STESSA 2015

Proceedings of the 8th International Conference on Behaviour of Steel Structures in Seismic Areas
Shanghai, China, 1-3 July 2015

Stessa 2015

BEHAVIOUR OF STEEL STRUCTURES IN SEISMIC AREAS

Editors

Federico M. Mazzolani
University of Naples "Federico II", Italy

Guo-Qiang Li, Suwen Chen, Xuhong Qiang
Tongji University, China

China Architecture & Building Press

图书在版编目（CIP）数据

第八届钢结构抗震国际会议 = Behaviour of steel structures in seismic areas：stessa 2015：英文/（意）马佐拉尼（Mazzolani，F. M.）等编著 . —北京：中国建筑工业出版社，2015.6
ISBN 978-7-112-18127-8

Ⅰ.①第… Ⅱ.①马… Ⅲ.①钢结构—抗震结构—国际学术会议—文集—英文
Ⅳ.①TU391.04-53

中国版本图书馆 CIP 数据核字（2015）第 102623 号

Executive Editors：Mengmei ZHAO，Dongxi LI

BEHAVIOUR OF STEEL STRUCTURES IN SEISMIC AREAS
（STESSA 2015）
Editors：Federico M. Mazzolani　Guo-Qiang Li　Suwen Chen　Xuhong Qiang

*

© 2015 China Architecture & Building Press，Baiwanzhuang，100037 Beijing，China
Published and distributed by China Architecture & Building Press in China
ISBN 978-7-112-18127-8（27354）
CIP
www. cabp. com. cn
Printed in China

PREFACE

The International Specialty Conference on Behaviour of Steel Structures in Seismic Areas, called STESSA, has reached its eighth edition.

The eighth edition of STESSA has been organized by the Department of Structural Engineering of Tongji University (China), in cooperation with the Department of Structures for Engineering and Architecture of the University of Naples "Federico II" (Italy).

The Conference is typically devoted to the behaviour of steel structures in seismic areas. Shanghai (China) was selected for the venue of the eighth edition of the Conference. Previous editions were held in Timisoara (Romania, 1994), Kyoto (Japan, 1997), Montreal (Canada, 2000), Naples (Italy, 2003), Yokohama (Japan, 2006), Philadelphia (United States, 2009) and Santiago (Chile, 2012).

The results of recent research from all over the world in the field of steel structures in seismic areas are represented through by experts from 25 countries. The papers included in the proceedings are subdivided into chapters with titles corresponding to the names of the working sessions of the STESSA'15 Conference. The Proceedings consists of 6 keynote lectures and two-pages extended abstracts for 198 papers in hardcopies, and the full texts of all the 204 papers in electronic format. The papers are categorized into the following areas:

1. *Performance-Based Design*: engineering descriptions of performance levels; conceptual design for multiple performance objectives; reliability-based design procedures; methods for analytical prediction of performance.

2. *Resilience Enhancement Technology*: self-centering; re-centering; behaviour of resilient structures.

3. *Earthquake, Wind and Exceptional Loads*: seismic load; wind load; fire; fire after an earthquake; explosions; impact.

4. *Member Behaviour*: rotation capacity; local buckling; overall buckling; classifications of sections; deterioration of strength, stiffness, ductility; bracing elements.

5. *Connection Behaviour*: cyclic behaviour of joints; analytical models; test results; pros and cons of welded and bolted connections; new innovations; data bank; influence of fully and partially restrained connections; seismic demands and capabilities of frames with welded, bolted or innovative connections.

6. *Global Behaviour*: moment resisting frames; braced frames; new and innovative structural systems; collapse mechanisms; redundancy of structures; P-Δ effects; modeling of deterioration; evaluation of reduction factors; damageability; large span structures; bridges; space frames; influence of non-structural elements.

7. *Analytical and Experimental Methods*: dynamic analysis; geometrical nonlinearity and material nonlinearity; optimization; static tests, shaking table tests; online real-time tests.

8. *Mixed and Composite Structures*: concrete filled tube (CFT) construction; steel encased in reinforced concrete (SRC) construction; mixed structures with reinforced concrete; mixed structures with timber; steel pile foundation and foundation problems.

9. *Passive, Semi-active, and Active Control*: behaviour of isolated structures; bridge bearings; energy dissipation; special devices; control algorithms; smart structures; design models; criteria for detailing.

10. *Codification, Design, and Practice*: up-dating of national and international codes; national practices case studies; cost-benefit ratio; design aids, fabrication and erection; aesthetics and habitability.

Organization

Honorary Chairman
Zu-Yan Shen

Chairman
Federico M. Mazzolani

Co-Chairman
Guo-Qiang Li

INTERNATIONAL SCIENTIFIC COMMITTEE

Carlos Aguirre, *Universidad Tecnica Federico Santa Maria, Valparaiso, Chile*
Jean Marie Aribert, *INSA Rennes, France*
Gustavo Ayala, *Universidad Nacional Autonoma de Mexico, Mexico City, Mexico*
MichelBruneau, *University of New York, Buffalo, USA*
Luis Calado, *Instituto Superior Tecnico of Lisbon, Portugal*
Suwen Chen, *Tongji University, Shanghai, China*
Gaetano Della Corte, *University of Naples "Federico II" Napoli, Italy*
Jie-Min Ding, *the Architectural Design and Research Institute of Tongji University, Shanghai, China*
DanDubina, *Politechnic University of Timisoara, Romania*
Gianfranco De Matteis, *University of Chieti-Pescara, Italy*
Amr Elnashai, *University of Illinois, Urbana-Champaign, USA*
Beatrice Faggiano, *University of Naples "Federico II" Napoli, Italy*
LarryFahnestock, *University of Illinois, Urbana-Champaign, USA*
MariaGarlock, *Princeton University, New Jersey, USA*
Ricardo Herrera, *University of Chile, Santiago de Chile, Chile*
Mamoru Iwata, *Kanagawa University, Yokohama, Japan*
Raffaele Landolfo, *University of Naples "Federico II", Italy*
StephenMahin, *University of California, Berkeley, USA*
Gregory McRae, *University of Canterbury, New Zealand*
Masayoshi Nakashima, *University of Kyoto, Japan*
Vincenzo Piluso, *University of Salerno, Italy*
Andre Plumier, *University of Liege, Belgium*
JamesRicles, *Lehigh University, Pennsylvania, USA*
Charles Roeder, *University of Washington, Seattle, USA*
Pedro Rojas, *ESPOL—Catholic University, Guayaquil, Ecuador*
MauricioSarrazin, *University of Chile, Santiago de Chile, Chile*
Bozidar Stojadinovic, *ETH, Swiss Federal Institute of Technology, Zurich, Switzerland*
Robert Tremblay, *Ecole Polytechnique of Montreal, Canada*
Ioannis Vayas, *National Technique University of Athens, Greece*
Carlos E. Ventura, *University of British Columbia, Canada*

Akira Wada, *Tokyo Institute of Technology*, *Japan*

In total31 members from 15 countries: Belgium, Canada, Chile, China, Ecuador, France, Greece, Italy, Japan, Mexico, New Zealand, Portugal, Romania, Switzerland and USA.

INTERNATIONAL ADVISORY COMMITTEE

Christoph Adam, University of Innsbruck, Austria

Hiroshi Akiyama, University of Tokyo, Japan

Enrique Alarcon, *Universidad Politecnica de Madrid*, *Spain*

Gulay Askar, *Bogazici University*, *Instanbul*, *Turkey*

Abolhassam Astaneh-asl, *University of California*, *Berkeley*, *USA*

Yi-Yi Chen, *Tongji University*, *Shanghai*, *China*

Dimitri E. Beskos, *University of Patras*, *Greece*

Constantin Christopoulos, *University of Toronto*, *Canada*

Charles Clifton, *University of Auckland*, *New Zealand*

Luis Da Silva, *University of Coimbra*, *Coimbra*, *Portugal*

Attilio De Martino, *University of Naples " Federico II "*, *Italy*

AhmedElghazouli, *Imperial College*, *London*, *UK*

Michael D. Engelhardt, *University of Texas*, *Austin*, *USA*

Mario Fontana, *ETH*, *Swiss Federal Institute of Technology*, *Zurich*, *Switzerland*

JeromeHajjar, *Northeastern University*, *Massachusetts*, *USA*

MohammedHjiaj, *INSA Rennes*, *France*

Kazuhiko Kasai, *Tokyo Institute of Technology*, *Japan*

Akihiko Kawano, *Kyusyu University*, *Fukuoka*, *Japan*

Roberto Leon, *Georgia Institute of Technology*, *Atlanta*, *USA*

AlbertoMandara, *The Second University of Naples*, *Italy*

Gian A. Rassati, *University of Cincinnati*, *USA*

Rodolfo Sarragoni, *University of Chile*, *Santiago de Chile*, *Chile*

Richard Sause, *Lehigh University*, *USA*

Motohide Tada, *Osaka University*, *Japan*

Lucia Tirca, *Concordia University*, *Montreal*, *Canada*

Keh-Chyuan Tsai, *National Taiwan University*, *Taiwan*, *China*

Gabriel Valencia, *Universidad Nacional de Colombia*, *Colombia*

Da-Sui Wang, *East China Architectural Design & Research Institute*, *China*

Satoshi Yamada, *Tokyo Institute of Technology*, *Japan*

Myung-ho Yoon, *Kongju National University*, *Korea*

In total 30 members from 18 countries: Austria, Canada, Chile, China, Colombia, France, Greece, Italy, Japan, Korea, Mexico, Portugal, Romania, Spain, Switzerland, Turkey, United Kingdom and USA.

ORGANIZING COMMITTEE

Chair: Guo-Qiang Li, *Tongji University*

Tongji University
Suwen Chen
Yuan-Qi Li
Yu-Shu Liu
Xu-Hong Qiang
Fei-Fei Sun
Wei Wang
Qiang Xie
Xian-Zhong Zhao
Feng Zhou

East China Architectural Design and Research Institute
Lian-Jin Bao
Cheng-Ming Li
Jian Zhou

Tongji Architectural Design (Group) Co. ,Ltd
Zhi-Jun He
Xin Zhao

CONFERENCE SECRETARIAT (*Tongji University*)
Xing Chen
Ya-Mei He

SCIENTIFIC SECRETARIAT (*University of Naples "Federico II"*)
Beatrice Faggiano
Antonio Formisano
Mario D'Aniello

Contents

Keynote Lectures

Performance-Based Design of Structures

Resilience Enhancement Technology

Contents

Member Behaviour

Contents

Connection Behaviour

Contents

Contents

Global Behaviour

Contents

Analytical and Experimental Methods

Contents

Mixed and Composite Structures

Passive, Semi-active and Active Control

Codification , Design , and Practice

Contents

Seismic, Wind and Exceptional Loads

Author Index

Keynote Lectures

MAJOR DEVELOPMENT OF RESEARCH AND PRACTICES ON SEISMIC DESIGN OF STEEL BUILDING STRUCTURES IN CHINA

Zu-Yan Shen

Department of Structural Engineering, Tongji University, Shanghai, 200092

E-mail: zyshen@ tongji. edu. cn

KEYWORDS: Steel building structures, Tall steel building, Seismic design, Codes and specifications, China

ABSTRACT

Steel structural systems have been widely used in different types of public, industry, even residential buildings in China in the past thirty years. Correspondingly, research and practices on seismic design of steel building structures in China have been keeping developing to offer strong technical support. In this paper, development and practices of different types of steel building structures were briefly reviewed at first. Then, research and standards on seismic design of steel building structures, as well as the basic theory and concept of seismic design for steel building structures in current codes and specifications in China, were discussed and introduced. Perspectives regarding seismic design of steel building structures in China were summarized finally.

1　INTRODUCTION

In the past thirty years, many types of steel building structures, including tall buildings, large-span spatial structures, heavy-type or light-weight industry factories buildings, even light-weight residential buildings, has been widely developed in China due to the rapid economy development as well as the correspondingly increase of raw steel production, as shown in *Fig. 1*.

On the other hand, seismic fortification to buildings is required based on the seismic fortification requirements almost in any places of China. Therefore, development of research and practices on seismic design of steel building structures is a primary issue in the codes and specifications related to designing of steel building structures. This paper will review the state-of-practice of seismic design of steel building structures, especially for high-rise buildings, in China.

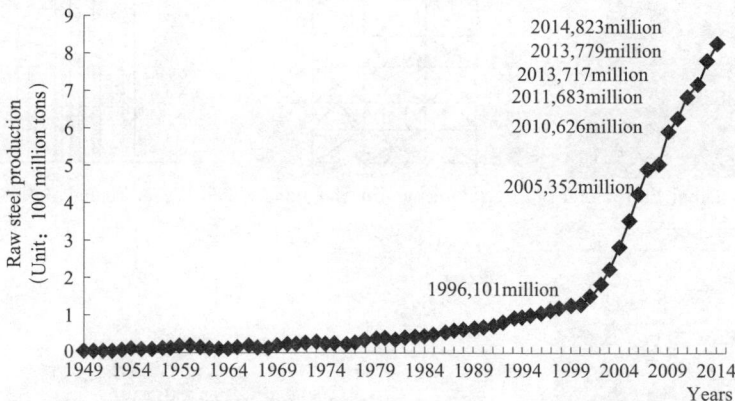

Fig. 1. Increase of raw steel production in China

2　DEVELOPMENT AND PRACTICES OF STEEL BUILDING STRUCTURES IN CHINA
2.1　Tall Steel Buildings

Up to now, many tall buildings have been constructed with steel structures and the height record has reached over 600m. The typical structural systems are summarized as follows: 1) Moment Frame. Shanghai Park Hotel (*Fig. 2*) is 83. 3m high with 2-story underground and 22 aboveground. Steel moment frame was adopted as well as reinforced concrete slab. 2) Braced Frame. Shanghai Jin Sha Jiang Hotel(*Fig. 3*) is 42m high with 2-story underground and 12 story aboveground. Its steel consumption was about 1300 t in total and 68kg per unit floor area. It was the first tall steel building which was designed by Chinese engineers and utilized domestic H profile steel. 3) Steel Frame-Bracing/Steel Plate Shear Wall. Shanghai Jin Jiang Tower is 153 m tall with 1-story underground and 43-story aboveground. Outriggers were used at the 23rd floor and the top floor to reduce horizontal displacement. Braces and steel plate shear walls were installed in the core tube for the stories above and below the 23rd floor, respectively (*Fig. 4*). Its total steel consumption was 8500t. 4) Mega Frame. Shanghai Stock Exchange Building(*Fig. 5*) is 121m high with 3-story underground and 30-story aboveground. It is composed of a 31m high,63m long overbridge and two 36m ×21m towers, which provides a sound background for the application of steel mega frame. Its total steel consumption was 9000t. 5) Frame with Viscous Dampers. TheSci-tech Complex of Tongji University is 48. 6m ×48. 6m in plan and 98 m tall with 1-story underground and 21-story aboveground. Its outline is a regular prism but its floors take an irregular configuration of "L" in plan, which rotates clockwise once for every three stories(*Fig. 6*). Outer braces with dampers were employed to control seismic effect of coupled lateral-torsional vibration. 6) Steel-concrete Hybrid Structure. Shanghai World Financial Center Tower is 492m tall with 3-story underground and 101-story aboveground, being presently the tallest building in China. It employs a lateral resistant structural system with the combination of a mega-frame structure, a reinforced concrete core and braced steel core connected by three 3-story outrigger trusses (*Fig. 7*). The mega-frame structure is composed of mega columns, mega braces and belt trusses.

152.000 43层

旱斜杆支撑

Outrigger Truss

K brace in core tube

伸臂支架

23F
钢板增板

Steel plate sheat Wall
in core tube

1层

Fig. 2. Shanghai Park Hotel *Fig. 3.* Shanghai Jin Sha Jiang Hotel *Fig. 4.* Shanghai Jin Jiang Hotel

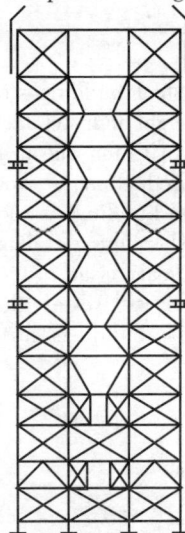

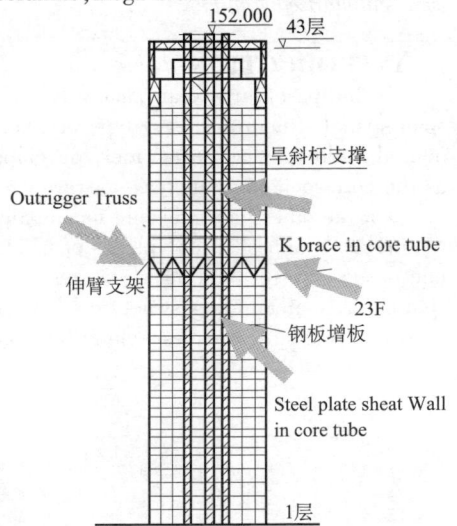

Fig. 5. Shanghai Stock Exchange Building   *Fig. 6.* The Sci-tech Complex of Tongji University

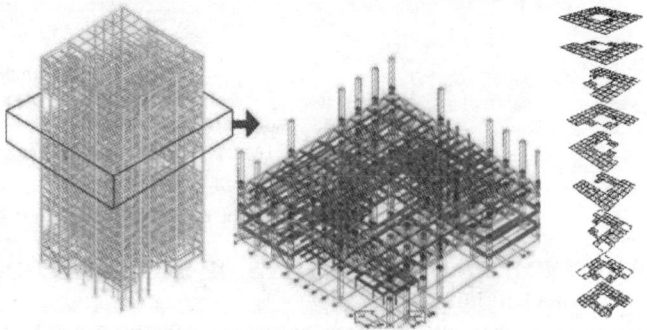

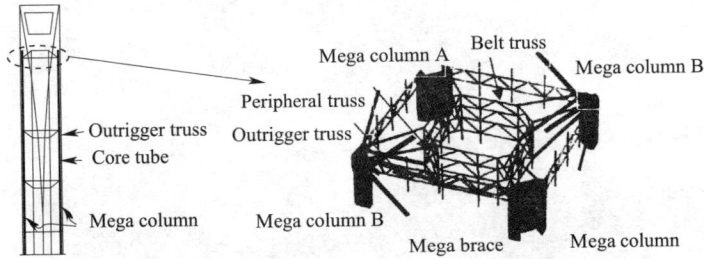

Fig. 7. Shanghai World Financial Center Tower

2. 2 Large-span Spatial Steel Structures

At the same time, large-span spatial steel structures with different systems, including spatial truss (*Fig. 8*) , single-layer or double-layer reticulated shell(*Fig. 9*) , prestressed or cable-stayed grid structures, cable structures, beam (truss) string structures (BSS/TSS, *Fig. 10* and *Fig. 11*) , suspen-dome structures(*Fig. 12*) , tensegrity system, etc. , are developed and more and more widely used in practices.

2. 3 Others

Besides tall building and large-span spatial buildings, many heavy-type or light-weight industry factories buildings(*Fig. 13*) , even low-rise or middle-rise light-weight residential buildings(*Fig. 14* and *Fig. 15*) , masts and towers, etc. , used steel structural system.

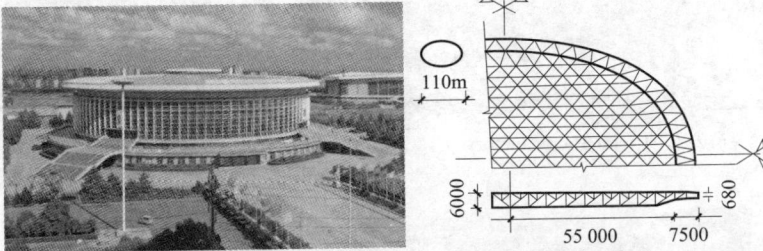

Fig. 8. Shanghai Gymnasium(110m , 1975)

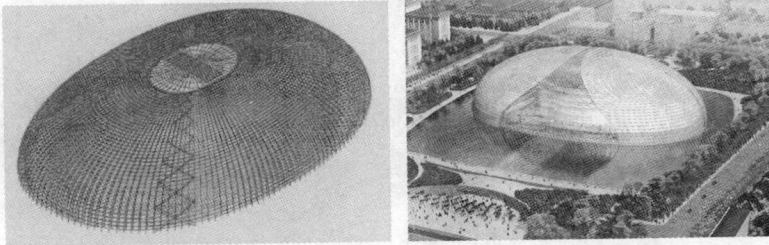

Fig. 9. National Grand Theater(146m × 212m)

Fig. 10. T1 terminal for Shanghai Pudong
International Airport

Fig. 11. T2 terminal for Shanghai Pudong
International Airport

Fig. 12. Jinan Olympic Gymnasium

Fig. 13. Portal frame factory

Fig. 14. middle-rise light-weight residential building

Fig. 15. Cold-formed thin-walled steel framing residential building

3 DEVELOPMENT OF RESEARCH AND STANDARDS ON SEISMIC DESIGN OF STEEL BUILDING STRUCTURES

For seismic design, due to the light-weight as roofs or walls, normally earthquake action is not the critical load for most large-span steel structures and light-weight steel industry factories and residential buildings. So the development of research and standards on seismic design of steel building structures mainly focuses on tall steel buildings.

Many research works regarding on seismic design of steel building structures have been conducted in the past thirty years, together with the necessary technical support requirement of rapid development of utilization of steel structural systems in different types of buildings. Based on the research work and corresponding practices, technical standard system regarding seismic design of steel building structures, including National Standard (approved by the authority appointed by the central government, which are nationwide unified technical standards), Professional standards(approved by industrial ministries, which are industrial unified technical standards), Local Standard(approved by the authority appointed by local government, which are provincial unified technical standards) and Standardization Associational Standard(approved by the Chinese Association of Engineering Construction Standardization, which are recommended by the Association for nationwide adoption), has been established.

The main codes and specifications for seismic design of tall steel buildings include: 1) Temporary Specification for Structural Design of Tall Steel Buildings(DJB08-32-92,1993), 2) Technical Specification for Steel Structures of Tall Buildings (JGJ99-98, 1998), and 3) Code for Seismic Design of Buildings(GB50011-2001,2001), Chapter 8, Multi-story and tall building steel structures.

While for technical inspection of seismic fortification of tall buildings designed beyond the code limits, three documents should be followed: 1) 1997, Temporary regulations for administration, 2) 2002, Regulations for administration, and 3) 2006, Technical outline.

Some main concepts and design methodology will be introduced based on above codes and specifications in the following parts of this paper.

4 BASIC THEORY OF SEISMIC DESIGN FOR STEEL BUILDING STRUCTURES IN CHINA

4.1 Design Specifications for Tall Steel Buildings in Shanghai

In order to meet the need of modernization construction, Shanghai local government sponsored 8 research projects in 1985. The research topics covered damping ratio, seismic design spectrum, panel zone of beam-column joint, mixed type of beam-to-column connection, welded thick plate column of box section, hysteretic model and composite column of tall steel buildings as well as mutual interference of wind effect between tall buildings. Based upon these projects and recent engineering practices, the first design specification for tall steel buildings in China-Standard DJB08-32-92 was published in 1993. After over 10 years of implementation, this specification was upgraded to the formal specification DG/TJ08-32-2008 in 2008, with significant adjustment and additional design provisions regarding ductility design of steel structures, transverse wind response, wind interference, semi-rigid beam-to-column connections and buckling restrained braces etc.

Meanwhile, Shanghai Code for Seismic Design of Buildings(DGJ08-9-2003) provided seismic design provisions for steel buildings not more than 12 stories or not higher than 40m.

A special type of steel-concrete hybrid structural system-steel frame-RC core structure was once suspected to be vulnerable to earthquake damage, as it lacks redundant lateral resisting system. However, over half tall steel buildings in earthquake-prone area of China employed this system. Based upon systematic research by Tongji University, the first design specification for the hybrid system in China-Standard DG/TJ08-015-2004 was put forward in Shanghai in 2004.

4.2 Technical Inspection of Seismic Fortification of Tall Buildings Designed beyond the Code Limits

Regarding structural systems of tall buildings can be too complex for design specifications to cover all the cases with accurate seismic design requirements, China Ministry of Construction issued Ministry of Construction Order No. 59 in 1997 and revised it as Ministry of Construction Order No. 111 in 2002. They were temporary and formal regulations on the requirement of technical inspection of seismic fortification of tall buildings designed beyond the code limits. Tall buildings beyond applicable height limit and those with non-applicable structural systems or especially irregularconfiguration are required to be subject to technical inspection by officially authorized committees. Later in 2006, the ministry issued a technical outline for such technical inspection(Ministry of Construction Order No. 220).

4.3 Seismic Design Methodology

1) Seismic Design Criteria

According to China National Standard GB50011-2001, seismic design criteria are defined as the following three levels: (1) No damage under frequent earthquakes with exceeding probability of 63.2% in 50 years, (2) Repairable damage under design earthquakes with exceeding probability of 10% in 50 years, and (3) Collapse prevention under rare earthquakes exceeding probability of $2 \sim 3\%$ in 50 years.

2) Definition of Seismic Action

In favor of response spectrum analysis, response spectra suggested by DGJ08-9-2003 is shown in Fig. 16 with α being earthquake influence coefficient, T_g characteristic period, γ attenuation coefficient, η_1 and η_2 modification factors and T structural natural vibration period. The values for earthquake maximum influence coefficient α_{max} are listed in Table 1. Regarding the equivalent base shear method, the base shear force F_{EK} is equal to $\alpha_1 G_{eq}$, where G_{eq} is the equivalent total mass and α_1 is the earthquake influence coefficient corresponding to the fundamental period of the structure.

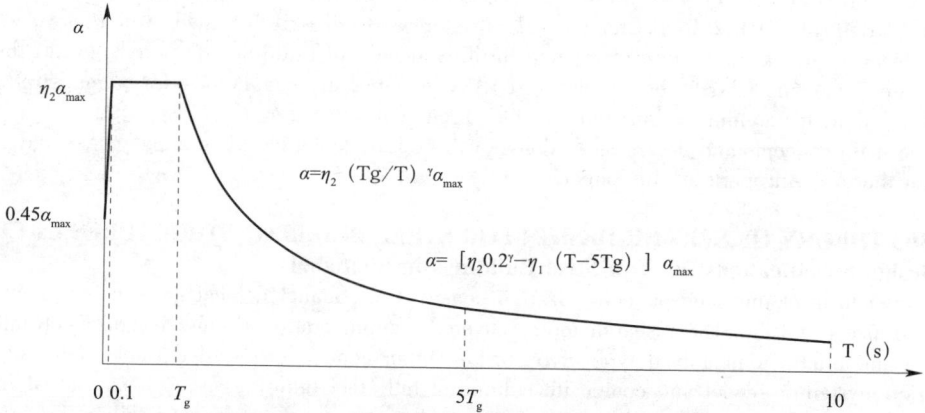

Fig. 16. Response spectra suggested by DGJ08-9-2003

Table 1. Earthquake maximum influence coefficient α_{Max} and amplitude of accelerogram

	Earthquake maximum influence coefficient			Amplitude of accelerogram (cm/s^2)		
Seismic intensity	6	7	8	6	7	8
Frequent earthquake	0.04	0.08	0.16	18	35	70
Rare earthquake	—	0.45	0.81	—	200	360

Regarding time history analysis, 4 accelerograms are suggested by DGJ08-9-2003 as shown in *Fig. 17* while accelerogram amplitude for each specific analysis can be determined with reference to *Table 1.*

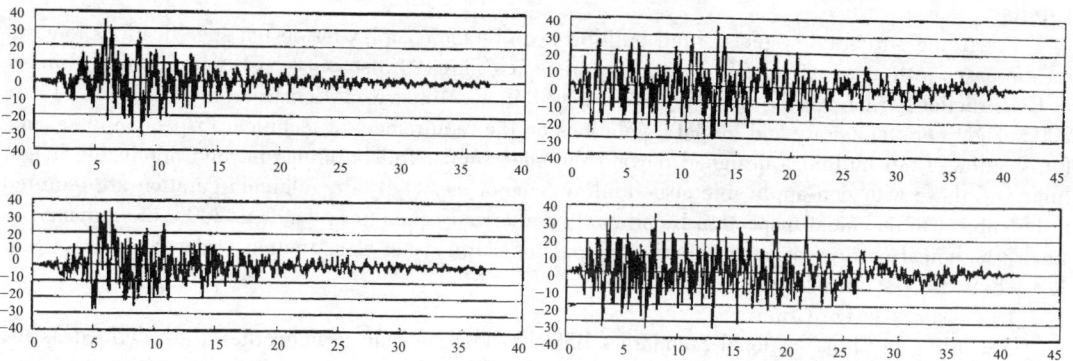

(a) and (b) were directly generated while   (c) and (d) were adapted from real accelerograms

Fig. 17. Artificial accelerogram suggested by DGJ08-9-2003

3) Ductility Design

Ductility design method under design earthquake was adopted in Standard DG/TJ08-32-2008 in place of elastic design method under frequent earthquake in Standard DJB08-32-92. However, design formula have been scaled to the form of frequent earthquake so as to make use of seismic action as defined in *Fig. 16* and *Table 1*, which is compatible to the definition in China National Standard GB50011-2001. Process of ductility design method suggested by DG/TJ08-32-2008 is as follows:

Step 1: Select an appropriate structural system according to expected ductility capacity. As listed in *Table 2*, four ductility types are defined according to ductility capability of structural systems with beam-to-column connections and member cross sections of different ductility levels.

Step 2: According to features and ductility requirement of the target structural system, determine if a structural member is allowed to be plastified under design earthquakes. As can be seen in *Table 2*, there are three section types of plastic members, which are defined as follows: section type A can form a plastic hinge with the rotation capacity required for plastic hinges, section type B can develop its plastic moment resistance, but has limited rotation capacity, section type C can have its extreme compression fibre reach its yield strength, but local buckling is liable to prevent development of the plastic moment resistance. Cross section of a plastic member should be proportioned not exceeding the limiting width-thickness ratio as listed in *Table 3*. Meanwhile, there are three types of beam-to-column connections as listed in *Table 2*. Improved Traditional Type and Improved Type are required to have a rotational capacity no less than 0. 02 and 0. 03 rad, respectively, while no such requirement is suggested for Traditional Type. Prequalified beam-to-column moment joints for the three connection types are provided by Standard DG/TJ08-32-2008 as depicted in *Fig. 18*.

Step 3a: Perform elastic structural analysis using the applicable one of equivalent base shear method and mode-superposition response spectrum method.

Step 3b: Do design checks for structural members and connections by means of equation 1.

$$S \leqslant R/\gamma_{\mathrm{RE}} \tag{1}$$

where R is design value of member capacity, γ_{RE} is seismic adjustment factor of member capacity, taking 0. 75,0. 80,0. 85 and 0. 90 for beam/column, brace, connection plate/bolt, and weld, respectively, S is design value of seismic load effect combination as $S = \gamma_{\mathrm{G}}S_{\mathrm{GE}} + \gamma_{\mathrm{Eh}}\gamma_{\mathrm{RS}}S_{\mathrm{Ehk}} + \gamma_{\mathrm{Ev}}\gamma_{\mathrm{RS}}S_{\mathrm{Evk}} + 0. 2\gamma_{\mathrm{w}}S_{\mathrm{wk}} \cdot S_{\mathrm{GE}}, S_{\mathrm{Ehk}}, S_{\mathrm{Evk}}, S_{\mathrm{wk}}$ are nominal value of load effect of permanent load, horizontal and vertical seismic actions, wind load, of which partial safety factors are $\gamma_{\mathrm{GE}}, \gamma_{\mathrm{Ehk}}, \gamma_{\mathrm{Evk}}, \gamma_{\mathrm{wk}}$, respectively. γ_{RS} is ductility factor of target structural system, taking 1. 0,0. 85,0. 70,0. 60 for Ductility Type I, II, III, IV of structural systems, respectively.

Step 3c: Do design check for story drift of the structure under frequent earthquakes using equation 2.

$$\gamma_{\mathrm{d}} \frac{\Delta}{h} \leqslant \left[\frac{\Delta}{h}\right] \tag{2}$$

where Δ/h is elastic story drift due to the same seismic load effect combination of equation 1, γ_{d} is displacement amplification factor, taking 1. 0,1. 18,1. 43,1. 67 for Ductility Type I, II, III, IV of structural systems, respectively. $[\Delta/h]$ is allowable story drift of structures under frequent earthquakes: (1)1/400 for buildings having non-structural elements of brittle materials attached to the structure, (2)1/300 for buildings having ductile non-structural elements, (3)1/200 for buildings having non-structural elements fixed in a way as not to interfere with structural deformations.

Table 2. Definition of ductility types of structural systems

Ductility type	Structural system	Type of beam-to-column connection	Section type of plastic members	Note
I	Ordinary frame	Traditional	C	
	Ordinary CBF	Traditional	C	Single
	Ordinary frame-RC panel	Traditional	C	Single
	Ordinary frame/truss tube	Traditional	C	

Ductility type	Structural system	Type of beam-to-column connection	Section type of plastic members	Note
II	Ductile frame	Improved Traditional	B	
	Ductile CBF	Improved Traditional	B	Single/dual
	Ductile frame-RC panel	Improved Traditional	B	Dual
	Ordinary frame-steel plate RC panel	Improved Traditional	B	Single/dual
	Ductile frame/truss tube	Improved Traditional	B	
	Ordinary tube-in-tube, bundled tubes	Traditional	C	
III	Highly ductile CBF	Improved Traditional	B	Dual
	Ductile EBF	Improved Traditional	B	
	BRBF	Traditional / Improved Traditional	B	
	Highly ductile frame-RC panel	Improved	B	Dual
	Ductile frame-steel plate RC panel	Improved	B	Single/dual
	Ductile frame-composite steel wall	Improved Traditional	B	Single/dual
	Ductile frame-steel plate wall	Improved Traditional	B	Single
	Highly ductile frame tube	Improved	B	
	Ductile tube-in-tube/bundled tubes	Improved Traditional	B	
IV	Highly ductile frame	Improved	A	
	Highly ductile EBF	Improved	A	Single/dual
	Highly ductile BRBF	Improved	A	Single/dual
	Highly ductile frame-steel plate wall	Improved	A	Dual
	Highly ductile tube-in-tube/bundled tubes	Improved	A	

Note: When the ratio of design shear of moment frame to base shear of overall structure is no less than 25%, a structure is recognized as a dual lateral system. Otherwise, it is a single lateral system.

Step 4: Do performance check for story drift of the structure under rare earthquakes. Allowable story drift of structures under rare earthquakes are prescribed as 1/50 for steel moment frames and 1/70 for all other structural systems. Non-linear time history analysis on the structure for at least 3 accelerograms of rare earthquakes is required by Standard DG/TJ08-32-2008 for structural systems of type III and IV and for extremely irregular structural systems. Simplified methods are applicable to take the place of non-linear time history analysis for other structural systems. Hysteretic models for different structural members are suggested by the standard as shown in *Fig. 19*.

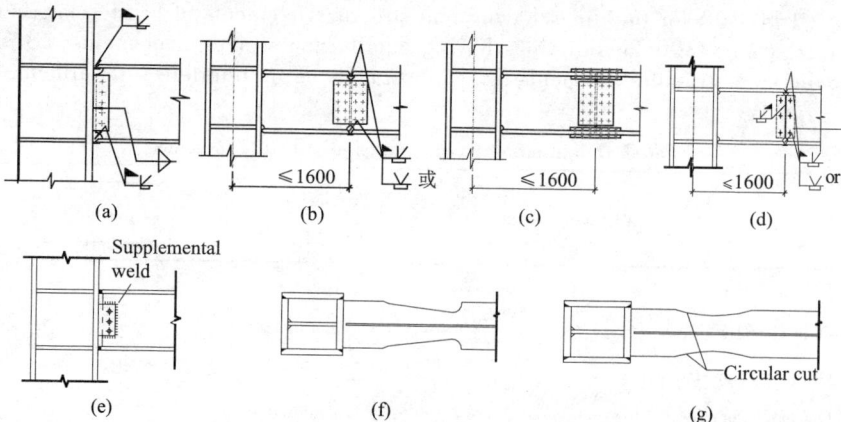

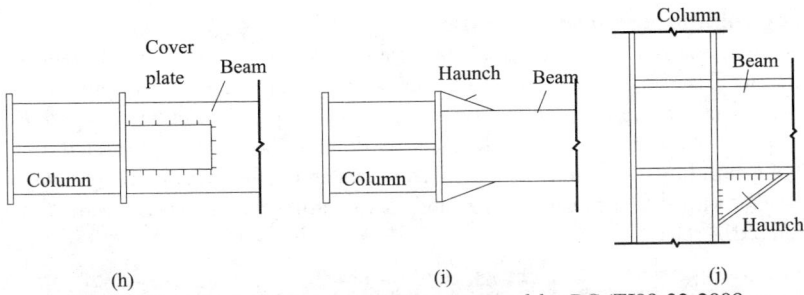

(h)   (i)   (j)

Fig. 18. Beam-to-column moment joint suggested by DG/TJ08-32-2008

(a) Traditional Type, (b)-(e) Improved Traditional Type, and (f)-(j) Improved Type

Table 3. Limiting width-thickness ratios for beams and columns

	Section type	A	B	C
Beam	free flange in I or box sections	9	11	15
	stiffened flange in box sections	33	38	42
	web plates in I or box sections	$72-100 N_b/(Af)$	$80-110 N_b/(Af)$	$85-120 N_b/(Af)$
Column	free flange in I or box sections	10	13	15
	stiffened flange in box sections	35	37	40
	web plates in I or box sections	43	43	45

Note: N_b, A and f are axial force, cross sectional area and design value of steel strength of the column, respectively.

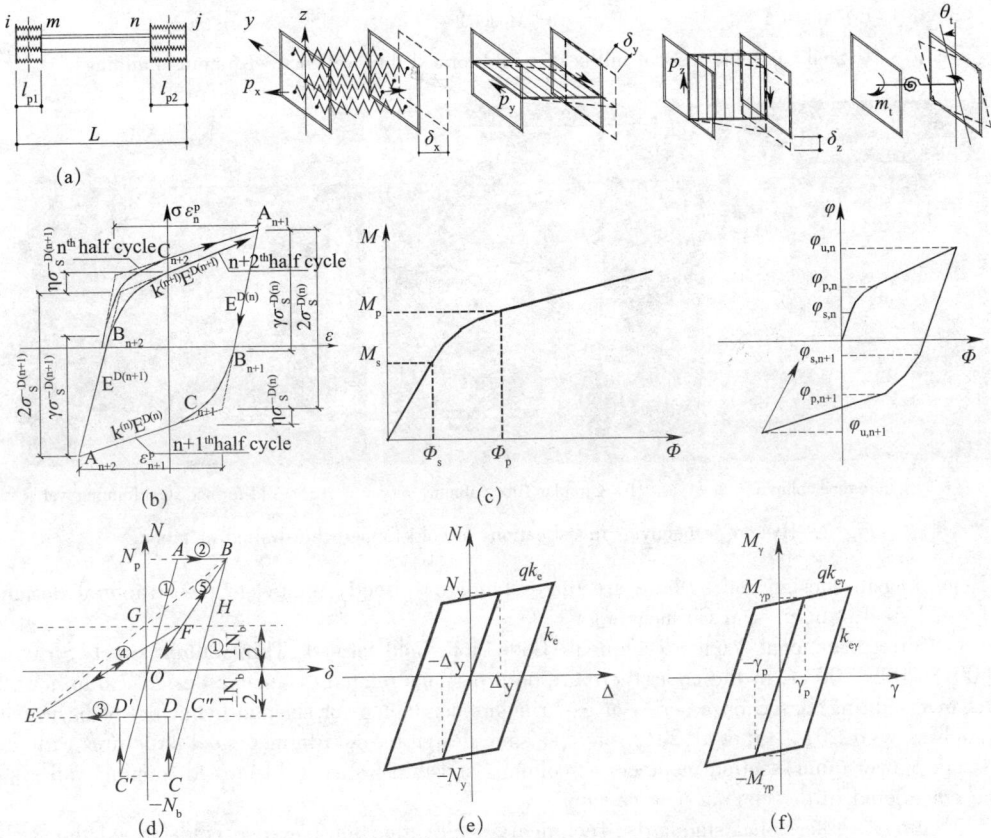

Fig. 19. Hysteretic models suggested by Standard DG/TJ08-32-2008

(a) Multi-spring model for beams and columns, (b) Hysteretic model for steel considering damage accumulation, (c) Yield surface model for beams and columns, (d) Axial hysteretic model for common braces, (e) Axial hysteretic model for buckling restrained braces (BRB), (f) Shear hysteretic model for joint web panels

4. 4　Regarding cold-formed steel structures

　　Due to the industrialization of building construction in China, cold-formed steel structures, with cold-formed open or close sections efficient to manufacturing and connecting in practices, are becoming widely used. Regarding seismic issues of cold-formed steel structures, a series of experiments have been conducted recently, including shaking Table tests of three full-scaled cold-formed thin-walled steel framing residential buildings, as shown in Fig. 20, and hysteretic Behavior investigations of cold-formed thin-walled columns with different geometrical and load-bearing parameters, as shown in *Fig. 21*. (Li, et al, 2013, Wen, 2014, etc.)

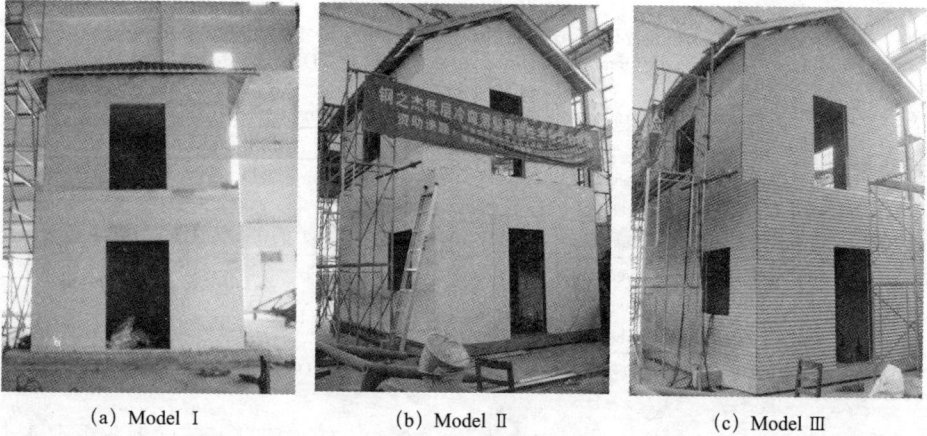

(a) Model I　　　　　　　(b) Model II　　　　　　　(c) Model III

Fig. 20. Shaking Table tests of full-scaled cold-formed thin-walled steel framing buildings

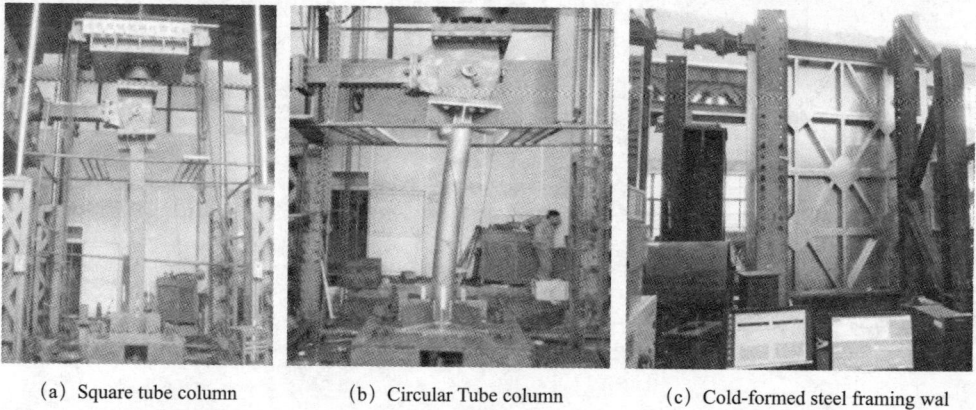

(a) Square tube column　　　(b) Circular Tube column　　　(c) Cold-formed steel framing wal

Fig. 21. Hysteretic behavior investigations of cold-formed thin-walled members

　　Regarding the design codes, there are three standards directly related to the structural design of cold-formed steel structures in China now:

　　1) Current national code, Technical Code for Cold-formed Thin-walled Steel Structures (GB50018-2003, 2003). In the code, there are no provisions related to seismic design. Up to now, this code is in re-editing. Based on a series of experiments conducted for seismic behaviors of basic members, such as Wen, 2014, Li, et al, 2013, etc. , seismic design for cold-formed steel structures with close thick-walled(over 6mm)section members as columns and thin-walled cold-formed framing wall system will be considered in the coming new version.

　　2) National Professional standards, Technical specification for Low-rise cold-formed thin-walled steel structures(JGJ227-2011, 2011). In the specification, if some construction details are satisfied the requirements, a sample base shear method, as provided in National Standard GB50011-2001 is recommended based on a series of shaking Table experiments(Li, et al, 2013).

3) Shanghai Engineering Construction Standard, Technical Specification for Light Weight Steel Building Structures (DG/TJ08-2089-2012 , 2012). Considering the current construction situation of low-rise thin-walled cold-formed framing wall systems in China, special construction details for earthquake action are emphasized, 0 ductility type as described out of Tab. 3 is defined for the low-rise thin-walled cold-formed framing wall systems, and a 1. 85 amplifying factor to often-occurred earthquake force is used based on the base shear method, as provided in National Standard GB50011-2001.

5 RESEARCH ON BUCKLING – RESTRAINED MEMBERS FOR SEISMIC RESISTANT STEEL BUILDING STRUCTURES IN CHINA

5. 1 Buckling restrained brace(BRB)

In mainland China, research on BRB was initiated in early 2000s. Among others, a series of new type BRBs have been systematically developed at Tongji University. In addition to common performance of sTable hysteretic behavior, TJ-series BRBs demonstrated superior performance in experimental results, e. g. (1) maximum average axial strain up to 1/16, (2) more than 60 cycles with amplitude of average axial strain up to 1/80, and (3) failure prevention capability of TJ-P type. A full-scale model of simple-connected steel frame with buckling restrained braces was tested to validate seismic performance of TJ-type BRB on the shaking Table. The story drift angles of the structure under El-centro wave are shown in Fig. 22. The BRB underwent obvious plastic deformation, but there was no buckling phenomenon appeared.

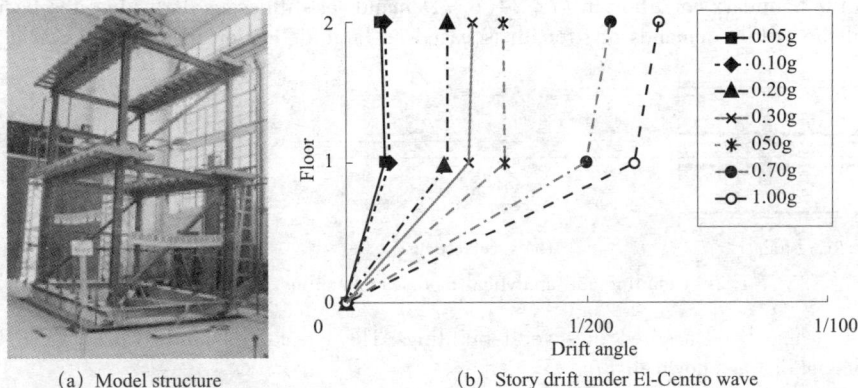

(a) Model structure   (b) Story drift under El-Centro wave

Fig. 22. Full scale shaking Table test on a simple-connected steel frame with BRB

Based upon a large amount of test data of hysteretic behavior of TJ-type BRB, theoretic and statistic study was conducted with the following output: (1) a hysteretic model for BRB in terms of axial force N and axial strain ε, considering Bauschinger Effect and strain hardening effect, (2) probability density functions for the compression strength adjustment factor and overstrength factor for low yield strength steel, (3) relation between strain hardening adjustment factor and plastic strain. Thereafter, probabilistic distributions of floor drift, inter-story drift angle, maximum inter-story drift ratio and roof drift of BRB-braced frames(BRBF) were all proved to satisfy the lognormal distribution. Based on this, a method of reliability evaluation and seismic design on BRBF under rare earthquakes was proposed.

In recent years, more and more buildings have been equipped with BRB. Up to now, the largest BRB was applied in 597m tall Tianjin Goldin Finance 117 Tower, as shown in Fig. 23. This mega BRB is 48 m in length, 260ton in weight, 1. 5m by 0. 9m in section, and 3600 ton in yielding capacity.

5. 2 Buckling restrained wall(BRW)

As a natural extension of BRB, buckling restrained wall(BRW) has also been intensively investigated. Three types of BRWs have been proposed: (a)I shape with large aspect ratio is sui Table for residential buildings with small space for walls, (b)I shape with low aspect ratio is sui Table for large shear capacity, (c)slit wall is suiTable for large elastic deformation. Each of them exhibited excellent hysteretic behavior. Simplified methods for BRW design and structural analysis have been proposed: (1)I-shaped wall can be idealized as an equivalent rectangular wall in favor of stiffness calculation,

(a) Structural model (b) Location in model   (c) Installation   (d) Global view

Fig. 23. Application of mega BRB in Tianjin Goldin Finance 117 Tower

(2) all the three types can be represented as a pair of braces for application in structural model, (3) connection analysis of slit wall can be performed with a simple extension form the above equivalent brace model, and (4) connection analysis of I-shaped wall can be modeled as flexural action in exterior regions and shearing action in interior region.

Two types of buckling restrainers for BRW were conceived as RC type in *Fig. 24* (a) and steel type in *Fig. 24* (b). Their demands in terms of stiffness and strength were put forward by means of classical theory with the boundary condition in *Fig. 24* (c). Demands of stiffness is derived on the basis of small deflection theory, while demands of strength is based on large deflection theory.

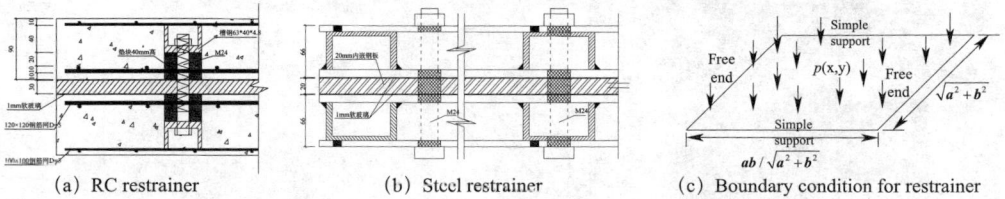

(a) RC restrainer   (b) Steel restrainer   (c) Boundary condition for restrainer

Fig. 24. Detailing and analytical model for buckling restrainer of BRW

BRW has also been applied in several buildings. The latest application China Expo Exhibition Complex, Shanghai was shown in *Fig. 25*.

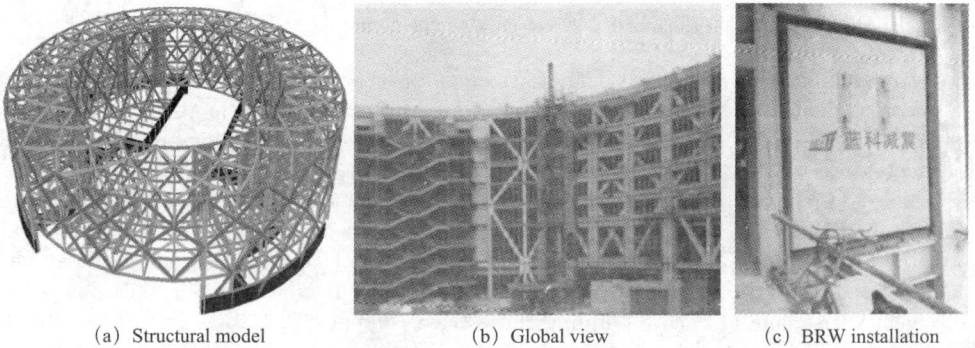

(a) Structural model   (b) Global view   (c) BRW installation

Fig. 25. Application of BRW in China Expo Exhibition Complex, Shanghai

6 PERSPECTIVES

Challenges for seismic design of steel building structures in China will always be twofold. On one hand, more and more steel skyscrapers beyond the scope of design codes are to be built, and new types of steel structural systems may be used due to the outbreak of span in large-span spatial structures, as well as the tendency of construction industrialization. Many over 600m-tall buildings, such as Shanghai

Center Tower, andmany over 300m large-span spatial structures will be built. On the other hand, there will be many important needs to improve the design specifications. Among others, research efforts have been devoted to accounting for risk of successive earthquakes in seismic design for over a decade. Last but not the least, it should be noted that many modern tall steel buildings and large-span spatial structures in China have not experienced the true severe earthquakes.

REFERENCES

[1] Shanghai Engineering Construction Standard DJB08-32-92, "Temporary Specification for Structural Design of Tall Steel Buildings", Shanghai Construction and Administration Commission, 1993. (in Chinese)

[2] Shanghai Engineering Construction Standard DGJ08-9-2003, "Code for Seismic Design of Buildings", Shanghai Construction and Administration Commission, 2003. (in Chinese)

[3] The National Professional standards JGJ99-98, "Technical Specification for Steel Structures of Tall Buildings", China Building Industry Press, 1998. (in Chinese)

[4] Shanghai Engineering Construction Standard DG/TJ08-015-2004, "Code for Design of Steel-concrete Hybrid Structures for High-rise Buildings", Shanghai Construction and Administration Commission, 2004. (in Chinese)

[5] Shanghai Engineering Construction Standard DG/TJ08-32-2008, "Specification for Steel Structure Design of Tall Buildings", Shanghai Construction and Administration Commission, 2008. (in Chinese)

[6] Shen Zuyan, Wen Donghui, Li Yuanqi. State-of-the-arts: Technical progress of steel building structures in china, *Building Structures*, 39(9):15-24, 14, 2009. (in Chinese)

[7] The National Standard GB50011-2001, "Code for Seismic Design of Buildings", China Building Industry Press, 2008. (in Chinese)

[8] Ministry of Construction Order No. 59, "Temporary Administrative Provisions for Technical Inspection of Seismic Fortification of Tall Buildings Designed beyond the Code Limits", Ministry of Construction P. R. China, 1997. (in Chinese)

[9] Ministry of Construction Order No. 111, "Administrative Provisions for Technical Inspection of Seismic Fortification of Tall Buildings Designed beyond the Code Limits", Ministry of Construction P. R. China, 2002. (in Chinese)

[10] Ministry of Construction Order No. 220, "Technical Outline for Technical Inspection of Seismic Fortification of Tall Buildings Designed beyond the Code Limits", Ministry of Construction P. R. China, 2006. (in Chinese)

[11] The National Standard GB50018-2003, "Technical code for cold-formed thin-walled steel structures", China Plan Press, 2003. (in Chinese)

[12] Wen Donghui. Research on seismic performance of cold-formed thick-walled steel tubular beam-columns, Ph. D. Thesis, Tongji university, 2014. (in Chinese)

[13] Li Yuanqi, Shen Zuyan, Yao Xingyou, Ma Rongkui, Liu Fei. Experimental investigation and design method research on low-rise cold-formed thin-walled steel framing buildings, *Journal of Structural Engineering*, ASCE, 139(5): 818-836, 2013.

[14] The National Professional standards JGJ227-2011 "Technical specification for Low-rise cold-formed thin-walled steel structures", China Building Industry Press, 2011. (in Chinese)

[15] Shanghai Engineering Construction Standard DG/TJ08-2089-2012, "Technical Specification for Light Weight Steel Building Structures" Shanghai Construction and Administration Commission, 2012. (in Chinese)

[16] Shen Zuyan, Sun Feifei. Discussion and recommendation on seismic design for steel structures, *Building Structures*, 39(11):115-122, 2009. (in Chinese)

[17] Sun Feifei, Li Guoqiang, Guo Xiaokang, Hu Dazhu, Hu Baolin. Development of new-type buckling-restrained braces and their application in aAseismic steel frameworks. *Advances in Structural Engineering*, 14(4):717-730, 2011.

[18] Jin Huaian, Li Guoqiang, Lu Ye, Sun Feifei, He Yamei. Buckling-restrained steel plate shear walls: demands on buckling restrainers. *EUROSTEEL*, 9:10-12, 2014.

HYBRID ANALYTICAL-EXPERIMENTAL SIMULATION AND APPLICATIONS TO STEEL FRAMES WITH SEMI-RIGID CONNECTIONS

Amr S. Elnashai* and Hussam N. Mahmoud**

* Pennsylvania State University, State College, PA, USA

e-mails: Elnashai@ engr. psu. edu

** Colorado State University, Fort Collins, CO, USA

e-mail: Hussam. Mahmoud@ colostate. edu

KEYWORDS: Seismic, Hybrid Simulation, Semi-rigid.

ABSTRACT

Hybrid simulation has emerged as a potentially-accurate and efficient tool for the evaluation of the response of large and complex structures under earthquake loading. As an application example, hybrid simulation is applied to semi-rigidly connected frames with different connection strength limits. In the sample study three hybrid simulations were conducted and included the most reliable, realistic, and computationally efficient experimental and analytical modules, which were developed and successfully integrated for a full system-level assessment. The simulations included a full-scale physical specimen for the experimental module and a 2D finite element model for the analytical module. It is demonstrated that hybrid simulation is a powerful tool for advanced assessment when used with appropriate analytical and experimental realisations of the components and that semi-rigid frames are a valuable, often preferable, option in earthquake engineering applications.

1 HYBRID SIMULATION METHODOLOGY

Seismic evaluation of structural systems has traditionally been explored using either experimental methods or analytical models. Issues of scale, equipment capacity and availability of research funding continue to limit the full-scale testing of complete structures. Analytical platforms on the other hand are limited to solving specific type of problems and in many cases fail to capture complex behaviours or failure modes in structural systems. One example of a limitation in analytical models would be in modelling and predicting failure in structural components. For example, current models, which employ crack initiation and extension often require the location of the crack to be defined and represented using a cohesive zone model prior to conducting the analysis. The concept of hybrid simulation hinges on combining experimental and analytical models in a single simulation while taking advantage of what each tool has to offer. In hybrid(experiment-analysis) simulation, part of a structural system is experimentally represented while the rest of the structure is numerically modelled. Typically, the most critical component, or the component whose behaviour cannot be well presented with a numerical model, is physically simulated. By combining a physical specimen and a numerical model, the system-level behaviour can be better quantified than modelling the entire system purely analytically or testing only a component. Hybrid simulation was first developed by Japanese researchers where a single-degree-of-freedom system was analysed under seismic loading[1]. The work used an analog computer to solve the equations of motion and an electromagnetic actuator to load the structure. Since then, simulation techniques have significantly evolved to include sub-structuring techniques with hybrid simulation[2,3], making it possible to consider distributed hybrid simulation and real-time hybrid simulation[4,5].

The simulation is advanced in timeusing a time-stepping numerical integration scheme and the interface between the substructures is controlled using hydraulic actuators. Significant developments in pseudo-dynamic and dynamic hybrid simulations have been achieved, although the applications of such have been primarily focused on pseudo-dynamic simulations. The slow, or pseudo-dynamic, hybrid simulation technique uses the implicit time-step integration schemes in two main forms; 1) iterative implicit method or 2) linearly implicit and nonlinearly explicit operating splitting(OS) method[6]. For systems with large inelastic deformations, an operator splitting method in conjunction with the α-modified Newmark scheme(α-OS method) has shown to be unconditionally stable[7]. One of the most developed and stable software used for orchestrating hybrid simulations is UI-SIMCOR, which is MATLAB-based. The software was developed at the University of Illinois at Urbana-Champaign(UIUC)and is capable of conducting the numerical integration as it steps through the seismic record. The numerical integration in UI-SIMCOR uses the OS method with a modified α-parameter through the Newmark integration scheme(α-OS method)which applies numerical damping to the undesired oscillations. In UI-SIMCOR, the earthquake force is calculated numerically using time step-integration of the equation of motion. The corresponding displacements are then applied simultaneously to the test specimens and the analytical models. The resulting restoring forces are measured for each module and used in a feedback loop for the calculation of the next displacement command corresponding to the next step. Fig. 1 shows the architecture of UI-SIMCOR and its ability to be integrated with various experimental and analytical modules.

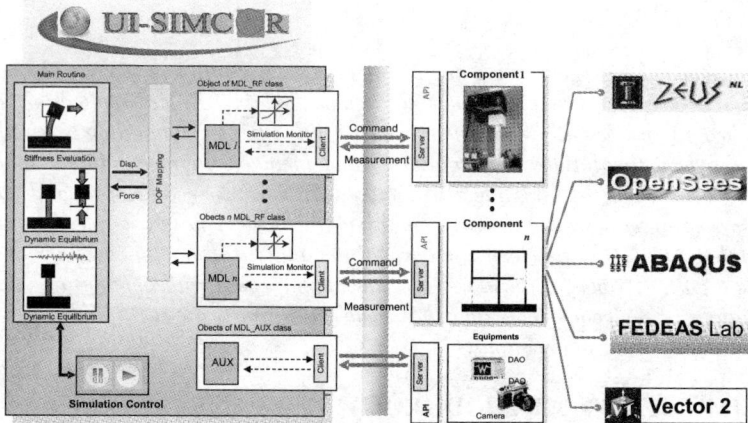

Fig. 1. Architecture of UI-SIMCOR hybrid Simulation framework[3]

2 APPLICATION TO SEMI – RIGID STEEL FRAMES
2.1 Background

The uncertainly in the performance of welded connectionsduring the Northridge (1994), and Hyogo-ken Nanbu(1995)earthquakes spiked interest in investigating the use of bolted connections in the construction of steel frames in seismic regions. With their lower construction costs and simple fabrication process, bolted partial-strength semi-rigid connections were evaluated as a viable alternative and their fundamental characteristics were assessed both experimentally and analytically. The cyclic behaviour of the connection was evaluated through testing of beam-column subassemblies. The experimental results demonstrated the large energy absorption capabilities of these connections under cyclic loading with stable hysteretic behaviour[8]. In addition, analytical models aimed at capturing the complicated behaviour of the connection such as slip, friction between surfaces in contact and prying action were heavily investigated. The previously conducted experimental and analytical studies were aimed at assessing the behaviour of the connection on component level bases. The assessment of the seismic performance of steel frames is then conducted using an idealized moment-rotation relationships obtained from the experimental results or the finite element models. The drawback of using such approach, in addition to idealizing the moment-rotation relationships, is that the interaction between the beam and column flanges and the angles comprising the connection is not captured. Such interaction is essential as it influences the spread of yielding in the beam and the overall joint rotation. In the discussion to follow, a new system-level hybrid simulation(experiment-analysis)application for the seismic assessment of steel frames with top-and-seat angle with double web-angle(TSADWA) connections is presented. In addition, major results of the simulations are discussed including global frame performance(drift and base shear)as well as localized connection behaviour(moment-rotation relationship). The hybrid simulation included realistic experimental and analytical modules, which were developed and successfully integrated in a closed-loop system-level simulation at the MUST-SIM facility at UIUC.

2. 2　Description of the Frames

The structure considered is a 2-story, 4-bay (longitudinal) and 2-bay (transverse) steel frame. The frame is assumed to be located in Los Angeles, California on soft soil. The height of the first and second story is 4. 57 m and 4. 11 m, respectively, and the bay width is 9. 14 m. The perimeter frames are special moment resisting frames (SMRF) designed according to the Structural Seismic Design Manual, Volume 3 of the International Building Code[9], while the inner frames are only responsible for carrying the gravity load. For the design of the frame, the importance factor (I) is 1, the force modification factor (R) is 8, the overstrength factor (Ω) is 3. The strong-column weak-beam design approach, with the panel zone remaining elastic, was used for the SMRF and resulted in beam and column sizes of W18x40 and W14 × 159, respectively. Plan view of the structure and an elevation of a typical SMRF are shown in *Fig. 2*. Following the sizing of the beams and columns, the assumed rigid connections in the frame were redesigned to reflect partial strength and semi-rigidity. Three different frames were considered with TSADWA connections designed according to the Eurocode 3[10]. The sizes of the angles and the bolts were optimized such that the resulting capacity of the connections employed in the three frames is 30%, 50%, and 70%, respectively, of the plastic moment capacity of the beam. The standardized parameters typically used for describing the geometry of these types of connections are shown in *Fig. 2*. The geometrical parameters for all three connections are also listed in the figure.

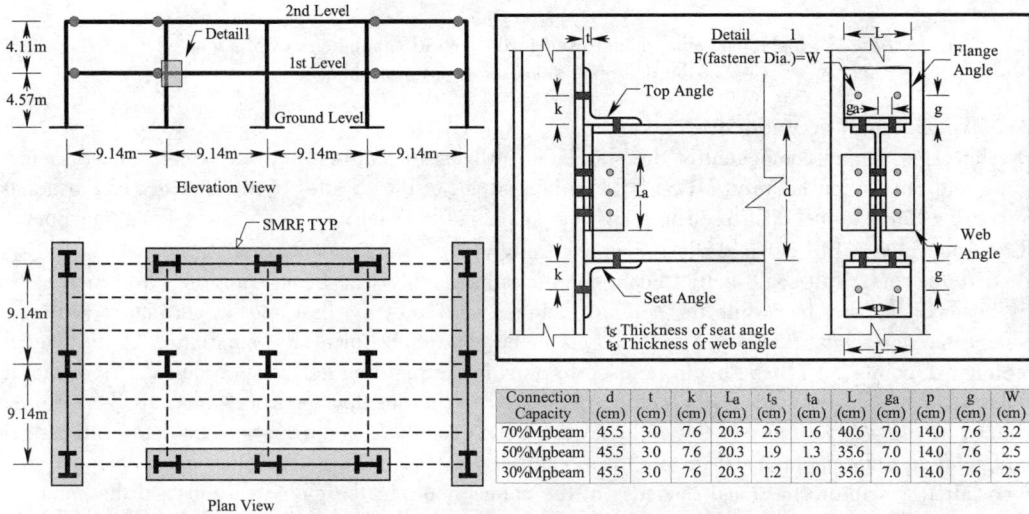

Connection Capacity	d (cm)	t (cm)	k (cm)	L_a (cm)	t_s (cm)	t_a (cm)	L (cm)	g_a (cm)	p (cm)	g (cm)	W (cm)
70%Mpbeam	45.5	3.0	7.6	20.3	2.5	1.6	40.6	7.0	14.0	7.6	3.2
50%Mpbeam	45.5	3.0	7.6	20.3	1.9	1.3	35.6	7.0	14.0	7.6	2.5
30%Mpbeam	45.5	3.0	7.6	20.3	1.2	1.0	35.6	7.0	14.0	7.6	2.5

Fig. 2. Geometric layout of the sample frames with dashed box on the plan view marking the physical component of the simulations

2. 3　Analytical component

Analytical models of frames have utilized line elements connected with springs representing the load deformation characteristics of the connection. Due to its minimal computational demands, this modelling approach has been viewed as the best alternative for hybrid simulation since the number of elements in this case is small and significant time is not required to complete a simulation step. However, the models typically represent idealized behaviour and in many cases cannot capture the local response of the various connection components. Furthermore, the deformation and the spread of yielding in the beam are not well represented since the prying action between the beam flanges and the top and seat angles is neither physically modelled nor accounted for. In light of these arguments, an inelastic finite element model is employed in the current investigation. The model comprises 2D plane strain elements for the beam-to-column connections and 1D beam elements between subsequent connections and is developed using ABAQUS which is a general purpose commercial package[11]. The model includes various behavioural features namely; 1) bolt preload, 2) friction between faying surfaces, 3) connection slip, 4) the effect of bolt-hole ovalization, and 5) hot-rolling residual stresses in the angles. In addition, the effect of the inner gravity frames on the stability of the moment resisting frame (i. e. large P-D effect) is included in the model through a leaner column modelled as two truss elements pinned at the base and at the first floor level. Tie multi-point constraints were used to provide rigid links between the semi-rigid frames and the leaner column. *Fig. 3* shows the described model with a zoom-in view of the deformed shape of the middle connection in the first floor.

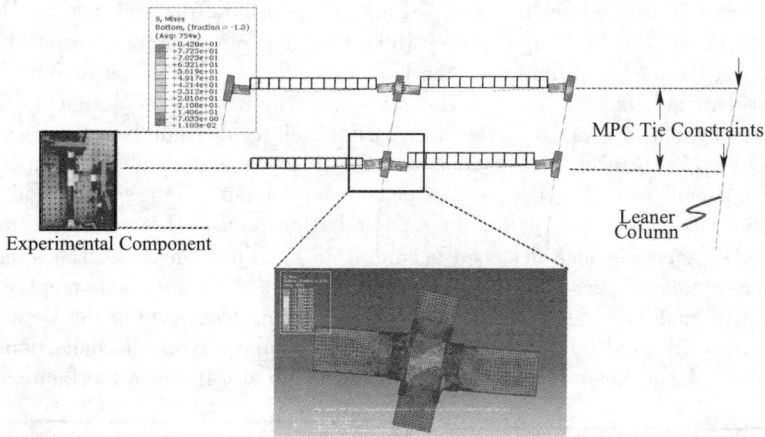

Fig. 3. Analytical frame model used in the hybrid simulations with a zoomed view of the deformed shape of an interior connection

2. 4　Experimental component

The experimental component of the simulation utilizes the Multi-Axial Full-Scale Sub-Structured Testing and Simulation Facility(MUST-SIM) which is part of the 15 sites of the Network of Earthquake Engineering Simulations(NEES). The main components of the facility include three Load and Boundary Condition Boxes(LBCBs) and the L-shaped strong wall[12]. The experimental component of the simulation represents a full-scale bolted beam-column subassembly. The beam comprises a portion of first story beam in the first bay while the column includes portion of the first and second story columns in the same bay for a total number of three points to be controlled during the simulation. At each of the three control points, an LBCB should be used to provide the required deformation commands during the simulation. However, the base of the column was fixed to the lab floor and only two LBCBs were used as shown in *Fig. 4*. The fixity of the column base was accounted for during the simulation through the use of relative deformation between all three control points. The instrumentation plan was developed and installed to capture the local response of the connection and the global response of the beam-column subassembly. In each test, a total of 175 channels were installed on the specimen and recorded using a National Instrument Data Acquisition(NI-DAQ) system. In addition, each LBCB houses 6 load cells and 6 LVDTs for displacement and load measurements, respectively, for each actuator. Fig. 4 shows the overall experimental setup and an example of linear pot arrangements used for measuring connection rotation.

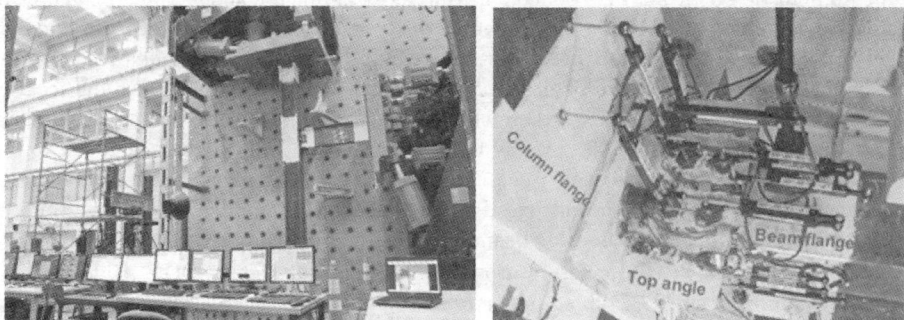

Fig. 4. Experimental setup(left) and detailed instrumentation of the connection(right).

2. 5 Integration Scheme for PSD Testing

The integration of the experimental and analytical modules in this study was conducted using UI-SIMCOR[3]. The simulation started with the stiffness evaluation step, followed by gravity loading, and finally the earthquake/dynamic simulation steps. In the earthquake steps, time stepping is achieved u-sing the α-Operator Splitting method. A schematic of the hybrid simulation application is shown in Fig. 5.

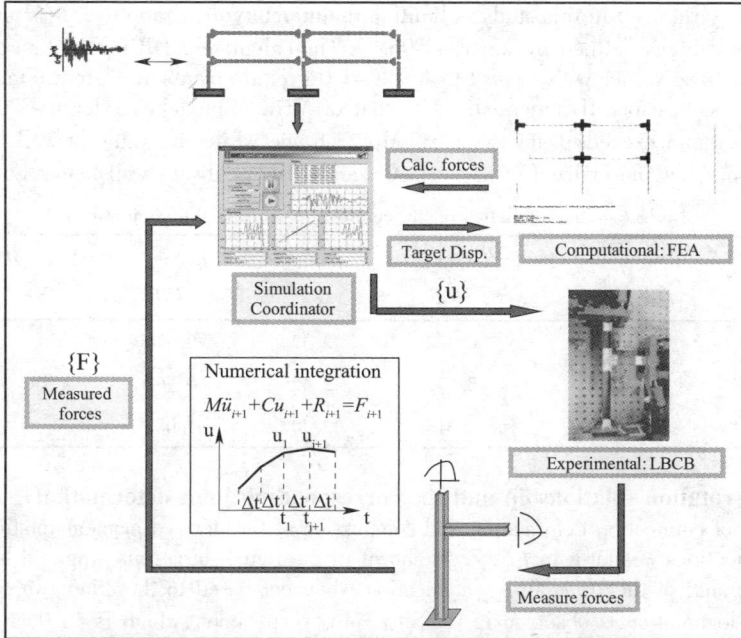

Fig. 5. Hybrid simulation framework for the investigated frames

3 SIMULATIONS RESULTS AND OBSERVATIONS

To ensure large demand on the structure during its period elongation, the 1989 Loma Prieta earth-quake was selected for the hybrid simulation. Specifically, the station used is USGS 1662 Emeryville, 77 km from the epicenter of the earthquake, on soft soil with peak ground acceleration of 0. 26 g.

3. 1 Interstory drift and Base shear

A comparison of the second-storyinterstory drift ratio(IDR) and the base shear normalized by the total weight for all three frames during the hybrid simulations is shown in Fig. 6. It is noted from the figure that the IDR response of the frames are very similar up to time 5 sec. In the time range of 5 sec

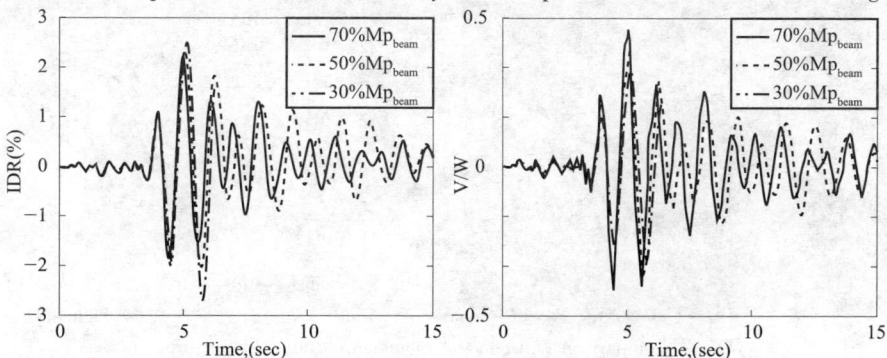

Fig. 6. Comparison of the interstroy drift ratio and base shear normalized by the total weight of the structure

to 6. 42 sec, the 30% Mp_{beam} frame starts to show signs of larger period elongation when compared to the other frames. This is a result of the nonlinearity experienced by the 30% Mp_{beam} frame at lower displacement. Noteworthy that for the 30% Mp_{beam}, the simulation was not completed and stopped after time 6. 42 sec of the earthquake motion due to conversion problems associated with the analytical model.

In addition to calculating the maximum IDR, the frames are assessed using two different performance levels, namely Design Basis Earthquake (DBE) and Maximum Considered Earthquakes (MCE). The acceptance criteria used in this study is limiting theinterstory drift ratio to 2. 5% and 5% for DBE and MCE, respectively, as defined by ASCE 41-06[13]. The calculated IDR for the first and second story and their normalized values with respect to ASCE 41-06 requirements are listed in table 1. As shown in the table, the second-story IDR normalized to that of MCE approaches 1 for the 70% Mp_{beam} and 50% Mp_{beam} frames and exceeds 1 for the 30% Mp_{beam} frame while the value is well below 1 for the DBE. The first-story IDR normalized to that of MCE and DBE is always well below 1.

Table 1. Characteristics of the connections during the simulations

Mp_{beam}	IDR^{1st}_{Max} (%)	$\dfrac{IDR^{1st}_{Max}}{IDR^{DBE}_{ASCE41}}$	$\dfrac{\mid IDR^{1st}_{Max} \mid}{IDR^{MCE}_{ASCE41}}$	IDR^{2nd}_{Max} (%)	$\dfrac{\mid IDR^{2nd}_{Max} \mid}{IDR^{DBE}_{ASCE41}}$	$\dfrac{\mid IDR^{2nd}_{Max} \mid}{IDR^{MCE}_{ASCE41}}$
70%	1. 61	0. 644	0. 322	2. 32	0. 928	0. 464
50%	1. 86	0. 744	0. 372	2. 42	0. 968	0. 484
30%	1. 58	0. 632	0. 316	2. 70	1. 080	0. 540

3. 2   3Moment-rotation relationship and the corresponding joint deformation

A depiction of connection deformation and comparison of the derived moment-rotation relationships for all three connections is shown in Fig. 7. As shown in the figure, large pinching and hardening is observed in the response of the 70% Mp_{beam} connection when compared to the other two connections. The highest stiffness degradation is observed in the 50% Mp_{beam} specimen, which is 46. 05% of the original stiffness, followed by the 30% Mp_{beam} specimen, which experienced degradation in its stiffness of 33. 59%, and finally the 70% Mp_{beam} specimen with stiffness degradation of 23. 47%. The characteristics of the connections during the simulations are listed in table 2. The table includes values for the initial stiffness (k_i), the unloading stiffness (k_u), the percent of stiffness degradation (k_{deg}), the maximum moment and rotations experienced by the connections, and the energy dissipated by each connection during the simulations. The maximum moment sustained by the 70% Mp_{beam}, 50% Mp_{beam}, and 30% Mp_{beam} connections is equal to 36, 404 kN. mm, 28, 879 kN. mm, and 19, 298 kN. mm, with corresponding rotations of 0. 0196 rad, 0. 0271 rad, and 0. 3400 rad, respectively. The sustained rotations are below the 0. 04 rad required by AISC since the demand did not reach such limit. At the maximum level of rotation no failures were observed in any of the subassembly components.

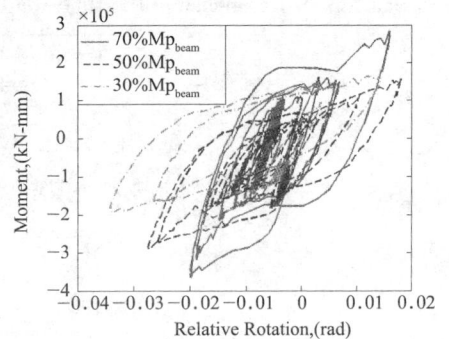

Fig. 7. Comparison between the moment-rotation relationships

Table 2. Characteristics of the connections during the simulations

Mp_{beam}	k_i (kN. mm/rad)	k_u (kN. mm/rad)	k_{deg} (%)	$|M|_{Max}$ (kN. mm)	% Mp_{beam}	$|\theta|_{Max}$ (rad)	Energy Dissipated (kN. mm. rad)
70%	5769941	4415751	23. 4	36404	82. 0	0. 0196	2205
50%	5584996	3013508	46. 0	28879	65. 2	0. 0271	2005
30%	3463221	2299975	33. 6	19298	43. 6	0. 034	1238

4 SUMMARY AND CONCLUSION

In summary, the concept of hybrid simulation is introduced and its advantages and disadvantages highlighted. In addition, a new sub-structured PSD hybrid simulation approach is developed for system-level evaluation of semi-rigid partial-strength steel frames. In this investigation, an experimental and an analytical component of a 2-bay 2-story semi-rigid partial-strength frame are used. Specifically, three connections with capacities equal to 30%, 50%, and 70% of the plastic moment capacity of the beam are employed in three different frames to investigate the response of the frames under seismic events under the 1989 Loma Prieta earthquake record. The following conclusions can be made on the discussed hybrid simulation framework and its application to the subject frames.

– Hybrid simulations allow for collaboration of various analytical platforms and experimental sites into a simulation of large, complex, and interacting system.

– The discussed platform is object-oriented so it can be easily extended to include various platforms.

– When comparing the results of all three hybrid simulations, it is observed that the maximum base shear is developed in the 70% Mp_{beam} frame followed by the 50% Mp_{beam} frame then the 30% Mp_{beam} frame.

– The maximum IDR is equal to 2. 7% in the second story of the 30% Mp_{beam} frame, which is slightly above the 2. 5% limit specified by ASCE 41-06.

– The moment sustained by the connections did not reach a plateau and the behaviour of all three connections was highlighted by stable hysteretic behaviour and high energy dissipation.

– The maximum moment sustained by the 70% Mpb_{eam}, 50% Mp_{beam}, and 30% Mp_{beam} connections is equal to 3, 222 kips. in, 2, 556 kips. in, and 1, 708 kips. in, respectively, with corresponding rotations of 0. 0196 rad, 0. 0271 rad, and 0. 3400 rad, respectively.

– Large pinching and hardening is observed in the 70% Mp_{beam} connection.

– The sustained rotations are below the 0. 04 rad required by AISC since the demand did not reach such limit. At the maximum level of rotation no failures were observed in any of the subassembly components.

– The highest stiffness degradation of the original stiffness is 46. 049%, 33. 59%, and 23. 47% for the 50% Mp_{beam}, 30% Mp_{beam}, and 70% Mp_{beam} respectively.

REFERENCES

[1] Hakuno, M., M. Shidawara, et al.. "Dynamic Destructive Test of a Cantilever Beam, Controlled by an Analog-computer. ", Trans. of Japan Society of Civil Engineering, 171: 1-9, 1969.

[2] Schellenberg A. and Mahin S., "Integration of hybrid simulation within the general-purpose computational framework OpenSees", Proceedings of the Eighth U. S. National Conference on Earthquake Engineering, San Francisco, U. S. A, 2005.

[3] Kwon, O., Elnashai, A. S., and Spencer, B. F., "A framework for distributed analytical and hybrid simulations. ", Structural Engineering and Mechanics, 30(3), 331-350, 2008.

[4] Nakashima, M., McCormick, J., and Wang, T., "Hybrid simulation: a historical perspective applications" published in "Hybrid simulation: theory, implementations and applications", Edited by V. E. Saouma and M. V. Sivaselvan, Taylor and Francis, 2008.

[5] Ricles, J., Sause, R., Karavasilis, T., and Chen, C., "Performance-based seismic design of building systems with dampers: experimental validation using real-time hybrid simulation", Proceed-

    ings 5th International Conference on Earthquake Engineering (5ICEE) , Tokyo Institute of Technology , Tokyo , Japan , 2010.

[6] Nakashima , M. and Kato , H. , "Experimental error growth behaviour and error growth control in on-line computer test-control method" , Building Research Institute , BRI-Report No. 123 , Ministry of Construction , Tsukuba , Japan , 1987.

[7] Combescure , D. and Pegon , P. , "a-Operator splitting integration technique for pseudo-dynamic testing. Error Propagation analysis" , *Soil Dyn. Earthq. Engr.* , 427-443 , 1997.

[8] Azizinamini , A. and Radziminski , J. B. , "Static and cyclic performance of semi-rigid steel beam-to-column connections. " , *J. Struct. Eng.* , 115 (12) , 2979-2998 , 1989.

[9] IBC Structural/Seismic Design Manual. (2006). "Volume 3 : building design examples for steel and concrete. "

[10] Eurocode 3 , ENV-1993-1-1 : 1992/A2. "Annex J. " , design of steel structures-Joint in building frames , CEN , European Committee for Standardization , Ref. No. ENV 1993-1-1 : 1992/A2 : 2 1998 E , Brussels. 1998.

[11] Simula (2007). "Abaqus/standard users' manual" Hibbit , Karlsson and Sorenson.

[12] Elnashai , A. , Spencer , B. , Kuchma , D. , Ghaboussi , J. , Hashash , Y. and Can , Q. , "Multi-Axial Full-Scale Sub-Structured Testing and Simulation (MUST-SIM) Facilityat the University of Illinois at Urbana-Champaign" , *Proceedings of the 13th World Conference on Earthquake Engineering* , Vancouver , B. C. , Canada , Paper No. 1756 , 2004.

[13] ASCE SEI 41-06. "Seismic rehabilitation of existing buildings. " American Society of Civil Engineers , 2007.

TEN YEARS OF E-DEFENSE ACTIVITIES— COLLAPSE, FUNCTIONALITY, AND RESILIENCE

Masayoshi Nakashima[*] and Taichiro Okazaki[**]
[*] Disaster Prevention Research Institute, Kyoto University
e-mails: nakashima@ archi. kyoto-u. ac. jp
[**] Hokkaido University
e-mail: tokazaki@ eng. hokudai. ac. jp

KEYWORDS: E-Defense, Shake-table test, Full-scale test, Collapse, Resilience.

ABSTRACT

Operation of E-Defense started in March 2005, ten years after the 1995 Kobe Earthquake. This paper reviews E-Defense research activities in steel building structures during the first ten years of operation. The initial objective of E-Defense was to understand how engineered buildings can collapse due to strong ground motions and to identify the reserve capacity beyond the capacity required by code provisions. Later, in recognition of the risk posed by ocean-ridge earthquakes, the safety of existing high-rise buildings became a focus. Different aspects of high-rise buildings have been addressed: functionality during long oscillation; performance under a very large number of loading cycles; and reserve capacity against collapse.

1 INTRODUCTION

As evidenced bythe 1995 Hyogoken-Nanbu(Kobe) earthquake and 2011 off the Pacific coast of Tohoku(Tohoku) earthquake, earthquakes pose a formidable challenge to Japan. We are frequently reminded that seismic hazard evolves with our society and that preparation is vital to the wellbeing of our nation. Fig. 1 shows a map of Japan with some past and expected future earthquakes. While the entire nation is under high seismic risk, a large earthquake produced by the ocean ridge called the Nankai trough is recognized as the source of substantial risk. The trough, running deep along the Pacific Coast of Japan, is divided into three segments, Tokai, Tonankai, and Nankai, named from the east. Each segment has caused large earthquakes in an interval of 100 to 150 years. Based on the pattern of previous occurrences, the next large earthquake is expected to occur by the middle of this century.

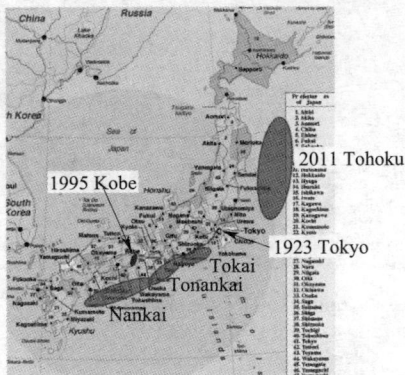

Fig. 1. Past and expected earthquakes

E-Defense is a large shake-table facility constructed in response to the 1995 Kobe earthquake with a mission to mitigate the substantial seismic risk faced in Japan. Since operation started in March 2005, a large number of projects have been conducted at E-Defense. Through the projects, E-Defense has played a key role in advancing earthquake engineering and in promoting countermeasures to prepare for future earthquakes. The initial objective of E-Defense was to understand how severe earthquakes can cause collapse of engineered buildings. Later, the risks posed by ocean-ridge earthquakes, particularly to high-rise and seismically-isolated buildings, became a major focus. More recently, in the wake of the 2011 Tohoku earthquake, resiliency of the urban built-environment has become a societal demand.

This paper will introduce the E-Defense shake table facility and subsequently highlight four projects on steel building structures.

2   E-Defense

E-Defense is the nickname for a large three-dimension shake table facility managed and operated by the National Research Institute for Earth Science and Disaster Prevention (NIED) of Japan. E-Defense was conceived as a facility to test full structures in full scale, under real earthquakes, to collapse. As shown in *Fig. 2*, the shake table is 20 by 15 meters in plan dimension, controlled by five actuators in each horizontal direction and fourteen actuators in the vertical direction. Originally, the table was designed to reproduce, under the full payload of 12 MN (1,200 tonf), motions up to a maximum velocity of 2.0 m/s. Originally, E-Defense focused on high velocity motions in the period range between 0.2 and 2.0 s and lasting less than one minute, which represent destructive near-field motions due to in-land earthquakes. Recently, in order to address the urgency to prepare for ocean-ridge earthquakessuch as the expected Nankai event and in the wake of the 2011 Tohoku earthquake, the shake table was upgraded to enable reproduction of long-period, long-duration motions. Consequently, E-Defense is a key facility to address a wide range of earthquake engineering concerns and to examine the real performance of structures under any expected earthquake ground motion and to develop new systems with advanced capabilities.

Since its inauguration in April 2005, E-Defense has conducted six to eight projects on an annual basis. The projects conducted to date range from residential timber buildings, steel buildings, reinforced concrete buildings, seismically isolated buildings, soil-pile-structure interaction during liquefaction, underground structures, reinforced concrete bridge piers, to equipments in nuclear power plants. The remainder of this paper will highlight selected projects in steel building structures.

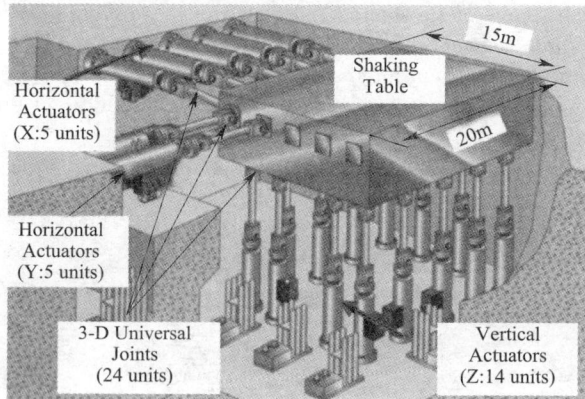

Fig. 2. E-Defense shake table

3   FOUR-STORY BUILDING COLLAPSE

The immediate recognition after the 1995 Kobe earthquake was the need to examine older buildings and infrastructures that do not meet current design requirements. The objectives of earlier E-Defense projects was to accurately understand the seismic capacity of older structures, and then to develop retrofit and rehabilitation methods in preparation for the next major earthquake. An example is the four-story steel moment-frame building tested to collapse[2] described below.

Fig. 3. Full-scale 4-story
specimen

The specimen was a four-story, two-bay by one-bay steel moment frame, shown in *Fig. 3*, with a plan dimension of 10 by 6 m and height of 14 m. The specimen was designed following the then-current specifications and practices. The moment frame comprised cold-formed HSS columns and rolled wide-flange beams connected using the through-diaphragm detail. External cladding panels were placed on three sides of the frame. Partition walls and gypsum ceiling boards were attached with windows and doors properly installed. The specimen was repeatedly subjected to the JR Takatori motion recorded from the 1995 Kobe earthquake, with progressively increasing amplitude from 20, 40, 60, to 100% of the record.

During the 20% motion, the specimen behaved elastically and the peak story drift ratio remained under 0.005 rad. This motion is equivalent to the Level I earthquake load in Japan under which allowable stress design is performed. During the 40 motion, minor yielding occurred at the column base and interior panel zones at lower floors. This motion is equivalent to Level 2 earthquake load under which the ultimate strength is checked. At the first and second stories, the drift ratio reached 0.01 rad. During the 60% motion, a sidesway energy-dissipation mechanism was formed over the first two stories. The drift ratio at the first story reached 0.02 rad. The ALC panels and partition walls were severally damaged. Residual drift in the first story was noted.

Within the first 7 seconds of the 100% motion, the specimen collapsed in the first-story. *Fig. 4*a shows the specimen resting on the safeguard system. Interestingly, the energy-dissipation mechanism transitioned from sidesway over multiple stories to sidesway of the first-story only. The primary reason for this transition was severe strength deterioration of the first-story columns caused by local buckling at the base and top, as shown in *Fig. 4*b and c, which, in turn, caused moment redistribution in the structural system. The data implies that, while the specimen was designed appropriately for codified design loads, the columns could not sustain the increased forces due to strain hardening after yielding of the panel zones. For this particular test, the effect of vertical motion was secondary: the vertical component of the JR Takatori motion is relatively small, and the variation in the axial force in the columns was primary due to overturning moments. Minimal torsion occurred in the specimen until collapse.

A blind analysis competition was conducted to predict the response of the specimen during the 60 and 100% motions[3]. Forty-seven groups of engineers and researchers around the world participated in this competition. The collected data was used to examine the ability of current analysis tools and to study the pros and cons of different analysis assumptions. No clear advantage was noted between 3D and 2D analysis for reliable estimation of the specimen, perhaps because the specimen was regularly planned and exhibited minimal torsion. Rayleigh damping with a damping ratio larger than 2% was adequate. Explicit consideration of P-Delta effects and modelling strength and stiffness degradation of components was essential to capture the behavior up to collapse. This is an example of how the extensive data collected from E-Defense tests can be used to advance numerical analysis techniques.

Fig. 4. Failure of 4-story specimen: (a) first-story collapse mechanism;
(b) local buckling at base of first-story column; (c) local buckling at top of first story column

4 FUNCTIONALITY OF HIGH-RISE BUILDINGS

Japan is preparing for the large ocean-ridge earthquake that is expected to occur by the middle of this century. One serious concern about such earthquakes is the long-period, long-duration ground motion caused in large cities including the Tokyo metropolitan area, in which several hundred high-rise buildings exist. Such motions can produce very large response velocity and displacement in the upper floors, which can damage nonstructural elements, furniture, and general contents. A test setup was devised to reproduce the large floor response in high-rise buildings[4].

Fig. 5 shows a rigid steel frame placed above two layers of concrete slab and rubber bearings. The system beneath the steel frame amplified the shake table motion to the floor responses expected in upper floors of high-rise buildings. The shake table motion was carefully adjusted so that motion of the

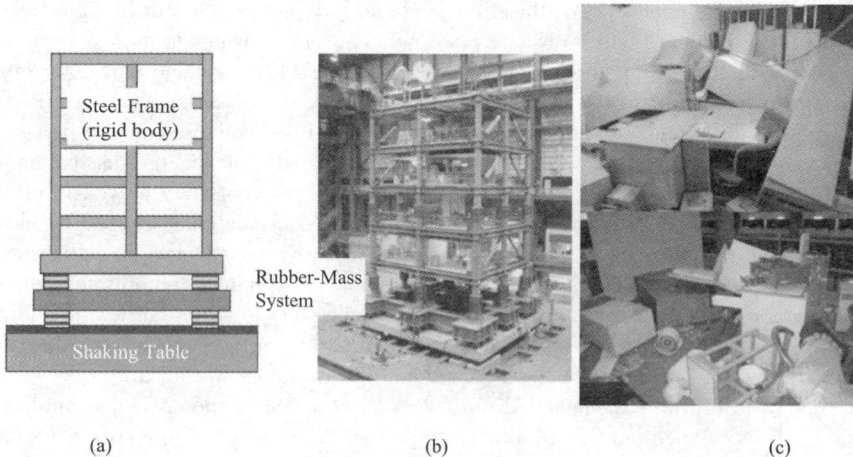

Fig. 5. Floor response of high-rise buildings: (a) concept of response amplification
system; (b) photo of test frame; (c) damage to furniture

steel frame coincided with the desired floor response. The steel frame measured a maximum displacement of 1.3 m and maximum velocity of 2.4 m/s. The test was conducted with various types of furniture placed on each of the four floors of the steel frame. The tests demonstrated the serious damage to contents, as shown in *Fig. 5c*, and the critical need to clamp furniture against sliding and overturning.

Anothertest setup, shown in *Fig. 6*, was devised to examine the structural damage to lower stories of high-rise buildings[1]. Opposite to the previous setup, the upper floors instead of the lower floors were represented by mass-and-damper layers. The combined system of the 4-story steel frame and mass-and-damper layers was designed to respond in the target periods and modes. The fundamental vibration period was 2.1 s. As the system was subject to a simulated long-period, long-duration ground motion, the steel frame experienced a very large number of small story drifts. In the first specimen, which was constructed according to practice in the 1970's, extensive fracture occurred in the beam bottom flanges near the CJP groove weld. The second specimen demonstrated that the damage may be mitigated effectively by proposed reinforcing schemes.

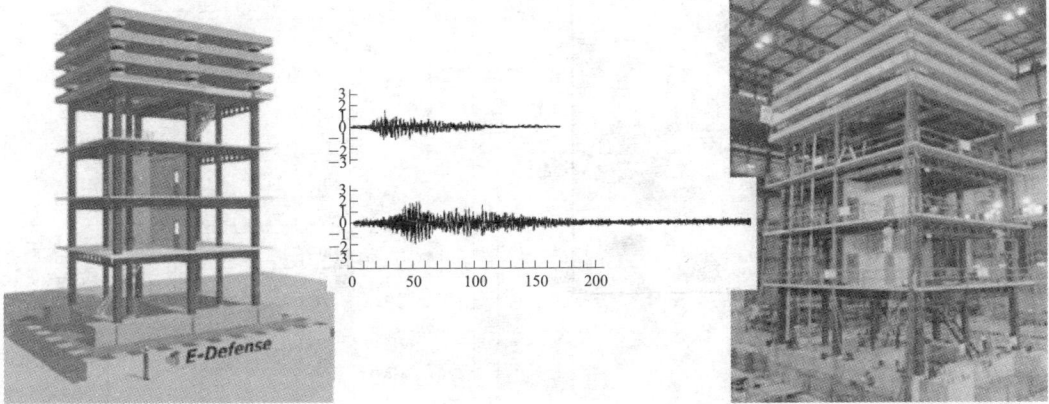

Fig. 6. Structural performance of high-rise buildings

5   COLLAPSE OF HIGH-RISE BUILDINGS

The 2011 Tohoku earthquake was the largest earthquake recorded in Japan in modern history. In contrast to the devastation caused by tsunami, the structural damage caused by ground shaking was rather limited. However, even in Tokyo which was more than 300-km away from the epicentre and where the ground motion was moderate, high-rise buildings vibrated for over 10 minutes. As suspected in recent years, earthquakes at oceanic troughs can produce long-duration ground motions whose dominant period coincides with the natural vibration period of high-rise buildings. While high-rise buildings are, in general, designed for a larger safety margin and constructed with greater care than the majority of buildings, observations from the Tohoku earthquake raise concerns for the safety and performance of high-rises. In particular, our building codes do not address such an extreme number of loading cycles. A project titled "Maintenance and recovery of functionality in urban infrastructures" was initiated to address such concerns for high-rise buildings. A 1/3-scale, 18-story, steel moment-resisting frame complete with composite slabs was tested to collapse at E-Defense[5].

As shown in *Fig. 7*, the specimen was 25.3-m tall and excited in one direction only. A protection frame was placed next to the specimen to prevent any damage to the facility when the specimen will have collapsed. The columns were built-up box sections of SMA490A steel in the lower six stories and BCR295 cold-formed HSS in all other stories. The primary beams were built-up I-sections of SMA490A steel. The beam-to-column connections adopted the internal diaphragm detail at box section columns and the through-diaphragm detail at HSS columns. The fundamental vibration period of the specimen was estimated to be 1.15 seconds. The specimen was subjected to a synthetic ground motion for a scenario earthquake where all three segments of the Nankai trough rupture simultaneously. The ground motion had a maximum acceleration of 4.2 m/s and lasted 460 seconds. The "average" ground motion was adjusted to time shortened to $1/3^{1/2}$ according to the scaling rule and to a peak spectral velocity, pSv, of 110 cm/s. The same ground motion was repeated with amplitude starting from small (pSv = 0.40 m/s), medium(0.81 m/s), average(1.1 m/s), large(1.8 m), very large(2.2 m/s), extremely large(3.4 m/s), to final(4.2 m/s). The final motion was repeated until the specimen collapsed.

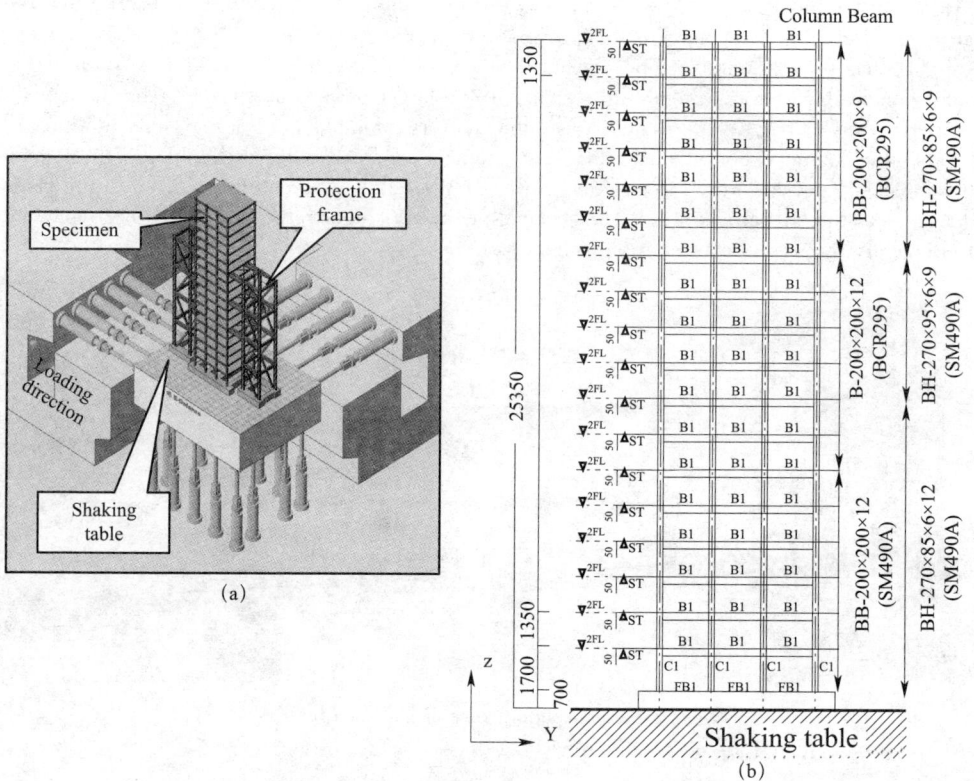

Fig. 7. Overview of high-rise collapse test: (a) Schematic; (b) Elevation of specimen

The specimen remained elastic during the small motion, but limited yielding was measured during the medium motion. Beam yielding was observed over 14 of the 18 floors during the large motion. Fracture of beams was observed in lower-floor columns during the very large to extremely large motions, at which stage the story drift ratio exceeded 1/30 rad. As the extremely large motion was repeated twice and final motion was repeated twice, the specimen gradually deformed in the same direction with story drift concentrating in the first five stories. Ultimately, as the final ground motion, which was 3. 8 times the average motion, was repeated for the third time, the specimen collapsed. *Fig. 8* shows the collapsed specimen leaning on the protection frame and close-up views ofrepresentative damage to the beams and columns.

6 CONCLUSION

The paperdescribed how E-Defense has played a critical role in seismic risk mitigation in Japan. Some of the research projects on steel building structures were highlighted. E-Defense was constructed in direct response to the 1995 Kobe earthquake. Therefore, an earlier project examined how a steel office building that met the then-current design requirements can collapse due to a ground motion recorded from the Kobe earthquake. Over the years, E-Defense has addressed a wide range of earthquake research needs. One of the largest seismicrisks recognized in Japan is the large earthquakes caused periodically by the Nankai trough, which can cause long-duration, long period motions in metropolitan areas. High-rise steel buildings can experience a large number of loading cycles and, at the upper stories, large floor responses that can lead to nonstrucutral damage and occupancy disruption. E-Defense tests have addressed these issues by devising special test setups that reproduce the response of high-rise buildings.

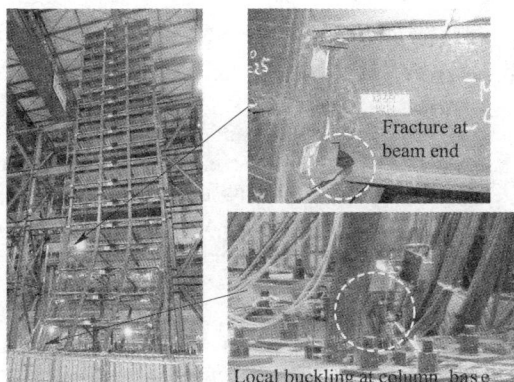

Fig. 8. Photograph taken after collapse

ACKNOWLEDGEMENT

The description presented in this paper was based primarily on various research projects conducted at E-Defense. Particular thanks go to Dr. Koichi Kajiwara, Director of E-Defense, for his willingness to share the test data with the writers. The writers also wish to extend their appreciation to numerous organizations and individuals who took part in the experimental projects introduced in the paper. The second writer is grateful to NSFC for their support under Grant No. 51408562. One of the ground motions used in this study was provided by the KiK-net of NIED.

REFERENCES

[1] Nakashima M. , "Roles of large structural testing for the advancement of earthquake engineering", *The* 14*th World Conference of Earthquake Engineering*, Beijing, China, 2008.

[2] Suita K, Yamada S, Tada M, Kasai K, Matsuoka Y, Shimada Y. , "Collapse experiment on 4-story steel moment frame: Part 2 detail of collapse behavior", *The* 14*th World Conf. on Earthquake Engineering*, Beijing, China, 2008.

[3] Lignos DG, Hikino T, Matsuoka Y, Nakashima M. , "Collapse assessment of steel moment frames based on E-Defense full-scale shake table collapse tests", *Journal of Structural Engineering*, ASCE, 139(1): 120-132, 2013.

[4] Ji X, Kajiwara, K, Nagae T, Enokida R, Nakashima M. , "A substructure shaking table test for reproduction of earthquake response of high-rise buildings", *Earthquake Engineering and Structural Dynamics*, 38: 1381-1399, 2009.

[5] Nakashima M and Luo YB. , "Research needs for earthquake engineering and current research and development efforts in Japan", *The* 5*th Asia Conference on Earthquake Engineering*, Taipei, Taiwan, 2014.

THE ACTIVITIES OF THE ECCS-TC13 SEISMIC COMMITTEE: BRIDGING THE GAP BETWEEN RESEARCH AND STANDARDS

Raffaele Landolfo

University of Naples Federico II, Department of Structures for Engineering and Architecture
e-mail: landolfo@ unina. it

KEYWORDS: Codification, Committee, Eurocode 8, Seismic design, Steel Structures.

ABSTRACT

The ECCS-TC13 Seismic Committee aims at promoting the use of steel structures in seismic areas. To this end, several activities have been carried out since its foundation. Among those, the Committee recently performed an assessment of the EN1998-1 (hereinafter also indicated as Eurocode 8 or EC8) provisions for the seismic design of steel structures, with the aim to find the main criticisms, and is actively involved in the preparation of the new version of the code, since the recent nomination as a specific working group of CEN/TC250/SC8. This paper, following a general overview on TC13 activities, discusses some issues that are crucial for a proper design of a steel structure in seismic areas and gives directions to bridge the gap between research and codification. Indeed, weaknesses of the current version of EN1998-1 are highlighted and some improvements are suggested. Finally, open research questions, which require further investigations, are presented with the aim to open new routes towards the next generation of the European seismic code.

1 INTRODUCTION

The European Convention for Constructional Steelwork (ECCS) is an international federation of national steelwork associations representing the steel construction industry in their respective countries. ECCS was established in 1955 and, nowadays, represents 18 European countries.

ECCS mission is to foster the use of steel in the construction field. To achieve this purpose, ECCS addresses its efforts on different action lines, ranging from the development of standards to promotional information. Publications, conferences, and active representation in European and International Committees dealing with standardisation, research and development and education are some of ECCS actions played to achieve its goals. All these activities can be carried out thanks to the particular organization of ECCS, which brings together the Steel Industry, the Fabrication and Contracting specialists, and the Academic world through an international network of construction representatives, steel producers, and technical centres.

Concerning the codification review and development, the ECCS activities are framed within several Technical Committees (TC) under the supervision of the ECCS Technical Management Board (TMB), involving more than 300 experts.

TCs are conceived as a forum between experts in specific topics that establishes consensus on European practice and provides the undisputed background for normalization, identifying the ongoing developments in specific fields and establishing priorities for research and developments of new products and components for the steel industry and steel construction market.

The Committee specifically devoted to the seismic design of steel structures is the TC13. As a matter of fact, the research on seismic design of steel structures is a very active branch in the field of engineering, but there is still the need to move onward in order to provide even more reliable and more competitive technical solutions. With that in mind, one key task for TC13 is to trace the guidelines for updated European seismic codes according to worldwide research trends. In addition, the Committee provides valuable support for facilitating and ensuring the transfer of knowledge to designers and structural steelwork fabricators through the publication of practical design guides and manuals.

2 THE ECCS TECHNICAL COMMITTEE 13 (TC13)

Since its establishment-about thirty years ago-TC13 addressed its activities towards both designers and standardization bodies, as reported in the mission, summarized below:

• to provide a comprehensive state-of-the-art on the ongoing research activities in the field of seismic design of steel structures;

    • to analyse the current status of European and worldwide codification;

    • to identify further research priorities and critical issues in the technical specifications;

    • to prepare background documents for the new generation ofEN1998-1[1] ;

    • to develop specific tools for designers and constructors in order to promote the use of steel structures in seismic areas.

In line with these objectives, during the '80s of the last century, TC13 worked as a consulting body of the drafting panel of EC8. Under the former Chairmanship of Federico M. Mazzolani, TC13 published the "European Recommendations for steel structures in seismic zones"[2] in 1988. These Recommendations were incorporated as the "Steel section" of the first edition of EN1998-1 (hereinafter also indicated as Eurocode 8 or EC8), and constituted the framework of the Chapter 6 of current version of the code. After this important result, the effort of TC13 was addressed to the publication of the "ECCS Manual on Design of Steel Structures in Seismic Zones"[3], which addressed practitioners with the basic principles of the codes and summarized the developments and the research results achieved within the period from the end of '70s to the beginning of '90s of the last century.

In 2007 all ECCS Technical Committees were restructured and also the organization and the membership of TC13 were renovated. Following this process, the Chairmanship changed and the Writer was appointed as the new Chair of the Committee. Consistently, also the technical secretariat changed and Aurel Stratan, from Polytechnic University of Timisoara, was appointed as the new technical secretary. The Committee membership changed also and, nowadays, 34 experts (24 full and 10 corresponding members), coming from all over the world, joined the Committee. *Fig. 1* gives an overview of the European countries represented by TC13 experts, while *Table 1* reports the composition per nationality.

Table 1. TC13 membership: Full and Corresponding Members (FM, CM).

European members	
Nation	Composition
Belgium	2 FM
Finland	1 CM
France	2 FM + 1 CM
Germany	1 FM + 1 CM
Hungary	1 FM
Italy	4 FM
Luxembourg	1 FM + 1 CM
Poland	1 FM
Portugal	3 FM + 1CM
Romania	3 FM + 1CM
Slovenia	1CM
Switzerland	1 FM + 1 CM
Turkey	2 FM + 1CM
United kingdom	2 FM
Extra European members	
Nation	Composition
USA	1 FM + 1 CM
Japan	1 FM

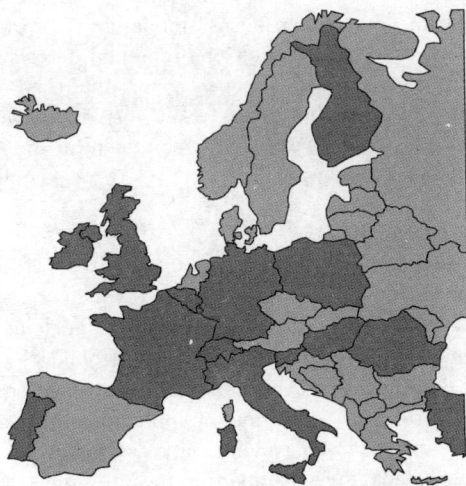

Figure1. EU Countries represented by TC13 members.

The new course of the Committee officially started with the kick-off meeting held in Naples (Italy) on 16 May 2008 (see *Fig. 2*). In that occasion, a new action plan was defined and a set of priorities was identified. Those include:

    • Identification of a list of weaknesses in the current codes;

Fig. 2. TC13 kick-off meeting in Naples(16 May 2008).

● Compilation of a list of questions from practicing engineers to create a list of priorities;
● Creation of some technical working groups to cope with the identified weakness and to address the specific priorities;
● Production of technical documents addressing these issues with the aim to contribute to the new generation of codes;

In line with TC13 mission and tradition, the work of the refurbished technical body was again effectively addressed to codification. Indeed, Committee members gathered a general complaint of the scientific community about the current limitation of design standards and codes that are not easily reactive to new research findings and to the latest developments in materials and construction practices. As a matter of fact, TC13 recognized that there is still a strong need to work so as to fill the gap between the progresses of knowledge and the professional practice and in this direction the Committee can play again a crucial role. Indeed, the European standard writing organization(e. g. the CEN Technical Committee 250(CEN/TC250) "Structural Eurocodes") has to develop consensus standards, where expert volunteers like TC13 members can guarantee the balanced representation of different interests, thus mediating between Producers and Users as well as Governmental and National agency interests.

Coming back to TC13 specific priorities, four technical working groups(TWG) have been established and subsequently the work programme of each TWG, together with the appointment of the TC13 members in charge of each task. The four TWGs are listed as follows:
– TWG1: Members and connections(coordinator D. Dubina)
– TWG2: Traditional typologies(coordinator A. Elghazouli)
– TWG3: Innovative systems(coordinator F. M. Mazzolani)
– TWG4: Low dissipative structures(coordinator A. Plumier)

The planned activities of TC13 have been organized in order to achieve short term and long term tasks. As for short term tasks, the activities were finalized to a critical review of EN1998-1[1] in its present form by means of proposals of simplifications, improvement of accuracy, reduction of requirements, extension of rules by more details and extension of rules by wider field of application. Moreover, it has been decided to develop Commentary documents in order to explain the reasons for the prescribed rules and to describe the scientific and technical background supporting the provisions. These documents are of great importance, because many of the rules prescribed in the current version of EC8 are obscure even to people working in this field either as researchers or practitioners.

In the long term the Committee aims(i) to prepare background documents for the new generation of Eurocodes namely EC8[1], and contribute in maintaining EC3[4]; (ii) to develop specific tools for designers and constructors in order to promote the use of steel structures in seismic areas; (iii) to disseminate the results.

In order to coordinate the activities, almost two TC13 meetings per year have been organized, each of them dealing with specific topics. Since 2008, the following meetings were held: Naples (16/5/2007); Timisoara (28/11/2008); London (29/5/2009); Ispra (6/11/2009); Aachen (7/5/2010); Salerno (19/11/2010); Rennes (20/5/2011); Ljubljana (18/11/2011); Budapest (11/5/2012); Esch-sur-Alzette (22/11/2012); Naples (27/6/2013); Istanbul (30/5/2014); Timisoara (7/11/2014); Porto (22/5/2015).

3 DISSEMINATION AND EDUCATION ACTIVITIES

According to its objectives, TC13 defined an action plan for the dissemination of the latest findings and outcomes in the field of seismic design of steel structures, also derived from R&D projects towards industry community and policy makers as well as general public of practitioners and Academia. According to a dynamic approach, the dissemination plan is continuously verified and updated, if needed. Dissemination activities include post-graduate training courses, support to the organization of conferences, newsletter, and publication of scientific work in different forms, i. e. state-of-art review, collaborative journal papers, reports, white papers, training handbooks.

With respect to training activities, under the aegis of ECCS, an education program has been launched and several short courses on seismic design of steel building according to EN1998-1 have been organized in different countries covering Europe from north to south. Indeed, two courses were held in Oslo (Norway) on 6-7/11/2013 and on 18/6/2014, one in Istanbul (Turkey) on 31/5-1/6/2014, and the last in Brussels (Belgium) on 16-17/10/2014. These courses were addressed to post-graduate students and practitioners working in the field of steel constructions but not confident with seismic design, as the case of low seismicity countries (e. g. the Norway) and/or not confident with the use of Eurocodes. This experience has been really encouraging as confirmed by the impressions of the attendants which positively responded, thus leading the trainers to write a design handbook with worked examples that was early provided for the last course taken in Brussels[5]. The development of this manual was encouraged and supported also by European Commission to foster the adoption and use of the Eurocodes in the EU Member States, where the most of practitioners still adopt their national codes. In line with that, TC13 is working on a more ambitious project that is the publication of a new ECCS design manual, namely "Design of steel structures for buildings in seismic areas" authored by some of TC13 experts[6] (see Fig. 3c).

As previously mentioned, the Committee is strongly involved in supporting thematic conferences by organizing specific special sessions where the members can show and discuss their research works developed within the framework of TC13 WGs. Accordingly, TC13 members are used to participating in international scientific conferences and workshops. An example in this sense is the work carried out for the STESSA Conference, which represents the direct field for the interests and the aims of TC13. Indeed, it is worth to be reminded the special session organized for the 7th Conference of STESSA in Santiago (Chile) on 9-11 January 2012, where 8 papers were presented covering the main topics addressed by the four TC13 WGs, namely members and connections, traditional and innovative structural typologies, low dissipative structures.

Recently, TC13 significantly contributed to the 7th Eurosteel conference held in Naples (Italy) on 10-12 September 2014, where a TC13 special session was organized and 10 papers were presented showing the latest outcomes of the research carried out by the committee members.

Within the field of scientific and technical dissemination, the Committee promoted also a special issue on the journal "Steel Construction" on TC13 activity (see Fig. 3a). Steel Construction is the official journal for ECCS-European Convention for Constructional Steelwork members since 2010 and it publishes peer reviewed papers covering the entire field of steel construction research. The TC13 special issue was published on the volume 4(2) in June 2011 and it consisted in an editorial preface and 4 papers, discussing on (i) the cyclic behaviour of dissipative moment-resisting beam-to-column joints, (ii) the seismic behaviour of concentrically braced frames, (iii) the use of dual steel concept to re-centre steel building after earthquake and (iv) a review on buckling restrained braces (which are not yet accounted for by EN1998).

After five years since its restructuring, one of most interesting outcome succeeded by the Committee is the publication of the book "Assessment of EC8 Provisions for Seismic Design of Steel Struc-

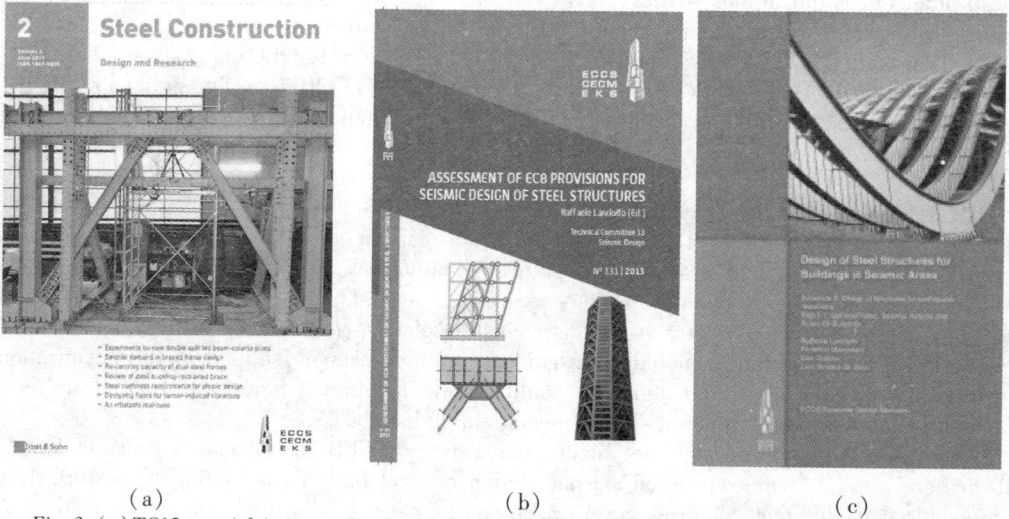

Fig. 3. (a) TC13 special issue on Steel Construction, vol. 4(2)2011; (b) ECCS/TC13 No. 131/2013:
Assessment of EC8 Provisions for Seismic Design of Steel Structures[7]; (c) ECCS handbook
in preparation "Design of steel structures for buildings in seismic areas"[6]

tures"[7], see *Fig.* 3b. This represents the finalization of the first short term objective, which was "to analyse the current status of European and worldwide codification and identify further research priorities and critical issues in the technical specifications". In this perspective, the publication summarizes all issues in the current version of EN 1998-1 needing clarification and/or development, aiming at contributing to a new generation of European codes. TC13 members actively contributed to the book providing their comments and integrations supported by their study carried out both within the TC13 action and their research programs. The report is organized in twelve Sections and one Annex.

This publication can be considered as the preparatory work to draft updated recommendation for the next version of Eurocode 8. Therefore, on the basis of the criticisms highlighted in this publication, the Committee is working to overcome the fallacies and weak points of the code thus proposing an updated alternative version of EC8. Different aspects are under investigation covering both global and local issues related to the seismic design of steel buildings. This activity is still going on and it will be collected in a new TC13 book "Toward improved European codification for the seismic design of steel structures"[8], which is in preparation.

It should be noted that all these outcomes have been obtained thanks to the voluntary and dedicated work of TC13 members, which constitutes the actual added value of the Committee and testifies the crucial and strategic importance of the addressed topics, as well.

In order to provide a brief overview of the investigated subjects, those criticisms of EN1998-1 detected by TC13 are described and discussed hereinafter.

4 ASSESSMENT OF EC8 PROVISIONS
4.1 Material overstrength
A well-established general concept in seismic design of structures is that non dissipative members must be designed on the basis of the expected material strength of the dissipative zones. The ratio between the average yield stress and the characteristic yield value for a given steel class is called γ_{ov} by EC8. There is no specific information about the values to be attributed to γ_{ov}, but National Authorities have the freedom to select the most appropriate ones. However, a constant value of $\gamma_{ov} = 1.25$ is suggested, which is contradictory with the available experimental evidence of the dependence on the yield strength of the steel, as also recently confirmed by the results of OPUS project[9]. In this research project statistical data on characteristics of steel products in Europe have been collected and processed in order to allow definition of normative values of material overstrength factors, which would account for

such aspects as steel grade, type of steel product, etc. The analysis of tensile yield strength of European steel by different producers clearly showed that the material overstrength factors specified in EN 1998-1 seems to be conservative only for higher-grade steels. Moreover, the results of tensile coupon tests sampled from hot-rolled sections showed that lower-grade steel is characterised by larger values of the actual yield strength, and consequently larger material overstrength factors[9]. On the contrary plated products have smaller overstrength in comparison with long hot rolled products.

Another criticism of Eurocode 8[1] about material overstrength is an unclear identification of the sources of overstrength, so that the reader is not always aware about the actual meaning of the coefficients suggested to cover the undesired effects due to overstrength. The code should provide the values of the overstrength coefficient with reference to each source of overstrength. Indeed, the following sources of material overstrength should be identified: 1) overstrength of delivered material; 2) code required overstrength; 3) strain-hardening overstrength; 4) random material variability overstrength; 5) overstrength due to strain rate.

Lastly, considering the recent trends of the steel market and the significant development in steel processing, the use of high strength steel(HSS) is very attractive and detailing rules on HSS deserve to be included in the next version of the code. Indeed, seismic applications potentially represent the rational field to exploit the high performance of HSS, because the combined use of HSS for non-dissipative members and of mild carbon steel(MCS) for dissipative members may allow an easier application of capacity design criteria. The expected design improvement would be obtained in terms of smaller member sizes than those obtained when using MCS only. Structures designed using the combination of HSS and MCS are termed "dual-steel" structures. The results of the international research project HSS-SERF[10] highlighted the advantages of dual-steel concept, especially for what concerns the control of seismic response of multi-storey buildings to achieve overall ductile mechanism.

4.2 Selection of steel toughness

Steel toughness is quantified by the fracture energy determined on standardised Charpy v-notch specimens, characterising the safety of steel structures to brittle failure. Regarding the selection of steel toughness EC8[1] states: "The toughness of the steels and the welds should satisfy the requirements for the seismic action at the quasi-permanent value of the service temperature (see EN 1993-1-10: 2005[11])". It is unclear from the information provided in EN 1998-1[1] and EN 1993-1-10[11] how to determine material toughness requirements in the seismic design situation. Firstly, no load combination is provided to be used for determination of the stress level σ_{Ed} in the seismic design situation. It can be considered, however, that in the seismic design situations the stress level in dissipative zones will reach the yield strength($\sigma_{Ed} = f_y(t)$). If this would be the case, the prescription given by EN 1993-1-10[11] cannot be applied, as it provides data only up to stress levels of $\sigma_{Ed} = 0.75 f_y(t)$, and no extrapolation is allowed. Secondly, one of the parameters contributing to the determination of the reference temperature T_{Ed} is the strain rate. No guidance is given in EN 1993-1-10[11] or EN 1998-1[1] on strain rate values to be considered in the seismic design situation. Anyway, some indications on the strain rate values can be found in the EN15129[12], which states that "... The differences due to strain rate shall be evaluated with reference to a frequency variation of $\pm 50 \%$".

Background information on toughness oriented rules in EN 1993-1-10[11] is available in a joint JRC-ECCS report[13]. According to this document, the $\sigma_{Ed} = 0.75 f_y(t)$ value corresponds to the maximum possible "frequent stress", where for the ultimate limit state verification yielding of the extreme fibre of the elastic cross-section has been assumed($\sigma_{Ed} = f_y(t)/1.35 = 0.75 f_y(t)$). Consequently, a tentative conclusion can be drawn, stating that the value $\sigma_{Ed} = 0.75 f_y(t)$ given by EN1993-1-10[11] would correspond to the possibly yielded cross-section, and presumably could be used for selection of material toughness and thickness in the seismic design situation.

However, other open issues remain, like the reference temperature in the seismic design situation and strain rate to be used for determination of the reference temperature, for which no guidance exists either in EN 1993-1-10[11], or in EN 1998-1[1].

4.3 Local ductility

In order to guarantee sufficient ductility, EN 1998-1[1] provide restrictions of width-thickness ratio

b/t of steel profiles according to the cross sectional classes specified in EN 1993-1-1[4]. However, the main criticisms to Eurocode classification are(i) the small number of parameters considered to characterize the beam performance and(ii) the Eurocode classification is based on monotonic loading while it must be different for seismic loading, because of strength deterioration induced by the repetition of inelastic deformations[14].

In the recent past, different classification criteria of bare steel profiles accounting for both cross section slenderness and member slenderness were early proposed by[15,16] and it has been pointed out that shifting from a section-based to a member-based classification would represent a significant advancement of the code. Indeed, the member-based classification given by[15] is based on a relationship between the member ductility and the stress ratio $s = f_c/f_y$ (i. e. the ratio between the peak collapse stress f_c and the yield stress f_y). The stress ratio s was derived from a regression analysis of experimental results on I-shaped beams, and obtained as function of the flange and web slenderness (λ_f and λ_w) , as well as the shear length L^* (namely the distance of the cross section subjected to the peak bending moment from the contra-flexure cross section) and the beam flange width(b_f). Recently, on the basis of a comprehensive experimental campaign, a more refined formulation of s and an extension to tubular member have been proposed by[16], which provided also reliable and accurate empirical equations to predict the rotation capacity of steel profiles accounting for the influence of cyclic loading. The tests showed that a cross section classification based on monotonic rotation capacity(as the case of EN 1993:1-1[4]) could be non-conservative for seismic applications. Fig. 4 clarifies this issue showing the reduction of rotation capacity for a steel beam due to cyclic loading as respect to monotonic loading.

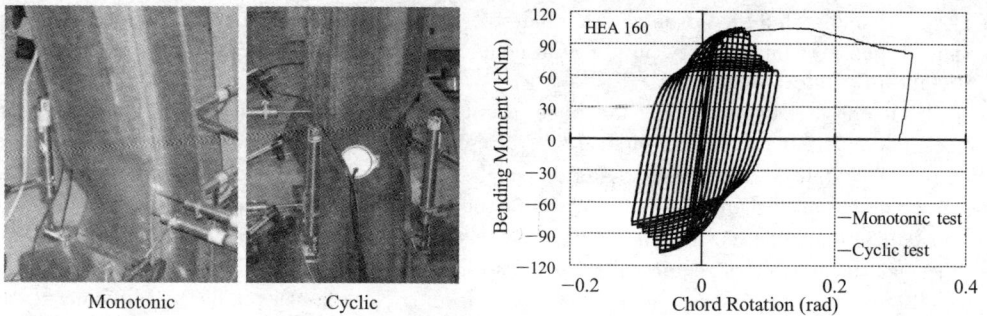

Fig. 4. Monotonic vs. cyclic response of a steel beam[16]

4. 4 Design rules for connections in dissipative zones

At the present time, one of the main criticisms about the connections in dissipative zones is that there are no reliable design tools able to predict the seismic performance of dissipativebeam-to-column joints to meet code requirements[14]. With this regard, EN 1998-1[1] prescribes design supported by experimental tests, resulting in impractical solutions within the time and budget constraints of real-life projects. This is one of the main causes penalizing the use of steel in seismic resistant structures at present time. On the other hand, also for full-strength joints reliable design tools are necessary. Owing to the variability of steel strength, these connections could not have enough overstrength (e. g. min 1. 1 × 1. 25 $M_{b.rd}$, being $M_{b.rd}$ the beam bending strength) , and in such cases their plastic rotation capacity must be prequalified by relevant test and numerical procedures.

A recently approved European research project "Equaljoints (European pre-QUALified steel JOINTS, RFSR-CT-2013-00021)[17] is currently working to introduce in European practice a qualification procedure for the design of moment resisting connection in seismic resistant steel frames, in compliance with EN1998-1 requirements. Further aims of the project are to qualify a set of standard for all-steel beam-to-column joints(see Fig. 5) , and to develop prequalification charts and design tools that can be easily used by designers. The project is also intended as a pre-normative research aimed at proposing relevant design criteria to be included in the next version of EN 1998-1[1].

(a)

(b)

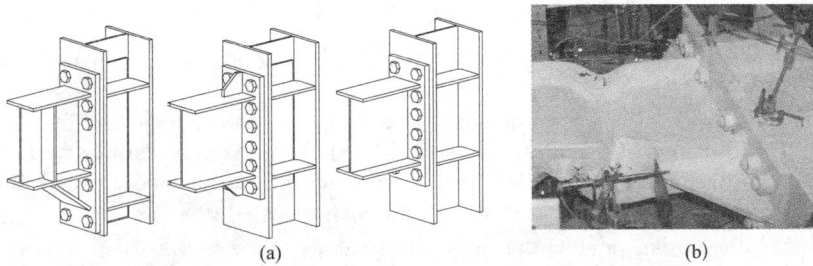

Fig. 5. Equaljoints(European pre-QUALified steel JOINTS)project[17] :
(a)bolted beam-to-column joints to be seismically qualified;(b)Qualification cyclic
test on a full strength extended stiffened endplate joint

4. 5   New links in eccentrically braced frames

Provisions for design of eccentrically braced frames(EBFs)in EC8[1] refer to links realised from I sections only. Other cross sections, such as tubular ones can also be used[18]. Tubular links are an attractive solution because they can provide a stable response without the need of bracings to restrain lateral-torsional buckling, as the case of traditional I-shape links. However, requirements for this type of link are currently missing in the EC8.

Conventional links are realised continuously with the beam containing it. However, replaceable links can be used providing effective performance[19,20]. Bolted flush end-plate connections at both link ends and the rest of the structure can be used, thus allowing replacing the dissipative elements damaged as a result of a moderate to strong earthquake, reducing the repair costs. However, experimental tests showed that bolted connections can be prone to premature failure. Indeed, as explained by[21], significant tensile axial forces may develop due to restraints to axial deformations and nonlinear geometric effects. These tensile forces acting in combination with the other two parameters may significantly modify the link shear overstrength, which can be significantly larger than the value currently recommended by EC8.

Therefore, extension of code provisions and development of design guidelines for new types of links are necessary and should be provided for the next version of EC8.

4. 6   Behaviour factors

The force-based design procedure implemented by EC8 relies on the correctness of the behaviour factors. While concerns could be raised about the real background information behind the values of the behaviour factors currently fixed by the code, the lack of information for some important more recent structural types is easily recognized[7].

As far as the aspects to be improved are concerned, one controversial issue is about the design of V bracings, where one unique value of the behaviour factor is specified for both ductility class "medium"(DCM)and ductility class "high"(DCH). Recent research has also proved that one simple but effective way to avoid damage concentration in V-braced frames is to use vertical ties connecting the braces over the frame height("zipper" bracing). This could also represent one area for the improvement of the code.

The behaviour factors of eccentric bracings(EBs)are set equal to the values of moment resisting frames(MRFs). Generally speaking, MRFs possess a larger plastic redistribution capacity than EBs. Furthermore, no distinction is made is short, intermediate or long links are used, even though the plastic deformation capacity of the frame is markedly affected by the type of link.

The mixed reinforced concrete(RC)walls and steel MRF structures are not explicitly dealt with, while they are one attractive solution to designers. Similar to the previous case is the one of MRFs coupled with bracing systems. There are many studies showing that the combination of the two systems may be advantageous, mainly because the large plastic redistribution capacity of the MRFs allows damage concentration in the braced bays to be strongly reduced or even completely avoided.

One recent and very successful application in the field of seismic resistant steel structures is represented by buckling restrained braces. They are not mentioned in EC8 and this is recognized as one main gap in the current version of the code.

4.7 Capacity design rules

The rules implemented for capacity design in case of steel structures are different from those implemented for other materials, and this deserves some comments.

In case of steel structures, capacity design of non-dissipative parts is regulated by a unique format applicable to all the different structural types covered by the code. Namely, earthquake-induced effects are increased by the factor 1.1 $\gamma_{ov}\Omega$, where γ_{ov} has previously been defined and $\Omega = \min(R_{pl,Rd,i}/R_{Ed,i})$, where $R_{pl,Rd,i}$ is the design strength of the i-th plastic zone and $R_{Ed,i}$ is the required strength. Therefore, the design value of the generic internal action for non-dissipative members is taken equal to $R_{Ed,i} = R_{Ed,G,i} + 1.1\ \gamma_{ov}\Omega R_{Ed,E,i}$ where subscripts "G" and "E" indicates the effect of gravity and earthquake loads, respectively. There are some controversial aspects in this approach. In case of small gravity load effects, amplifying the earthquake-induced counterpart of the required strength by the factor $\gamma_{ov}\Omega$ means that the internal actions corresponding to the first real plastic hinge formation are being calculated. However, in case of large gravity load effects, the proposed Ω factor markedly underestimates the real overstrength. One proposal of correction has been recently reported by[22] . The meaning of the multiplicative coefficient 1.1 is not clearly stated in the code. According to[22] it was introduced to take into account strain hardening of steel and strain rate effects. However, the exact meaning of this coefficient 1.1 and, especially, the rational background behind the assumed value, remains unknown.

4.8 Dual structures: MRFs combined with CBFs/EBFs

Dual frames are obtained by combination of moment resisting frames (MRFs) and concentrically braced frames (CBFs) or eccentrically braced frames (EBFs). However, only the MRF + CBF combination is mentioned in the title of clause 6.10.2 of EN 1998-1[1] directly. Nevertheless, dual frames obtained by combination of MRFs and EBFs are mentioned in paragraph 6.10.2(2), which makes reference to clause 6.8 "Design and detailing rules for frames with eccentric bracings" as a component subsystem of dual frames. A clarification on this issue is thus needed in EN 1998-1[1]. Recently, the European project "DUAREM" experimentally investigated the performance of Dual-EBFs designed to provide re-centering capability to the structure[23] , see *Fig. 6*. The experimental tests showed that the effectiveness of dual systems and the need to improve the EC8 provisions for these systems.

(a)   (b)

Fig. 6. DUAREM (FP7/2007-2013 SERIES n° 227887) project:
(a) full scale mock-up; (b) removable link response

The requirements for design of dual frames in EC8 are: (1) a single behaviour factor is used, (2) seismic forces should be distributed between the different frames according to their elastic stiffness and (3) each structural subsystem (MRFs, CBFs, EBFs) should be designed according to their specific provisions. In US AISC341-10[24] an additional requirement can be found: the MRFs should be able of resisting at least 25% of the seismic force. Such a requirement makes sense, because a too weak MRF would provide no benefit to the dual system in terms of increased strength, stiffness and redundancy. The assumptions/requirements for the selection of the q-factor are not clear. In reality the development of plastic mechanism will be sequential, not simultaneous. In this context the assumption of the distribution of lateral forces according to elastic stiffness appears questionable.

4.9 Drift limitations and second-order effects

A very general problem with the current codified EN 1998-1 rules for the design of MRFs is the compatibility of maximum drifts imposed at the damage limitation limit state and the large behaviour factors for the ultimate limit state. It has long been recognized that the design of MRFs is dictated by drift limitations of EN 1998-1, which produces strong overstrength and, consequently, reduced ductility demand as well as increase of costs[22,25]. Limitations on P-Delta effects could also be a source of significant frame overstrength[22,25].

A study on second-order effects in seismic design of steel structures was recently undertaken by[26]. It was shown that the interstorey drift sensitivity coefficient from EN 1998-1, which is based on the maximum inelastic displacements, appears too severe in the design of steel structures. They showed that the $P - \Delta$ effects may affect the behaviour factor q and the displacement behaviour factor significantly. A simplified formulation was proposed for taking into account of these modified values, based on the elastic interstorey drift sensitivity. It was proposed to replace the usual upper limit 0.10 of this coefficient allowing neglecting the $P-\Delta$ effects with the 0.045 value. This value appears appropriate when the moment resisting frames are preliminary designed for ductility class DCH using a q factor of about 6 and an overstrength factor Ω of the most stressed beam close to 1.0.

4.10 New structural types

The rules for non-conventional structural typologies are missing in the current version of EC8[1]. Therefore, there is a need to incorporate design provisions for concentrically braced frames with buckling restrained braces (BRBFs), steel plate shear walls (SPSWs), truss moment frames (TMFs) and concentrically braced frames with dissipative connections in EN 1998-1.

Design provisions for the above mentioned structural systems (BRBFs, SPSWs and TMFs) already exist in other seismic design codes, such as AISC[24]. Availability of design provisions for these structural systems in EN 1998-1 would enlarge the opportunities of European designers.

Another structural system that currently lacks code support is concentrically braced frames with dissipative connections. Experimental and theoretical investigations on concentrically braced frames with dissipative connections exist[27] that indicate an excellent seismic performance in terms of strength, stiffness and ductility. Design guidelines for concentrically braced frames with dissipative connections can be prepared and implemented in future versions of EN 1998-1.

4.11 Low-dissipative structures

The analysis of the current EC8 shows that it is necessary to develop simplified rules and design aids for design of structures in low-seismicity regions. Besides, the development of intermediate rules, less severe than DCM and DCH rules, seems desirable thus allowing the justification of a low-to-moderate ductility of the structure. Moreover, the work undertaken by the Committee showed that background studies are still necessary to detect weaknesses of DCL design in moderate to high seismicity regions in case of an earthquake with PGA higher than expected and definition of required elastic overstrength providing the same safety level as DCM or DCH design.

5 THE CURRENT ACTIVITY OF TC13

Consistently with its mission, TC13 is still working on proposals of improvement of sections 6 and 7 of EC8 (namely seismic design of steel and steel composite structures, respectively) in the light of the identified criticisms, deeply discussed in[7] and briefly summarized in the previous section. This purpose is also in line with the activity of the CEN/TC250/SC8, which is the official body devoted to the maintenance and development of all parts of Eurocode 8. In order to share the TC13 past and ongoing work on codification with the CEN/TC250/SC8 Committee, TC13 recently expressed its willingness to become an official Working Group within CEN/TC250/SC8, also considering that TC13 gather together the majority of European experts in the field of steel and composite structures and it is representative of almost all European countries. This proposal was accepted by the CEN/TC250/SC8 that, during the meeting held in London on 8 9 January 2015, approved a resolution in which the Working Group 2 (WG2)-Steel and Composite Structures-has been established with the aim to deal with the seismic design of steel and composite structures (i. e. chapter 6 and 7 of EC8). With the same resolution, the

Writer has been appointed as Convenor of this WG for the next three years (i. e. triennium 2015-2018). This decision has a great value for the Committee, because it ratifies the spontaneous effort of its expert members and also allows finalizing the mission of TC13. This important recognition will motivate even more TC13 Committee to continue working to improve the current version of EC8.

6 CONCLUSION

This paper summarises the main activities carried out within the ECCS-TC13 Committee on Seismic Design since its recent restructuring in 2008. According to its mission, TC13 is devoted to the promotion of the use of steel structures in seismic areas and to this end, two main action lines were pursued. Indeed TC13 has been working on the dissemination of the steel culture across Europe and beyond and it is also giving its contribution in the field of scientific research. With respect to the first item, TC13 set up an ambitious and effective dissemination plan in order to widespread the latest research findings and the specialized knowledge in the field of seismic design of steel structure to a large audience. Indeed, the Committee established an educational training program for post-graduate students and practitioners, gave its support to Conferences and pursued high-level publications. As for research activities, TC13 analysed the current status of European standard and codes, identifying critical issues in Eurocode 8. Among the main outcomes succeeded by the TC13 is the publication of the book "Assessment of EC8 Provisions for Seismic Design of Steel Structures"[6], which summarizes all issues in the current version of EN 1998-1 needing clarification and/or development. Nowadays, TC13 members are also actively involved in the maintenance operations of the current version of the EC8, being parts of the WG2 of TC250/SC8. The Writer, which has been serving as Chair the Committee since its restructuring, would like to express his deepest gratitude to all the TC13 members that provided, provides and undoubtedly will provide their enthusiastic, valuable and precious contribution to achieve TC13 goals and objectives, confident that the this demanding work could contribute to the overall goal of a safer and more resilient European Society.

REFERENCES

[1] EN 1998-1, Design of structures for earthquake resistance-Part 1: General rules, seismic actions and rules for buildings. CEN 2005.
[2] European Recommendations for Steel Structures in seismic Zones. *Technical Committee* 13 —*Seismic Design*, No. 54/1988. ECCS, 1988.
[3] Mazzolani F. M., Piluso V., "ECCS manual on design of steel structures in seismic Zones". *Technical Committee* 13—*Seismic Design*, No. 76/1994. ECCS, 1994.
[4] EN 1993-1. Eurocode 3: Design of steel structures—Part 1-1: General rules and rules for buildings. CEN; 2005.
[5] Eurocodes-background and applications: "Design of Steel Buildings with Worked Examples". JRC publication (in press).
[6] Landolfo R., Mazzolani F. M., Dubina D., da Silva L., D'Aniello M., Design of steel structures for buildings in seismic areas. *ECCS* (in preparation).
[7] Landolfo Raffaele, editor. "Assessment of EC8 provisions for seismic design of steel structures". *Technical Committee* 13—*Seismic Design*, No 131/2013. ECCS, 2013.
[8] Landolfo Raffaele, editor. "Toward improved European codification for the seismic design of steel structures". *Technical Committee* 13 —*Seismic Design*, (in preparation).
[9] OPUS (2007-2010). OPUS: Optimizing the seismic performance of steel and steel concrete structures by standardizing material quality control. RFSR-CT-2007-00039.
[10] Application of High Strength Steels in Seismic Resistant Structures. DanDubina, Raffaele Landolfo, Aurel Stratan and Cristian Vulcu Editors. "Orizonturi Universitare" Publishing House. ISBN 978-973-638-552-0.
[11] EN 1993-1-10:2005 -Eurocode 3: Design of steel structures-Part 1-10: Material toughness and through-thickness properties. CEN 2005.
[12] EN 15129. Anti-seismic devices. European Committee for Standardization. CEN 2010.
[13] Sedlacek G., Feldmann M., Kühn B., Tschickardt D., Höhler S., Müller C., Hensen W., Stranghöner N., Dahl W., Langenberg P., Münstermann S., Brozetti J., Raoul J., Pope R., Bij-

laard F. ,COMMENTARY AND WORKED EXAMPLES to EN 1993-1-10 "Material toughness and through thickness properties" and other toughness oriented rules in EN 1993. JRC 47278 EUR 23510 EN,European Communities,2008. ISSN 1018-5593.

[14] Plumier A. ,General report on local ductility. *Journal of Constructional Steel Research* 55: 91-107,2000.

[15] Mazzolani F. M. , Piluso V. , Member behavioural classes of steel beams and beam-columns. *Proc. of First State of the Art Workshop*,*COSTI*,Strasbourg,1992.

[16] D'Aniello M,Landolfo R,Piluso V,Rizzano G. Ultimate behavior of steel beams under non uniform bending. *Journal of Constructional Steel Research*,78:144-58,2012.

[17] http://dist. dip. unina. it/2013/12/09/equaljoints/

[18] Berman J. ,Bruneau M. ,Tubular links for eccentrically braced frames. II: Experimental verification,*Journal of Structural Engineering*,ASCE,134(5),2008.

[19] Dubina D. ,Stratan A. ,Dinu F. ,Dual high-strength steel eccentrically braced frames with removable links. *Earthquake Engineering & Structural Dynamics*,37(15): 1703-172,2008.

[20] Mazzolani F. M. ,Della Corte G. ,D'Aniello M. ,Experimental analysis of steel dissipative bracing systems for seismic upgrading. *Journal of Civil Engineering and Management*,15(1): 7-19, 2009.

[21] Della Corte G. ,D'Aniello M. Landolfo R. ,Analytical and numerical study of plastic overstrength of shear links. *Journal of Constructional Steel Research* 82:19-32,2013.

[22] Elghazouli A. Y. ,Seismic design of steel framed structures to Eurocode 8. *Proceedings of the 14th World Conference on Earthquake Engineering*,Beijing,China,12-17 October 2008.

[23] http://www. series. upatras. gr/DUAREM

[24] American Institute of Steel Construction,Inc. (AISC)(2010). Seismic provisions for structural steel buildings. Standard ANSI/AISC 341-10,Chicago(IL,USA).

[25] Tenchini A. ,D'Aniello M. ,Rebelo C. ,Landolfo R. ,da Silva L. S. ,Lima L. ,Seismic performance of dual-steel moment resisting frames. *JCSR*,101:437-454,2014.

[26] Aribert JM,Vu HT. ,New criteria for taking into account P-Δ effects in seismic design of steel structures. *Proc. of the 6th STESSA Conference*,2009.

[27] Vayas,I. ,Thanopoulos,P. ,Castiglioni,C. ,Plumier,A. ,Calado,L. ,Behaviour of seismic resistant braced frames with innovative dissipative (INERD) connections. *Proc. of the 4th European Conference on Steel and Composite Structures-Eurosteel*, 2005.

NZ RESEARCH ON STEEL STRUCTURES IN SEISMIC AREAS

GregoryA. MacRae* and G. Charles Clifton**

* Civil and Natural Resources Engineering, University of Canterbury, New Zealand
gregory. macrae@ canterbury. ac. nz
* * Civil and Environmental Engineering, University of Auckland, New Zealand
c. clifton@ auckland. ac. nz

KEYWORDS: Steel structures, Seismic, Floor diaphragm, Low Damage, Decision

ABSTRACT

This paper describes recent and ongoing research at the civil engineering universities(Auckland and Canterbury) to produce better steel structures and to reduce the vulnerability of the built infrastructure. The types of structure considered are (i) new structures designed considering current code performance objectives, (ii) existing structures, (iii) nascent structural systems, and (iv) damaged structures. Effective decision support tools to assist in determining what approach is best are also discussed. As the different structural types above are described, research efforts, especially related to the work of the authors on the seismic performance on these structures at NZ universities are explained.

1 INTRODUCTION

The 2010-2011 Christchurch earthquake sequence showed the performance of steel structures was generally better than anticipated especially given the levels of ground shaking which were more than twice that considered explicitly in design. Furthermore, unlike many reinforced concrete structures, a lot of damage was reparable. These earthquakes have resulted in steel being the construction material of choice in the Christchurch rebuild.

Due to this success, it could be considered that steel structures are a good proven solution and that further research on steel structures is not required. In fact, since the mid-1980s, when the provision of ductility and the use of capacity design became a requirement, member sizes and details in the majority of structures around the world have not changed much. This is in spite of considerable research into efforts on topics such as performance-based earthquake engineering. There have been some changes to design of steel structures though. These include the development of more ductile steel structures after the 1994 Northridge earthquake through the SAC Steel Project, and the more explicit provisions for very tall buildings especially in regard to in-service performance.

It may be argued thatsince recent developments in steel structures, and in earthquake engineering in general, have relatively made small impacts, that further significant changes in approach are unlikely, and that earthquake engineering research should not be a high priority for national funding. In this case, work related to structures in earthquake zones should emphasize education of practitioners to correctly use current codes, ensuring that the intent of standards is not eroded by competing interests or the passage of time since severe earthquake events, etc. Here, structural earthquake engineering may be regarded as mature, and the work of structural engineer is much like that of any tradesman, such as an electrician or a mechanic, who applies technology in a clear standardized way to obtain a standard of construction which meets the performance expectations of society-at least while those expectations are life safety. It is interesting that many institutions that were famous for their ground-breaking work in earthquake engineering in the US and Japan in the 1970s and 1980s are emphasizing this area less. These institutions are now employing academics without a strong earthquake engineering background and whose research is often focused on other topics. The consequence of the status quo is that we are likely to see more Christchurch-type damage in future earthquakes around the world where the life-safety performance objective was met.

While the status-quo may be acceptable to some, for the people associated with the Christchurch event, it was considered that the damage level was unacceptable(Buchanan et al. ,2011)[1], and that we should do better. What engineers accepted as "earthquake resistant" had a very different meaning to building owners and occupiers, most of whom following the Christchurch earthquake of 22 February

2011 had to leave their damaged buildings either temporarily or, for many commercial building owners and occupiers, permanently. Society as a whole suffered considerably from the severe disruption to the fabric of city life, which is only now starting to be reinstated.

In contrast to the argument that structural earthquake engineering methodologies are mature, there is another argument. This is that structural earthquake engineering research and techniques are essential to modify and improve structures and make them more effective and more economical. The need for better structures can be considered for the following structural types:

(i) existing structures, for better retrofit interventions;

(ii) new structures designed according to the current life safety philosophy, to ensure intended response by addressing inadequacies in current standards;

(iii) new structures designed for performance levels above the minimum defined in current standards, and

(iv) damaged structures, to allow reinstatement and reuse.

This paper outlines some issues associated with the above four types of structure, and some recent and ongoing research at the Universities of Canterbury and Auckland and elsewhere to develop better performance of steel building systems subject to earthquake shaking for a specified cost.

2　OLDER STRUCTURES

Many existing buildings were constructed before current standards were enforced. Often they use materials or techniques that are not included in current design approaches. A decision needs to me made about any vulnerable older buildings, sometimes called earthquake-prone buildings, as to whether or not people should be permitted inside, or near, them. The need for the decision, and the decision itself, are affected by the structural assessment and estimated total costs associated with structural retrofit. Different stakeholders and interest groups, such as the owner, occupier, local business groups and councils, risk analysis groups, relatives of those killed in recent events, the construction industry, and politicians may have very different vested interests and opinions about what to do with such structures (MacRae, 2015) [2]. The recommendation for a specific structure may range from doing nothing to full retrofit. There are large economic, safety, and societal implications of this. Strongly differing views of the different stakeholder groups are normal for such high impact low probability events (Mezaros, 2005) [3]. There is a role for engineering here to minimize these polarized views. This can be done by (i) performing more accurate assessments, and (ii) developing cheaper retrofit options for the life safe performance objective.

An example of one structural form where there is significant discrepancy between standard calculations and observed performance is low-rise industrial steel portal frame structures. Some of these have been assessed to have strengthsas low as 25% of that to the new building standard. However, of the many frames that went through the Christchurch earthquakes, where there was over 200% of the design ultimate limit state shaking, very few came near collapse. A number or reasons have been cited for this approximately 800% difference between observed and computed behaviour. These include the influence of soil structure interaction, "non-structural" elements contributing to the response, different end fixities, the lack of consideration of nonlinear elastic buckling response of the members, and conservative assumptions of engineers undertaking the studies. In a recent undergraduate study, Makwana and Nanayakkara(2014) [4] used the computer software ABAQUS to capture elastic and inelastic out-of-plane buckling deformations and considered likely rotational stiffness and explained the observed behaviour. Practitioners generally use simple simple frame analysis software for their assessments so cannot capture many important effects. This lack of consideration of the likely response results in possible unnecessary retrofit of many structures.

3　CURRENT DESIGN

The primary performance objective of most current standards is to limit the possibility of life loss in a structure during one design level shaking event. While this aim is clear, the degree of confidence that this will always be achieved may be low. Some examples of inadequacies in current design are the following:

3. 1   Floor Diaphragms

In undergraduate structural classes around the world, the importance of load paths is emphasized. The capacity of an element meant to remain elastic is required to be greater than the expected force demand. However, the load paths between the seismic resisting system and the floor diaphragm and those through the diaphragm itself are not well considered for the following reasons:

i) *Demands are difficult to estimate*

While diaphragm demands may be estimated using 3-D frame non-linear dynamic analysis, practitioners generally do not have the will and/or capability to use such techniques.

A number of methods have been proposed to estimate demands using 3-D elastic analysis software. These include:

(i) thepESA method (Bull, 2004[5], Gardiner, 2011[6]) which has been used in NZ frame design. It uses lateral forces associated with the frame overstrength shear. It considers both inertia and transfer force effects with one lateral force distribution. This is however only effective for buildings up to about 9 stories in height (Tiong and Lyes, 2014)[7].

(ii) A method by Cowie et al. (2013)[8] that explicitly calculates inertial and transfer forces. Force transfer between a composite floor diaphragm and the seismic resisting system frame, and design/detailing requirements are included.

(iii) Sabelli et al. (2002)[9] has also proposed a similar method which does not consider frame overstrength.

(iv) An alternative method (Tiong and Lyes, 2014[7]) was also considered. It includes transfer forces and intertial forces. It is conducted by applying the overstrength force distribution to the frame, but making sure that the forces at the level considered are no less than those resulting from the anticipated in the Parts and Components section of the loadings standard. This approximation is more conservative than some other methods.

ii) *Diaphragm modelling requires an additional step*

In most software used to design buildings a rigid diaphragm is assumed, and the forces within the diaphragm are difficult to obtain. In NZ design, deep beam modelling (Cowie et al 2013)[8] and various types of strut-and-tie/truss analysis (Bull 2004[5] and 2014[10], Gardiner 2011[6], Scarry 2013[11] and 2014[12]) are starting to be used for the diaphragm itself. The truss analysis generally requires a separate analysis of the diaphragm itself which may be time consuming.

Further studies are required to reasonably estimate diaphragm demands, and computer software needs to be developed to easily obtain diaphragm demands, and to analyse the diaphragm itself.

3. 2   Beam Axial Forces

Beams in moment-frames are often considered to carry bending only and it is generally assumed that the slab will transfer the lateral forces to the column through compression on the column face. However, since slab inertia forces act in the same direction as the frame sways, a gap opens at location "A" shown in *Fig. 1*. Because of this, slab inertia forces cannot go directly into the column. Instead the forces must move into the steel beam through friction and mechanical transfer using studs. This means that:

a) reinforcement is required in the direction perpendicular to the shaking direction to enable the strut and tie forces in *Fig. 2* to be developed, and

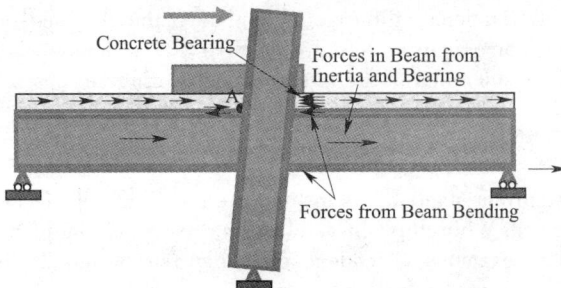

Fig. 1. Inertia and Slab Bearing forces Actions

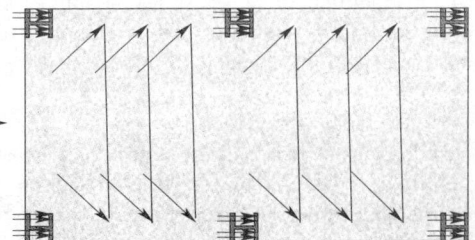

Fig. 2. Inertia and Slab Bearing forces Actions

b) beam plastic hinge regions, and connections at the beam ends, should be considered to carry the beam axial forces into the column.

Also, if there is no construction gap between the slab and column, as the column sways, it bears a-gainst the concrete slab on the far side of the column causing a slab-interaction effect that increases the forces that the slab must transfer into the beam and the beam must transfer back to the columns.

In addition to (i) beam axial forces from inertial effects, and (ii) slab interaction effects, beam axi-al forces also occur due to transfer force effects, which are a function of the building configuration.

When considering the slab and beam effects above it is useful to understand whether a construc-tion separation gap should be placed around the column or not. In the current NZ steel specification (SNZ 2007), slab effects are not considered when sizing the beam, but they are considered when com-puting the overstrength moments from the beams into the columns. Typically this increases the moment demand on a column by at least 30%.

It would be advantageous to consider the slab in the beam strength for design, but this can only be done if the beam strength is able to be maintained through large deformations. This has been done (MacRae, Hobbs et al, 2013[13]) by special detailing around the column, and more economical ways of providing this detailing are being studied (Chaudhari et al. ,2015[14]).

An alternative to this is to fully isolate the column and elements connected to it from the slab. In a welded beam connection, only one layer of isolation may be required on each side of each flange con-sidering strong axis bending. For a bolted end-plate connection it is necessary to isolate the column flange, end-plate, bolts/nuts, and any haunch. This may be done with a number of layers of isolating material. Alternatively the slab may be have a diamond gap cut all around the column, end-plate, haunch, etc. The gap may initially be made using polystyrene cut following the profile of the decking a-round each column for casting. The polystyrene is removed later and the gap filled with a fire proof matting. A further alternative, used on two medium rise steel framed buildings recently at the Universi-ty of Auckland, involves wrapping a ceramic fibre blanket in polyethylene (to prevent water ingress during concrete placement), pin this around the column and cast the concrete against it. Full isolation can provide good performance (Chaudhari et al. ,2015[14]), but care is needed to minimize the chance of column instability due to the decrease in restraint associated with having no slab, and the frame is likely to have greater displacements as its stiffness is reduced and fundamental period longer.

Whatever method is chosen has implications for the column, for the slab and for beam axial forces.

3.3 Column Splices

Columns can generally be trucked to site in lengths of no more than about 3 stories, and therefore require connecting to the column above and below. This connection is typically referred to as a splice, and it is placed near the middle of the interstorey height in accordance with the requirements of NZS 3404. Also, because of past issues with field welding, and the desire to construct quickly and independ-ently of the weather, bolted connections are desirable in NZ. When such connections use vertical plates, holes are made in the column thereby reducing the flexural and shear capacity of the column member at the splice. While efforts have been made to do away with bolted connections in recent ver-sions of US codes, the economic costs of this are high. In NZ, the splice strength in a major column is permitted to be as low as 45% of the member capacity for ductile systems, which is lower than the val-ue of 55% in older US codes. Recent analyses indicate that splice strength may be exceeded in some cases (Tork Ladani et al. ,2014[15]). Since many floors may be supported above these splices, it is es-sential that splice failure resulting in collapse not occur. The provisions of NZS 3404[16] have not changed since the 1992 edition of the standard and were in many modern steel framed buildings be-tween 3 and 22 storeys in height that were pushed into the inelastic range during the Christchurch earthquake series; no adverse effects of splice behaviour were reported.

A programme is underway at the University of Canterbury to assess the likely performance of par-tial strength/reduced stiffness splices in frames under seismic loading.

3.4 Gravity Frame Issues

Gravity frames are often designed to carry generally axial forces only. However, they need to follow through seismic frame displacements (E. g. SNZ 2007[16]) and may be subject to moments for the fol-lowing reasons:

(i) while beam connections to such columns are nominally pinned, there is generally some moment transfer(Liu and Astaneh,2004[17]) ,

(ii) relative floor drifts also cause moments(MacRae et al,2004[18] ; MacRae,2011[19]). These relative floor drifts occur in both horizontal directions.

Furthermore,it is possible in slender columns that the peak moment may not occur at the member end causing the possibility of yielding at a location where the likely response is not known(MacRae, Lu et al. ,2010[20] ; MacRae et al. 2004[8]).

Such moment demands are seldom considered explicitly in design. In addition to the demands on the column itself,there are flexural and shear demands on column splices and on the base connections (Borzouie et al. 2014[21]). These elements must be considered for satisfactory behavior in design.

3. 5   Proprietary System Issues

These can be understood with relation to the following discussion about William Tell. Mr Tell is known as a famous archer because he shot an apple off a boy's head. However,while he was successful in that particular instance,it is not clear if he could have done it again. Perhaps he was just lucky! Also,if the boy were running around,rather than standing still,his chance of success may also have been lower.

There are the same issues with many patented products,such as BRBs. Tests have shown that they can perform very well. However,there is no guarantee that they would behave well if one of the dimensions or material properties were altered by a small percentage due to a change in design or statistical differences related to the same design. The overall sensitivity to construction differences is not generally known or understood(Jones et al. 2014[22]). Also, braces from well-known manufacturers fail when they have been put in a building frame as shown in tests of the shaking(e. g. Tsai et al,2008[23]).

Furthermore, the vast majority of BRB testing around the world considers axial load only. Sometimes in-plane bending is considered,as may occur due to in-plane frame displacements when the brace does not have a perfect pin at the end. Design and detailing procedures for this are developed (Wijanto,2012[24])based on very limited testing. In addition, an earthquake acts in more than one horizontal direction causing out-of-plane bending of the BRB connections into the structural system which is almost never considered in testing or design. For these reasons,it is not clear that such proprietary elements are likely to behave well in all earthquake situations. Other BRB issues are related to gusset plates as described below.

3. 6   Gusset Plate Issues

Gusset plates are attached to the ends of compression/tension members in frames subject to seismic and non-seismic loading. A number of design recommendations-including the Uniform Force Method(Thornton,1991[25]) , the proportioning method (Clifton, 2000[26]) , and the Generalized Uniform method(Muir, 2008[27]) , and some simplified methods-are available to consider direct forces. Frame action effects may be considered(e. g. Clifton,2000[26] ,Lin et al. 2013[28]) ,standardized generally accepted methods are not available. Methods to explicitly consider the following actions on the gusset plates are also not generally available:

i) In-plane moments from brace in-plane bending when brace connections are not totally pinned, and

ii) Out-of-plane moments(such as may occur due to frame out-of-plane deformation, or due to brace buckling.

Also,most procedures advocated for gusset plate design consider the gusset plate has an effective length for buckling of 0. 50-0. 70 even though sway is the predominant failure mode! Other types of axial cleat connection have exhibited problems with swayand mitigation methods have been developed (E. g. Clifton et al,2009[29] ; MBIE,2010[30]). Some researchers are making recommendations for gusset plate design based on tests with a gusset plate restrained so that sway cannot occur(E. g. Yam and Cheng 2002[31]). While the gusset plate effective length and width may be calibrated with these short effective length factors to realistically estimate overall capacity in some cases,poor behaviour can occur.

Clifton(2000) [29] and Lin et al. (2005) [32] have shown that for unstiffened gusset plates the effective length factor should be increased from 0. 65 to 2. 0. A recent study at the University of Canterbury(Westeneng and Crake,2014[33])used stability functions to obtain the actual gusset plate effective length factor corresponding to overall buckling failure of an elastic gusset plate connected to an elastic BRB brace member as shown in *Fig. 3*a. They showed that the effective length factor is

(i) dependent of the stiffness of the brace,

(ii) greater than unity as would be expected for a sway element, and

(iii) may be greater than 2. 0 for high gusset plate stiffnesses.

An implication of this is that gusset plates designed according to standard methods, with small effective length factors, may fail under compressive load alone at strengths lower than that specified in the standards. This has been shown for a typical configuration by Tsai andHsaio(2008)[23] and replicated by Westeneng and Crake(2014)[33] in *Fig. 3*b. Furthermore, while gusset plate stiffeners can increase strength, care must be taken to prevent beam twist reducing gusset plate end restraint affecting brace strength.

Frame action is not considered explicitly in most design methods, although Lin et al. (2013)[28] have shown that gusset plate stiffeners assist the behaviour under cyclic loading.

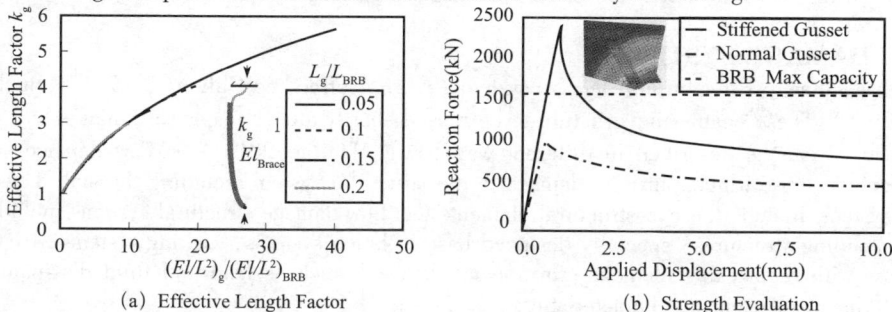

(a) Effective Length Factor   (b) Strength Evaluation

Fig. 3. Gusset PlatePerformance(Westeneng and Crake,2014)[33]

Design of welds is not always for the plate strength even though many design procedures consider the plastic strength of the plate. This means that non-uniform distributions of force in the gusset plate due to element flexibility, frame(beam-column) opening effects, brace in-plane moments, brace force eccentricity, or any assumptions that are different from that used to size the plate, as well as out-of-plane bending of theplate, from member out-of-plane deformation etc. , can result in weld fracture.

3. 7 Column Continuity Issues

Similar to CBFs, EBFs and steel plate shear walls, there are no explicit requirements for BRBFs to have sufficient column stiffness/strength to prevent a soft storey mechanism. It is theoretically possible to design somebraced frames to be totally pinned at each member end according to most current standards. However, such a design will exhibit a soft storey as soon as brace yielding occurs. The continuous column concept(MacRae and Kimura,2004[18] ; MacRae,2011[19])seems to be a good way to manage this as it provides the stiffness over the height to mitigate a large drift concentration. However, most standards worldwide do not have requirements for this.

3. 8 Overturning and Torsional Effects

After the recent Canterbury earthquakes, changes to the loading standard (SNZ, 2004[34]) were suggested by the Royal Commission for the Canterbury Earthquakes.

One of these was related to the likely greater response of structures with cantilevers. This resulted in the building having a significantly different strength in each direction. This difference is also seen in symmetrically loaded structures with different strengths in different directions. This issue has been recognized(Yeow et al. ,2013[35]) and some design recommendations do exist. For example, for structures with BRBs, the maximum strength difference permitted in two different directions is 10% (SNZ, 2007[16]). However, implications of the amount of strength difference on frame design have not been studied to the extent that clear design recommendations are available for incorporation in the loadings standard.

Another change was requested for buildings that have lack of torsional restraint as soon as the first lateral force resisting system yielded. While, a large amount of research has been conducted on torsion worldwide there seems to be little guidance to explicitly consider such effects. Two methods are referenced in MacRae andDeam(2009)[36], but further work needs to be conducted to obtain something simple and robust for building codes.

3.9 Other Details

Many other issues are not considered by current codes and it is not always clear what minimum provisions are required to develop systems to satisfy life safety criteria. Some of these are with the use of different combinations of materials. In the Christchurch rebuild, two-way moment frames are being used with rectangular concrete-filled tubular columns. While many techniques for constructing such columns exist in Asia, these generally often use sophisticated automatic processes to obtain good quality welds, and such automation is not available in NZ. Economical means of connecting beams to such columns have been considered(Chunhaviriyakul et al. ,2015[37]) and recent structures have been constructed using external diaphragm connections with low damage friction connections. Research work has also followed to develop simple and robust design recommendations for such structures(Tjahjanto et al. ,2015[38]) and to see if better and more economical connections could be made.

4 LOW DAMAGE SYSTEMS

Much has been written on low-damage systems(E. g. MacRae and Clifton(2013)[39], Buchanan et al. (2012)[1]). These enable the structure to be reused shortly after a major earthquake, and performance objectives can be described in different ways(E. g. MacRae,2013[40]). Low damage structural systems include those which cause no damage to the structural system including the slab. Low damage building systems include the non-structural elements too. Low damage structural systems include: elastically responding structures; specially designed base isolation systems; rocking systems; friction systems; those with replaceable elements; or special devices; such as lead and fluid dissipators. Some but not all, have self-centring characteristics.

Not all so-called low damage systems are equal, and they may have significantly different costs. For example, post-tensioned beam systems, which were advocated strongly in the early 1990s have not become popular, because, while they behaved well in beam-column tests, they do not behave well when a slab exists, as it needs to damage the slab to perform as expected unless special and expensive mitigation measures are taken. Furthermore, these post-tensioned beam systems tend to push columns apart causing extra demands. These effects are most significant in frames with more bays and deeper beams(MacRae and Clifton,2013[39]).

At the University of Canterbury, recent studies have been conducted on friction braces(Chanchi et al. 2015[41]), and base connections(Borzouie et al. 2015[21]) and desirable behavior has been obtained. Similar work at the University of Auckland has been conducted on frictional rotational links (Leung et al. ,2015[42]), rocking frames(Djojo et al. 2015[43]) and frictional connections(Ramhormozian et al. 2015[44]). Current design recommendations for asymmetric friction connections with hard (e. g. Bisalloy 400 or harder) shims is to use a sliding force equal to 0. 25 multiplied by the number of surfaces in sliding(normally 2) multiplied by the number of bolts multiplied by the proof load per bolt. A strength reduction factor, ϕ, of 0. 7 for friction is used, and the overstrength for the connection, ϕ_o, considering bolt, and surface variations is 1. 40.

A device to prevent buckling of tension dissipative devices in compression was recently described by Gunning and Weston(2013)[45]. The device was inspired by plastic cable ties which can carry tension force, but carry no force in compression as they are pushed through the orifice. It has similar, but opposite, characteristics to a car axle jack which is a compression only device. The tension only device has the behavior described in *Fig. 4* and *Fig. 5*. The device itself is shown with the teeth in blue. A small lateral compressive force is required to encourage the two parts not to fall away from each other, so that the teeth engage. The dissipate element is shown in brown. Dissipation may occur due to yielding, frictional sliding or other means. Initially the device is loaded elastically in tension(A-B) then yielding/frictional sliding occurs in the dissipative element increasing its length(B-C). When the force is taken off(C-D)there is some elastic shortening of the dissipative element. When compression force is applied, the device carries very little compression but slides(D-E). When tension force is applied again (E-F), the device slips until the teeth are engaged but the dissipative element does not change in length since the axial force in this stage is very small. The maximum possible E-F distance is the tooth pitch. For greater tensions(F-G),displacement increases in the elastic range and then causes dissipation in the dissipative element as before. Preliminary tests of small scale devices indicate excellent behaviour with monotonic dissipation only of the yielding element(Cook et al. 2015)[46].

Such a device has the potential to be used on the outside of rocking walls,in brace,and in other applications requiring energy dissipation. The dissipative element would need to be replaced,and the device reset,after every major event.

Typical low-damage systems offer a significant performance enhancement for additional costs generally less than 5% (MBIE,2015)[47]. However,other disincentives also exist. Many low-damage systems do not satisfy the acceptable solution of the NZ Building Code. They are therefore considered as alternative solutions to meet the minimum performance requirements of the Building Code. This can involve increased design and consenting costs,with PEER review as well as testing or test records demanded by the Building Consent Authority(MBIE,2015)[47]. Also,unfamiliarity of different construction processes by the construction community can lead to higher costs.

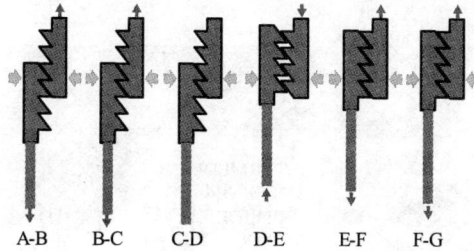

Fig. 4. Tension-Only System Push-Pull Behaviour

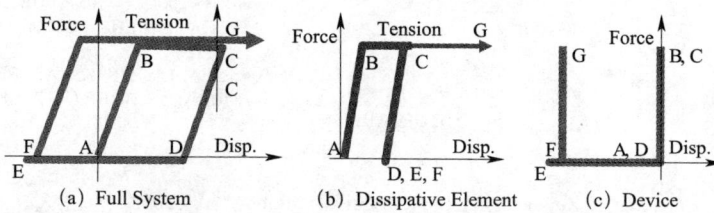

Fig. 5. Tension-Only System Hysteretic Behaviour

5   REPAIR

After a major event it is generally desirable to reinstate the building as fast as possible. In the best case,if there is no structural damage,nothing needs to be replaced. In other cases,it may be necessary to replace parts of the structural system. It is preferably if these parts are small,not part of the gravity force resisting system,and are at one location or level in the structure to minimize access time.

When an item is damaged,a decision needs to be made as to whether the damage is so minimal that it can be left in place to resist a larger event,whether it can be repaired,or if replacement is needed. This decision requires assessment tools. After the Canterbury earthquakes it was found that appropriate assessment tools are not available for many of practical situations.

In reinforced concrete buildings subject to the Christchurch earthquakes often showed a few large cracks near the column face. These large cracks were different from the large number of distributed cracks often seen in laboratory tests because of the dynamic pulse type loading in one direction, the presence of the slab,and the concrete strength being significantly greater than that specified at 28 days. Since the cracks cause large strains in the reinforcing steel which passes into the column joint, repair is difficult. As a result many of the Park and Paulay type reinforced concrete structures,which satisfied the life safety performance objective well,were replaced after the Canterbury earthquakes.

For steel structures,success has been obtained in determining the level of damage ineccentrically braced frame active links by Nashid et al. (2013)[48] using the field based hardness. This method takes into account the considerable variation of hardness readings through requirements for surface preparation of the surface to be hardness tested, where and how to take the measurements, how to establish a baseline for the unyielded material, the assessed loading regime etc. It has been applied to the assessment of a 2010 built 12 storey EBF structure in Christchurch to determine what yielded active links could be left in place. It is only applicable to the webs of shear links that have yielded in a principally shear mode.

An example of an EBF link replacement for Pacific Tower is given in *Fig. 6*.

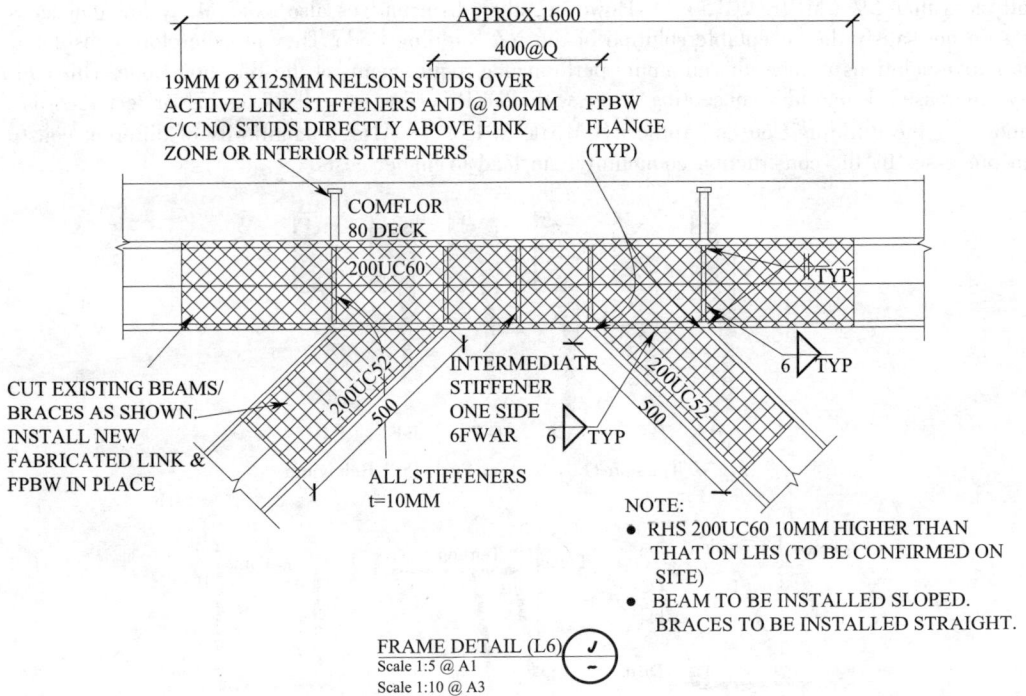

Fig. 6. Replacement Link Detail(Gardiner et al. ,2013)[49]

It is possible that structures without any permanent or non-replaceable damage may have residual displacements after a major event. These displacements may result in further damage during aftershocks (e. g. Abdolahirad et al. 2015)[50]. When straightening is conducted manually, some sequences of placing straightening cables/members, disconnecting elements, and performing the straightening, may leave the structure more vulnerable to an aftershock than others. Also, immediately after a structure has been pushed over, some temporary stabilization may be required to limit the possibility of the structure deforming further in the same direction during an aftershock. Such stabilization may be performed using very simple techniques, such as tension cables, or compression elements. If it is done well, when an aftershock comes, the aftershock may be used to straighten the structure. Some techniques may also stop the structure moving too far in the opposite direction. One of these simple techniques being investigated by Abdolahirad involves the use of Tension Only Devices in two directions. Those in one direction are done up tight to limit the possibility of further significant movement in that direction, while those in the opposite direction may be provided with initial slackness so that do not grip until the structure starts to move in the opposite direction.

6   DECISION

The best structure, or best retrofit, in a given situation may depend on many factors such as the viewpoint of the stakeholder, their vested interests and their insurance policy.

Two methods to support decision-making are described below. They are(i)probabilistic loss assessment(PLA), and(ii)subjective quantitative analysis(SQA).

In the first, all probabilistic information is combined using convolution integrals to obtain scenarios losses(for a particular event), or probabilistic loss(estimating the loss over time). While this type of analysis can be used to quantify dollar losses due to damage, death(and injury), and downtime, many assumptions are required as input information of good quality is seldom available and it is difficult to include all factors. When more accuracy is desired, and more parameters and factors are considered in the analysis, there is also an increase in uncertainty. This uncertainty may swamp the analyses. PLA may be useful though. For example, a break-even analysis (MacRae 2006[51], Bradley et al. 2009[52], Yeow et al. 2012[53]), such as that shown in *Fig. 7* below, can be used to evaluate the "best option" considering both initial(or retrofit) cost, as well as loss smeared out over time including discount rate as a result of natural hazards or other effects. This is the line with lowest total loss at the time of interest.

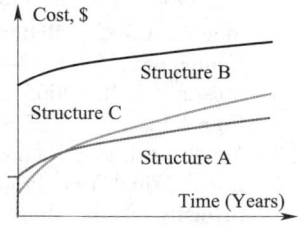

Fig. 7. Comparison of Different Retrofit Approaches-Break-Even Analysis (MacRae, 2006)[51]

In SQA, the decision is made based on the outcomes that are seen to be important. For each option, a rating is given for each outcome, and then these are combined to obtain the final rating in a way that seems appropriate to the decision makers using a decision matrix. The "best option" is the one with the highest rating. In a recent study of some building structures(Chanchi et al. 2012[54])factors considered included(i)frame damage, (ii)slab damage, (iii)element replaceability and(iv)permanent displacement. While this approach is subjective, it allows factors which are not quantified easily in a probabilistic approach to be directly included empowering the decision makers. It also includes the information in a way that allows it to be easily communicated to other audiences. A show of hands at the 2014 ASEC conference(MacRae, 2014)[55] indicated that the vast majority of consultants prefer this SQA approach to the PLA approach which is regarded as being a black box, difficult to perform, and difficult to check.

PLA and SQA may be used together in the decision making process.

7 CONCLUSIONS

This paper describes recent and ongoing research on the seismic performance and design of steel structures at NZ universities. It is shown that while steel structures generally exceeded their life-safety performance objectives in recent earthquakes, significant research is required to address:

(i)inadequacies with current standards that require addressing, including issues with diaphragm design, beam axial force design, column splices, gravity frames, proprietary systems(such as BRBs), gravity frames, gusset plates for BRBs, column continuity, frame ratcheting, frame torsion, and details for composite construction,

(ii)a lack of information about the real likely behaviour of older steel structures, including older portal frame structures,

(iii)a need for robust design techniques that can provide structures with a higher performance than that provided in current standards. A new tension-only system is introduced here.

(iv)a need for better assessment and repair techniques for damaged structures.

Research relating to all these topics in NZ is discussed. In addition, decision support tools to assist in selecting the best structural system, is described. Further research into these areas will result in better steel structures and more seismically sustainable cities.

ACKNOWLEDGEMENTS

Much of the work referenced here was developed through funding provided by the MBIE National Hazards Research Platform, the NZ Heavy Engineering Research Association, the Earthquake Commission, and the Quake Centre. Much of the work was undertaken by students at the Universities of Canterbury and Auckland, and in discussion with people in many professional organizations in NZ and around the world. The authors are is grateful to these people and organizations. The majority of the contents of this paper were published initially by MacRae(2015).

REFERENCES

[1] Buchanan A. H. , Bull D. , Dhakal R. P. , MacRae G. A. , Palermo P. , Pampanin S. , Base Isolation and Damage-Resistant Technologies for Improved Seismic Performance of Buildings, Report to the Royal Commission for the Canterbury Earthquakes, New Zealand, August 2011. http:// canterbury. royalcommission. govt. nz/

[2] MacRae G. A. , "Better Structures", SEFC 2015 PROCEEDINGS, Structural Engineering Frontier Conference, March 18-19, 2015, Tokyo Institute of Technology, Yokohama, Japan

[3] Meszaros J. R. (2005). "High-impact, Low-likelihood Risks: Engineers are from Oz, Owners are from Kansas", ASCE Structures Congress, New York.

[4] Makwana J. and Nanayakkara C. (2014), "Assessment of Seismic Capacity of Existing Structures", Final Year Project Report, Supervised by G MacRae and J Bothara, Civil Engineering, University of Canterbury.

[5] Bull D. K. (2004). Understanding the Complexities of Designing diaphragms in Buildings for Earthquakes, Bulletin of the New Zealand Society for Earthquake Engineering, 37(2), 70-88.

[6] Gardiner, D. R. (2011), Design Recommendations and Methods for Reinforced Concrete Floor Diaphragms Subjected to Seismic Forces, PhD Thesis, Univ. Cant. NZ.

[7] Tiong D. and Lyes D. (2014), Minimum Slab Thickness for Composite Floor Diaphragms", Final Year Project Report, Supervised by MacRae, Bull and Cowie, Dept. of Civil and Natural Resources Engineering, Univ. of Canterbury.

[8] Cowie, K. A. , A. J. Fussell, G. C. Clifton, G. A. MacRae and S. J. Hicks 2014. Seismic Design of Composite Metal Deck and Concrete-Filled Diaphragms-A Discussion Paper, NZSEE Conference, 2014.

[9] Sabelli, R. , T. Sabol and W. Easterling (2011), NEHRP Seismic Design Technical Brief No. 5-Seismic Design of Composite Steel Deck and Concrete-filled Diaphragms. National Institute of Standards and Technology.

[10] Bull, D. K. (2014), Module-2 ENCI429:2014 [PowerPoint slides]. University of Canterbury, Christchurch, New Zealand.

[11] Scarry, J. (2013), Floor Diaphragms and a Truss Method for Their Analysis, NZSEE Conference, 2014.

[12] Scarry, J. M. (2014), Floor diaphragms-Seismic bulwark or Achilles' heel, New Zealand Society for Earthquake Engineering 2014 Annual Conference. Christchurch, New Zealand, NZSEE.

[13] MacRae G. A. , Hobbs M. , Bull D. , Chaudhari T. , Leon R. , Clifton G. C. , and Chase C. , "Slab Effects on Beam-Column Subassemblies-Beam Strength and Elongation Issues", Composite Construction VII Conf. , Palm Cove, Queensland, Australia, 28-31 July 2013.

[14] Chaudhari T. D. , MacRae G. A. , Bull D. , Chase G. , Hicks S. , Clifton G. C. and Hobbs M. . (2015). Composite Slab Effects on Beam-Column Subassembly Seismic Performance, 8th International Conference on Behavior of Steel Structures in Seismic Areas, STESSA, Shanghai, China, July 1-3, 2015.

[15] Tork Ladani F. , MacRae G. , Chase G. & Clifton C. , "Effects Of Column Splice Properties On Seismic Demands In Steel Moment Frames", New Zealand Society of Earthquake Engineering Conference, Aotea Centre, Auckland, 21-23 March 2014. Paper P33.

[16] SNZ(2007), NZS 3404-1997 + Amendment 2-2007, Steel Structures Standard, Standards New Zealand, Wellington, New Zealand.

[17] Liu J, Astaneh-Asl A. Moment-rotation parameters for composite shear tab connections. J Struct Eng 2004;130(9):1371-80.

[18] MacRae G. A. , "The Continuous Column Concept-Development and Use", Proceedings of the Ninth Pacific Conference on Earthquake Engineering, Building an Earthquake-Resilient Society, 14-16 April, 2011, Auckland, New Zealand. Oral presentation. Paper 083.

[19] MacRae G. A. , Kimura Y. , and Roeder C. W. "Effect of Column Stiffness on Braced Frame Seismic Behavior", Journal of Structural Engineering, ASCE. Journal of Structural Engineering, March 2004, 130(3), pp. 381-391.

[20] MacRae G. A. , Lu A. Y. C. , Peng B. H. H. , Hahn C. , Ziemian R. and Clifton G. C. (2010), Plastic Hinge Locations In Steel Columns, Bull. of NZ Society of Equake Eng. , 43(1), March,

pp7-12,ISSN No. 1174-9857.

[21] Borzouie J. ,MacRae G,Chase G. J. ,and Clifton C. , "Experimental Studies On Cyclic Behaviour Of Steel Base Connections Considering Anchor Bolts Post Tensioning", New Zealand Society of Earthquake Engineering Conference, Aotea Centre, Auckland, 21-23 March 2014.

[22] Jones A. ,Lee C. ,MacRae G. & Clifton C. "Review Of Buckling Restrained Brace Design And Behaviour", New Zealand Society of Earthquake Engineering Conference, Aotea Centre, Auckland, 21-23 March 2014. Paper O87.

[23] Tsai, K. -C. and Hsiao, P. -C. (2008), Pseudo-dynamic test of a full-scale CFT/BRB frame—Part II: Seismic performance of buckling-restrained braces and connections. Earthquake Engineering and Structural Dynamics, 37(7), 1099-1115.

[24] Wijanto, S. 2012. Behaviour and Design of Generic Buckling Restrained Brace Systems. Masters thesis. Civil and Environmental Engineering. University of Auckland. Auckland. New Zealand.

[25] Thornton WA. On the Analysis and Design of Bracing Connections, National Steel Construction Conference Proceedings, AISC, Chicago, IL, 1991.

[26] Clifton, G. C. (2000) Revised Design Guidance for Proportioning Design Actions from the Braces into the Supporting Members of Brace/Beam/Column Connections. HERA Steel Design and Construction Bulletin, Issue No 47.

[27] Muir, L. S. , Designing Compact Gussets with the Uniform Force Method, Engineering Journal, American Institute of Steel Construction, First Quarter, 2008, p13-19. http://www. larrymuir. com/Documents/UFM. pdf

[28] Lin P. C. ,Tsai K-C,An A-C,Chuang M-C. (2013), "Seismic Design and Test of BRB Connections", 10 CUEE Symposium, 1-2 March 2013, Tokyo, Japan.

[29] Clifton, G. C. , N. Mago, et al. (2009). Eccentric Cleats in Compression and Columns in Moment-Resisting Connections, New Zealand Heavy Engineering Research Association: 1-70.

[30] MBIE(2010), NZ Ministry of Business, Industry and Enterprise, Practice Advisory #12, "Unstiffened eccentric cleat connections in compression", http://www. dbh. govt. nz/practice-advisory-12.

[31] Yam M. C. H. ,Cheng J. J. R. ,2002. Behaviour and design of gusset plate connections in compression, Journal of Constructional Steel Research 58, 1143-1159.

[32] Lin M-L,Tsai K-C. , Hsiao P-C and Tsai C-Y. , (2005). Compressive Behaviour of Buckling-Restrained Brace-Gusset Connections, The First International Conference on Advances in Experimental Structural Engineering, AESE 2005, July 19-21, Nagoya, Japan.

[33] Westeneng B. , Crake M. , Lee C-L. , MacRae G. A. , Jones A. (2015). Gusset Plate Effective Length Factor and the Effect of Stiffened Gusset Plate Connection on Buckling Restrained Brace Frames, New Zealand Society of Earthquake Engineering Annual Conference Proceedings, Rotorua, April.

[34] SNZ(2004), Earthquake Actions Standard, NZS1170. 5-2004, Standards New Zealand, Wellington, New Zealand.

[35] Yeow T. , MacRae G. A. , Kawashima, K and Sadashiva V. K. Dynamic Stability and Design of C-Bent Columns, Journal of Eqke Eng. ,17(5), 2013, pp. 750-768.

[36] MacRae G. A. and Deam B. "Building Regularity For Simplified Modelling", Chapter 6, EQC Project No. 06/514, Department of Civil and Natural Resources Engineering, University of Canterbury, Christchurch 8140, New Zealand, June 2009. 118pp. Commissioned by the NZ Earthquake Commission.

[37] Chunhaviriyakul P. , MacRae G. A. , Anderson D. , Clifton G. C. , Leon R. T. "Suitability of CFT Columns for New Zealand Moment Frames", Bulletin of the NZ Society for Earthquake Engineering, 48(1), March 2015, pp 63-79.

[38] Tjahjanto H. H. ,MacRae G. A. ,Abu A. K. ,Clifton G. C. ,Beetham T. and Mago N. , (2015), Behaviour of External Diaphragm Connections for Square CFST Columns under Bidirectional Loading, 8th International Conference on Behavior of Steel Structures in Seismic Areas, STESSA, Shanghai, China, July 1-3.

[39] MacRae G. A and Clifton G. C, (2013). "Low Damage Steel Construction", Steel Innovations Conference, Steel Construction New Zealand, Wigram, Christchurch, 21-22 February, Paper 1.

[40] MacRae G. A. , (2013), "Low Damage Construction-Some Systems Issues", 10CUEE Conference Proceedings, 10th International Conference on Urban Earthquake Engineering(10CUEE), March

1-2,Tokyo Institute of Technology,Tokyo,Japan.

[41] Chanchi Golondrino J. ,Xie R. ,MacRae G. A. ,Chase G. ,Rodgers G. and Clifton C. . "Braced Frame Using Assymmetrical Friction Connections AFC,8th International Conference on Behavior of Steel Structures in Seismic Areas,STESSA,Shanghai,July 1-3,2015.

[42] Leung H-K,Clifton G. C. ,Khoo H-H. and MacRae G. A. (2015),"Experimental Studies of Eccentrically Braced Frame with Rotational Bolted Active Links",8th International Conference on Behavior of Steel Structures in Seismic Areas,Shanghai,China,July 1-3.

[43] Djojo G. S. ,Clifton G. C,Henry R. S. and MacRae G. A. (2015). Experimental Testing of a Double Acting Ring Spring System for Use in Rocking Steel Shear Walls,8th International Conference on Behavior of Steel Structures in Seismic Areas,STESSA,Shanghai,China,July 1-3.

[44] Ramhormozian S. ,Clifton G. C. ,MacRae G. A. ,Khoo H. -H. ,"The Optimum Use of Belleville Springs in the Asymmetric Friction Connection",8th International Conference on Behavior of Steel Structures in Seismic Areas,STESSA,Shanghai,China,July 1-3,2015.

[45] Gunning,M. & Weston,D. 2013. Assessment of Design Methodologies for Rocking Systems,EN-CI493 Report,Supervised byMacRae G. A. ,Dept. of Civil and Natural Resources Engineering, University of Canterbury.

[46] Cook J. ,Rodgers G. W. ,MacRae G. A. & Chase J. G. 2015. "Development of a ratcheting,tension-only fuse mechanism for seismic energy dissipation",NZSEE Conference Proceedings,Rotorua.

[47] MBIE(2015),NZ Ministry of Business,Industry and Enterprise,Practice Advisory-Using Low-Damage Building Technologies in Seismic Design & Construction.

[48] Nashid H. ,Ferguson W. G. ,Clifton G. C. ,Hodgson M. ,Seal C. ,MacRae G. A. ,(2013). "Investigate the relationship between hardness and plastic strain in cyclically deformed structural elements",New Zealand Society of Earthquake Engineering Conference,Wellington,26-28 April. P7.

[49] Gardiner S. ,Clifton G. C. andMacRae G. A. "Performance,Damage Assessment and Repair of a Multistorey Eccentrically Braced Framed Building following the Christchurch Earthquake Series",Steel Innovations Conference,Steel Construction New Zealand,Wigram,Christchurch,21-22 February 2013,Paper 19.

[50] Abdolahirad A. ,MacRae G. A. ,Yeow T. Z. and Bull D. (2015). "Steel Building Behaviour under 9/2010 and 2/2011 Excitations in the Christchurch Earthquake Sequence ", 8th Int. Conf. on Behavior of Steel Structures in Seismic Areas,STESSA,Shanghai,July 1-3.

[51] MacRae G. A. ,"Decision Making Tools for Seismic Risk",Proceedings of the New Zealand Society of Earthquake Engineering Annual Conference,Paper 28,Napier,2006.

[52] Bradley BA,Dhakal RP,Cubrinovski M,MacRae GA,and Lee DS. Seismic loss estimation for efficient decision making. Bulletin of the New Zealand Society for Earthquake Engineering 2009; 42(2): 96-110.

[53] Yeow T. Z. ,MacRae G. A. ,Dhakal R. P. & Bradley B. A. ,"Seismic Sustainability Assessment of Structural Systems: Frame or Wall Structures?",World Conference on Earthquake Engineering,Lisbon,Portugal,Paper number 1242. August 2012.

[54] Chanchí,J. C. ,MacRae,G. A. ,Chase,J. G. ,Rodgers,G. W. ,and Clifton,G. C. ,Methodology for quantifying seismic sustainability of steel framed structures,STESSA Conference,Santiago, Chile,January 2012.

[55] MacRae G. A. ,Chanchi J. ,and Yeow T. ,"What Structure is Best?",Australian Structural Engineering Conference(ASEC),9 July to Friday 11 July 2014,Auckland,PN:177.

THE APPLICATION AND DESIGN OF VISCOUS DAMPERS IN SUPER HIGH-RISE BUILDING

Da-Sui Wang

East China Architectural Design & Research Institute

E-mail: dsw1973@ ecadi. com

KEYWORDS: Viscous damper, Energy dissipation, Outrigger truss, Damping coefficient.

ABSTRACT

The wind load and the seismic load are two main lateral load for high rise structures, which would cause a power series increase in structural response with the increase of structure height. So the traditional method, to improve the capability of wind and seismic resistance by improving the structure stiffness, strength and ductility is facing great challenge. The application of energy dissipation technique has been found to be very effective at mitigating the structure response subjected to wind load and earthquake load. The damping device can dissipate energy effectively in dynamic structural response, thus the structures are well protected.

Compared with the metal damper and tuned mass damper(TMD), which are commonly used in high-rise buildings, the viscous dampers have many excellent characteristics such as early intervention, good applicability, high redundancy and superior energy dissipation capacity. In this paper, the Green Centre Project in Urumqi is illustrated to introduce the application of viscous dampers. The performance objective of key elements in structure has been improved a lot compared with the traditional design method. Meanwhile, the reinforcement ratio and steel ration of main load-bearing elements such as core shear wall and frame column are reduced significantly, resulting in obvious economic benefits. With the application of viscous dampers, the additional damping ratio is 4. 5% in frequent earthquake, 2. 8% in design earthquake and 1. 9% in rare earthquake respectively, which shows a very obvious damping effect.

1 INTRODUCTION OF VISCOUS DAMPERS

Viscous damper mainly consists of cylinder, piston, damping channel, damping medium (viscous fluid) and the guide bar and other parts(Fig. 1). The piston in the cylinder will be activated under the external stimulus, forcing pressurized fluid passing through pores or cracks and a damping force is produced, thus the input energy can be dissipated and the structural response will be hampered.

Fig. 1(b) shows the relationship of viscous pamper damping force F and piston velocity v:

$$F = Cv\alpha \tag{1}$$

where C and α are the damping and velocity coefficientcorrespondingly.

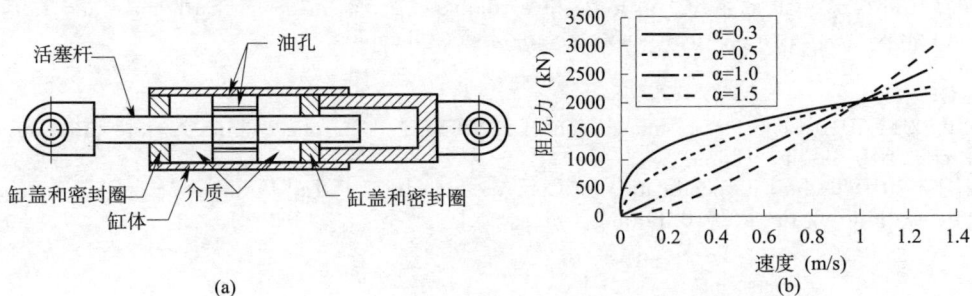

Fig. 1. (a) Construction of Viscous Dampers; (b) Damping Force and Velocity Relation

2 URUMAI GREEN CENTER DESIGN OVERVIEW

The Green Center project is located in Urumqi, including two super-tall twin towers. Tower structure has a height of 245m and the aspect ratio is 5. 8. The basic wind pressure is 0. 60kN/m^2 and seismic intensity is 8 degree, and the basic seismic acceleration is 0. 20g. Seismic design category is C according to GB50011-2010[1]. The structure has multi-lateral resistance systems: reinforced concrete core tube and shaped steel reinforced concrete core wall. Three viscous dampers are located in the three device stories respectively, as shown in *Fig. 2*.

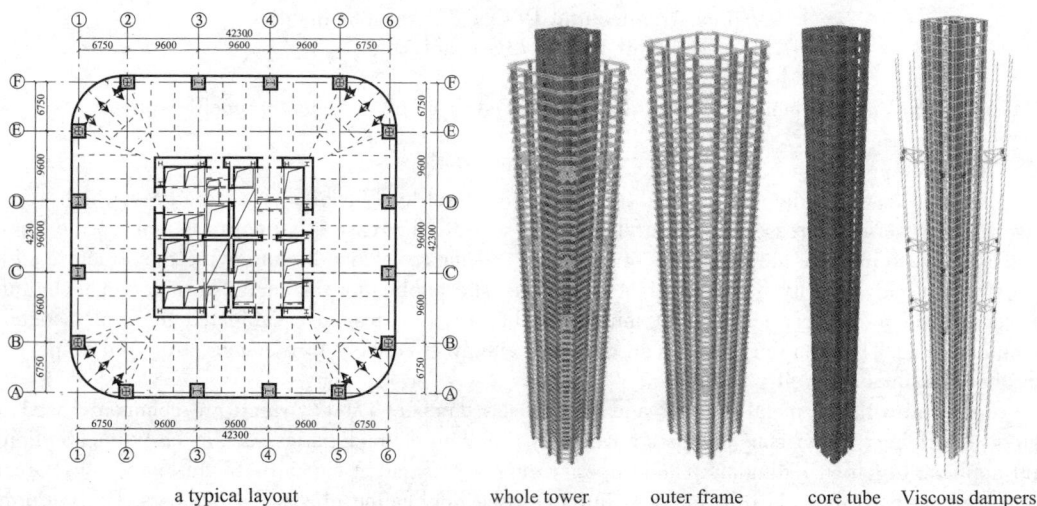

a typical layout   whole tower   outer frame   core tube   Viscous dampers

Fig. 2. Structure System of Green Center Project

3   CONCLUSIONS

The first order translationalperiod is 6. 09s in X direction and 6. 04s in Y direction correspondingly; and the first order torsional period is 3. 79s. The maximum inter-story displacement angle is 1/ 730, which satisfies the specification limits 1/510. The application of dampers reduces the request of the lateral stiffness of the structure so that the natural vibration period of the structure is increased. Furthermore, the damping effect also mitigates the base shear effectively. The shear-weight ratio is 0. 025, which is within the specification limits. Meanwhile, the stiffness weight ratio is 1. 76, which has exceeded the specification limit 1. 4.

The time history analysis shows that the average base shear decreases by 21% in frequent earthquake, 15% in design earthquake and 9% in rare earthquake compared with the model without dampers, the base overturning moment decreases by 20% , 12% , 7% and the inter-story displacement angle decreases by 25% , 16% , 10% respectively. With the application of viscous dampers, the additional damping ratio is as high as 4. 5% in frequent earthquake, 2. 8% in design earthquake and 1. 9% in rare earthquake respectively, which shows a very obvious damping effect.

REFERENCES

[1] GB50011-2010 Code for seismic design of building[S]. Beijing: China Architecture&Building Press, 2010, Beijing, China.

[2] JGJ 3-2010 Technical Specification for Concrete Structures of Tall Buildings[S], China Architecture & Building Press, 2010, Beijing, China.

Performance-Based Design of Structures

AN ENERGY-BASED NONLINEAR STATIC PROCEDURE FOR ESTIMATING THE SEISMIC RESPONSE OF HYBRID STEEL MOMENT RESISTING FRAMES

Ke Ke[a], Yi-Yi Chen[a], Guang-Hong Chuan[a]

[a] State Key Laboratory of Disaster Reduction in Civil Engineering, Tongji University, Shanghai 200092, China

1987keke@ tongji. edu. cn, yiyichen@ tongji. edu. cn, 1130335ChuanGH@ tongji. edu. cn

KEYWORDS: hybrid steel moment frames, energy approach, nonlinear static procedure

ABSTRACT

The innovative moment resisting frames defined as the hybrid steel moment resisting frames showing multiple yielding stages during ground motions is demonstrated to be of noteworthy features in recent research works. To date, while the nonlinear dynamic analysis is the most rigorous approach to compute the response of the system, the nonlinear static procedure, or the pushover analysis, is widely used and recommended by design documents such as EC8 and FEMA-273 for its reasonable accuracy and satisfactory computational efficiency. However, it is noted that most current approaches are proposed focusing on conventional structures which can be idealized with an elasto-plastic model or bilinear model with negligible post-yielding stiffness. For hybrid steel moment resisting frames showing multiple yielding stages, the conventional idealized method might conceal the behavior of the system and lead to inaccurate estimations. In addition, it is not informative to neglect the yielding stages which are directly related to the design systems. Therefore, an applicable approach to evaluate the behavior of hybrid steel moment resisting frames considering the yielding stages explicitly is in need.

In the present work, motivated by the modified energy balance concept, a nonlinear static procedure for estimating the seismic response of hybrid steel moment resisting frames showing multiple yielding stages (bilinear feature or tri-linear feature) is proposed. The theory is explored and underlying assumptions are clarified. For purpose of demonstration, the procedure is validated by the nonlinear dynamic analysis of a prototype structure. To validate the improved accuracy, current nonlinear static procedures are also performed. The results show that the proposed procedure can evaluate the responses of systems with attractive accuracy compared with the other approaches, which can be used to evaluate the systems effectively.

CONCLUSIONS

To evaluate the seismic response of hybrid steel MRF system, an energy-based nonlinear static procedure is proposed. The theory and assumptions are clarified in detail, and the application of the example structure is also presented. This research indicates that the structural energy balance of hybrid steel MRF systems is dependent on the yielding stages, and the balance mode is altered in different stages. The post-yielding stiffness, the sequence factor and the property of ground motions all influence the structural energy balance. The analysis and comparison with the conventional approach show that the proposed procedure considering the yielding stages can lead to more accurate estimates regarding the target displacement and story drift distributions compared with the energy-based procedure which is based on elasto-plastic models. The lateral forces distributions determined by elastic analysis are applicable in estimating the response of hybrid steel MRF systems.

The proposed procedure can estimate the seismic response of hybrid steel MRF systems with reasonable accuracy. Currently, research works regarding the applicability of the procedure in hybrid steel MRF considering more yielding stages are in progress, and they will provide full validation of the procedure.

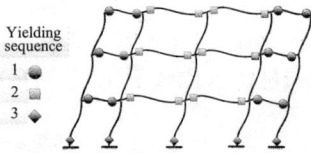

Fig. 1 The yielding sequence of a hybrid steel moment frame

Yielding sequence
1
2
3

Table 1. Dynamic properties of the example structure

Structure	subject	1st Mode	2nd Mode	3rd Mode
9-story	Period(s)	2. 25	0. 84	0. 49
	Modal participation factor	1. 37	0. 55	0. 25

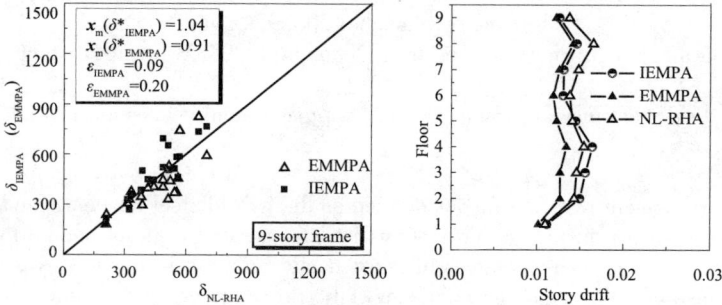

Fig. 2 The roof displacements response comparison and drift response comparison

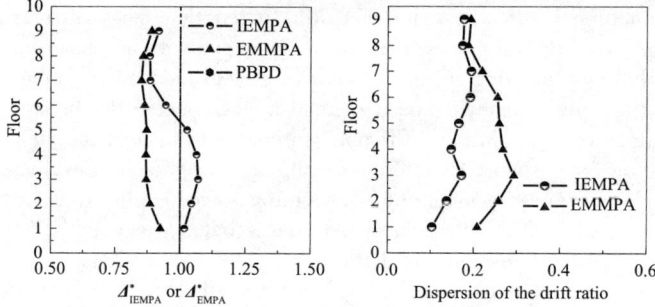

Fig. 3 The drift ratios and dispersion of the drift ratios

ACKNOWLEDGMENTS

This research is supported by the National Science Foundation of China(Grant No. 51038008). Any opinions, findings, and conclusions presented in this paper are those of the authors and do not necessarily reflect the views of the sponsors.

REFERENCES

[1] Charney FA, Atlayan O, "Hybrid moment-resisting steel frames", *Engineering Journal*, Vol. 48, pp. 169-182.

[2] Hernandez-Montes E, Kwon OS, Aschheim MA, "An energy-based formulation for first-and multiple-mode nonlinear static (pushover) analyses", *Journal of Earthquake Engineering*, Vol. 8, pp. 69-88.

[3] Manoukas G, Athanatopoulou A, Avramidis I, "Static pushover analysis based on an energy-equivalent SDOF system", *Earthquake Spectra*, Vol. 27, pp. 89-105.

[4] Leelataviwat S, Saewon W, Goel SC, "Application of energy balance concept in seismic evaluation of structures", *Journal of Structural Engineering*, Vol. 135, pp. 113-121.

[5] Jiang Y, Li G, Yang D, "A modified approach of energy balance concept based multimode pushover analysis to estimate seismic demands for buildings", *Engineering Structures*, Vol. 32, pp. 1272-1283.

[6] Chopra AK, Goel RK, Chintanapakdee C, "Evaluation of a modified MPA procedure assuming higher modes as elastic to estimate seismic demands", *Earthquake Spectra*, Vol. 20, pp. 757-778.

[7] Shome N, Cornell C A, Bazzurro P, et al, "Earthquakes, records, and nonlinear responses", *Earthquake Spectra*, Vol. 14, pp. 469-500.

SEISMIC DESIGN OF MULTISTORY TENSION-ONLY CONCENTRICALLY BRACED BEAM-THROUGH FRAMES AIMED AT UNIFORM INTER-STORY DRIFT

Chao Zou[a,b], Wei Wang[a,b], Yiyi Chen[a,b], Yunfeng Zhang[c]

[a] Tongji University, State Key Laboratory of Disaster Reduction in Civil Engineering, China

[b] Tongji University, Department of Structural Engineering, China

zouxiaochao2009@ 163. com, weiwang@ tongji. edu. cn, yiyichen@ tongji. edu. cn

[c] University of Maryland, Department of Civil & Environmental Engineering, USA

zyf@ umd. edu

KEYWORDS: Seismic design, Beam-through, Braced steel frame, Uniform inter-story drift.

ABSTRACT

In this paper, a seismic design procedure aiming for achieving uniform inter-story drift in multi-story tension-only concentrically braced beam-through frames (TOCBBTFs) is presented. The TOCBBTF primary structural system with its special strong-beam-weak-column configuration is designed so that the columns and beams remain elastic for their negligible lateral stiffness while tension-only concentric braces resist all lateral loads as a reliable way of absorbing seismic energy under seismic loading. Three performance objectives of seismic design procedure aiming for uniform inter-story drift along the building height are established first. A seismic design procedure is proposed and applied to a 4-and 6-story model structures to demonstrate the design procedure. Time-history analyses are carried out to verify whether the pre-specified performance objectives are achieved or not. The analysis results show that the seismic performance of the 4-and 6-story model structure designed with the proposed design procedure agrees very well with the pre-specified design objectives.

CONCLUSIONS

1. The mode shapes of the multi-story TOCBBTFs according to the design procedure are close to linear lines, and the first modal participating mass ratios are large enough to neglect the higher modes effect, which is the basic premise for the establishment of the design procedure.

2. The multi-story TOCBBTFs designed according to the design procedure can achieve uniform inter-story drift when the braces yield for the first time in both pushover analysis and time-history analysis, and the maximum story displacements of theirs are closer to linear lines in time-history analysis than those designed with other methods.

3. The proposed design procedure provides a convenient tool for seismic design of multi-story TOCBBTFs to achieve the goal of uniform inter-story drift.

ACKNOWLEDGMENT

The research presented in this paper was supported by the Open Fund of State Key Laboratory of Disaster Reduction in Civil Engineering through Grant No. SLDRCE14-04.

REFERENCES

[1] Wada A, Iwata M, Huang Y H, "Seismic design trend of tall steel buildings after the Kobe earthquake", Passive Energy Dissipation and Control of Vibrations of Structures, 7:251-269, 1997.

[2] Wang W, Zhou Q, Chen Y, et al, "Experimental and numerical investigation on full-scale tension-only concentrically braced steel beam-through frames", Journal of Constructional Steel Research, 80:369-385, 2013.

[3] Redwood R G, Channagiri V S, "Earthquake resistant design of concentrically braced steel frames", Canadian Journal of Civil Engineering, 18(5):839-850, 1991.

[4] Smith B S, Coull A, "Tall building structures: analysis and design". University of Texas Press, 1991.

[5] Rosenblueth E, Herrera I, "On a kind of hysteretic damping". *Journal of Engineering Mechanics Division*, 90(1):37-48, 1964.

[6] INOUE K, "Hysteresis-type vibration dampers-Design of steel frames incorporating hysteretic vibration dampers, tow-part series", *Steel Construction Today & Tomorrow*, 8:4-6, 2004.

[7] Kim J, Seo Y. "Seismic design of low-rise steel frames with buckling-restrained braces", *Engineering Structures*, 26(5):543-551, 2004.

COMPARISON BETWEEN CRITERIA FOR SELECTING THE PARAMETERS OF HYSTERETIC ENERGY DISSIPATORS FOR SEISMIC PROTECTION OF STEEL BUILDING STRUCTURES

David Domínguez[a], Francisco López-Almansa[b], Amadeo Benavent-Climent[c]

[a] Construction Engineering and Managing Department, University of Talca, Chile
ddomínguez@ utalca. cl

[b] Architecture Structures Department, Technical University of Catalonia, Barcelona
francesc. lopez-almansa@ upc. edu

[c] Structures Mechanics Department, Technical University of Madrid, Spain
amadeo. benavent@ upm. es

KEYWORDS: Energy Dissipators, Steel Structures, Steel Buildings, Design of Dissipators, Dynamic Analysis.

ABSTRACT

This work deals with seismic protection of steel building structures using hysteretic energy dissipators. The sensitivity of the performance of these systems to the design parameters of the dissipative devices is numerically investigated. Particularly, the influence of the vertical distribution of the yielding forces of the dissipators is deeply examined; two major approaches are compared: the complex formulation described in [Benavent-Climent 2011, 2014] and the simpler strategy presented in [Foti et al. 1998]. Comparison is established in terms of the response of a 15-story steel (*Fig. 1*) to the Lorca earthquake. The 2011 earthquake in Lorca (11-05-2011) is the most destructive event ever recorded in Spain, causing nine fatalities and other severe consequences. Initial results seem to indicate little sensitivity of that response to the design yielding forces of the dissipators.

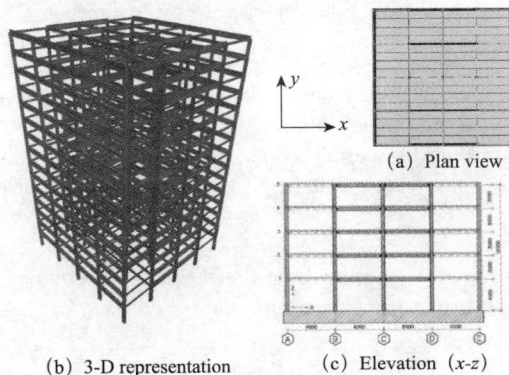

(a) Plan view

(b) 3-D representation   (c) Elevation (*x-z*)

Fig. 1.   Selected 15-story prototype building

The proposed strategy consists of incorporating chevron steel bracing members and to installing energy dissipators in the connections between each bracing and the above beam (*Fig. 2*); energy dissipators are represented as. *Fig. 3* displays the top floor time – history responses in the *x* direction (*Fig. 1.* a) of the considered building under the strongest component of the Lorca record. *Fig. 3* represents three responses: frame without any bracing (bare frame), frame with conventional concentric chevron bracing (braced frame), and frame with energy dissipators installed as indicated in *Fig. 2* (protected frame). Comparison among plots in *Fig. 3* confirms the capacity of energy dissipators to reduce the structural response far beyond the reduction provided by conventional concentric bracing. Design yielding forces of the dissipators have been determined according the two aforementioned criteria listed. Only minor differences have been found; this indicates little sensitivity of the performance of the dissipation system to the vertical distribution of the yielding forces.

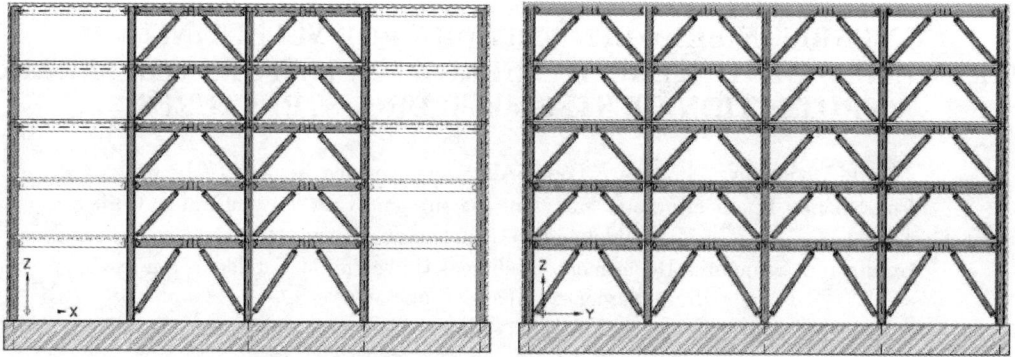

(a) x direction   (b) y direction

Fig. 2.   Proposed seismic protection using energy dissipators

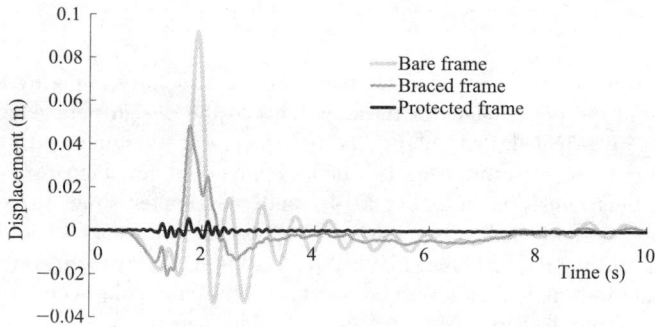

Fig. 3.   Time-history responses in x direction

CONCLUSIONS

This paper presents a numerical study on the capacity of hysteretic energy dissipators to reduce the time-history response of a 15-story steel building under the strongest component of the Lorca earthquake. Preliminary resultsshow that dissipators reduce significantly the maximum displacement and that results are highly insensitive to the design criteria of the yielding forces of the dissipators.

ACKNOWLEDGMENT

This work has received financial support from the Spanish Government under projects CGL2011-23621 and BIA2011-26816 and from the European Union(Feder).

REFERENCES

[1] Benavent-Climent A , "An energy-based method for seismic retrofit of existing frames using hysteretic dampers" , Soil Dynamics and Earthquake Engineering, 31 : 1385-1396 , 2011.

[2] Benavent-Climent A , "Energy - based design of non-traditional structures incorporating hysteretic dampers: experimental validation with shaking table tests" 15ECEE(15th European Conference of Earthquake Engineering) , Istanbul , Turkey , 2014.

[3] Foti D , Bozzo LM , López Almansa F , "Numerical Efficiency Assessment of Energy Dissipators for Seismic Protection of Buildings" , Earthquake Engineering & Structural Dynamics, 27 : 543-556 , 1998.

PERFROMANCE BASED DESIGN OF MR-FRAMES BY TPMC AND ENERGY APPROACH

Elide Nastri[a] and Vincenzo Piluso[a]

[a] Department of Civil Engineering, University of Salerno, Italy

e-mails: enastri@ unisa. it, v. piluso@ unisa. it

KEYWORDS: Theory of Plastic Mechanism Control, global mechanism, Energy approach.

ABSTRACT

The importance of the collapse mechanism typology of seismic resistant structures is universally recognized. In particular, the best seismic performances are obtained when a collapse mechanism of global type is developed, because of the maximization of dissipative zones. Two design procedures leading to such collapse mechanism are herein investigated: the Performance Based Design approach byGoel and Lee[1] and the Theory of Plastic Mechanism Control (TPMC)[2]. To this scope reference is made to a MR-Frame structure designed by Goel and Lee[1]. By using the same design seismic forces and the same gravity loads, the structure has also been designed by means of TPMC. The seismic design forces are derived by means of the energy balance equation, as proposed by Goel and Lee[1] where the earthquake input energy is evaluated as γ times the input energy occurring in the elastic system and evaluated. The γ factor accounts for the expected ductility demand and the reduction factor according to the well-known Housner formulation. The maximum energy that the structure is able to absorb in elastic range is estimated by means of the Akiyama approach. As a result, the energy to be dissipated by means of hysteresis is obtained as the difference between the seismic input energy and the maximum stored elastic energy. The energy to be dissipated is related to the plastic collapse mechanism by means of the classical work equation by assuming an appropriate distribution of the seismic horizontal forces. The internal work is, therefore, strictly related, on one hand, to the collapse mechanism typology and, on the other hand, to the cumulated plastic rotation demand expected for a given level of the seismic intensity measure. As a result, the seismic design horizontal forces are derived as a function of the seismic intensity measure, i. e. the spectral acceleration value, and of the plastic rotation capacity.

Regarding the beam dimensioning two cases can be identified. The first case corresponds to beams whose size is governed by gravity loads. The second case correspondsto beams, parallel to the direction of secondary beams of building deck, whose size is governed by the seismic load combination. The design example herein presented refers to beam sections governed by seismic actions, i. e. parallel to the direction of the deck slab corrugation. Therefore, being known the seismic forces at each storey, provided by means of the energy approach, beam sections can be dimensioned by assuming a distribution of their size according to the storey shear distribution and by imposing the prevention of storey mechanism at first storey. The fulfilment of these two design conditions results in a system of two equations where the two unknowns of the problem are the plastic moment of beams at top storey and the plastic moment of columns at first storey. This criterion for beam design, proposed in[1], is herein applied also for the preliminary design of beam sections needed to apply TPMC. Concerning the design of columns at the upper storeys the two design approaches are significantly different. In fact, within the framework of rigid-plastic theory, while TPMC is based on a kinematic approach, the Goel and Lee procedure is based on a static approach using free-body diagrams to derive the distribution of column bending moments along their height. Moreover, the influence of second order effects is rigorously accounted for in TPMC by means of the concept of mechanism equilibrium curve and of the application of kinematic theorem of plastic collapse. Conversely, in the procedure suggested by Goel and Lee, second order effects are considered by applying the moment amplification method to the columns bending moments which are derived by means of a rigid-plastic analysis, rather than to moments derived by means of elastic analyses as commonly made. Serviceability requirements have been checked with reference to NEHRP provisions. The seismic performance of a MR-Frame designed according to Performance Based Design approach by Goel and Lee has been compared, by means of both push-over and dynamic non-linear analyses, with the one resulting from the application of TPMC.

CONCLUSIONS

With reference to moment resisting frames, a very brief summary of two plastic design procedures aimed to assure a collapse mechanism of global type has been presented. The Goel and Lee design method is characterized by the use of a static approach where, starting from the plastic moments of beams and the plastic moment of first storey's columns, the bending moments to design the column sections along the building height are derived by means of free body diagrams. In addition, second order effects are not directly taken into account within rigid-plastic analysis, but are successively considered by means of the application of the moment amplification method, commonly applied for elastic analyses. TPMC is, conversely, based on a kinematic approach where the desired collapse mechanism is assured by imposing that the mechanism equilibrium curve corresponding to the global mechanism has to be located, up to the design displacement, below those corresponding to all the undesired mechanisms. The results coming from the application of the examined design approaches have been presented with reference to a 5bay-9storey MRF. Despite it is only a preliminary investigation, some observations can be outlined:

● The Goel and Lee design procedure does not lead to a maximum base shear consistent with its design goal as pointed out by the results of pushover analysis;

● The use of the moment amplification method to amplify bending moments which are derived by means of a rigid-plastic analysis and a static approach does not seem correct from a theoretical point of view. In spite of such a use of the moment amplification method seems to lead to an overestimation of second order effects, it is not able to compensate the underestimation of the plastic moment actually needed to reach a maximum base shear consistent with the target design;

● Also the frame designed by means of TPMC is not able to develop the maximum base shear assumed in the design process, but in this case it is due to the adoption, for comparison purposes, of beam sections equal to those of Goel and Lee design example. However, this observation needs to be investigated and, eventually, confirmed by designing a structure where both the beams and the columns are derived by TPMC;

● Both Goel and Lee design method and TPMC based design method allow to reach the main design goal. This has been confirmed by the pattern of yielding exhibited as result of both push-over and dynamic non-linear analyses;

● Given the beam sections, TPMC leads to a lighter structures guaranteeing the global mechanism;

● Due to the oversizing of the column sections with respect to those required to assure a collapse mechanism of global type, the Goel and Lee design procedure leads to better seismic performances, but with an increased structural weight(23% increase in the examined case).

A more deep investigation of the differences between the two design approaches is needed before general conclusions can be gained. In the meantime, it is worthwhile to underline that, among different design approaches suggested in the technical literature, they are the only one able to assure a collapse mechanism of global type.

REFERENCES

[1] Goel S. C., Lee S-S, "Performance-Based Design of Steel Moment Frames using Target Drift and Yield Mechanism", Research Report UMCEE 01-17, December 2001.

[2] Piluso, V., Nastri, E., Montuori, R. "Advances in Theory of Plastic Mechanism Control: Closed Form Solution for Mr-Frames", accepted for publication on Earthquake Engineering and Structural Dynamics, 2015.

FRAGILITY AND SEISMIC BEHAVIOUR OF PRE- AND POST-RETROFIT CONCENTRICALLY BRACED FRAMES

Lucia Tirca[a], Ovidiu Serban[a], Mingzheng Wang[a]

[a]Concordia University, Dept. of Building, Civil and Environmental
Engineering, Montreal, Canada
lucia. tirca@ concordia. ca, o_serban@ encs. concordia. ca, m_wang@ encs. concordia. ca

KEYWORDS: Concentrically Braced Frames, Fragility, Seismic deficiencies, Retrofit, Damage

ABSTRACT

In general, Eastern Canada ischaracterized by M6 and M7 shallow earthquakes and the risk of losses could be high. The later seismic map was developed for a probability of exceedance of 2% in 50 years or 2500 years return period, while the 2nd generation of seismic map based on a probability of exceedance of 50% in 50 years was released in 1970 and the 3rd generation was released in 1985. Moreover, seismic design and detailing provisions were first included in the 1989 edition of CSA S16 standard and buildings designed between 1970 and 1985 were proportioned to withstand lower seismic forces in comparison with those required by the current code. Thus, for seismic assessment of buildings, the year of building construction and knowledge of changes in code and regulations are essential parameters. Due to the large CBF building stock designed and built in Quebec, Canada in the decade 1980-1989, this study focuses on the seismic fragility assessment of pre-and post-retrofit low-rise(3-storey) and middle-rise(6-storey) CBFs office buildings located in Montreal and Quebec City. These CBF(concentrically braced frame)systems were designed according to the NBCC 1980[1] and CSA/S16. 1-78[2] standard and were assessed against the current design requirements[3], [4].

Employing incremental dynamic analysis(IDA)[5] and using OpenSees [6-7], the IDA curves computed as pairs of earthquake intensity measure parameter(e. g. spectral acceleration at the first mode)versus the engineering demand parameter(e. g. peak interstorey drift)were obtained for all studied CBF buildings before and after retrofit. As reported from IDA curves, all existing buildings experienced collapse when subjected to ground motion intensities associated to the current code demand. In this study, a conventional rehabilitation strategy consisting in local strengthening of CBF beams and columns and replacement of gusset plates used for brace-to-frame connections was considered. According to ASCE/SEI 41-13 provisions[8], for office buildings, the rehabilitation objective class is *Basic Safety*(BSO) and the targeted performance levels are *Life Safety*(*LS*) when subjected to frequent, moderate earthquakes(10% / 50 yrs. hazard) and *Collapse Prevention*(*CP*) when subjected to more severe rare events(2% / 50 yrs.). Quantitative approaches were considered to define each performance level associated to each ground motion on the computed IDA curve based on acceptable ranges of deformation demand on structural and non-structural components. Thus, damage levels (*DL*) namely: *Very Light*, *Light* (*LD*), *Moderate* (*MD*) and *Severe* (*SD*) vary as a continuous function of building deformation and were associated to the defined building performance levels[9]. Additionally, ASCE/SEI 41-13 provides recommended values for the maximum interstorey drift, δ_{max} and maximum residual interstorey drift, δ_{max_res} for each performance level and type of building structure. In this study, fragility functions were derived from parameters estimated using IDA curves and fragility curves are assumed to follow a lognormal cumulative distribution, which is defined by a median value and a dispersion. In addition, both aleatoric and epistemic uncertainties were considered[10].

CONCLUSIONS

In this study, fictitious 3-and 6-storey CBF office buildings located in Montreal and Quebec City designed according to the 1980 provisions were assessed against the current code though nonlinear dynamic analysis. Critical seismic deficiencies were detected at the brace-to-frame connections. These connections, consisting of a gusset plate welded to the slotted HSS brace, have short welding lengths, do not present the 2tg clearance and cannot sustain buckling of braces due to the lack of strength. Among these deficiencies, the shearing of welds governed. It is noted that factored forces developed in members were used in design instead of brace capacity. Furthermore, beams do not have sufficient strength to carry the tributary gravity load component in addition to the axial force developed due to the effect of braces and lack of compression strength was encountered by the middle column of the two adjacent CBFs with tension/compression diagonal braces. These deficiencies are common for CBF buildings designed according to the 1980 code provisions. When brace-to-frame connections fail at the same floor, the storey mechanism is formed due to hinging of columns. When this mechanism occurs at the bottom floors it leads to building failure, when it happens at the upper floors it leads to partial failure.

The selected conventional retrofit strategy was able to respond to the *Basic Safety Objective* which has the following targeted performance levels: *Life Safety* and *Collapse Prevention*. From nonlinear dynamic analysis using OpenSees it was found that all post-retrofit buildings were able to reach *LS* under 10% / 50 yrs. hazard and *CP* when subjected to 2% / 50 yrs. earthquake.

After seismic assessment was completed, building fragility associated to each performance level was calculated through a lognormal cumulative distribution function. The analytical fragility curves were derived from the computed IDA curves on which the performance levels were defined such as: *Immediate Occupancy* at the end of the elastic segment when the first brace experienced buckling, *Life Safety* when the maximum residual interstorey drift reached 0. 5% hs and *Collapse Prevention* when the slope of the IDA curve approaches zero (flattening effect). Three damage levels were associated to each performance level such as: *light*, *moderate* and *severe* damage. In the calculation of fragility curves both epistemic and aleatoric uncertainties were considered. For the pre-retrofit buildings damage state jumped from *light damage* to *sever damage* while for post-retrofit buildings the probability of severe damage was less than 30% for buildings located in Montreal, while for buildings in Quebec City the probability was less than 10%.

ACKNOWLEDGMENT

The financial support provided by the NSERC is gratefully acknowledged by the authors.

REFERENCES

[1] NBCC 1980. *National Building Code of Canada*, The 8th ed. , National Research Council of Canada.

[2] CSA (Canadian Standards Association), 1980. "Limit states design of steel structures". CSA/ S16. 1-*M78 Standard*, Toronto.

[3] NBCC 2010. *National Building Code of Canada*, The 13th ed. , National Research Council of Canada, Ottawa.

[4] CSA (Canadian Standards Association), 2009. "Design of steel structures". CSA/S16-09 *Standard*, Toronto, ON.

[5] Vamvatsikos D. , Cornell C. A. , 2002. "Incremental dynamic analysis". *Earthquake Eng. Struct. Dyn.*, J. Wiley & Sons, Ltd. , ISSN 1096-9845, Vol. 31, No. 3, pp. 491-514.

[6] McKenna F. , Fenves G. L. , 2004. "Open system for earthquake engineering simulation (OpenSees)". OpenSees software 2013, *Pacific Earthquake Engineering Research Center (PEER)*.

[7] Tirca L. and Chen L. , 2014. "Numerical simulation of inelastic cyclic response of HSS braces upon fracture", *Advanced Steel Construction Journal*, Vol. 10, No. 4, pp. 442-462.

[8] American Society of Civil Engineers (ASCE), 2013. "Seismic rehabilitation of existing buildings". *ASCE/SEI Standard* 41-13, Reston, VA.

[9] Tirca L. , Serban O. , Lin L. , Wang M. , Lin N. , 2015. "Improving the seismic resilience of existing braced frame office buildings". *Journal of Structural Engineering (ASCE)*, in press.

[10] Ellingwood B. , Celik O. C. , Kinali K. , 2007. "Fragility assessment of building structural systems in Mid-America", *Earthquake Engineering and Structural Dynamics*, Vol. 36, pp. 1935-1952.

SEISMIC BEHAVIOR OF CONCENTRIC BRACED FRAMES DESIGNED USING DIRECT DISPLACEMENT-BASED DESIGN METHOD

Dipti Ranjan Sahoo[a] and Ankit Prakash[b]

[a]Department of Civil Engineering, Indian Institute of Technology Delhi, New Delhi, India
e-mail: drsahoo@ civil. iitd. ac. in

[b]National Thermal Power Corporation Limited, New Delhi, India
e-mail: ankitprakashntpc@ gmail. com

KEYWORDS: Concentric braced frame, Design basis earthquake, Nonlinear dynamic analysis, Seismic.

ABSTRACT

The performance based design of structures has gained importance and is under development ever since release of the earliest work 'Vision 2000' after the 1994 Northridge earthquake. The new design philosophy aims at achieving the pre-determined performance goals for a structure when subjected to the specified seismic hazard-level events. Considering the inherent weakness of conventional concentric braced frames (CBFs) in terms of the unpredictable and adverse performance during large seismic e-vents, the development of such a performance-based design methodology is extremely necessary. Through the DDBD approach the design of structures to meet performance objectives defined by drift or displacement limits can be accomplished. The procedure developed by Wijesundara[1] is based on classical DDBD approach by Priestley[2] and uses the equivalent viscous damping (EVD) of the system based on the brace slenderness ratio. This procedure has been adopted in the present study. Although the capability of the DDBD procedure has been validated for CBF structures in achieving design drift limits when subjected to corresponding ground motions, further evaluation of the design procedure is required in terms of comprehensive performance objectives of the performance matrix. Two chevron braced-type CBFs of 3-and 6-story buildings were taken up for this investigation. Forty ground motions originally developed by Somerville[3] for SAC project were considered in this study. The ground motions for DBE have been considered as those corresponding to 10% in 50 years probability (LA01 to LA20) while for MCE as those corresponding to 2% in 50 years probability (LA21 to LA40). 3-storey four-bay and 6-storey five-bay CBFs designed using DDBD procedure were modelled in OpenSEES finite element program[4] using a simplified 2-D non-linear line-frame model. Simulation of inelastic deformation during of buckling of braces were realised by using ten number non-linear force-based beam-column elements per brace. P-delta geometric transformation is used for global co-ordinate system. Gusset plates were modelled as zero-length non-linear beam-column elements made of steel02 material with two integration points and corotational geometric transformation. Rigid portions of the members at the gusset plates i. e. end of braces, beams and columns were defined as uniaxial rigid elastic elements modelled with large values of cross-sectional area and moment of inertia. To capture P-delta effects due to gravity load, leaning columns with rigid frame links were modelled. Nonlinear time-history (NLTH) analyses have been carried out for the selected ground motions. Stiffness and mass proportional Rayleigh damping was applied with damping coefficients calculated using the first and third natural frequencies. The value of tangent stiffness proportional damping assumed was 3% and Newmark acceleration time integration scheme with beta and gamma as 0.5 and 0.25 was adopted. The main parameters investigated are peak interstory drift ratio, residual inter-story drift ratio, hinge mechanism, and low-cycle fatigue damage of braces. *Fig. 1.* shows the comparison of interstory drift response of 3-story and 6-story DDBD frames with respective NEHRP frames.

CONCLUSIONS

Performance-based design procedure for conventional concentric braced frames (CBF) using direct-displacement based design (DDBD) approach by Wijesundara[1] was presented. Performance objectives for the CBF structures were defined for two-level seismic hazard comprising of peak and residual inter-story drifts and structural damage in terms of low-cycle fatigue damage of braces, yielding of

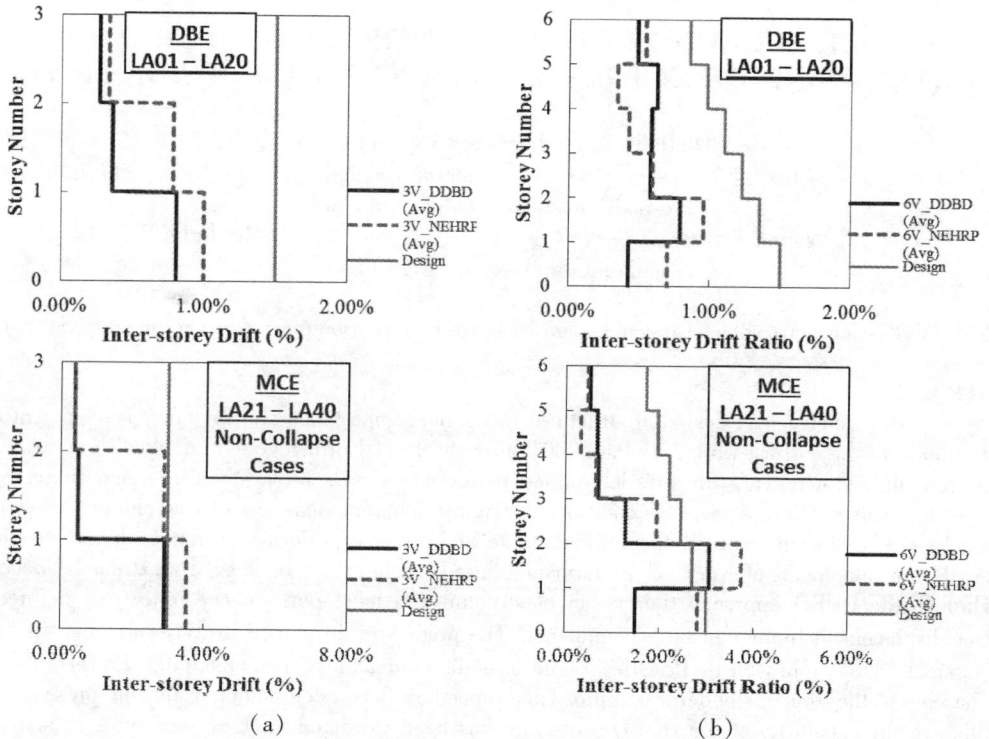

Fig. 1. Comparison of interstory drift response of (a) 3-story and (b) 6-story frames

columns and beams and structural failure. Two example CBFs 3-and 6-storey were designed using the proposed methodology as well as current force-based code provisions. These frames were also subjected to same sets of ground motions and the NLTH results obtained were compared with those of DDBD frames. The analyses results obtained confirmed capability of the design process in imparting enough strength to the CBFs to meet target performance in DBE scenario and also improved performance than current code-based designed structure. For MCE hazard level events, improved performance was marked by significant reduction in number of collapse cases. Though the collapse cases could not be entirely eliminated for DDBD structures in MCE scenario largely because of some extremely strong ground accelerations in the suit of time histories, it is obvious that these incidents are exceptionally rare. Nevertheless the performance of DDBD frames was better than NEHRP frames in this case too.

REFERENCES

[1] Wijesundara K. K. Seismic design of steel concentric braced frame structures using direct displacement based design approach, 2nd International Conference on Sustainable Built Environments (ICSBE 2012) , Sri Lanka, 2012.

[2] Priestley M. J. N. , Calvi G. M. , and Kowalsky M. J. Displacement-based seismic design of structures, IUSS Press, Pvia, Italy, 2007, 721 pp.

[3] Somerville P. G. , Smith M. , Punyamurthula S. , and Sun J. Development of ground motion time histories for phase 2 of the FEMA/SAC Steel Project, Rep. No. SAC/BD-97/04, Sacramento, CA. 1997.

[4] McKenna F. , Fenves G. L. , Jeremic B. , and Scott M. H. Open System for Earthquake Engineering Simulation (OpenSEES) , Berkeley, CA, 2007.

DIRECT DISPLACEMENT BASED DESIGN: APPLICATION FOR STEEL MOMENT RESISTING FRAMES WITH CLT INFILL WALLS

Matiyas Ayalew Bezabeh[a], Solomon Tesfamariam[a], Siegfried F. Stiemer[b]

[a] The University of British Columbia, School of Engineering, Kelowna, Canada
matiyas. bezabeh@ ubc. ca, Solomon. Tesfamariam@ ubc. ca
[b] The University of British Columbia, Department of Civil Engineering, Vancouver, Canada
sigi@ civil. ubc. ca

KEYWORDS: Cross laminated timber, direct displacement based design, equivalent viscous damping, and steel moment resisting frames.

ABSTRACT

Innovative hybrid structure has been developed as an alternative lateral-load resisting system, at the University of British Columbia (UBC) and FPInnovation[1]. The hybrid structure incorporates Cross Laminated Timber (CLT) shear panels as an infill in steel moment resisting frames (SMRFs) (*Fig. 1*). L-shaped steel brackets were used as connectors that are bolted to the steel frame and nailed to the CLT panel. Thorough experimental[2] and analytical studies[3] have been undertaken on the cyclic behaviour of the bracket connections. The brackets ensure full confinement between the CLT and SMRF systems, and contribute to energy dissipation during seismic loads. The interface between the CLT infill wall and the steel frame is provided with a small gap to allow the connection brackets to deform under lateral load. In this paper, for this hybrid structure, a new iterative Direct Displacement Based Design (DDBD) method is developed[4].

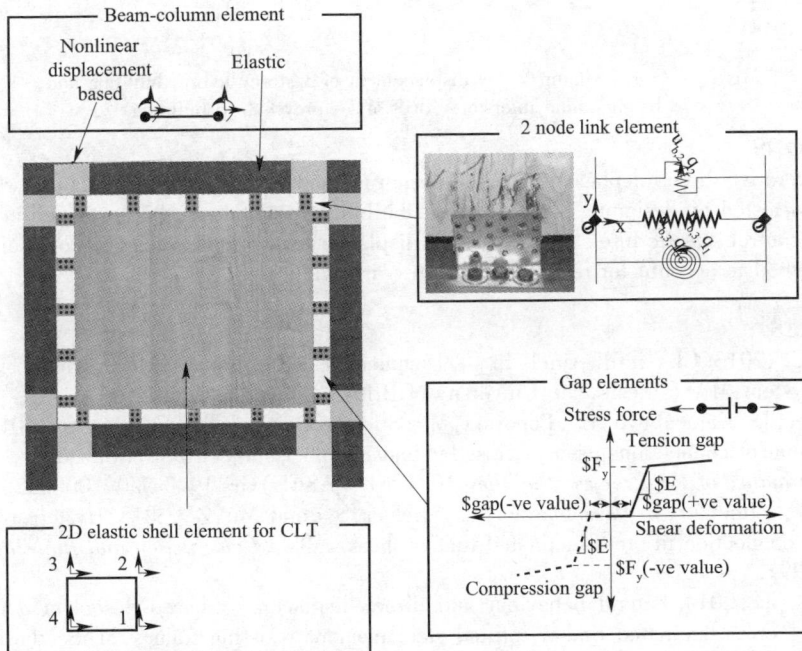

Fig. 1. Steel moment resisting frames with CLT infill walls[5]

The design procedure was started by assuming the following initial modeling variables: gap between CLT panel and steel frame, bracket (connection) spacing, CLT panel thickness and strength, and post yield stiffness ratio of steel members. Subsequently, the design displacement profile was developed by assigning an initial relative strength between the CLT wall and frame elements. This profile was used to obtain the characteristics of an equivalent single degree of freedom(SDOF) system. A system ductility value was established based on the proportions of the overturning moment resistance of the CLT wall and steel moment frame. In order to represent the energy dissipative capacity of the hybrid system, an equivalent viscous damping(EVD) equation was developed using response surface method. To formulate the EVD equation, 243 single-storey single-bay CLT infilled SMRF models were developed and subjected to monotonic static and semi-static cyclic analysis. The EVD of each model was calculated from the hysteretic responses based on Jacobsen's area based approach and later calibrated using nonlinear time history analysis. By using the proposed EVD-damping law, effective period and secant stiffness of the system were calculated to obtain the final design base shear. A 3-storey and three bays hybrid building(Fig. 2a), with middle bay CLT infill was designed using the proposed method. Nonlinear time history analysis using twenty earthquake ground motion records was used to validate the performance of the proposed design methodology. A story drift(Fig. 2b) and maximum interstorey drift(Fig. 2c) of the 3-storey hybrid building show the effectiveness of the proposed design method, where the average responses of nonlinear time history analyses meet the target drift values.

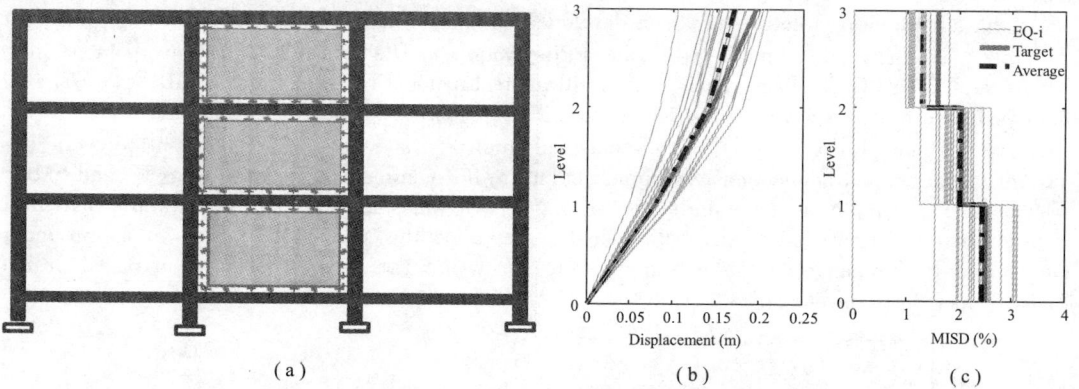

(a) (b) (c)

Fig. 2. (a) Maximum storey displacement of 3 storey hybrid building and
(b) Maximum interstorey drift of 3-storey hybrid building

CONCLUSIONS

A new iterative direct displacement based design methodfor SMRFs with CLT infill walls has been developed and tested by designing 3-storey hybrid building. In summary, the developed method proved to effectively control seismic interstorey drifts and displacements. Future research should aim at investigating the method to account for residual interstorey drift values.

REFERENCES

[1] Dickof, C., 2013. CLT infill panels in steel moment resisting frames as a hybrid seismic force resisting system, MASc thesis, The University of British Columbia, April 2013.
[2] Schneider, J., Karacabeyli, E., Popovski, M., Stiemer, S., and Tesfamariam, S., 2013. "Damage assess-ment of connections used in cross laminated timber subject to cyclic loads". ASCE Journal of Per-formance of Constructed Facilities, 10. 1061/(ASCE)CF. 1943-5509. 0000528.
[3] Shen, Y., Schneider, J., Tesfamariam, S., Stiemer, S., and Mu, Z., 2013. Hysteresis behavior of brack-et connection in cross-laminated-timber shear walls". Construction and Building Materials, 48, 980-991.
[4] Bezabeh, M., 2014. Lateral behaviour and direct displacement based design of a novel hybrid structure: cross laminated timber infilled steel moment resisting frames, MASc thesis, School of Engineering, The University of British Columbia, August 2014.
[5] Tesfamariam, S., Stiemer, S., Dickof, C., and Bezabeh, M., 2014b. "Seismic vulnerability assessment of hybrid steel-timber structure: Steel moment resisting frames with CLT infill." Journal of Earthquake Engineering, 18(6), 929-944.

Resilience Enhancement Technology

SELF-CENTERING STEEL PLATE SHEAR WALLS FOR IMPROVING SEISMIC RESILIENCE

Patricia M. Clayton[a], Daniel M. Dowden[b], Chao-Hsien Li[c],   Jeffrey W. Berman[d],
Michel Bruneau[b], Laura N. Lowes[d], and Keh-Chyuau Tsai[e]

[a] University of Texas at Austin, Dept. of Civil, Architectural, and Environmental Engineering
clayton@ utexas. edu

[b] University at Buffalo, Dept. of Civil, Structural, and Environmental Engineering
dmdowden@ buffalo. edu, bruneau@ buffalo. edu

[c] National Center for Research on Earthquake Engineering, Taipei, Taiwan
chli@ ncree. org

[d] University of Washington, Dept. of Civil and Environmental Engineering
jwberman@ uw. edu, lowes@ uw. edu

[e] National Taiwan University, Dept. of Civil, Structural, and Environmental Engineering
kctsai@ ntu. edu. tw

KEYWORDS: Self-centering; Steel plate shear walls; Large – scale experiment; Performance-based design; Nonlinear response history analyses.

ABSTRACT

As part of a Network for Earthquake Engineering Simulation(NEES) small group research project led by researchers at the University of Washington (UW) with collaborators at University at Buffalo (UB), University of Illinois at Urbana-Champaign(UIUC), and Taiwan NationalCenter for Research on Earthquake Engineering(NCREE), a self-centering steel plate shear wall(SC-SPSW) system has been developed to achieve enhanced performance objectives following earthquakes, including recentering. The SC-SPSW consists of thin steel infill panels, referred to as web plates that serve as the primary lateral load resisting and energy dissipating element of the system. Post-tensioned(PT) beam-to-column connections(*Fig. 2*) are employed to provide system recentering capabilities. A performance-based design procedure has been developed for the SC-SPSW, and a series of nonlinear response history analyses have been conducted to verify intended seismic performance at multiple hazard levels[1]. Quasistatic subassembly and scaled three-story tests have been conducted at UW(*Fig. 1*a) and UB, shake table tests of scaled three-story specimens have been conducted at UB(*Fig. 1*b), and pseudo-dynamic tests of two full-scale two-story SC-SPSWs were conducted at NCREE(*Fig. 1*c). These test programs were used to better understand system behaviour and to verify seismic performance at multiple hazard levels.

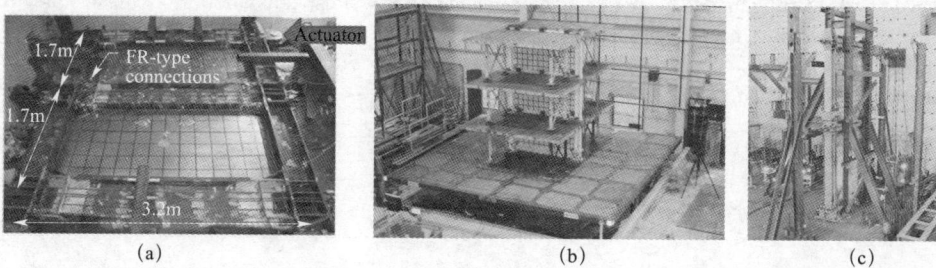

Fig. 1.   (a) Subassembly test at UW; (b) Shake table test at UB; (c) Pseudo-dynamic test at NCREE

CONCLUSIONS

This research project addressed several key issues in the topic of self-centering and resilient systems, including development of a new PT beam-column connection that only rocks about its top flange

(*Fig.* 2b) to eliminate frame expansion[2]. Methods of connecting web plates to the boundary frame, including connecting the web plates only to the beams and using diagonal web plate strips rather than full infill plates, were investigated to mitigate damage and improve repairability. A PT column base connection was also incorporated into the SC-SPSW performance-based design procedure to eliminate column hinging and improve structural recentering capabilities. SC-SPSW behaviour was simulated with numerical models of varying complexity, including simple line-element models and more complex shell-element models. These numerical studies found that the idealized tension-only behaviour that is commonly assumed for steel plate shear walls[3] neglects reloading stiffness and unloading resistance that are present in real web plates that can significantly impact simulated seismic performance. As a culmination of this multi-year, multi-institutional project, this paper presents an overview of the SC-SPSW numerical and experimental research programs. This paper will also discuss innovative PT connection and web plate designs that were investigated to improve constructability, resilience, and seismic performance and that can be applied to other self-centering and steel plate shear wall systems.

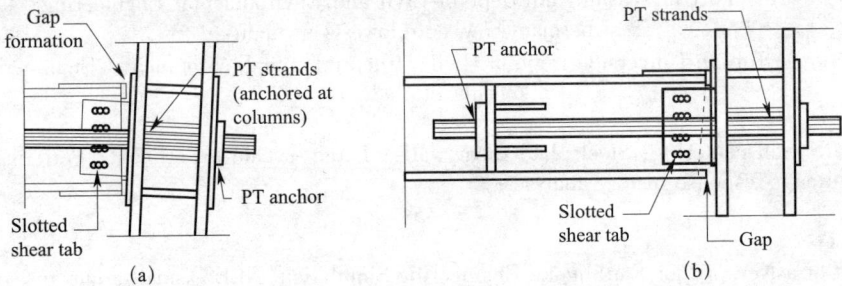

Fig. 2.   PT connections that (a) rock about both flanges and (b) rock only about the top flange

ACKNOWLEDGMENT

Financial support for this study was provided by the National Science Foundation as part of the George E. Brown Network for Earthquake Engineering Simulation under award number CMMI – 0830294 and by NCREE. The authors would also like to acknowledge material donations from the American Institute of Steel Construction.

REFERENCES
[1] Clayton, P. M. , Berman, J. W. , and Lowes, L. N. (2012a). Performance Based Design and Seismic Evaluation of Self-Centering Steel Plate Shear Walls. ASCE Journal of Structural Engineering, 138:1, 22-30.
[2] Dowden, D. , Bruneau, M. , (2011). NewZ-BREAKSS: Post-tensioned Rocking Connection Detail Free of Beam Growth. AISC Engineering Journal. Second Quarter 2011, 153-158.
[3] Sabelli, R. and Bruneau, M. (2007). Design Guide 20: Steel Plate Shear Walls, American Institute of Steel Construction, Chicago, IL.

INFLUENCE OF MEMBER INELASTICITY ON THE PERFORMANCE OF CONTROLLED ROCKING STEELBRACED FRAMES

Taylor C. Steele and LydellD. A. Wiebe

Department of Civil Engineering, McMaster University, Hamilton, ON, Canada

e-mails: steeletc@ mcmaster. ca, wiebel@ mcmaster. ca

KEYWORDS: performance – based earthquake engineering, self – centering systems, controlled rocking steel braced frames, higher mode effects, capacity design.

ABSTRACT

Higher mode effects increase the capacity design forces in controlled rocking steel braced frames (CRSBFs) relative to the first-mode lateral forces[1,2,3,4]. Underestimating the capacity design forces may result in members yielding and buckling, compromising performance or even leading to collapse. Studies to date have assumed that frame members remain elastic even during ground motions larger than the design level[3,4,6]. This study investigates the influence of member inelasticity on the performance of CRSBFs through a series of non-linear time-history analyses on a six-story structure designed using $R = 8$. A design based on only the first-mode lateral forces is compared to three additional designs including estimates of the higher mode effects at the design basis earthquake (DBE), maximum considered earthquake (MCE) and 1.5 times the MCE level. When the frames were not designed for at least the MCE level, collapse occurred during one record and undesirable brace buckling occurred during other records.

CONCLUSIONS

Four controlled rocking steel braced frames (CRSBFs) were designed for an archetypical six-story structure in an area of high seismicity. The base rocking joint was designed using the method proposed by Wiebe and Christopoulos[7]. The first frame was designed to resist the capacity design forces from only the first-mode lateral distribution of forces (FMD). Subsequent frames were designed for capacity design forces that included an estimate of the higher mode effects at the DBE, MCE and 1.5 times the MCE level. The higher mode effects were estimated using a modified response spectrum analysis[8] implemented in ETABS[9]. Using OpenSees[10], the four frames were modelled to capture yielding and buckling in the braces, beams, and columns, and to capture yielding and fracture of the post-tensioning and energy dissipating elements. Nonlinear push-pull analyses of the four frames showed that the first-mode responses are nearly identical because this response is dominated by rocking, and the rocking joint design was the same for each frame.

These four frames were subjected to a suite of seven ground motions scaled to the MCE. Designing the CRSBF for the FMD load case resulted in collapse for oneof the seven records without fracture of the post-tensioning or energy dissipating elements, and undesirable brace buckling the top story during the other records; this buckling resulted in a non-uniform interstory drift profile and residual drifts of up to 0.8%. It is preferable to prevent such behaviour in CRSBF members to prevent collapse and reduce residual drifts. Including an estimate of the higher mode contribution at the DBE level still resulted in collapse during one record, but reduced the residual drifts and limited the yielding of the frame members for the other six records. Increasing the higher mode estimates to the MCE level and 1.5 times the MCE level prevented collapse and slightly improved the drift performance of the system by minimising frame member yielding and buckling. For the three designs that included an estimate of the higher mode effects, one record did not cause collapse but had non-zero residual drifts because the post-tensioning yielded, reducing the system's ability to self-center.

This study was limited to a six-story archetypical structure designed using a force reduction factor of $R = 8$ and analysed for a 2% in 50 year event. Using a larger force reduction factor would be possible without excessive demands on the post-tensioning, energy dissipation element, or gravity system, but it would also make it more critical to account for higher mode effects in design. Failure to do so would increase the probability of collapse or of larger residual displacements. Further analysis is ongoing to

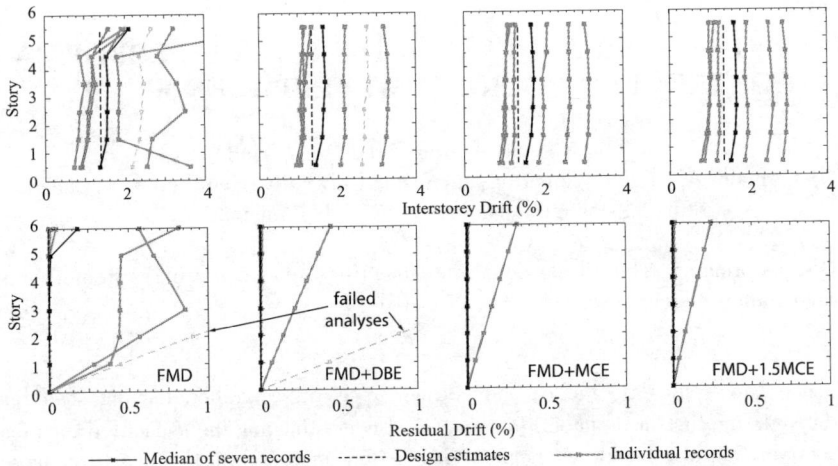

Fig. 1. Displacement results for the four frames from nonlinear time history analysis at the MCE level

quantify the collapse fragility in CRSBFs where buckling is modelled.

REFERENCES

[1] Wiebe L, Christopoulos C, Tremblay R and Leclerc M, 2013. "Mechanisms to limit higher mode effects in a controlled rocking steel frame. 1: concept, modelling, and low-amplitude shake table testing," *Earthquake Engineering and Structural Dynamics*, John Wiley & Sons Limited, doi: 10. 1002/eqe. 2259, Vol. 6, pp. 1069-1086.

[2] Roke D, Sause R, Ricles JM and Chancellor NB, 2010. "Damage-free seismic-resistant self-centering concentrically-braced frames," ATLSS Report No. 10-09, Bethlehem, PA, United States.

[3] Eatherton MR and Hajjar JF, 2010. "Large-scale cyclic and hybrid simulation testing and development of a controlled-rocking steel building system with replaceable fuses," NSEL Report No. NSEL-025, University of Illinois at Urbana-Champaign, United States.

[4] Ma X, Krawinkler H and Deierlein GG, 2011. "Seismic design, simulation and shake table testing of self-centering braced frame with controlled rocking and energy dissipating fuses," The John A. Blume Earthquake Engineering Center Report No. 174, Stanford University, CA, United States.

[5] Tremblay R, Poirier L-P, Bouaanani N, Leclerc M, Rene V, Fronteddu L, and Rivest S, 2008. "Innovative viscously damped rocking braced frame," *Proc. of the 14th World Conf. on Earthquake Engineering*, Beijing, China.

[6] Wiebe L and Christopoulos C, 2014. "Performance-based seismic design of controlled rocking steel braced frames. II: Design of capacity-protected elements," *Journal of Structural Engineering*, American Society of Civil Engineers, doi: 10. 1061/(ASCE) ST. 1943-541X. 0001201.

[7] Wiebe L and Christopoulos C, 2014. "Performance-based seismic design of controlled rocking steel braced frames. I: Methodological framework and design of base rocking joint," *Journal of Structural Engineering*, American Society of Civil Engineers, doi: 10. 1061/(ASCE) ST. 1943-541X. 0001202.

[8] Steele TC and Wiebe L, 2014. "A practical approach to capacity design of controlled rocking steel braced frames," *Proc. of the 4th International Structural Specialty Conf.* , Halifax, NS, Canada.

[9] Computer and Structures Inc. , 2013. Extended Three-Dimensional Analysis of Building Systems (ETABS)2013 [program], Version 13. 1. 0, Computer and Structures Inc. , Berkeley, CA.

[10]McKenna F, Fenves GL, Scott MH and Jeremic B, 2013. Open system for earthquake engineering simulation(OpenSees) [program], Version 2. 4. 5, Pacific Earthquake Engineering Research Center, University of California at Berkeley, United States.

EXPERIMENTAL STUDY OF RESTORING FORCE MECHANISM IN SELF-CENTERING BEAM(SCB)

Abhilasha Maurya[a] and Matthew R. Eatherton[a]

[a] Department of Civil and Environmental Engineering, Virginia Tech

e-mails: abhilasha. maurya@ vt. edu, meather@ vt. edu

KEYWORDS: Self-centering, seismic design, hysteretic behavior, restoring force, gap opening mechanism

ABSTRACT

In the past, several self-centering(SC) seismic systems have been developed. However, examples of self-centering systems used in practice are limited due to unusual field construction practices, high initial cost premiums and deformation incompatibility with the gravity framing. A self centering beam moment frame(SCB-MF) has been developed that mitigates several of these issues while adding on to the advantages of a typical SC system.

The SCB consists of aI-shaped steel beam augmented with a restoring force mechanism attached to the bottom flange and can be shop fabricated. Additionally, the SCB has been designed to eliminate the deformation incompatibility associated with self-centering mechanism. The strength and stiffness of the beam are decoupled resulting in a cost-effective system that may be competitive with conventional moment frames.

This paper describes the SCB concepts and the experimental program on two-thirds scale SCB specimens. Key parameters will be varied to investigate their effect on global system hysteretic response and their effect on system components. Parameters that will be varied include SCB depth, initial post-tensioning stress, ratio of initial post-tensioning force to fuse yield capacity, and nominal moment strength of the system. The main design equations which define the SCB will be presented. The testing results will include the global force-deformation behavior of the SCB. The main focus of the paper will be the experimental results associated with the restoring-force mechanism of the SCB.

CONCLUSIONS

Based on the work presented in this paper, several observations and conclusions can be made.

- The SCB moment frame is a high-performance self-centering lateral force resisting system which reduces residual drifts and prevents inelastic damage to the main structural components of the system.
- In addition to the advantages associated with a typical SC system, the SCB can be shop fabricated which allows for conventional field construction methods. The design of a SCB also eliminates deformation incompatibility with the gravity framing system.
- Based on the experimentalbehavior, the SCB was shown to have a large deformation capacity. All three SCBs were successfully tested up to a story drift of 6% without causing any damage to the end connections, SCB body or columns.
- The strength equations developed for the SCB predicted the moment capacity well, with the mean difference between the predicted and experimental values of 9%.
- The results obtained from equations developed to predict beam moment at gap opening, M_{PTi} and post-gap opening stiffness contribution by the PT strands(K_{PT}) were found to be in good agreement with the experimental results with mean percent difference of 12.6% and 5% respectively.
- The behavior of PT strands were examined with respect to the gap opening ratio. Depending on the initial post-tensioning stress, the PT strands were shown to lose a certain percentage of PT stress due seating losses and inelastic deformations sustained during the loading. This loss in PT stress leads to a decrease in self-centering capacity of the SCB.

REFERENCES

[1] Ricles, J. M., Sause, R., Garlock, M. M., and Zhao, C., (2001) "Post-Tensioned Seismic Resist-

ant Connections for Steel Frames," *Journal of Structural Engineering.* , 127 (2) , *pp.* 113-121 ,2001.

[2] Ricles, J. , Sause, R. , Wolski, M. , Seo, C-Y. , and Iyama, J. (2006). "Post-Tensioned Moment Connections with a Bottom Flange Friction Device for Seismic Resistant Self-Centering Steel MRFs". *Proceedings ,4th International Conference on Earthquake Engineering. Taipei ,Taiwan.*

[3] Garlock, M ,and Li, J. (2008). "Steel Self-Centering Moment Frames with Collector Beam Floor Diaphragms" , *Journal of Constructional Steel Research , Elsevier ,pp.* 526-538.

[4] Maurya, A. , Eatherton, M. R. (2014). "Self-Centering Beams with Resilient Seismic Performance" *Structures Congress , Boston , MA.*

[5] Eatherton, M. and Hajjar, J. F. (2011) "Residual Drifts of Self-Centering Systems Including Effects of Ambient Building Resistance" *Earthquake Spectra , , Vol.* 27 , *No.* 3

[6] Bruce, T. L. (2014). "Behavior of Post-Tensioning strand systems subjected to inelastic cyclic loading" , *M. S. Thesis , Virginia Tech.*

LARGE-SCALE TESTS ON A RE-CENTRING DUAL ECCENTRICALLY BRACED FRAME

Aurel Stratan[a], Adriana Ioan[a], Dan Dubina[a,b], Martin Poljanšek[c], Javier Molina[c],
Pierre Pegon[c], Fabio Taucer[c] and Gabriel Sabău[a]

[a] Politehnica University of Timisoara, Department of Steel Structures and Structural Mechanics,
Timisoara, Romania
aurel. stratan@ upt. ro, adriana. ioan-chesoan@ student. upt. ro, dan.
dubina@ upt. ro, gabriel. sabau@ student. upt. ro
[b] Romanian Academy, Fundamental and Advanced Technical Research
Centre, Timisoara, Romania
[c] European Commission, Joint Research Centre(JRC), Institute for the Protection and
Security of the Citizen(IPSC), European Laboratory for Structural Assessment Unit,
Via Enrico Fermi 2749, 21027 Ispra VA, Italy
martin. poljansek@ jrc. ec. europa. eu, francisco. molina@ jrc. ec. europa. eu, pierre.
pegon@ jrc. ec. europa. eu, fabio. taucer@ jrc. ec. europa. eu

KEYWORDS: re-centring, bolted links, eccentrically braced frames.

ABSTRACT

Conventional seismic design philosophy is based on dissipative response, which implicitly accepts damage of the structure under the design earthquake and leads to significant economic losses. Repair of the structure is often impeded by the permanent(residual) drifts of the structure. The repair costs and downtime of a structure hit by an earthquake can be significantly reduced by adopting removable dissipative members and providing the structure with re-centring capability. These two concepts were implemented in a dual structure, obtained by combining steel eccentrically braced frames with removable bolted links and moment resisting frames. The paper presents the results of a large-scale experimental program on a dual eccentrically braced frame with replaceable links performed at the European Laboratory for Structural Assessment(ELSA) at the Joint Research Centre(JRC) in Ispra within the framework of Transnational Access of the SERIES Project. Its objectives were to: (1) validate the re-centring capability of dual structures with removable dissipative members; (2) assess overall seismic performance of dual eccentrically braced frames and(3) obtain information on the interaction between the steel frame and the reinforced concrete slab in the link region.

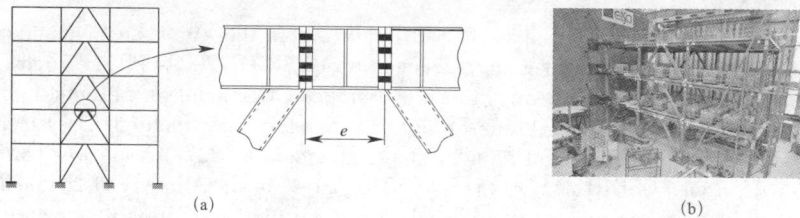

Fig. 1.   Bolted link concept(a) and the experimental mock-up in front of the reaction wall(b).

Testing sequence on the mock-up in the reaction wall facility of ELSA consisted in modal evaluation, snap-back andPsD tests. *Table 1* summarises the sequence of PsD tests and link replacements performed on the test structure. Links were replaced two times: after DL test and after SD test.

Table 1. Sequence of PsD and link removal tests.

Links	Test	Scope
	Full operation(FO1)	Assess elastic response of the structure.
Set 1	Damage Limitation(DL)	Observe structural response to a moderate earthquake.
	Replacement of set 1 links(LR1)	Investigate removal of links by unbolting and their replacement.

Links	Test	Scope
Set 2	Full operation(FO2)	Assess elastic response of the structure.
	Significant Damage(SD)	Observe structural response to a design-level earthquake.
	Pushover(PO1)	Induce large permanent drifts.
	Replacement of set 2 links(LR2)	Investigate removal of links by flame cutting and their replacement.
Set 3	Full operation(FO3)	Assess elastic response of the structure.
	Near collapse(NC)	Observe structural response to a severe-level earthquake.
	Pushover(PO2 and PO3)	Investigate ultimate capacity of the structure.

CONCLUSIONS

Dual frames with removable dissipative members can be used to provide a structure with a re-centring capability, and can greatly reduce the costs and manpower required for post-earthquake repair. This paper has described a large-scale experimental programme that has been carried at the European Laboratory for Structural Assessment(ELSA) on a three-storey dualEBF with replaceable shear links. The testing sequence on the mock-up in the reaction wall facility of ELSA consisted of modal evaluations, snap-back, and pseudo-dynamic tests.

The dual eccentrically braced structure exhibited excellent performance when subjected to DL and SD earthquakes. Small residual deformations were recorded for both seismic intensity levels, which were within the erection tolerance limits. Such small permanent deformations effectively mean that the structure is self-centring to a certain degree, which allows an easy repair of the structure through replacement of damaged links. Residual drifts are further reduced by removing and replacing the bolted links. Re-centring was better for the frame with links disconnected from the slab. Moreover, damage to the concrete was avoided in this case. Nevertheless, good re-centring was observed even for the frame with the slab cast over the links, while damage to the reinforced concrete slab was low at the DL and SD tests.

Provided the residual deformations after an earthquake are small, the links can be removed simply by unbolting, as was done after the DL test. If larger residual drifts occur, flame cutting of links is recommended to allow for smooth release of forces, as was done after the SD/PO1 tests.

The experimental investigation validated the re-centring capability of dual eccentrically braced frames with removable links, which was accomplished without major technological difficulties.

ACKNOWLEDGMENT

The research leading to these results received funding from the European Community's Research Fund for Coal and Steel(RFCS) under grant agreement n° RFSR-CT-2009-00024 "High strength steel in seismic resistant building frames" and from the European Community's Seventh Framework Programme [FP7/2007-2013] for access to the European Laboratory for Structural Assessment of the European Commission-Joint Research Centre under grant agreement n° 227887 and was partially supported by the strategic grant POSDRU/159/1.5/S/137070(2014) of the Ministry of National Education, Romania, co-financed by the European Social Fund-Investing in People, within the Sectorial Operational Programme Human Resources Development 2007-2013.

REFERENCES

[1] Dubina D, Stratan A, Dinu F. "Dual high-strength steel eccentrically braced frames with removable links." *Earthquake Engineering & Structural Dynamics*, 37(15):1703-1720, 2008.

[2] Stratan A, Ioan A, Dubina D, Taucer F, Poljanšek M, Molina J, Pegon P, D'Aniello M, Landolfo R. "Experimental program for large-scale tests on a re-centring dual eccentrically braced frame." *7th European Conference on Steel and Composite Structures EUROSTEEL* 2014, Napoli, Italy: European Convention for Constructional Steelwork, ECCS; 2014, p. paper no. 37-300, 8 p.

SEISMIC DESIGN OF NOVEL STEEL RESILIENT STRUCTURES

T Y Yang[a], D P Tung[a], Yuanjie Li[a]

[a]University of British Columbia, Department of Civil Engineering, Canada

yang@ civil. ubc. ca, dorian_tung@ hotmail. com, yuanjieli01@ gmail. com

KEYWORDS: Earthquake, resilience, Structural Fuse, Energy-based, Performance-based.

ABSTRACT

Recent earthquakes in Japan and New Zealand have proven that even developed countries with modern building codes still experience unrecoverable economic and social losses after a significant earthquake shaking. The issue lies in the fundamental approach in designing structures to deform inelastically during earthquakes. This design approach results in damages and leads to hefty financial losses. Such losses can be minimized by earthquake resilient structures which use sacrificial structural fuses to dissipate the sudden surge of energy to the system during strong earthquake shaking. These fuses are de-coupled from the gravity system. The location of fuses is designed to be accessible so they can be inspected, repaired, and replaced to achieve rapid return of occupancy performance objective. This paper introduces two new earthquake resilient structures named spine fused frames(SFFs) and fused truss moment frames(FTMFs) as shown in *Fig. 1*.

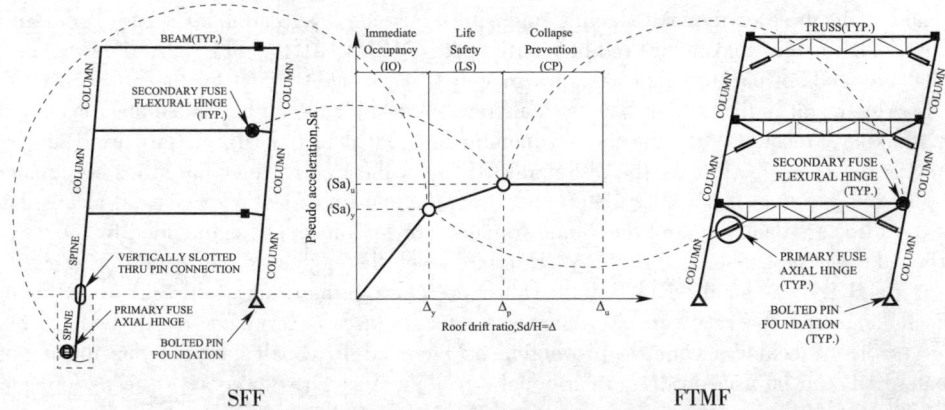

Fig. 1. Diagrammatic illustration of proposed earthquake resilient structures

SFF uses a spine column to engage the superstructure and structural fuse at the basement. At the ground level, the spine column is restrained laterally with a vertically slotted hole. This detail allows the spine column to pivot about the ground level. The other columns in SFF are pinned at the ground level to ease foundation design and avoid column base damages. To ensure SFF has additional resistance after the structural fuse yields, the beam ends are designed as the secondary fuses. FTMF introduces structural fuses at the bottom chords of long-span trusses. In addition, the top chords are moment connected to columns to act as the secondary fuses. Both proposed systems rely on structural fuses as the primary seismic force resisting system(SFRS). Secondary SFRS is included in the systems to mitigate structural collapse after the primary fuses yield. Both systems are developed to achieve multiple performance objectives under different seismic hazard levels.

To facilitate the design of SFFs and FTMFs, an equivalent energy-based design procedure(EEDP) is presented. This novel design method can be readily adapted by consulting offices to design alternative fused SFRSs. EEDP allows designers to choose performance targets, namely roof drift ratios, for various seismic hazard levels at the onset of a project. EEDP is developed based on the relationship between the energy stored by a linear single degree-of-freedom system to the energy absorbed by a nonlinear system under cyclic loading. Designers are able to determine structural sizes without iterations. To demonstrate that the new fused SFRSs can be practically designed and detailed in a consulting office and that the systems can achieve the expected performances, a three-storey office prototype building in Los Angeles is designed by EEDP. *Fig.* 2 presents the structural members of the prototype building using SFFs and FTMFs. Non-linear static and dynamic analyses are conducted to understand their seismic behaviour.

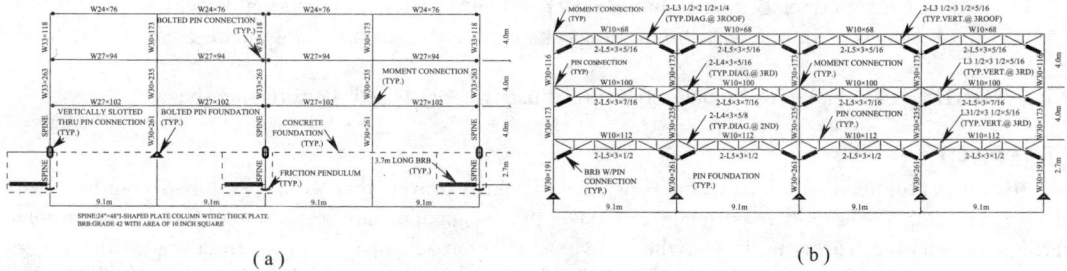

(a)   (b)

Fig. 2.   (a) SFF; (b) FTMF

CONCLUSIONS

Non-linear static analyses are performed on the SFF and FTMF. It is found that the systems perform as intended where the primary fuses start to yield when the roof drift ratio reaches 0.65% for the SFF and 0.35% for the FTMF. The secondary fuses start to yield at the drift ratio of about 1.80% for both systems. *Fig. 3* shows the median roof drift ratios of both systems when subjected to twenty scaled ground motions at three earthquake shaking intensities: service level earthquake (SLE), design based earthquake (DBE), and maximum credible earthquake (MCE). At the SLE hazard level, the result shows that the median roof drift ratio is approximately 0.42% and 0.35% for the SFF and FTMF, respectively. This result indicates that both systems remain elastic and can be immediately occupied after the SLE shaking intensity. When the ground motions are scaled to the DBE hazard level, the median roof drift ratio is about 1.31% for the SFF and 1.19% for the FTMF. Since the ratios are greater than the 0.65% and less than the 1.80% determined from the non-linear static analyses, it means that only the primary fuses are damaged, and the systems indeed provide life safety as intended by the EEDP design. When the motions are scaled to the MCE hazard level, the median roof drift ratio is about 1.73% for the SFF and 1.69% for the FTMF. Both values are close to the design value calculated using EEDP. This indicates the reserved capacity from the secondary fuses is very effective in delaying building collapse and hence achieves collapse prevention as intended. In conclusion, the non-linear analyses show that EEDP can be used to design earthquake resilient structures to achieve the performance objectives as selected by designers.

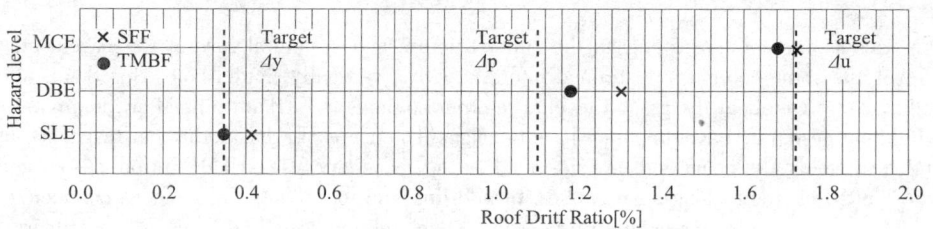

Fig. 3.   Median roof drift ratio due to various seismic hazards

REFERENCES

[1] Yang T. Y., Tung D. P., Li Y., 2015. "Performance-based Plastic Design and Evaluation of Linked Column Frames", *Progress in Steel Building Structures*, 17(1).
[2] Chao S. H., Goel S. C., Lee S. S., 2007. "A Seismic Design Lateral Force Distribution Based on Inelastic State of Structures", *Earthquake Spectra*, 23(3), pp. 547-569.
[3] OpenSees, 2010. "Open System for Earthquake Engineering Simulation [OpenSees] Framework-Version 2.5", *Pacific Earthquake Engineering Research Center*.

DEVELOPMENT AND VALIDATION OF A STEEIDUAL-CORE SELF-CENTERING BRACE FOR SEISMIC RESISTANCE: FROM BRACE MEMBER TO ONE-STORY ONE-BAYBRACED FRAME TESTS

Chung-Che Chou[a], Ping-Ting Chung[b], Tsung-Han Wu[b], Alexis Rafael Ovalle Beato[b]

[a]Professor, National Taiwan University, Department of Civil Engineering, Taiwan
cechou@ ntu. edu. tw

[b]Research Assistant, National Taiwan University, Department of Civil Engineering, Taiwan

KEYWORDS: Dual-core self-centering brace(DC-SCB), Braced frame tests, Residual deformation.

ABSTRACT

The steel dual-core self-centering brace(DC-SCB) is an innovative structural member that provides both energy dissipation and self-centering properties to structures, reducing residual drifts of structures in earthquakes. Chou et al. [1-3] develops a new steel dual-core self-centering brace (DC-SCB), which utilizes three conventional steel bracing members, two friction devices, and two sets of tensioning elements that are in a parallel arrangement. The three bracing members and the two sets of PT elements in the DC-SCB double the axial elongation capacity of the self-centering energy-dissipating(SCED)brace[4] if the same PT elements are used in both braces.

The seismic performance of a steel frame with DC-SCBs has never been studied experimentally, especiallytogether with inelastic responses or buckling in the beam, column and brace in multiple tests. Therefore, the objective of the work is to validate the seismic behavior of a steel frame with the DC-SCB as an earthquake-resisting mechanism. The DC-SCB in this study uses high-strength steel tendons as tensioning elements to eliminate sudden failure of FRP tendons when the brace is overloaded beyond the design limit[1]. The brace also uses the second core as an intermediate member for half the tendon anchorages to reduce half the initial post-tensioning work and to simplify the deformation mechanism of the brace compared to the previous DC-SCB[3].

This paperfirst presents the mechanics and hysteretic responses of the cross-anchored DC-SCB. A 7950-mm long cross-anchored DC-SCB, which was designed with ASTM A572 Gr. 50 steel bracing members and ASTM A416 Gr. 270 steel tendons, was tested six times to evaluate its cyclic performance[3]. A prototype three-story steel DC-SCBF was then designed and analyzed; a full-scale one-story, one-bay DC-SCBF specimen that represented the prototype first-story braced frame was tested under multiple loading protocols. Fig. 1 shows responses of the one-story, one-bay DC-SCBF specimen in Phase 3 test; the residual drift of the braced frame was caused by local buckling of the steel beam after 1.5% drift. The objectives of the test program were to(1) validate the system response of the DC-SCBF, (2) study force distributions in framing members as damage progresses in the DC-SCB, beam or columns, and(3) investigate the repair and replacement characteristics of the braced frame, as the same frame, brace and PT elements will be reused in eight tests.

CONCLUSIONS

Traditional seismic resisting systems in earthquakes dissipate energy through beams or braces, leading to structural damage or residual drifts that are difficult and expensive to repair. A steel dual-core self-centering brace(DC-SCB) has been proposed to provide both the energy dissipation and self-centering properties to structural systems. To investigate the seismic performance of a steel frame with DC-SCBs, a prototype three-story DC-SCBF was designed based on ASCE standard and available research works. A DC-SCB subassemblage was first tested to evaluate its seismic behavior; a full-scale one-story one-bay DC-SCBF subassembly specimen was then tested eight times to evaluate its system performance, damage progresses in the DC-SCB, beam or column, and post-earthquake behavior of the system under multiple earthquake motions[5]. Subassemblage and braced frame tests confirmed that the DC-SCB performs as the mechanics; except for beam local buckling in the one-story, one-bay braced frame tests, no damage was found in the DC-SCB member throughout the test. Although the maximum axial force of the DC-SCB was around 1700 kN in the subassemblage tests and frame tests in this study, the maximum axial force of the DC-SCB that was ever tested was around 6000 kN[6], indicating a reliable force transfer mechanism in multiple seismic loadings. The initial PT force in the DC-SCB decreased 10% after Phase 2 test; no reduction of the initial PT force was observed from Phase 3 to 8 tests. Friction force of the DC-SCB decreased 14% after all eight tests due to gradual abrasion of the asperity of brass shim plates; this reduction might be acceptable because the one-story, one-bay DC-SCBF subassembly specimen had experienced several increasing cyclic loading tests and multiple DBE and MCE loading tests. Except for energy dissipation, no differences could be found from the hysteretic responses of the DC-SCB in Phase 3 and 8 tests.

(a) One-Story, One-Bay Braced Frame Test   (b) Hysteretic Response

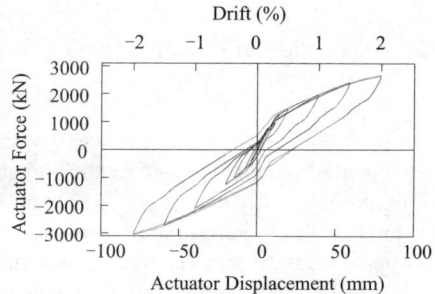

Fig. 1. DC-SCBF subassembly frame specimen at 2% drift (Phase 3 test)

ACKNOWLEDGMENT

The authors would like to thank theMinistry of Science and Technology, Taiwan for financially supporting this research under Contract No. 102-2221-E-002-101-MY3.

REFERENCES

[1] Chou C-C, Chen Y-C. 2013. Development of steel dual-core self-centering braces: quasi-static cyclic tests and finite element analyses. *Earthquake Spectra.* (available online, September 2013)

[2] Chou C-C, Chen Y-C, Pham D-H, Truong V-M. 2014. Steel braced frames with dual-core SCBs and sandwiched BRBs: mechanics, modeling and seismic demands. *Engineering Structures*, 72, 26-40.

[3] Chou C-C, Chung P-T. 2014. Development of cross-anchored dual-core self-centering braces for seismic resistance. *J. Constructional Steel Research*, 101, 19-32.

[4] Christopoulos C. , Tremblay R. , Kim H. J. , Lacerte M. 2008. Self-centering energy dissipative bracing system for the seismic resistance of structures: development and validation. *J. Structural Engineering*, ASCE, 134(1), 96-107.

[5] Chou C-C, Wu T-H, Beato Ovalle A. R. , Chung P-T, Chen Y-H, Chou C-H. 2014. Seismic design and tests of a novel steel dual-core self-centering braced frame (SCBF). *Report No: NCREE* 14-029, National Center for Research on Earthquake Engineering, Taiwan. (in Chinese)

[6] Chou C-C, Chung P-T, Cheng Y-T. 2014. Seismic tests of large-scale energy dissipating braces: dual-core self-centering brace and sandwiched buckling-restrained brace. *5th Asia Conference on Earthquake Engineering*, October 16-18, Taiwan.

FULL-SCALE CYCLIC TESTING OF A LOW-DUCTILITY CONCENTRICALLY-BRACED FRAME

Joshua G. Sizemore[a], Larry A. Fahnestock[b], Eric M. Hines[c], and Cameron R. Bradley[d]

[a]Graduate Research Assistant, Department of Civil and Environmental Engineering, University of Illinois at Urbana-Champaign; email: sizemor2@ illinois. edu

[b]Associate Professor, Department of Civil and Environmental Engineering, University of Illinois at Urbana-Champaign; email: fhnstck@ illinois. edu

[c]Professor of Practice, Department of Civil and Environmental Engineering, Tufts University; Principal, LeMessurier Consultants; email: ehines@ lemessurier. com

[d]Graduate Research Assistant, Department of Civil and Environmental Engineering, Tufts University, email: cbradley@ lemessurier. com

KEYWORDS: Low-ductility, Steel Frames, Braced Frames, Moderate Seismic, Reserve Capacity

ABSTRACT

Steel concentrically-braced frames (CBFs) are used extensively as lateral-force-resisting systems for low to mid-rise buildings in moderate seismic regions of the United States, such as the East Coast and Midwest. Although good structural performance of CBFs in moderate seismic regions for typical gravity and wind loading is well-established, there is essentially no data for earthquake loading. As a result of this situation, a research project was initiated to investigate the seismic performance of CBFs in moderate seismic regions. This paper summarizes one aspect of the project: a full-scale cyclic test of a one-bay two-story CBF designed assuming $R = 3$ and not specifically detailed for seismic resistance-focusing on the sequence of limit states and associated system behavior. The frame experienced brittle brace buckling in both upper story braces at $\pm 0.35\%$ frame drift. Brace-to-gusset weld fracture was subsequently induced in the lower story to observe the influence of brace re-engagement on system strength.

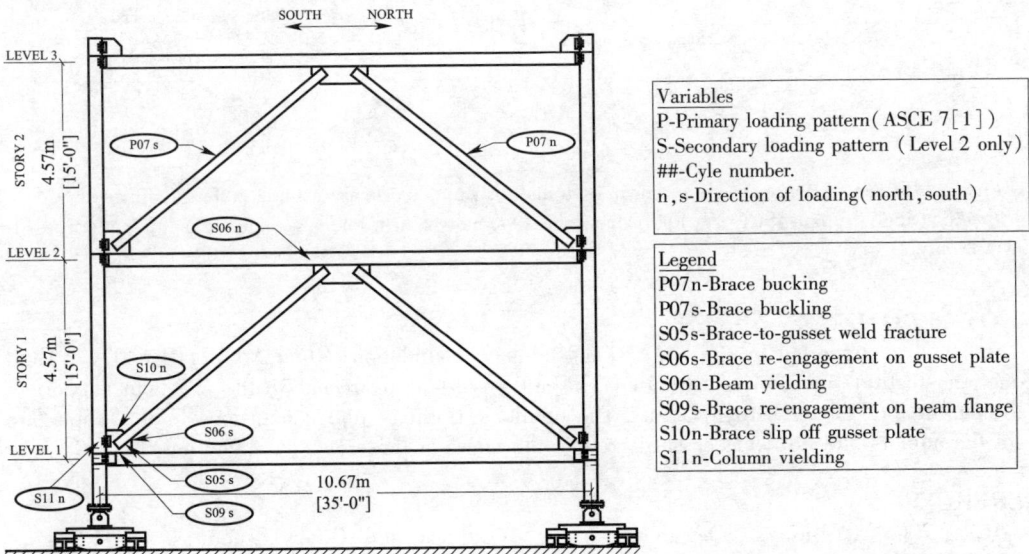

Fig. 1.   Frame elevation with observation locations.

CONCLUSIONS

Full-scale testing of a two-story $R = 3$ chevron concentrically-braced frame (CBF) has provided valuable new data on the cyclic behaviour of braced frames not specifically detailed for seismic resist-ance. During this test, both Story 2 braces experienced a brittle buckling mechanism with significant loss of strength. This is in contrast to the other full-scale test performed by the authors on an $R = 3.25$ ordinary concentrically-braced frame(OCBF), which experienced more ductile brace buckling behaviour as expected due to its seismic detailing requirements[2]. Despite the relatively brittle brace buckling in the $R = 3$ test described here, the frame still maintained a capacity of 445kN [100 kip] to Story 2 drifts upwards of 2%, identifying the reserve capacity achievable from the connections and the buckled braces. As shown in *Fig.* 2a, the maximum base shear achieved under a loading protocol mimicking the equivalent lateral force distribution used in design was 2060 kN. Hysteretic behavior of Story 1 following an induced brace-to-gusset weld fracture revealed that the story regained nearly all of its original stiffness when the brace was bearing on the gusset plate(*Fig.* 2b). Overall, the $R = 3$ frame performed reasonably well, achieving a capacity of nearly 1.4 times the design base shear and displaying appreciable reserve capacity. In the case of brace buckling, the frame exhibited reserve strength primarily from the post-buckling brace behavior. In the case of weld fracture, however, the frame exhibited reserve strength that varied based on the direction of load. In the positive direction, the frame exhibited reserve strength from friction between the fractured brace and gusset plate, beam flexural hinging, and column bending. In the negative direction, the frame exhibited reserve strength predominantly from the fractured brace re-engaging onto the gusset plate and bearing in compression. The combination of this EBF-like behaviour with brace re-engagement created a promising source of reserve capacity which seemed to outperform brace buckling on its own. However, insuring the formation of such a mechanism can be difficult, as shown in this experiment, and further studies are required before it can be suggested as a design strategy.

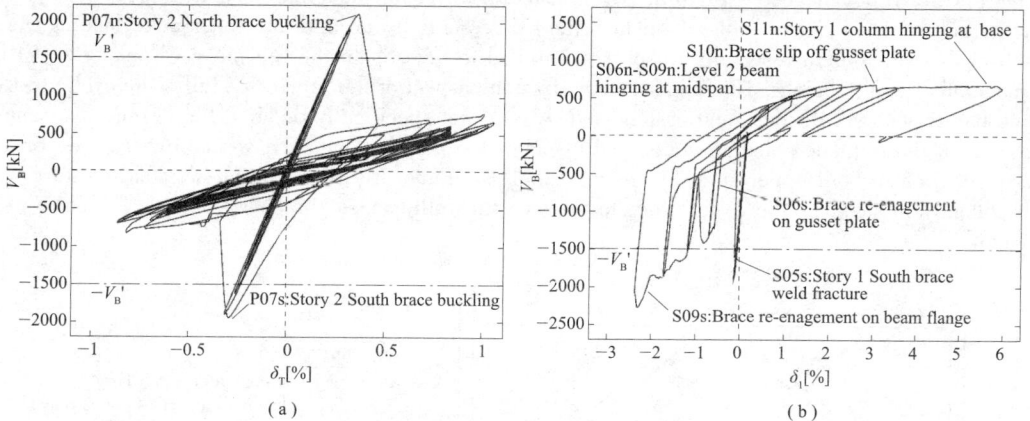

Fig. 2.   Frame response: (a) Base shear vs. frame drift under primary loading pattern(equivalent lateral force distribution from ASCE 7[1]); (b) Base shear vs. Story 1 drift under secondary loading pattern(loading Level 2 only).

ACKNOWLEDGMENTS
This study is supported by the US National Science Foundation(Grant No. CMMI-1207976) and the American Institute of Steel Construction. The authors gratefully recognize the excellent support of the lab staff in theNEES@ Lehigh facility. The opinions, findings, and conclusions in this paper are those of the authors and do not necessarily reflect the views of those acknowledged here.

REFERENCES
[1] ASCE (2005). Minimum Design Loads for Buildings and Other Structures. ASCE/SEI 7-05. Reston, VA: American Society of Civil Engineers.
[2] AISC(2010). Seismic Provisions for Structural Steel Buildings. ANSI/AISC 341-10. Chicago, IL: American Institute of Steel Construction.
[3] Bradley, C. et al. (2014). "Full-Scale Cyclic Testing of an Ordinary Concentrically-Braced Frame". In: Proceedings of the 2015 Structures Congress. (Oregon Convention Center, Portland, OR). ASCE.

EXPERIMENTAL LNVESTIGATION OF SEISMIC BEHAVIOR OF STEEL BUIDING STRUCTURE WITH NONLINEAR VISCOUS DAMPWRS USING REAL-TIME HYBRID EARTHQUAKE SIMULATION

Baiping Dong[a], Richard Sause[a] and James M. Ricles[a]

[a] ATLSS Engineering Research Center, Department of Civil and Environmental
Engineering, Lehigh University, USA
bad209@ lehigh. edu , rs0c@ lehigh. edu, jmr5@ lehigh. edu

KEYWORDS: Experimental, Seismic behaviour, Steel building, Nonlinear viscous damper, Real – time hybrid simulation.

ABSTRACT

Experimental investigation of a large-scale 3-story steel building structure is presented. The seismic lateral force resisting system of the building includes a moment resisting frame(MRF) and a frame with bracing and nonlinear viscousdampers(DBF). The building is assumed to be located on a stiff soil site in Southern California, an area of high seismicity in the United States. The experimental investigation uses the real-time hybrid earthquake simulation(RTHS) experimental method. In the RTHS, the complete structure system is divided into an experimental substructure and an analytical substructure. The experimental substructure is the DBF with nonlinear viscous dampers, and the analytical substructure is comprised of the remaining parts of building, including a steel MRF, gravity load frames, the associated seismic mass, and the inherent damping of the building. RTHS at the frequently occurring earthquake(FOE), design basis earthquake(DBE), and maximum considered earthquake(MCE) hazard level were conducted. The results from the RTHS are presented and the observed behaviour of the experimental substructure and the complete structural system are discussed.

CONCLUSIONS

This paper presents an experimental investigation ofthe seismic behaviour of a large-scale three-story structure with nonlinear viscous dampers under FOE, DBE, and MCE level ground motions. The earthquake simulations were performed using the real-time hybrid simulation(RTHS) method at the NEES Real-Time Multi-Directional(RTMD) facility located at Lehigh University[1,2]. The test structure for the RTHS, as shown in *Fig. 1*, includes a three-story, one-bay moment resisting frame(MRF), a three-story, one-bay frame with one nonlinear viscous damper and associated bracing in each story (DBF), and a gravity load system. During the simulations, the MRF was modelled numerically as the analytical substructure while the DBF was tested in the lab as the experimental substructure.

Table 1. shows the peak story drift ratios from the RTHS at the FOE, DBE, and MCE levels. The FOE, DBE, and MCE is a ground motion intensity with a 50%, 10%, and 2% probability of exceedance in 50 years, respectively. The maximum peak story drift ratio of the test structure from the RTHS is 0.34%, 0.73%, and 1.35% rad under the FOE, DBE, and MCE, respectively. The test structure remained elastic under the FOE and DBE, and showed minor yielding under the MCE. Based on the story drift limits in ASCE/SEI 41-06[3] for the performance levels of steel MRFs, the performance level of the test structure is "Immediate Occupancy" performance level under the FOE, is between the "Immediate Occupancy" performance level and the "Life Safety" performance level under the DBE, and is "Life Safety" performance level under the MCE.

The seismicbehaviour of the dampers in the DBF and the interaction of the DBF with the MRF were investigated in the RTHS. The damper forces are were found to be partially in-phase with the story drifts, and the DBF story shear forces are partially in-phase with the MRF story shear forces and the story drifts stiffens the test structure and helps to reduce the story drift response of the test structure under earthquake loading.

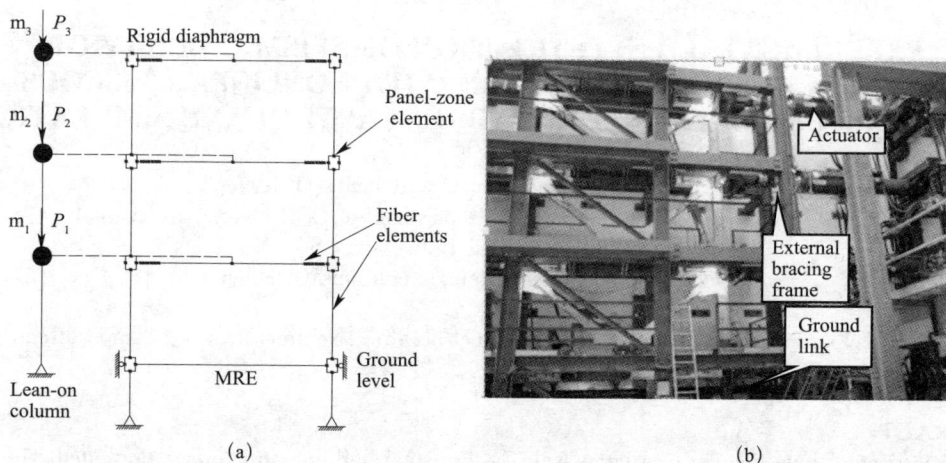

Fig. 1. RTHS substructures: (a) analytical substructure; (b) experimental substructure

Table 1. Peak story drift ratios of test structure.

Hazard level	PBD prediction of story drift ratio(% rad)			RTHS story drift ratio(% rad)		
	1st story	2nd story	3rd story	1st story	2nd story	3rd story
FOE	-	-	-	0.30	0.34	0.25
DBE	0.76	0.81	0.64	0.63	0.73	0.52
MCE	1.33	1.41	1.12	1.09	1.35	1.02

REFERENCES

[1] Lehigh RTMD Users Guide. http://www. nees. lehigh. edu/resources/users-guide,2013.

[2] Dong,B. ,Sause,R. ,and Ricles,J. M. ,"Accurate Real-time Hybrid Earthquake Simulations on Large-scale MDOF Steel Structure with Nonlinear Viscous Dampers. " *Earthquake Engineering and Structural Dynamics*,DOI:10. 1002/eqe. 2572,2015.

[3] ASCE/SEI 41-06, *Seismic Rehabilitation of Existing Buildings*. American Society of Civil Engineers,Reston,VA,2007.

BRACED FRAME USING ASYMMETRICAL FRICTION CONNECTIONS (AFC)

J. Chanchi Golondrino[a], R. Xie[b], G. A. MacRae[b], G. Chase[b], G. Rodgers[b], C. Clifton[c]

[a]University of Canterbury-New Zealand, National University of Colombia, Colombia
jcchanchigo@ unal. edu. co

[b]University of Canterbury-New Zealand
robin. xie@ pg. canterbury. ac. nz, gregory. macrae@ canterbury. ac. nz,
geoff. chase@ canterbury. ac. nz,
geoffrey. rodgers@ canterbury. ac. nz

[c]University of Auckland-New Zealand
c. clifton@ auckland. ac. nz

KEYWORDS: Low Damage Structures, Energy Dissipation, Moment Resisting Frame, Singly Braced Frame, Asymmetric Friction Connections (AFC).

ABSTRACT

Asymmetrical Friction Connections (AFC) were proposed and developed in New Zealand as means to dissipate seismic energy in different structural systems. AFCs can be assembled with a slotted plate sandwiched by two plates with standard holes. One of these plates is termed fixed plate because its movement is restricted, and the other plate is a floating plate termed cap plate and that is attached to the slotted and fixed plate by means of high strength bolts. Thin plates termed shims can be inserted at the interface between the slotted plate and the cap plate, and at the interface between the slotted plate and the fixed plate in order to improve the hysteretic behaviour of the AFCs. Use of shims with a hardness greater than the hardness of the slotted plate is recommended in order to develop a stable hysteretic behaviour with low degradation. Testing of braces equipped with an AFC detail at one end (AFC brace) have been undertaken recently at University of Canterbury-New Zealand showing that this type of braces can be used as fuses for reducing damage on moment resisting frames during a seismic event. This paper reports on the quasi-static testing of a single storey moment resisting frame with one bay and equipped with a concentric AFC brace. The asymmetrical friction connection brace (AFC brace) was assembled using one 250PFC channel, two Bisalloy 500 shims, and two M16 grade 8. 8 galvanized bolts. Results show that by equipping the frame with the AFC brace, the frame can undergo drifts up to 3. 0% without yielding any frame member, and more importantly with low damage on the AFC detail. Results also show that the amount of load that the frame can absorb during a seismic event can be controlled by the AFC detail strength.

CONCLUSIONS

This paper describes the hysteretic behavior of a singly braced frame equipped with an AFC brace (AFC frame). It was shown that:

ⅰ) The hysteretic behaviour of AFC frames is stable and almost rectangular. This hysteretic behaviour can be described by an initial stage termed pre-sliding where the frame absorbs lateral load until the sliding mechanism of the brace is activated, and a second stage termed sliding where the frame can undergo high drifts with very low increments on the lateral load.

ⅱ) By adding AFC braces to moment resisting frames the lateral capacity of the frame can be controlled and assumed equal to the force that fully activates the sliding mechanism of the slotted plate on the AFC brace. AFC braces also increase the ductility of moment resisting frames allowing them to undergo high drifts without yielding any frame member and with low damage on the brace.

ⅲ) AFC frames can be considered as low damage structural systems given that the hysteretic behaviour is stable over an amount of cycles comparable to those typical in a severe earthquake, and also because AFC frames can dissipate the seismic energy via friction and exhibiting large drifts rather than yielding or degrading severely any member or component of the structural system.

ACKNOWLEDGMENT

The authors would like to acknowledge the MBIE Natural Hazards Research Platform(NHRP) and John Jones Steel for their support of this study as part of the Composite Solutions project. All opinions expressed remain those of the authors.

REFERENCES

[1] Butterworth, J. W. (1999). Seismic Damage Limitation in Steel Frames Using Friction Energy Dissipators. 6th *International Conference on Steel and Space Structures* -September 1-2. Singapore.

[2] Chanchí, J. C. , Xie, R, MacRae, G. A. , Chase, J. G. , Rodgers, G. W. , & Clifton, G. C. (2014). Low damage braces using Asymmetrical Friction Connections(AFC). *NZSEE*, April, Auckland.

[3] Chanchí, J. C. , MacRae, G. A. , Chase, J. G. , Rodgers, G. W. , Clifton, G. C, & Munoz, A. (2012). Design considerations for braced frames with asymmetrical friction connections(AFC). *STESSA* 2012, *Santiago de Chile*, January.

[4] Chanchí, J. C. , MacRae, G. A. , Chase, J. G. , Rodgers, G. W. , & Clifton, G. C. (2011). Behaviour of Asymmetrical Friction Connections(AFC) using different shim materials. *NZSEE-annual conference*. New Zealand.

[5] Grigorian, C. E. & Popov, E. P. (1994). *Experimental and Analytical Studies of Steel Connections and Energy Dissipaters. Report UCB/EERC-95/13, Engineering Research Center*. San Francisco (USA).

[6] Clifton, G. C. (2005). Semi-Rigid Joints for Moments Resisting Steel Framed Seismic Resisting Systems. *Published PhD Thesis, Department of Civil and Environmental Engineering*. University of Auckland-New Zealand.

[7] MacRae, G. A. (2008). A New Look at Some Earthquake Engineering Concepts. M. J. *Nigel Priestley Symposium* Proceedings, IUSS Press, 2008.

[8] MacRae, G. A. , Clifton, C. G. , MacKinven, H. , Mago, N. , Butterworth, J. , Pampanin, S. (2010). *The Sliding Hinge Joint Moment Connection. Bulletin of the New Zealand Society for Earthquake Engineering*. Vol. 43, No 3, September.

Member Behaviour

OUT-OF-PLANE STABILITYASSESSMENT OF BUCKLING-RESTRAINED BRACES WITH CHEVRON CONFIGURATIONS

Toru Takeuchi[a], Ryota Matsui[a] and Saki Mihara[a]

[a] Tokyo Institute of Technology

ttoru@ arch. titech. ac. jp, matsui. r. aa@ m. titech. ac. jp

KEYWORDS: Buckling-restrained brace, Stablity, Connections, Chevron, Cyclic Loading Tests

ABSTRACT

Buckling-restrained braces (BRBs) are expected to exhibit stable hysteresis under cyclic axial loading; however, one of their key limit states is global flexural buckling including connections. The authors have previously proposed a unified simple equation set for ensuring BRB stability that includes their bending-moment transfer capacity at the restrainer ends for various connection stiffness values with initial out-of-plane drifts with symmetric conditions at both ends as in *Fig. 1*(a). In the present study, these equations are extended to BRBs with asymmetric boundary conditions as chevron configurations (*Fig. 1*(b)). Cyclic loading tests with initial out-of-plane drifts (*Fig. 1*(c)) are conducted, and the results are compared with those obtained using the proposed equations.

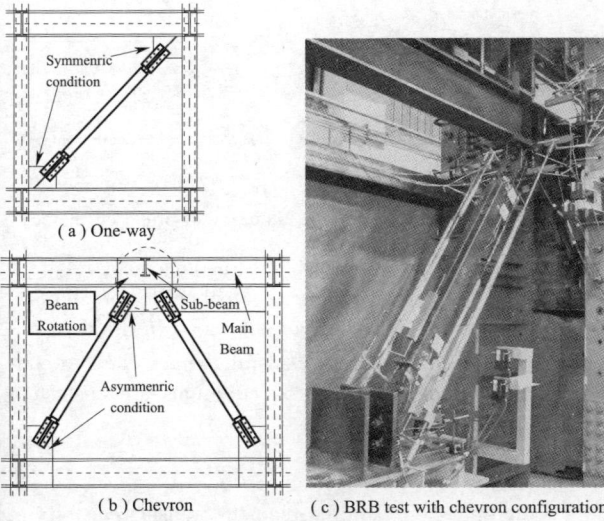

Fig. 1. BRB configuration

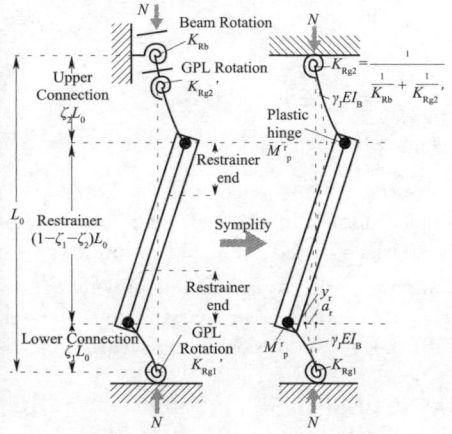

Fig. 2. Collapse model of BRB for chevron layout

The authors have proposed the following equations for evaluating the stability limit of BRBs that include connections considering bending-moment transfer capacity at the restrainer ends. The limit axial force with elastic springs for gusset plates, $N_{\text{lim}1}$, is expressed as follows:

$$N_{\text{lim}1} = \frac{(M_p^r - M_0^r)/a_r + N_{cr}^r}{(M_p^r - M_0^r)/(a_r N_{cr}^B) + 1} > N_{cu} \tag{1}$$

In the above equations, under asymmetrical conditions at connections as shown in *Fig. 2*, N_{cr}^r can be obtained by the equivalent slenderness ratio given as

$$\lambda_r = \frac{2L_0}{i_c} \cdot \sqrt{\frac{B}{A}} \quad (asymmetrical \quad mode) \;, \; A = \frac{{}_\xi \kappa_{Rg1}}{\xi_1^{\,3}({}_\xi \kappa_{Rg1} + 3)} + \frac{{}_\xi \kappa_{Rg2}}{\xi_2^{\,3}({}_\xi \kappa_{Rg2} + 3)} \;,$$

$$B = \frac{{}_\xi \kappa_{Rg1} \pi^2 \{ (1 - \xi_1 - \xi_2) + 2\xi_1 \} + 24(1 + \xi_1 - \xi_2)}{\pi^2 \xi_1 (1 - \xi_1 - \xi_2)({}_\xi \kappa_{Rg1} + 3)}$$

$$+ \frac{{}_\xi \kappa_{Rg2} \pi^2 \{(1 - \xi_1 - \xi_2) + 2\xi_2\} + 24(1 + \xi_2 - \xi_1)}{\pi^2 \xi_2 (1 - \xi_1 - \xi_2)({}_\xi \kappa_{Rg2} + 3)} \tag{2}$$

$$\lambda_r = \frac{2\xi_2 L_0}{i_c} \sqrt{\frac{(1 - \xi_1)({}_\xi \kappa_{Rg2} + 24/\pi^2)}{(1 - \xi_1 - \xi_2){}_\xi \kappa_{Rg2}}} \quad (one - sided \ mode) \tag{3}$$

Furthermore, the stability limit with plastic hinges at the gusset plates, N_{lim2}, can be expressed as

$$N_{\text{lim2}} = \frac{(M_p^r - M_0^r + C_4)/a_r}{(M_p^r - M_0^r + C_4)/(a_r N_{cr}^B) + 1}, \quad C_4 = \left(\frac{M_p^{g1}}{\xi_1} + \frac{M_p^{g2}}{\xi_2}\right)$$

$$\frac{1}{1/\xi_1 + 1/\xi_2 + 4/(1 - \xi_1 - \xi_2)} \quad (asymmetrical \ mode) \tag{4}$$

$$N_{\text{lim2}} = \frac{((M_p^r - M_0^r)(1 - \xi_1)/(1 - \xi_1 - \xi_2) + M_p^{g2})/a_r}{((M_p^r - M_0^r)(1 - \xi_1)/(1 - \xi_1 - \xi_2) + M_p^{g2})/(a_r N_{cr}^B) + 1} (one - sided \ mode)$$

For confirming the proposed stability conditions, cyclic loading tests were performed on BRBs with a chevron configuration including out-of-plane drifts, as shown in *Fig. 1(c)*.

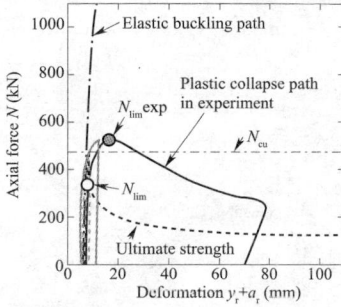

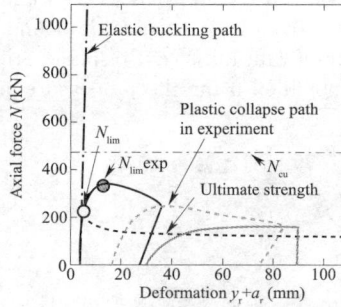

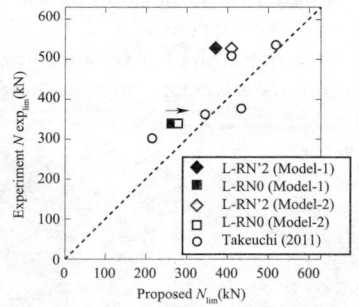

Fig. 3. Relationship between axial force and out-of-plane displacement   *Fig. 4.* Accuracy of proposed method

Fig. 3 shows a comparison between the measured axial force-out-of-plane displacement relationships obtained in the tests and those obtained using the proposed equations. Further, *Fig. 4* shows a comparison of the estimated failure axial forces N_{lim} with the peak axial force obtained in the tests, together with the results of our previous study[1]. It can be seen that the test results with a chevron configuration have higher margins, and the results obtained using the proposed equations are consistent with the test results with some variation.

CONCLUSIONS

The concept of stability evaluation of BRBs including their bending-moment transfer capacity at the restrainer ends and various connection stiffness values proposed previously by the authors was extended in this study to BRBs with asymmetrical conditions at both ends as chevron configurations, and a compact equation set for evaluating global BRB stability under consideration of asymmetrical conditions was proposed. A series of cyclic loading tests were conducted on BRBs with chevron configurations, and it was confirmed that the test results are consistent with those obtained using the proposed equations.

ACKNOWLEDGMENT

The authors would like to acknowledge the support of Nippon Steel and Sumikin Engineering Co. , Ltd. , for this study.

REFERENCES

[1] Takeuchi T. , Ozaki H. , Matsui R. , Sutcu F. , "Out-of-plane Stability of Buckling-Restrained Braces including Moment Transfer Capacity" *Earthquake Engineering & Structural Dynamics*, Vol. 43, Issue 6, pp. 851-869, 2014. 5

EVALUATION METHOD OF PLASTIC DEFORMATION CAPACITY OF STEEL BEAM GOVERNED BY DUCTILE FRACTURE AT THE TOE OF THE WELD ACCESS HOLE

Satoshi Yamada[a], Yu Jiao[b], Shoichi Kishiki[a]

[a]Tokyo Institute of Technology, Structural Engineering Research Center, Japan

naniwa@ serc. titech. ac. jp, kishiki@ serc. titech. ac. jp

[b] Tokyo Univ. of Science, Dept. of Archtecture, Japan

yujiao@ rs. tus. ac. jp

KEYWORDS: Beam, Loading History, Plastic Deformation Capacity, Ductile Fracture, Numerical Analysis

ABSTRACT

Ductile fracture occurs when the steel beam reaches its ultimate plastic deformation capacity. In order to evaluate the plastic deformation capacity of beam governed by ductile fracture, it is necessary to clarify the plastic strain capacity of the beam-end flange. In this paper, the in-plane beam analysis was conducted to achieve the strain history at the fracture zone. A hysteresis model of the structural steel effective in evaluating material damage was introduced into this analysis. Moreover, loss of beam section due to the weld access holes and the local out-of-plane bending of the tube wall on the RHS column were taken into account to the analytical model. The analytical method was proved by comparing the analytical results to the experimental results from a database. The plastic strain capacity of beam-end flange till ductile fracture was evaluated by studying the analytical flange strain histories.

CONCLUSIONS

The plastic deformation capacity of the structural components till ductile fracture plays a significant role in seismic design because it is one of the indexes of structural performance. In this paper, the fracture condition of the steel beam flange was discussed by studying the average flange surface stress-strain history of the beam section at the toe of the weld access hole. The in-plane analyses of the steel beams subjected to cyclic loading histories were conducted to obtain the above mentioned average stress-strain history. A hysteresis model of the structural steel consideringBauschinger effect, which is effective in evaluating material damage, was introduced into this analysis. Moreover, the analytical model takes into account the decrease of the joint efficiency, which directly affects the strain capacity of the beam flange. By comparing the analytical load-deformation relation as well as the local strain histories to the experimental results from a database, the analytical simulation was proved to be authentic.

The plastic strain capacity of beam flange till ductile fracture was evaluated using the evaluation method. This method focuses on the relation between the cumulative plastic strain ratio of both the skeleton curves and the Bauschinger parts derived from the average beam-end flange stress-strain histories obtained in the in-plane analysis. At the fracture zone of the beam flange, there is a linear relation between the ultimate normalized cumulative plastic strain ratio of the skeleton curve on the tension side (η_S^+/η_0) and the total normalized cumulative plastic strain ratio(η_T/η_0). Ductile fracture of the beam flange occurs when the cumulative plastic strain ratio at the critical area satisfies *Equation*(8).

$$\eta_T/\eta_0 = 8 - 7.5\eta_S^+/\eta_0$$
$$(0.234 \leqslant \eta_S^+/\eta_0 \leqslant 0.941)$$

REFERENCES

[1] Suita K. , and Tada M. "Full-scale Tests to Determine Plastic Rotation Capacity of Steel Wide-flange Beams Connected with Square Tube Columns" Proceedings on Welded Construction in Seismic Areas, International Conference on Welded Constructions in Seismic Areas, 289-301, 1998.

[2] Jiao Y. , Yamada S. , Kishiki S. , and Shimada Y. , "Evaluation of plastic energy dissipation ca-

pacity of steel beams suffering ductile fracture under various loading histories", Earthquake Engineering and Structural Dynamics, 40: 1553-1570, 2011. 3, DOI 10. 1002/eqe. 1103.

[3] Yamada M, Sakae K, Tadokoro T, Shirakawa K. , "Elasto-plastic bending deformation of wide flange beam-columns under axial compression, Part I: Bending moment-curvature and bending moment-deflection relations under static loading" (In Japanese). Journal of structural and construction engineering, Transactions of the AIJ ; No 127: 8-14, 1966.

[4] Yamada S, Imaeda T, and Okada K. , "Simple hysteresis model of structural steel considering the Bauschinger effect " (In Japanese). Journal of structural and construction engineering. Transactions of AIJ ; No 559, 225-32, 2002.

[5] Architectural Institute of Japan, "Recommendation for design of connections in steel structures" (In Japanese) , 2006.

[6] Suzuki T, Ishii T, Morita K, and Takanashi K. , "Experimental study on fracture behavior of welded beam-to-column joint with defects" (In Japanese). Steel construction engineering; Vol 6, No 23: 149-64, 1999.

[7] Kobayashi A. , "Evaluation of ultimate energy dissipation capacity of steel under cyclic loading" (In Japanese) , Master Thesis submitted to Tokyo Institute of Technology, 2005.

[8] Okada K, Matsumoto Y, Yamada S. , "Evaluation of effect of joint efficiency at beam-to-column connection on ductility capacity of steel beams" (In Japanese). Journal of Structural and Construction Engineering, Transactions of AIJ, 568: pp. 131-8, 2003.

[9] Yamada S. , Jiao Y. and Kishiki S. , "Cyclic Loading Test on Steel Beam under Various Loading Histories " (in Japanese) , Proc. of annual meeting, JAEE, 286-287, 2011.

[10] Architectural Institute of Japan, " Japanese architectural standard specification JASS 6 steel work" (In Japanese) , 1996.

[11] Jiao Y. , Yamada S. , and Kishiki S. , "Plastic deformation capacity of structural steel under various axial strain histories" , Proc. of 8th International Conference on Urban Earthquake Engineering, 935-40, 2010.

APPLICATION AND MODIFICATION OF SHIBATA-WAKABAYASHI MODEL TO SIMULATION OF BUCKLING HYSTERESIS LOOP OF STEEL BRACES

Ryota Matsui[a], Toru Takeuchi[a]

[a]Tokyo Institute of Techonology, Architecture and Building Engineering, Japan

matsui. r. aa@ m. titech. ac. jp, ttoru@ arch. titech. ac. jp

KEYWORDS: Braces, Post-buckling, hysteresis

ABSTRACT

A phenomenological model proposed by M. Shibata, called the Shibata-Wakabayashi(SW) model, is one of the most commonly used rules for simulating the buckling hysteresis loop of steel braces[1]. The SW model(*Fig. 1*(a)) divides the buckling hysteresis loop of the steel braces into four stages, and the rule of this model is somewhat simpler than those of other phenomenological or physical models. In[1], the authors validated the SW model in each stage by using the test results of several flat steel plates. Several researchers have further proposed modified SW models for simulating the degradation of the buckling strength[2]. Examples of such models include the modified models shown in *Figs. 1*(b) and 1(c), termed SW 01 and SW 02, respectively. In SW 01(*Fig. 1*(b)), the initial buckling strength is determined by the elastoplastic column curve and the buckling strength in the subsequent cycles is equal to that of the original SW model. Similarly, in SW 02(*Fig. 1*(c)), the initial buckling strength is determined as SW 01. However, the buckling strength from the second cycle is determined by the cumulative compressive plastic strain of the steel brace. Thus far, validation of these modified models through a comparison of a wide range of experimental results has not been attempted.

The present authors previously conducted several series of tests on steel braces in the form of circular hollow sections(CHSs), H-sections, L-shaped angles, double L-shaped angles, double T-shaped angles, and double channels[3-5]. In these series of tests, 31 specimens were used. In the present paper, each SW model is validated using these test results. *Fig. 2* shows a comparison of the hysteresis loops of the three considered SW models(original SW model, SW 01, and SW 02) with the corresponding results of one of the tests. *Fig. 2* illustrates that SW 02 largely agrees with the test results and provides higher accuracy than the other two SW models. Here, the accuracy of each SW model in terms of the buckling strength is determined using the average error r_{mean} and the sample variance V(*Eq.* (1)) of the error between the test results and the results of the SW models.

$$V = \sqrt{\frac{1}{N}\sum_{i=1}^{N}\left\{\frac{\sigma_i - \sigma_{i,\text{test}}}{(\sigma_i + \sigma_{i,\text{test}})/2} - r_{mean}\right\}^2}\left(r_{mean} = \frac{1}{N}\sum_{i=1}^{N}\left|\frac{\sigma_i - \sigma_{i,\text{test}}}{(\sigma_i + \sigma_{i,\text{test}})/2}\right|\right) \tag{1}$$

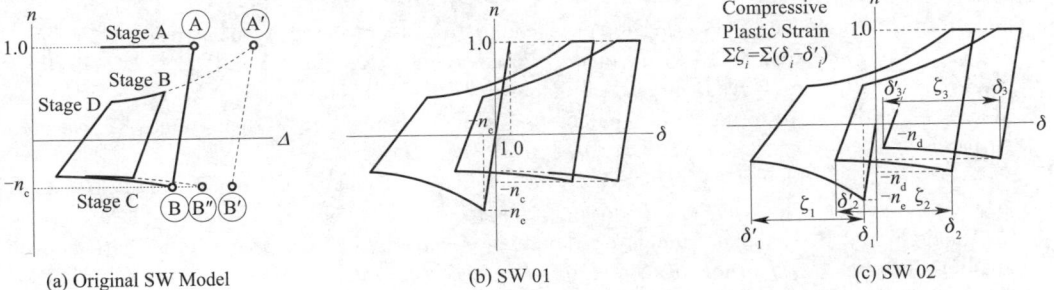

(a) Original SW Model (b) SW 01 (c) SW 02

Fig. 1. Shibata-Wakabayashi Models(SW Models)

Here, σ_i is the buckling stress calculated using each SW model, $\sigma_{i,\text{test}}$ is the buckling stress measured in the test, and N is the number of specimens. *Fig. 3* shows the relationship between the test re-

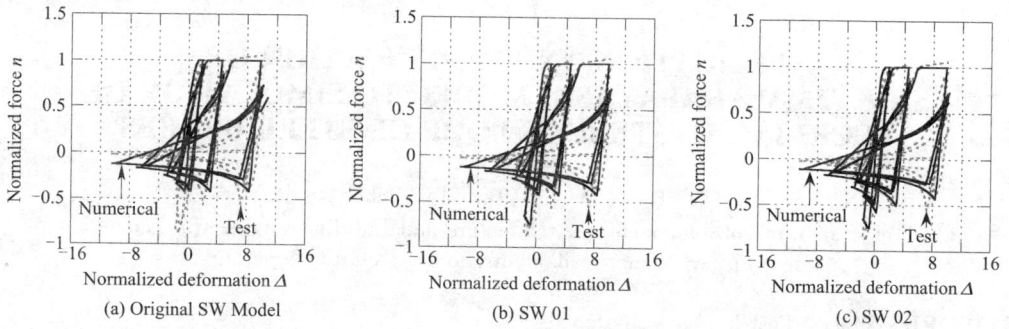

Fig. 2. Comparison of Hysteresis Curves between Test Results and Numerical Results of Considered SW Models
(Dia. :89. 1 mm;Thickness:3. 2 mm;Slenderness Ratio:70;Diameter-to-Thickness Ratio:32)

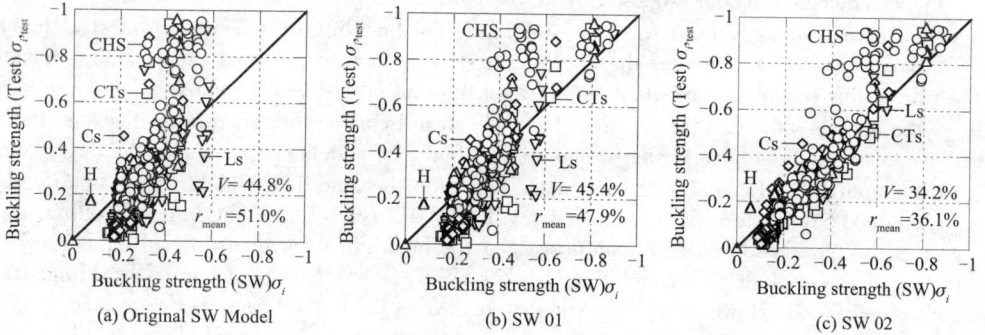

Fig. 3. Accuracy of SW Models

sults and the numerical results of the SW models for the buckling strength, along with the values of the average error r_{mean} and sample variance V of the error. SW 02 provides the highest accuracy from among the three considered SW models.

CONCLUSIONS

The accuracy of modified Shibata-Wakabayashi models was investigated using the test results of 31 specimens. One of the modified Shibata-Wakabayashi models was shown to provide higher accuracy than the original Shibata-Wakabayashi model. The average error and the sample variance between the test results and the numerical results of the best modified Shibata-Wakabayashi model were approximately 2/3rd those of the other Shibata-Wakabayashi models.

REFERENCES

[1] M. Shibata. " Influence of restoring force characteristics of braces on dynamic response of braced frame. ",7th World Conference on Earthquake Engineering Proceedings,Vol. IV,185-192,1981

[2] H. Taniguchi,B. Kato,N. Nakamura,Y. Takahashi,T. Saeki,T. Hirotani,and Y. Aikawa. "Study on restoring force characteristics of X-shaped brace steel frame. ", Journal of Structural Engineering,AIJ,Vol. 37B,303-316,1991(in Japanese)

[3] T. Takeuchi and R. Matsui. "Cumulative cyclic deformation capacity of circular tubular braces under local buckling. ",Journal of Structural Engineering,ASCE,Vol. 137,1311-1318,2011

[4] T. Takeuchi and R. Matsui. "Cumulative deformation capacity of steel braces under various cyclic loading histories. ", Journal of Structural Engineering, ASCE, 10. 1061/(ASCE) ST. 1943-541X. 0001146,04014175

[5] T. Takeuchi,Y. Kondo,R. Matsui,and A. Imamura. "Post-buckling hysteresis and cumulative deformation capacity of built-up member braces with local buckling. ", Journal of Structural and Construction Engineering,Transactions of AIJ,Vol. 77,No. 681,1781-1790,2012(in Japanese)

LOADING PROTOCOLS FOR EVALUATING THE SEISMIC BEHAVIOR OF STEEL BEAMS IN WEAK-BEAM MOMENT FRAMES

Yu Jiao[a], Shoichi Kishiki[b], Satoshi Yamada[c]

[a] Tokyo University of Science, Faculty of Engineering, Japan
yujiao@ rs. tus. ac. jp
[b] Osaka Institute of Technology, Dept. of Architecture, Japan
kishiki@ archi. oit. ac. jp
[c] Tokyo Institute of Technology, Structural Engineering Research Center, Japan
yamada. s. ad@ m. titech. ac. jp

KEYWORDS: Loading protocol, Steel beam, Weak-beam moment frame, Seismic behaviour, Response analysis

ABSTRACT

In steel moment frames with strong-column weak-beam mechanism, beams are the dominant seismic components that dissipate the majority of the earthquake input energy. Therefore, the plastic deformation capacity of beams is one of the most important topics in seismic design of steel structures.

The loading protocols employed in beam tests to evaluate its seismicbehavior have been studied by researchers universally for decades. Basically, each country has its own standard loading protocols for steel beam testing. For example, the AISC 2005[1] protocol and SAC 2000[2] protocol are often used in the USA, and the JISF 2002[3] loading protocol is commonly accepted in Japan. However, seismic effects of significantly different characteristics are applied to beams during distinct earthquakes. Earthquake ground motions tend to have longer durations and long-period characteristics when the magnitude scale of the earthquake is large. During the 2011 Great Tohoku Earthquake, many highrise buildings in Tokyo and Osaka area, which are far away from the epicenter, also experienced long period of shaking. Moreover, the number and amplitudes of the loading cycles the beam experiences during earthquakes also depend on the configuration, strength, stiffness, etc. of the structure. It is questionable whether the current loading protocols are comprehensive enough to simulate real beam responses in structures during earthquakes. This paper temps to investigate the currently used loading protocols and gives suggestions on the selection of the loading protocols in beam testing.

The first part of this paper studies beam behavior in low to mid-rise weak-beam moment frames during different earthquakes. Response analyses of three plane moment frames with different number of stories were conducted. The plastic deformation capacity of the beams in those frame models were investigated. The second part of this study compares the beam's seismic behavior with the presently used loading protocols. In-plane beam analyses of beams subjected to different loading histories were carried out. The beams' behavior under the present loading protocols was compared with that under real beam responses. The appropriate combinations of the loading protocols employed in beam tests for the purpose of evaluating the seismic behavior of steel beams were suggested.

CONCLUSIONS

Fig. 1 shows the deform performance of the beam under all loading protocols investigated in this paper. The SAC, FEMA 461[4], and JISF loading protocols show very similar results, which can reproduce the beam's performance under generic earthquakes. When the amplitude of the loading protocol decreases, the growing processes lean to the upper-left side. The loading protocols in this area reproduce the beam's performance under long duration ground motions. On the other side, when the amplitude increases, the growing processes lean to the lower-right corner, which is the domain of near-fault ground motions.

Based on the above discussion, suggestions are given as follows to the selection of loading protocols in beam tests:

Single specimen testing program

One loading protocol from the SAC 2000/ASIC, FEMA 461, or JISF loading protocols.

*For the beams possibly suffering brittle fracture, SAC 2000/ASIC loading protocol is recommended.

Multiple specimens testing program

Testing program with at least three specimens is recommended.

1) One loading protocol from the SAC 2000/ASIC, FEMA 461, or JISF loading protocol.

*For the beams possibly suffering brittle fracture, SAC 2000/ASIC loading protocol is recommended.

2) One constant amplitude loading protocol with small loading amplitude, for example: $\pm 3\theta_p$.

3) Either the monotonic loading protocol or the SAC-Near-fault loading protocol.

*When there is a limitation of the facility, constant amplitude loading protocol with the largest loading amplitude of the facility is recommended.

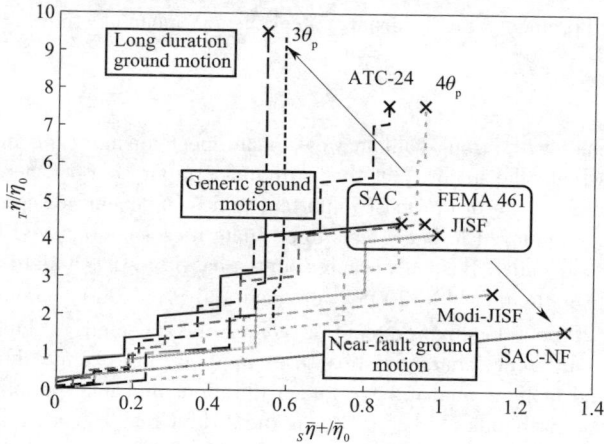

Fig. 1. Equivalent cumulative deformation ratio of beams under recommended loading protocols

REFERENCES

[1] American Institute of Steel Construction. Seismic provisions for structural steel buildings (Including Supplement No. 1) ;2005.

[2] Krawinkler H. , Gupta A. , Medina R. , and Luco N. Loading histories for seismic performance testing of SMRF components and assemblies, SAC Joint Venture, Report No. SAC/BD-00/10 ;2000.

[3] Building Research Institute/Japan Iron and Steel Federation, Testing methods of the evaluation of structural performance for the steel structures, 2002.

[4] Federal Emergency Management Agency. Interim testing protocols for determining seismic performance characteristics of structural and nonstructural components. FEMA 461 Draft Document, Redwood City, CA : Applied Technology Council ;2007.

LATERAL-TORSIONAL BUCKLING CAPACITY OF TAPERED-FLANGE MOMENT FRAME SHAPES

Leah S. O'Neill[a], Trevor A. Jones[a], and Paul W. Richards[a]

[a]Brigham Young University, Department of Civil and Envonmental Engineering, USA

leah@ leahoneill. me, trevoralexanderjones@ gmail. com, prichards@ byu. edu

KEYWORDS: moment frames, lateral-torsional-buckling, beams.

ABSTRACT

Steel moment frames are popular for buildings because of their architectural flexibility, but use more material than steel braced frames. Under lateral loads, steel moment frames experience bending in the beams and columns that varies linearly along the length of the shapes. When shapes with uniform cross-section are used for these beams and columns, the steel is non-uniformly stressed leading to material inefficiencies.

Shapes with tapered flanges could result in more uniformly stressed material in steel moment frames, but are difficult to analyze. *Fig. 1* illustrates two beams with the same cross-section, one straight-flange and one tapered-flange. The tapered-flange beam has cross-sectional properties that vary in proportion to the expected bending moments in moment frames and would result in more efficient moment frames. Like regular moment frame beams, intermediate lateral bracing would be required. However, while lateral-torsional buckling (LTB) can be predicted for straight-flange shapes, closed-form solutions for LTB have not been achieved for tapered-flange beams with intermediate bracing and moment gradients.

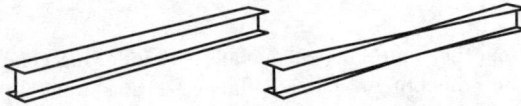

Fig. 1. Straight-flange and tapered-flange shapes for moment frame beams

The objective of this study was to find a simple method to determine the lateral-torsional-buckling capacity of tapered-flange shapes with moment gradients and intermediate lateral bracing.

The approach consisted of performing bifurcation analysis for thousands of straight-flange and tapered-flange finite element beams, and comparing the buckling stresses of beams with the same end-cross section and length, but different flange configurations (straight or tapered). The models were developed and analysed using Abaqus. A validation study, where the results for the straight-flange beam were compared with closed-form solutions, indicated that the finite element models were reasonably accurate for predicting lateral-torsional-buckling.

Results indicate that lateral-torsional-buckling is not an issue that would preclude the used of tapered-flange shapes in moment frames. A representative result is shown in *Fig. 2* for tapered-flange beams with an end cross-section of a W21 × 111. For the typical moment frame beam lengths, 8 to 10m, third point bracing is required to have a buckling stress greater than 350 MPa (*Fig. 2*), but such lateral bracing is frequently used for special moment frames even for straight-flange shapes so it does not represent an onerous requirement.

When the buckling stress for tapered-flange shapes are normalized by the buckling stress of straight-flange shapes with the same end cross-section, length, and lateral bracing, a convenient relationship emerges. A typical result is illustrated in *Fig. 3*. The buckling stress for tapered-flange shapes tends to be around 20 percent of the buckling stress of a comparable straight-flange shape for typical moment frame beam lengths. The same relationship was found for hundreds of end cross-sections.

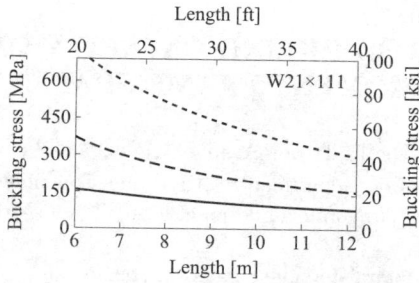

Fig. 2. Buckling stress curves for tapered flange shapes with: end lateral restraint(solid line), midpoint lateral restraint(long dashed) , and third-point lateral restraint(short dashed)

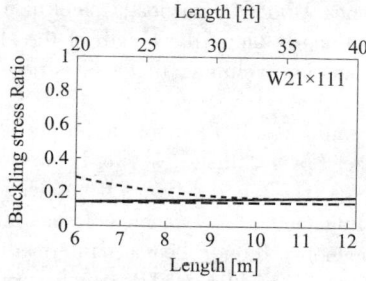

Fig. 3. Ratio of tapered-flange to straight-flange buckling stress

CONCLUSIONS

Based on the results from the finite element models, the following conclusions can be made:

1. Overall, lateral-torsional buckling of tapered-flange I-beams is not a problem that would prohibit wide-scale use of this configuration in moment frames.

2. The lateral-torsional buckling capacity of a tapered-flange shape, with moment gradient and intermediate bracing, is approximately 20% of the lateral-torsional-buckling capacity of a straight-flange shape with the same end cross-section.

ACKNOWLEDGMENT

The use of tapered-flange beam shapes is protected by pending patents(U. S. and International).

The authors gratefully acknowledge the support from Brigham Young University that made this work possible, but retain responsibility for the work and conclusions.

REFERENCES

[1] Yang YB, Yau JD. , "Stability of beams with tapered I-sections" , *Journal of Engineering Mechanics*, 113. 9:1337-1357, 1987.

[2] Bradford MA, Cuk PE. , "Elastic buckling of tapered monosymmetric I-beams" , *Journal of Structural Engineering*, 116(7):1893-1906, 1988.

SIMULATION OF HYSTERETIC BEHAVIOR OF RHS COLUMNS UNDER BI-DIRECTIONAL HORIZONTAL FORCES AND CONSTANT AXIAL FORCE

Takanori ISHIDA[a], Yuko SHIMADA[b], Satoshi YAMADA[a]

[a]Tokyo Institute of Technology, Structural Engineering Research Center, Japan
ishida. t. ae@ m. titech. ac. jp, naniwa@ serc. titech. ac. jp
[b] Chiba University, Graduate School of Enginnering, Japan
yshimada@ faculty. chiba-u. jp

KEYWORDS: Rectangular Hollow Section (RHS) column, Local-buckling, Deterioration behavior, Multi Spring(MS) model, Random bi-directional horizontal loading test.

ABSTRACT

Cold formed rectangular hollow section(RHS) steel tubes are generally used as columns for middle-rise and low-rise steel moment-resisting frames(MRFs) in Japan. Columns are subjected to biaxial bending, since buildings behave 3-dimensionally under seismic excitations. In order to clarify the 3D collapsebehavior of steel MRFs, it is necessary to simulate the behavior of RHS column under bi-directional horizontal forces and axial force, including the post-local-buckling and strength deterioration. The objective of this study is to develop a hysteretic model to simulate post-local-buckling and deterioration behavior of RHS columns under bi-directional horizontal forces and constant axial force.

A multi spring (MS) model[1], as shown in *Fig. 1*, which consists of elastic and inelastic elements was applied in the simulation. The hysteretic model of the element springs which compose the inelastic element was established based on the hysteretic model of RHS columns under uniaxial bending, proposed by Yamada te. al. [2]. That is, before the maximum strength which is governed by local buckling, the hysteretic model consists of the skeleton curve which corresponds to the load-deformation relationship under monotonic loading, Bauschinger part, and the elastic unloading part, as shown in *Fig. 2*(a). After the maximum strength, the hysteretic model of compression side consists of the degrading part which corresponds to the extended skeleton curve, the upgrading part and the unloading part as shown in *Fig. 2* (b). The extended skeleton curve was expressed in terms of the ratio of the maximum strength to yield strength S', three deterioration slopes E_{d1}', E_{d2}', E_{d3}', and two deterioration stiffness changing points T', T_2' as shown in *Fig. 3*. Also, as Bauschinger part, Akiyama and Takahashi's model[3] was applied. As the upgrading part, the bilinear model proposed by Yamada et al. [2] was applied.

In addition, a series of cyclic loading tests of RHS columns under random bi-directional horizontal forces and constant axial force were conducted toverify the proposed analytical model. In this cyclic loading tests, the random horizontal loading protocol, established based on the random horizontal displacement orbits obtained by elasto-plastic response analyses of multi-story steel frames subjected to bi-directional horizontal ground motions, was employed. The maximum strength of all specimens was governed by local buckling in this test.

An example of experimental and analytical result was shown in *Fig. 4*. The analytical results using the MS model closely approximated the hysteresis of RHS columns including the deterioration range under bi-directional horizontal forces and constant axial force.

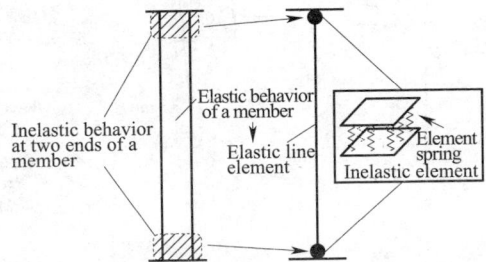

Fig. 1. MS Model

CONCLUSIONS

This studyes tablishes an analytical model of RHS columns that can simulate the members' behaviors, including the post-local-buckling and strength deterioration under bi-directional horizontal forces and axial force. The simulated RHS column behaviors, including the post-local-buckling and deterioration range, agreed with the experimental results.

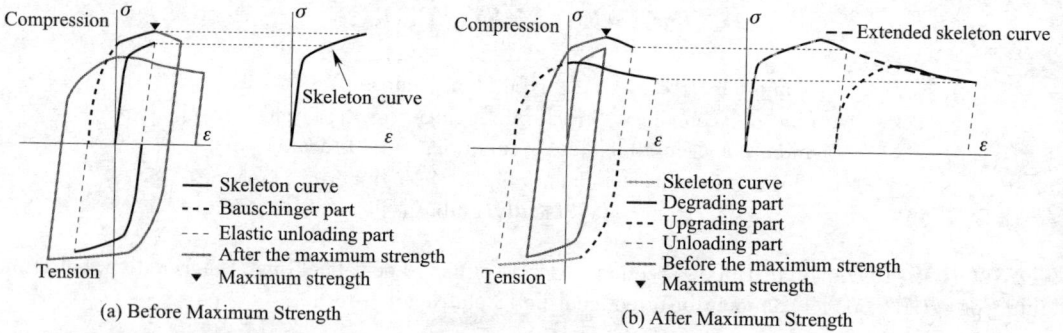

(a) Before Maximum Strength   (b) After Maximum Strength

Fig. 2. Decomposition of Hysteretic Curve and Skeleton Curve · Extended Skeleton Curve

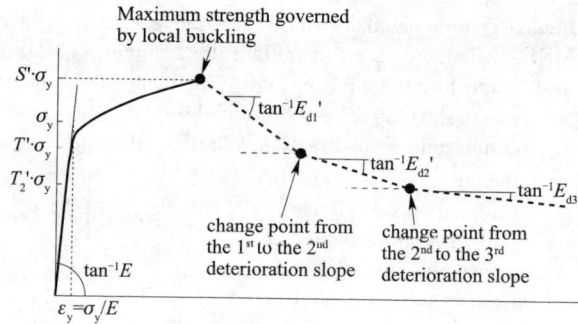

Fig. 3. Deterioration Model of Extended Skeleton Curve

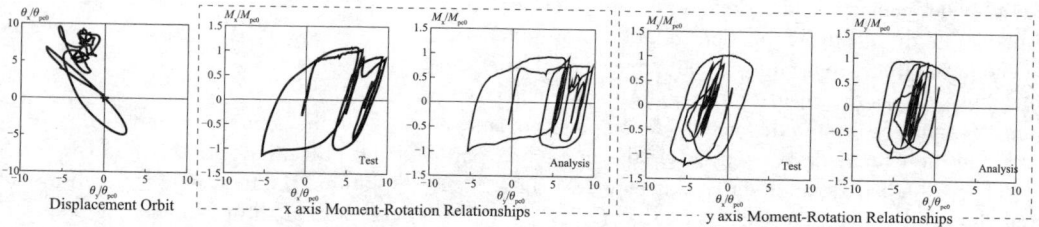

Fig. 4. Comparisons of Experimental and Analytical Results
(Width/Thickness Ratio : 22. 2 , Axial Force Ratio : 0. 2)

REFERENCES

[1] S. S. Lai, G. T. Will, S. Otani, 1984. "Model for Inelastic Biaxial Bending of Concrete Members", *Journal of Structural Engineering*, ASCE, Vol. 110, No. ST11, pp. 2563-2584.

[2] S. Yamada, T. Ishida, Y. Shimada, 2011. "Hysteresis Model of RHS Columns in the Deteriorating Range Governed by Local Buckling", *Journal of Structural and Construction Engineering*, AIJ, No. 674, pp. 627-636. (in Japanese)

[3] H. Akiyama, M. Takahashi, 1990. "Influence of Bauschinger Effect on Seismic Resistance of Steel Structures", *Journal of Structural and Construction Engineering*, AIJ, No. 418, pp. 49-57. (in Japanese)

STUDY ON X-SHAPE BUCKLING RESTRAINED STEEL PLATE SHEAR WALL WITH TWO-SIDE CONNECTIONS

Wen-Yang Liu[a,b], Guo-Qiang Li[a,c]

[a] Tongji University, College of Civil Engineering, China
wyliu81@ 126. com, gqli@ tongji. edu. cn
[b] Heilongjiang Bayi Agricultural University, College of Engineering, China
[c] Tongji University, State Key Laboratory of Disaster Reduction in Civil Engineering, China

KEYWORDS: Buckling restrained steel plate shear wall, Two-side connections, X-shape steel plate shear wall, Stiffness, Capacity.

ABSTRACT

Buckling restrained steel plate shear wall, which consists of a steel plate and restraining plates on both sides, is recognized as a kind of structural element with excellent energy dissipation capacity. This paper presents X-shape buckling restrained steel plate shear wall (*Fig. 1*) based upon rectangular shape[1] and I-shape[2] buckling restrained steel plate shear wall connected at top and bottom sides. Finite element analysis indicates that the ductility and energy dissipation capacity of X-shape steel plate shear wall is better than that of rectangular shape or I-shape steel plate shear wall with similar sizes, while the stiffness and capacity are almost the same as rectangular shape steel plate shear wall. So X-shape steel plate shear wall has better earthquake-resistant performance than rectangular shape or I-shape steel plate shear wall. Furthermore, formulas to calculate stiffness and capacity of X-shape buckling restrained steel plate shear wall are presented.

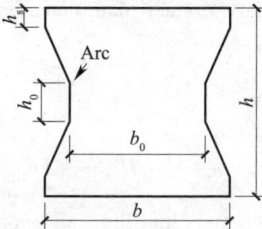

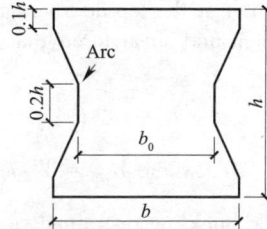

Fig. 1. X-shape buckling restrained steel plate shear wall   Fig. 2. Geometry of X-shape steel plate

CONCLUSIONS

Based upon the study on stressing mechanism of buckling restrained steel plate shear wall with two-side connections[3], X-shape steel plate can be formed by linear reduction from the point 0. 1h away from the top and bottom edges to the point 0. 1h away from the middle height of steel plate, as shown in *Fig. 2*. The width of middle part of steel plate can be determined by the criterion that the cross-section at the top and bottom edges yield considering moment and shear force when the cross-section at the middle height of steel plate yield only considering shear force, that is

$$b_0 = \frac{b}{\sqrt{0.85(\frac{h}{b})^2 + 1}} = \frac{b}{\sqrt{0.85\lambda^2 + 1}} \tag{1}$$

To study the ductility and energy dissipation capacity of X-shape buckling restrained steel plate shear wall, three kinds of buckling restrained steel plate shear wall areanalyzed by finite element method using Abaqus software. As shown in *Table 1*, X-shape steel plate has better energy dissipation capacity than rectangular shape steel plate and I-shape steel plate with similar sizes, and has lower PEEQ (Equivalent plastic strain, which is the cumulative result of plastic strain of the whole deformation process) peak value. The greater of PEEQ the more easily fracture occurs. So X-shape steel plate with reasonable sizes can bear more loading cycles than rectangular shape steel plate and I-shape steel plate before fracture occurs, which can greatly enhance energy dissipation capacity of buckling restrained steel plate shear wall for earthquake-resistance.

Table 1. Comparison of three kinds of steel plate shear wall with sizes 3m × 3m under cyclic loading

No.	Type	b (m)	h (m)	b_0 (m)	h_s (m)	h_0 (m)	Area (m^2)	PEEQ	Plastic energy dissipation(J)	Energy dissipation in unit area(J/m^2)
1	Rectangular shape	3	3	3	–	–	9	1. 297	971320	107924
2	X-shape	3	3	2. 2	0. 3	0. 6	7. 80	0. 943	1047400	134282
3	I-shape	3	3	2. 2	0. 8	0. 6	8. 02	0. 950	981359	122364
4	X-shape	3	3	2. 2	0. 3	1. 2	7. 56	0. 803	941825	124580
5	I-shape	3	3	2. 2	0. 5	1. 2	7. 54	0. 906	910922	120812

Note: The thickness of steel plate is 6mm. Yield strength of steel is $235N/mm^2$. Radius of the arc is 0. 4m. The cyclic loading is controlled by displacement at the top. The target displacement is set as inter-story drift of 2%. Cyclic loadings were repeated three times at the target displacement.

The stiffness of X-shape buckling restrained steel plate shear wall can bedetermined by assuming X-shape steel plate to be an equivalent rectangular shape steel plate with the same height and equivalent width $b_1 = (b + b_0)/2$, that is

$$K = \frac{Et}{2. 6\lambda + \lambda^3} \tag{2}$$

where $\lambda = h/b_1$ is the height-width ratio of equivalent rectangular shape steel plate.

The capacity of X-shape buckling restrained steel plate shear wall can beobtained by the capacity of cross-section at the top and bottom edges of the middle part of steel plate considering combined action of moment and shear force, that is

$$V_{y0} = \frac{M_{p0} V_{p0}}{\sqrt{V_{p0}^2 H_{y0}^2 + M_{p0}^2}} \tag{3}$$

where $M_{p0} = W_{p0} f_y = \frac{tb_0^2}{4} f_y$, $V_{p0} = A_0 f_{vy} = b_0 t \frac{f_y}{\sqrt{3}}$, $H_{y0} = 0. 5h_0$.

X-shape buckling restrained steel plate shear wall has better ductility and energy dissipation capacity than rectangular shape buckling restrained steel plate shear wall with less steel consumption. In actual design X-shape steel plate with sizes determined using the method proposed in this paper is a better choice to replace rectangular shape steel plate with better mechanical performance. The structural design parameters such as stiffness and capacity of X-shape buckling restrained steel plate shear wall can be determined using *Eq.* (2) and *Eq.* (3).

REFERENCES

[1] Sun Fei-fei, Gao Hui, Li Guo-qiang. "Experimental research on two-sided composite steel plate walls". Shanghai: Tongji University Press, 2006: 281-291(in Chinese)

[2] Lu Ye, Li Guo-qiang, Sun Fei-fei. "Experimental and theoretical study on slim I-shape buckling-restrained steel plate shear walls". *CHINA CIVIL ENGINEERING JOURNAL*, 2011, 44(10):45-52(in Chinese)

[3] Li Guo-qiang, Liu Wen-yang, Lu Ye, etc. "Mechanical behavior and equivalent brace model of buckling restrained steel plate shear wall with two-side connections". *Journal of Building Structures*, submitted(in Chinese)

CYCLIC BEHAVIOR OF REPLACEABLE STEEL COUPLING BEAMS

Xiaodong Ji[a], Yandong Wang[a], Qifeng Ma[a], Jiaru Qian[a]

[a] Department of Civil Engineering, Tsinghua University, Beijing, P. R. China
jixd@ mail. tsinghua. edu. cn, jltxwyd008@ 163. com,
maqf@ foxmail. com, qianjr@ tsinghua. edu. cn

KEYWORDS: replaceable steel coupling beam, seismic behavior, connection, replaceability, inelastic rotation capacity

ABSTRACT

For improving the resilience capacity, a replaceable steel coupling beam is developed, which comprises a central "fuse" shear link connected to normal steel segments at its two ends. This paper presents a series of quasi-static tests used to examine seismic behavior of the replaceable steel coupling beams with various types of connections between shear link and normal segments.

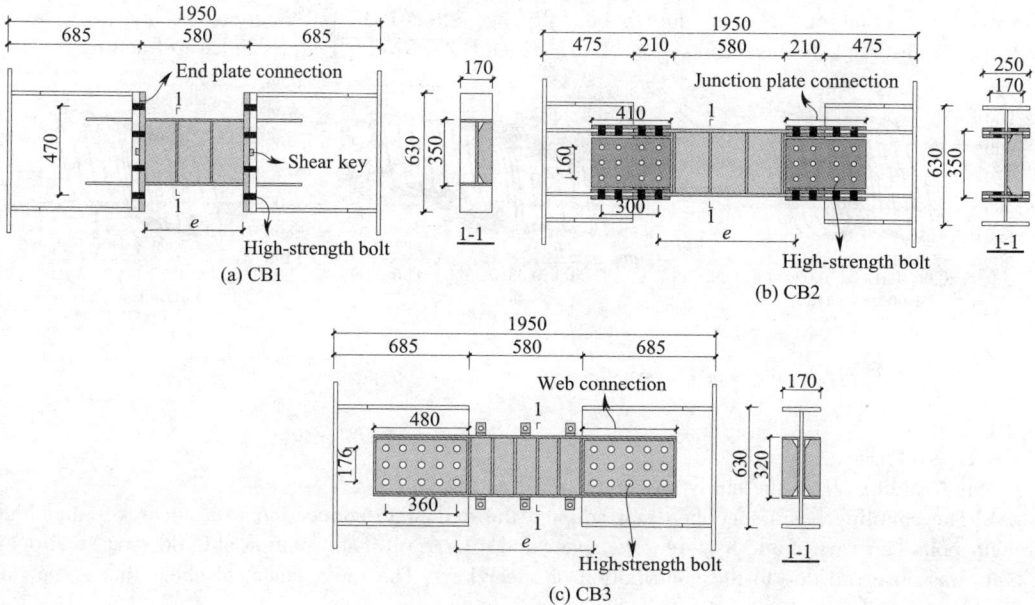

Fig. 1. Test specimens

A total of three specimens, with various types of connections, were designed and tested (see *Fig. 1*). The normal segments of all specimens were I-shaped steel. The shear links of Specimens CB1 and CB2 were built-up I-shapes, while that of CB3 was double channel-shapes. Specimens CB1 adopted end plate connection, where a shear key transfers the shear force and high-strength bolts transfer the bending moment. Specimen CB2 used junction plate connection. For Specimen CB3, the double channel steels were connected to the webs of normal segments by high-strength bolts. All links were designed to yield in shear. The strengths of normal segments and connections were designed to be higher than the maximum shear strength of shear link.

Fig. 2 shows the test setup and instrumentation. In Phase I loading, the specimen was loaded to 0. 02 rad rotation, approximately the demand for coupling beams under the maximum considered earthquake. Afterwards, the shear link was replaced. In Phase II loading, the specimen with reinstalled shear link was loaded till failure.

In Phase I loading, "fuse" links yielded in shear for all specimens. In Phase II loading, Specimen CB1 showed very stable hysteretic loops under large inelastic rotation (see *Fig. 3*). Specimens CB2 and

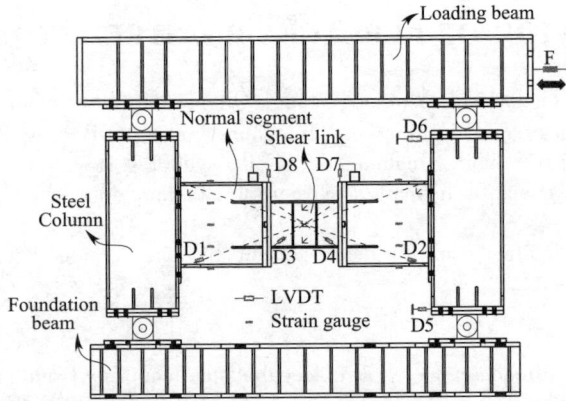

Fig. 2. Test setup and instrumentation

CB3 exhibited pinching behavior due to bolt slippage after 0. 03 rad rotation. Two types of failure modes were observed in shear links, i. e. , flange-to-end plate weld fracture and web fracture.

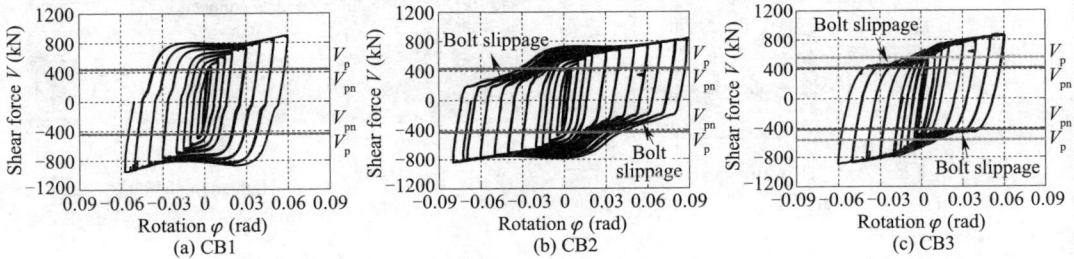

Fig. 3. Hysteretic loops of shear force versus rotation of specimens in Phase II loading

CONCLUSIONS

Major findings from the study are summarized as follows:

1) The coupling beam specimen that adopted the end plate connection with shear key and high-strength bolts had very stable hysteretic behavior and a large ultimate rotation of 0. 06 rad. No slippage of bolts was observed due to the contribution of shear keys. The replacement of shear link spent only 0. 4 h by two workers.

2) The specimens that adopted the junction plate connection and web connection also had an ultimate rotation no less than 0. 06 rad. However, slippage of high-strength bolts occurred in the connections during tests. The time for replacing shear links for these two specimens was five times that of the specimen with the end plate connection.

3) The I-shaped shear links with a length ratio of 0. 7 had an inelastic rotation capacity of approximately 0. 15 rad and an overstrength factor of 2. 0. The double channel-shaped shear link with a length ratio of 1. 24 had an inelastic rotation capacity of 0. 12 rad, and an overstrength factor of 1. 5.

4) The coupling beam that adopted the end plate connection with shear key and high-strength bolts was found to be superior in both hysteresis behaviour and rapid replacement.

REFERENCES

[1] Fortney P J, Shahrooz B M, Rassati G A. , "Large-scale testing of a replaceable "fuse" steel coupling beam", *Journal of Structural Engineering*, 133(12):1801-1807, 2007.

LATERAL BUCKLING BEHAVIOR OF WIDE-FLANGE BEAMS WITH CONCRETE FLOOR SLAB SUBJECTED TO CYCLIC BENDING MOMENT: Part 1 Experiment

Yuji Koetaka[a], Haruna Iga[a], Jun Iyama[b] and Takashi Hasegawa[c]

[a] Kyoto University, Dept. of Architecture and Architectural Engineering, Japan
koetaka@ archi. kyoto-u. ac. jp, iga. haruna. 35w@ st. kyoto-u. ac. jp
[b] The University of Tokyo, Dept. of Architecture, Japan
iyama@ arch. t. u-tokyo. ac. jp
[c] Building Research Institute, Dept. of Structural Engineering, Japan
hase@ kenken. go. jp

KEYWORDS: Wide-flange beam, Lateral-torsional buckling, Concrete floor slab, Loading test, Plastic deformation capacity.

ABSTRACT

In terms of rationalizing the structural design practice, effect of lateral support by RC slab to a beam has been recently gathering attention. It is expected to reduce the number of lateral stiffeners, simplify the connection details, and reduce the construction cost by means of evaluating the effect of the RC slab. In order to verify the effect of RC slab on lateral-torsional buckling of the wide-flange beam, cyclic loading test was conducted in our research. This paper presents test results, e. g. hysteresis loop, out-of-plane deformation of beam, maximum strength and plastic deformation capacity, by means of varying length of the beam, bending moment distribution, and constraint of axial deformation of the beam. And prediction method of maximum strength and plastic deformation capacity of steel beams with RC slab is proposed based on test results.

Test specimen, as illustrated in *Fig. 1*, is a half scale single-bay plane frame subassembly, whose span is from 6. 0 to 9. 5 meters. Test parameters are bending moment distribution of the beam, length of the beam, with or without RC slab, constraint of axial deformation of the beam, consequently the number of specimen is 10. Loading is controlled by story drift angle, and each amplitude increasing gradually is repeated in two times.

Fig. 2 shows typical examples of relationship between bending moment and rotation angle at the end of the beam. Compressive stress is acted on the lower flange in positive loading of the hysteresis loops. Bending moment degraded and the slope of hysteresis loop in elastic range decreased while out-of-plane deformation of the beam increased due to lateral-torsional buckling after reaching the maximum strength. The amplitude at the maximum strength of the specimen with RC slab is larger than that without RC slab.

In order to evaluate the maximum strength and the plastic deformation capacity of wide-flange beams with RC slab, normalized slenderness ratio is modified and is applied to the existing evaluation method without RC slab[1]. For calculating the normalized slenderness ratio of the beam with RC slab, elastic lateral-torsional buckling moment[2] of steel beams with the continuous complete restraint upper flange is adopted. And 0. 5 times of beam length is applied for the buckling length in consideration with restraint about weak axis at both ends of the beam. In *Fig. 3*, comparison between evaluation (lines) and test results (plots) of plastic deformation capacity is shown. All test results are close to the evaluation results, and the estimation about the increment of plastic deformation capacity by RC slab has reasonable accuracy from a practical standpoint.

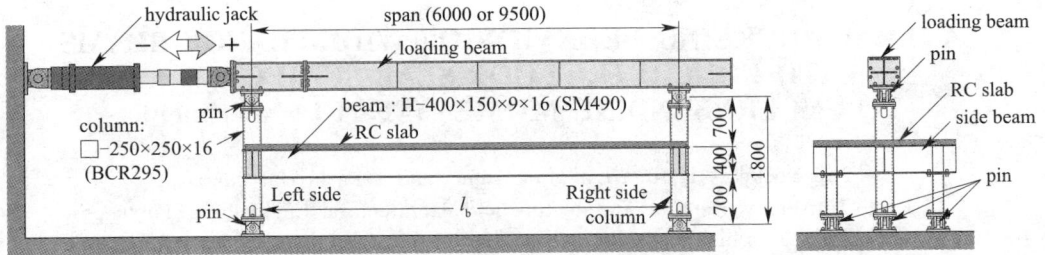

Fig. 1. Test specimen and set up

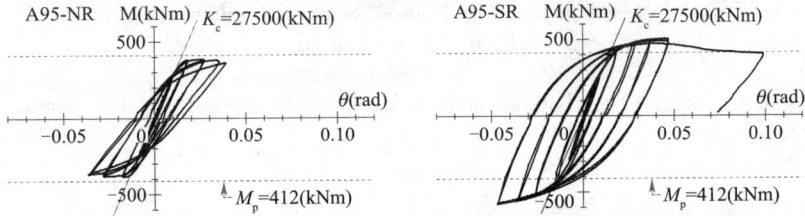

Fig. 2. Relationship between bending moment and rotation at the beam end

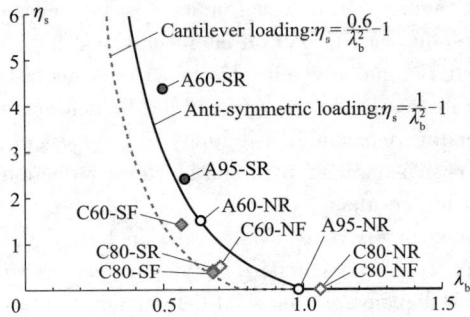

Fig. 3. Evaluation of plastic deformation capacity

CONCLUSIONS

From cyclic loading test results, it is clarified that hysteresis loop can be stable and the maximum strength and the plastic deformation capacity increase by means of connecting RC slab to the beam. The effects of beam length and bending moment distribution on the plastic deformation capacity of the beams with RC slab indicate a similar tendency of the beams without RC slab by previous researches. The difference of mechanical behavior between axial deformation of the beam restrained and released is remarkable in case of the beam without RC slab, but is quite little in case of the beam with RC slab.

Furthermore, the prediction method of the maximum strength and the plastic deformation capacity of the steel beams with RC slab was proposed by means of modification of normalized slenderness ratio. According to the comparison between the evaluated results and test results, it is verified that the prediction method can be utilized for practical structural design.

REFERENCES

[1] Architectural Institute of Japan, *Recommendation for Limit Design of Steel Structures*, 2010. 2
[2] Kikuo Ikarashi and Yuki Ohnishi, "Elastic buckling strength of H-shaped beams with continuous complete restraint on upper flange", *Journal of Structural and Construction Engineering*, Vol. 79, No. 706, pp. 1899-1908, 2014. 11 (in Japanese)

EXPERIMENTAL STUDY ON THE TORSIONAL RESTRAIN EFFECT OF THE CONCRETE SLAB TO IMPROVE DUCTILITY OF H-SHAPED STEEL BEAMS SUBJECTED TO BENDING MOMENT

Tsuyoshi Koyama[a], Jun Iyama[a], Satoru Inamoto[a], Yuka Matsumoto[b] and Tomoki Tamura[b]
[a]the University of Tokyo, Department of Architecture,
Graduate School of Engineering, Japan
koyama@ arch. t. u-tokyo. ac. jp, iyama@ arch. t. u-tokyo. ac. jp,
inamoto@ stahl. arch. t. u-tokyo. ac. jp
[b] Yokohama National University, Department of Architecture and Building Science,
College of Engineering, Japan
yk-mtsmt@ ynu. ac. jp, tamura-tomoki-tp@ ynu. ac. jp

KEYWORDS: STEEL COMPOSITE BEAM, CONCRETE FLOOR SLAB, TORSIONAL RESTRAINT, LATERAL BUCKLING, FINITE ELEMENT ANALYSIS.

ABSTRACT

To attain sufficient plastic ductility of wide flange steel beams, the current design code requires lateral supports to prevent early lateral-torsional buckling of the beam. Such supports complicate construction, leading to an increase in cost. The reinforced concrete(RC) slab constructed on top of steel wide flange beams has recently been gathering attention as a structural element capable of restraining the lateral-torsional buckling of the beam by constraining the motion of the top flange[1]. This paper especially focuses on the torsional constraint that the RC slab enforces on the motion of the attached wide flange steel beam.

Bending tests of the cantilever wide flange steel beams without RC slab(S0b) and with RC slab (S800b), shown in *Fig. 1*, are reported. To properly assess the interaction between the concrete slab, headed stud, and steel beam, full scale experiments are conducted. The beam-slab interaction is simplified, by placing a gap between the concrete slab and fixed end to remove any internal compressive force transfer, so that force from the slab is transmitted only through the headed studs and slab-beam flange contact in the form of a torsional restraint. A displacement is enforced at the tip of beam, which is free to rotate in any direction but constrained laterally, so that the beam bends around the strong axis of the beam section. The load-tip displacement curve of the specimens are shown in *Fig. 2*, where S800b is loaded so that the slab is in tension. The results show that the slab suppresses the lateral-torsional buckling of the beam, increasing the deformation at maximum load to 3 times that of a beam without a slab. The restraining effect can also be seen in the relationship between the tip vertical displacement and lower flange lateral displacement at 2/3 the length of the beam from the fixed end.

Two types of slab interaction models, shown in *Fig. 3* a), are proposed to understand the effect of the slab on the beam and simulate itsbehavior in FEA. In one model, the slab is modeled as an elastic shell(M800S) and the other as an equivalent beam(M800B). The method in which the slab-beam interaction is modeled is also different. The load displacement curves obtained from the simulation (*Fig. 3* b))show that the shell model shows a ductility larger than the simulation of the steel beam (M0), but smaller than the experimental results.

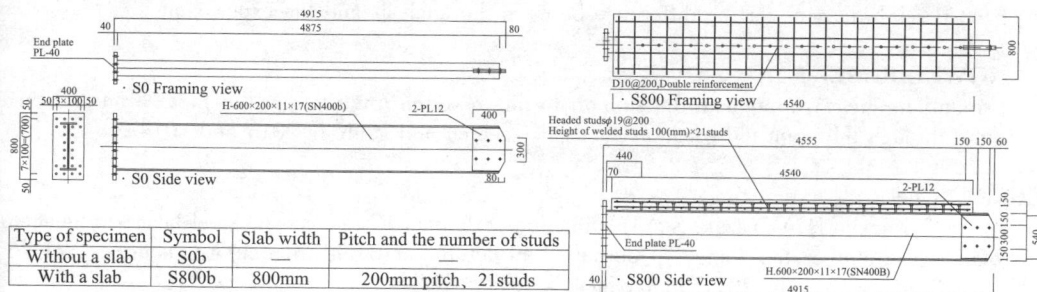

Type of specimen	Symbol	Slab width	Pitch and the number of studs
Without a slab	S0b	—	—
With a slab	S800b	800mm	200mm pitch, 21studs

Fig. 1. Test specimen

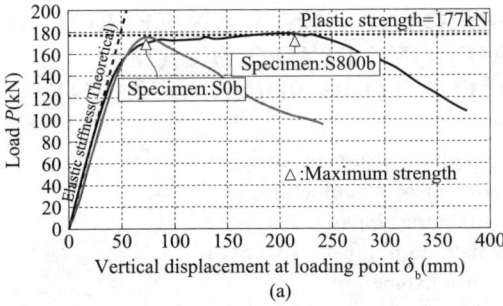

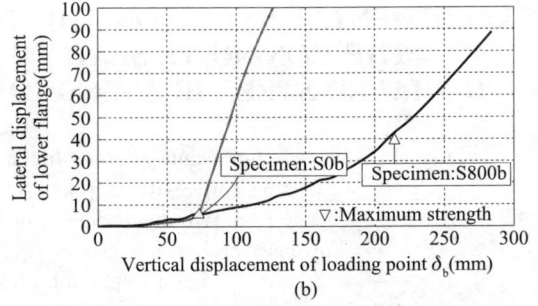

Fig. 2.　(a) Load displacement curves of experiment
(b) Relationship between tip vertical displacement and lower flange lateral displacement at 2/3L

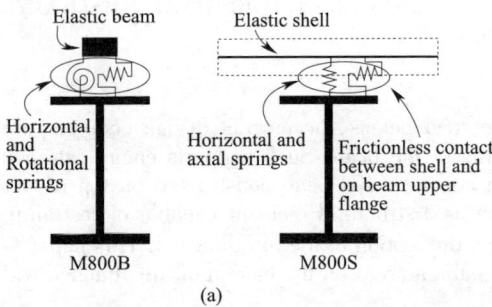

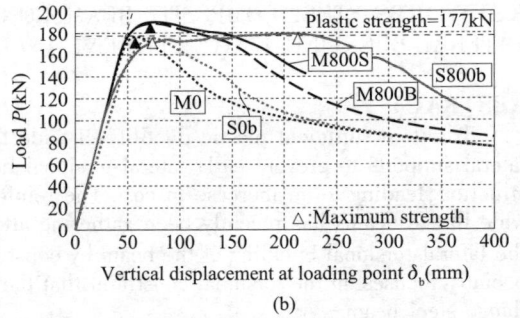

Fig. 3.　(a) Slab interaction model　(b) Load displacement curves of experiment of simulation

CONCLUSIONS

Bending experiments on cantilever wide flange steel beams with and without a RC slab are conducted to understand the effect of the slab intorsionally restraining the lateral-torsional buckling of the beam. The test results show that the specimen with the RC slab exhibits similar maximum strength to that of the specimen without, but this strength is attained at a displacement approximately 3 times the yield deformation. The lateral displacement and rotation of the beam is greatly suppressed by the slab. The strain gauge history of the headed studs show that the torsional restraint is most active near the fixed end.

In order to understand thebehavior of the slab torsional restraint, two finite element models are proposed. The one in which the slab is modeled with shell elements and the slab-interaction behavior through axial and shear springs, and contact is able to better simulate the behavior compared to the one in which the slab is modeled with beam elements and the slab-interaction behavior with axial, shear, and rotational springs. Though the model shows improved ductile behavior compared to the case of a beam without a slab, it is still not able to reproduce the extreme ductile behavior observed in the experiments. The strain distribution in the headed studs measured in the experiments implies that there may be significatnt composite beam effect, which was ignored in the finite element model. This would be one of the possible reasons for the difference between the analysis and the experiment

ACKNOWLEDGMENT

The authors greatly appreciate the support to this research project from the Japan Iron and Steel Federation through the grant-aid program for steel research and study in 2013 and 2014.

REFERENCES

[1] Usami T, Kaneko H, Yamazaki K, "Deformation performance of composite steel beam: influence of rotation restraint of top flange", Journal of Structural and Construction Engineering, Vol. 76, No. 668, pp. 1847-1854, 2011 (in Japanese).

SUBASSEMBLAGE TESTING OF ALL-STEEL WEB-RESTRAINED BRACES

Johnn Judd[a], Adam Phillips[a], Matthew Eatherton[a],
Finley Charney[a], Igor Marinovic[b], Clifton Hyder[b]
[a]Department of Civil and Environmental Engineering,
Virginia Tech, Blacksburg, Virginia, USA
jjohnn11@ vt. edu, arp12@ vt. edu, meather@ vt. edu, fcharney@ vt. edu
[b]BlueScope Buildings North America, Inc. , Memphis, Tennessee, USA
igor. marinovic@ bluescopebuildingsna. com, skip. hyder@ bluescopebuildingsna. com

KEYWORDS: Seismic resistance, subassemblage tests, energy dissipation, buckling restrained braces.

ABSTRACT

This paper discusses cyclic testing of two full-scale web-restrained brace frame subassemblages. The web-restrained brace(WRB) is an all steel buckling restrained brace(Fig. 1). A brace-and-column test apparatus was used to simulate the combined axial and rotational deformations expected in a braced-frame structure. The WRB sub-assemblages were subjected to the qualification loading sequence stipulated in the AISC Seismic Provisions[1] to evaluate the hysteretic behavior of the WRB.

CONCLUSIONS

Both specimens exhibited stable and repeatable load-displacement behavior with no degradation of the cyclic envelope(Fig. 2). They possessed high initial stiffness, predictable yield strength, and full cyclic behaviour. Both specimens successfully survived the AISC 341-10 displacement protocol without rupture, brace instability, brace end connection failure, or any other undesirable limit states. As such, both specimens satisfied the BRB subassemblage acceptance criteria described in AISC 341-10

For each cycle corresponding to a deformation greater than the yield displacement, Δ_{by}, the maximum compression forces were less than 1. 3 times the maximum tension forces for both specimens. This demonstrates that the WRB controls compression overstrength to be comparable with typical BRBs. Managing the over-strength results in smaller design forces for the boundary elements which can produce more economical member selection. Except during the first tension cycle for Specimen 1, the maximum tension and compression forces in the brace were greater than the nominal strength of the

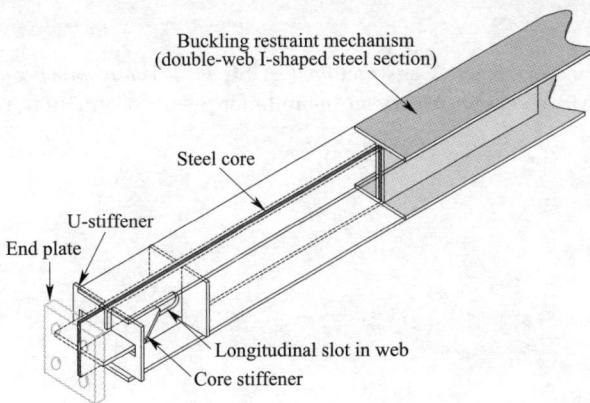

Fig. 1. Primary components of all-steel web-restrained brace(WRB)

core. Furthermore, Specimen 2 successfully underwent a cumulative inelastic axial deformation of 436 times the yield deformation without rupture, brace instability, brace end connection failure, or any other undesirable limit states. Specimen 2, therefore, also satisfied the acceptance criteria in AISC 341-10 for a BRB brace test. The WRB exhibited significant ductility without any performance degradation, which makes it an ideal seismic lateral load resisting system option.

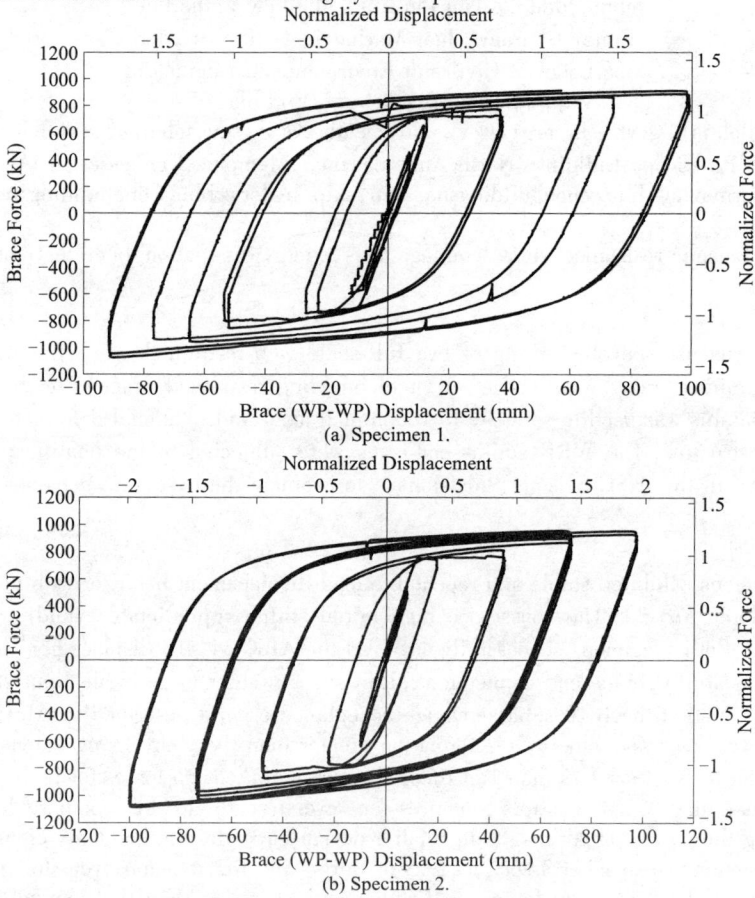

(a) Specimen 1.

(b) Specimen 2.

Fig 2. Brace axial force versus axial displacement

REFERENCES

[1] American Institute for Steel Construction (AISC). *Seismic provisions for structural steel buildings*. ANSI/ AISC 341-10, American Institute for Steel Construction, Chicago, Illinois; 2010.

EFFECTIVENESS OF BUCKLING RESTRAINED BRACES ON AN INDUSTRIAL STRUCTURE

Ricardo A. Herrera[a], Karina Santelices[b]

[a]University of Chile, Department of Civil Engineering, Chile
riherrer@ ing. uchile. cl
[b] Arcadis Chile, Chile
karinasantelices@ gmail. com

KEYWORDS: Buckling restrained brace, Energy dissipation, Industrial building, Structural steel.

ABSTRACT

Frame (CBF). This type of frame has advantages over other structural systems in terms of stiffness and strength. However, the performance under cyclic loads of the braces is affected by the occurrence of buckling in compression, followed shortly after by fracture of the brace during the tension part of the cycle. The performance objectives sought by current industrial design codes result in structures with large brace sizes required to limit damage. In this context, buckling restrained braces (BRBs) appear as an attractive alternative because they have reduced sizes as compared to conventional braces without affecting the structural performance, increase the energy dissipation capacity of the structure, and can be more easily replaced after an extraordinary event.

This work presents a comparison of the performance of an industrial building (shown in *Fig.* 1) with conventional braces and the same structure with BRBs.

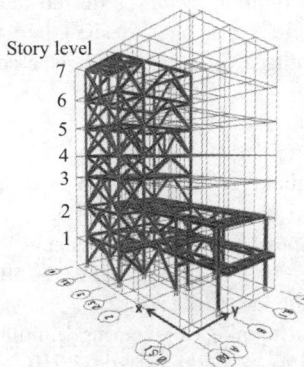

Fig. 1. Industrial mill building

Initially, the conventional braces are replaced by BRBs with similar strength in compression. Both configurations are modeled considering geometric and material nonlinearities and the resultant models are subjected to lateral pushover and time-history analyses with actual ground motion records. The following results are compared: maximum story drift, maximum story shear, damage on structural elements, residual deformations and energy dissipated. In addition, the response modification factor (R) overstrength factor (Ω), and displacement amplification factor (C_d) are back calculated and compared to the design values recommended by the code. *Fig.* 2 shows representative results of the static pushover analyses.

The structure with BRBs has a lower elastic stiffness and yield strength, but larger ductility and a smooth load-displacement response than the conventional CBF.

Regarding the time-history analysis results, under the application of the unscaled records the structure behaved elastically. An incremental dynamic analysis was conducted to try to determine the ultimate state of the BRBF structure. The elastic response was then obtained for the amplified records and the seismic response parameters were calculated. It can be seen that the nonlinear displacement do not grow unbounded and that the base shear is always equal or smaller than the base shear for the linear elastic structure.

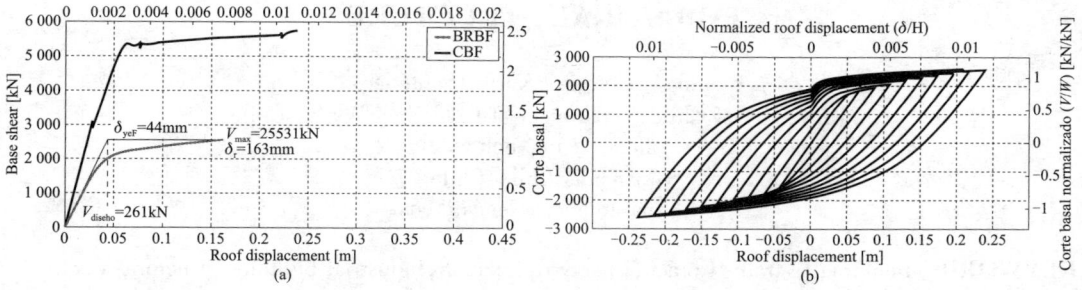

Fig. 2. (a) Monotonic static pushover results; (b) Cyclic pushover results

CONCLUSIONS

The structure with BRBs achieved a better performance than the conventional CBF in terms of stability of the hysteresis cycles and ductility. Most of the energy dissipation is done by the BRBs, keeping the other structural members nearly undamaged. However, the use of BRBs does not prevent the formation of soft stories in braced frames.

The response modification factors are significantly different for the two directions of analysis considered, which can be explained by the differences in structural configuration between both directions. An average value of 9 is obtained from the analyses, three times larger than the design value used, and even larger if the effective reduction factors, required to comply with the minimum base shear requirement, are considered. The structure has been designed to remain nearly elastic for the design level event. For this reason, the advantages of using BRBs over conventional braces, although visible, cannot be fully appreciated.

REFERENCES

[1] INN, "NCh2369. Of2003: Earthquake-resistant design of industrial structures and facilities", Instituto Nacional de Normalización, Santiago, Chile, 2003.
[2] Astica G, "Evaluación de las disposiciones de diseño sísmico para marcos arriostrados en edificios industriales". Thesis to obtain the title of Civil Engineer, College of Physical Sciences and Mathematics, Universidad de Chile, Santiago, Chile, 2012.
[3] OpenSees, "Open System for Earthquake Engineering Simulation". Pacific Earthquake Engineering Research Center. http://opensees. berkeley. edu, 2010.
[4] Takeuchi T, Hajjar JF, Matsui R, Nishimoto K, Aiken ID, "Effect of local buckling core plate restraint in buckling restrained braces", Journal of Constructional Steel Research, 66: 139-149, 2010.
[5] Ariyaratana C, Fahnestock LA, "Evaluation of buckling-restrained braced frame seismic performance considering reserve strength", Engineering Structures, 33: 77-89, 2011.
[6] Fermandois G, "Marcos con riostras de pandeo restringido: Comportamiento y factores de modificación de la respuesta sísmica". Thesis to obtain the degree of M. Sc. in Civil Engineering, Universidad Técnica Federico Santa María, Valparaíso, Chile, 2009.
[7] FEMA P695, "Quantification of Building Seismic Performance Factors". Federal Emergency Management Agency, Washington, DC, 2009.
[8] ASCE, "Minimum Design Loads for Buildings and Other Structures", ASCE Standard ASCE/SEI 7-10, American Society of Civil Engineers, Reston, Virginia, 2010.

LATERAL BUCKLING BEHAVIOR OF WIDE-FLANGE BEAMS WITH CONCRETE FLOOR SLAB SUBJECTED TO CYCLIC BENDING MOMENT: Part 2 Finite element analysis

Jun IYAMA[a], Yuji KOETAKA[b] and Takashi HASEGAWA[c]

[a] The University of Tokyo
iyama@ arch. t. u-tokyo. ac. jp

[b] Kyoto University
koetaka@ archi. kyoto-u. ac. jp

[c] Building Research Institute
hase@ kenken. go. jp

KEYWORDS: Lateral-torsional buckling, Concrete floor slab, Finite element analysis, Plastic deformation capacity.

ABSTRACT

In Part 1, the test results of bending test of wide flange steel beam with and without reinforced concrete(RC) slab was presented, where the specimens with RC slab showed larger ductility compared to those without RC slab. This is considered to be due to the effect of slab to restrain out-of-plane horizontal displacement and torsional deformation of the steel beam. In Part 2, finite element analysis is performed using a model considering these effect of RC slab. Through the comparison between the analysis result from this model and the test result, the accuracy of the model is validated, in terms of the maximum strength, plastic deformation capacity, and out-of-plane displacement of the beam.

The overview of the finite element model is shown in *Fig. 1*. The model shape is the same as the test specimen described in Part 1. The wide-flange steel beam and the rectangular hollow section column are modeled with shell element. The RC slab is modeled by an elastic wire element, and the shear connectors are modeled with horizontal and rotational springs. The top and bottom of each column are connected with a rigid link element, and the rigid links at both ends of the beam are rotated to apply bending moment to the beam.

Ten models are analyzed using ABAQUS 6.12. As an example, the moment-rotation skeleton curves of A60-NR(no RC slab) model and A60-SR(with RC slab) model obtained by the analysis and the test are shown in *Fig. 2*. It is shown that the FEA result agrees with the test result with a good accuracy and that the skeleton curve of A60-SR shows a significant improvement in terms of plastic deformation capacity compared to A60-NR.

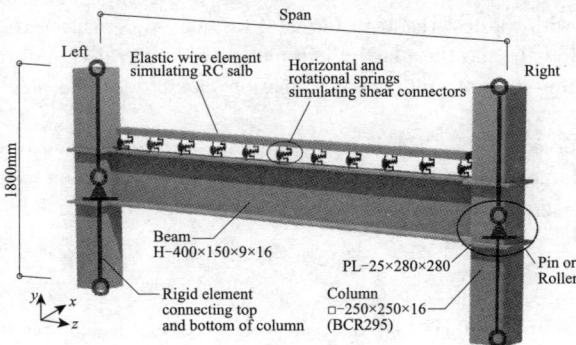

Fig. 1. Overview of finite element model

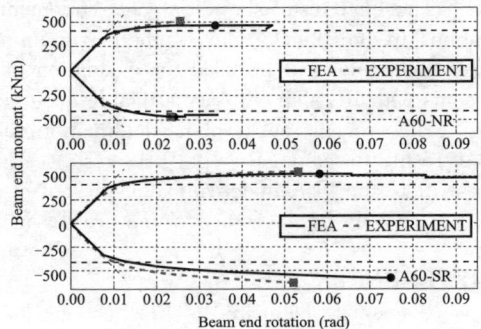

Figure 2. Skeleton curves

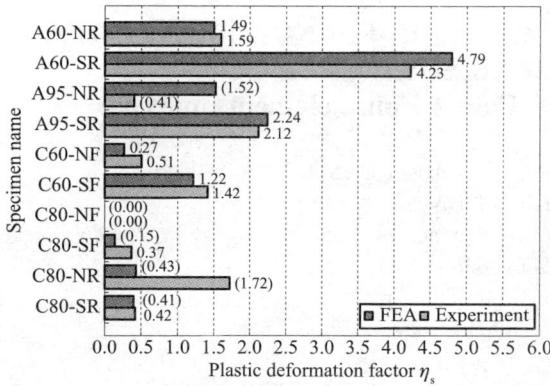

Fig. 3. Comparison of plastic
deformation capacity from FEA and test

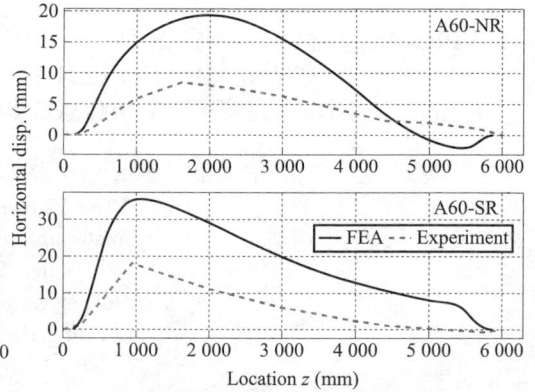

Figure 4. *Out-of-plane displacement of
bottom flange at the maximum load*

The plastic deformation capacity obtained by the FEA result and the test result is summarized in *Fig. 3*. In this graph, the FEA result and the test result shows a good agreements, implying that this model could properly represent the effect of RC slab and shear connectors for evaluation of plastic deformation capacity of the steel beam with RC slab.

The out-of-plane deformation of the beam from the FEA is also compared to the test result. As an example, the out-of-plane displacement distributions of the bottom flange of A60-NR and A60-SR, at the time that the load reaches the maximum, are shown in *Fig. 4*. The rough shape of the out-of-plane deformation distribution is similar to the test result, however, significant errors in the magnitude of the displacement are observed. This means that it is still difficult to simulate the complete behavior of steel beams with RC slab, and that it is necessary to improve the model.

CONCLUSIONS

Finite element analysis(FEA) is performed to validate the results of bending test of wide-flange steel beams with and without reinforced concrete(RC) floor slab reported in Part 1. The load displacement relationships, distributions of out-of plane horizontal and torsional deformations are compared to those obtained from the tests.

The finite element model used in this study consists of a wide flange steel beam modeled with shell element, a wire element simulating the slab bending and torsional stiffness, and the lateral and rotational springs simulating shear connectors connecting the steel beam and the wire. In this model, the wire element and spring elements are supposed to be elastic. The residual stress or strain in the steel beam is not considered.

The FEA result showed a good agreement with the test result in terms of the load-deformation relationship, the skeleton curves, the maximum strength, and the plastic deformation capacity. It is shown that this model can be used for evaluation of the strength and the deformation capacity of the steel beams with RC slab with a reasonable accuracy.

However, the out-of-plane deformation observed in the FEA had a significant difference from the test result. In many cases, the FEA results are larger than the test results, however, there are some cases where the FEA underestimated the out-of-plane deformation. In order to obtain more accurate results, it would be necessary to improve the model.

REFERENCES

[1] Ikarashi K, Ohnishi Y, "Elastic buckling strength of H-shaped beams with continuous complete restraint on upper flange", *Journal of Structural and Construction Engineering*, 79(706), 1899-1908, 2014. 11(in Japanese)

APPLICATION OF COUPLED SHEAR WALLS WITH BUCKLING-RESTRAINED STEEL PLATES IN HIGH-RISE BUILDINGS

Guo-Qiang LI[a], Hai-Jiang WANG[b], Xiao-Kun HUANG[c]

[a] State Key Laboratory of Disaster Reduction in Civil Engineering, China
gqli@ tongji. edu. cn

[b] College of Civil Engineering, Tongji University, Shanghai, 200092, China
haijiang_wang@ 163. com

[c] China Academy of Building Research, Beijing, 100013, China
huangxiaokun@ cabrtech. com

KEYWORDS: Coupled shear wall, Buckling-restrained steel plate, Energy-dissipation element, High-rise building, Seismic design.

ABSTRACT

Concrete wall structures are widely used in high-rise building designbecause they could provide desirable lateral load resistant stiffness, since drift control is very important for high-rise buildings. However, cantilever shear walls tend to fail in shear or the boundary elements of bottom walls possibly suffer severe damages under seismic loads, both of which are apparently undesirable. Shear walls coupled with concrete coupling beams, which have advantages over traditional cantilever shear walls, also have some drawbacks. For example, the key elements in the structural system-concrete coupling beams are lack of sufficient ductility and energy-dissipation capacities in most cases. Besides, they inevitably suffer damages in earthquakes and most of which are impossible to be fully repaired. Compared to cantilever shear walls, they often possess decreased structural stiffness with identical wall thickness. This will without doubt bring adverse effect to the drift control of tall buildings.

In this research paper, a state of the art structural system- coupled shear wall with buckling-restrained steel plates(CSW-BRSPs) is proposed for the engineering community. In this system, a pair of concrete walls is inter-connected by the steel plates, which are equally distributed along the structural height. Obviously, under lateral loads, the steel plates will suffer large in-plane shears. Characteristic of large height-to-thickness ratio, they will experience out-of-plane buckling. In order to restrain the undesirable buckling, precast concrete plates are provided on each side of the thin steel plate. Also, a sufficient gap is provided between the concrete bracing plates and the wall boundaries. This ensures that the concrete plates will not bear against the boundary elements when the system deforms in lateral loads. The infilled steel plates are welded to the steel columns in wall boundary elements through fin plates. The configuration of this system is shown in *Fig. 1*.

Since out-of-plane buckling is effectively restrained by the concrete plates, the thin steel plates are able to provide superior energy-dissipationcapacities. Their hysteretic loops are always statured. Furthermore, compared to cantilever shear walls, the stiffness of CSW-BRSPs are able to decrease in a limited scope. This is probably the predominate reason why this proposed system is especially applicable in high-rise buildings. Besides, the buckling-restrained steel plates could be readily repaired or replaced if they were damaged in earthquakes.

A typical wall-frame structure isproposed and extensively studied in order to compare the seismic behaviour of ordinary cantilever shear walls and the above-mentioned CSW-BRSPs system. Elastic analysis results show that with proper design parameters(the steel plate thickness, width and height) the CSW-BRSPs is able to possess similar stiffness with original cantilever shear walls. Based on the elasto-plastic time-history analysis, the proposed CSW-BRSPs system exhibits superior seismic performance, such as reduced bases shear and story drift ratio as well as relieved concrete damage at wall bottom.

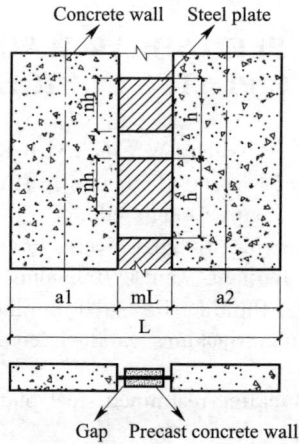

Fig. 1. Configuration of CSW-BRSPs system

CONCLUSIONS

A state-of-the-art structural system, in which shear walls are coupled with buckling-restrained steel plates, is proposed to increase the structural seismic behaviour. Compared to reinforced concrete cantilever walls, the proposed coupled shear walls with buckling-restrained steel plates are able to have similar lateral load resisting stiffness, which is beneficial for the drift control.

Besides, their seismic behaviour is extremely excellent under severe earthquakes. Because of the existence of energy-dissipation elements-buckling-restrained steel plates, they will experience decreased seismic responses. This kind of structure is undoubtedly beneficial for the engineering community in the seismic design.

REFERENCES

[1] Wallace, J. W. February 27, 2010 Chile Earthquake: Preliminary Observations on Structural Performance and Implications for U. S. Building Codes and Standards. *Structures Congress.* 1672-1685, 2011.

[2] Telleen, K. , Maffei, J. , Willford, M. etc. Lessons for Concrete Wall Design from the 2010 Maule Chile Earthquake. Proceeding of the International Symposium on Engineering Lessons Learned from the 2011 Great East Japan Earthquake, Tokyo, Japan. 1766-1777, 2012.

[3] Astaneh-Asl, A. , "Seismic Behavior and Design of Composite Steel Plate Shear Walls", Steel TIPs Report, Structural Steel Educational Council, Moraga, CA, 2002

[4] Lu, Y. , Li, G. Q. , "Slim Buckling-Restrained Steel Plate Shear Wall and Simplified Model", *Advanced Steel Construction*, 8(3): 282, 2012.

NUMERICAL INVESTIGATION ON THE EFFECT OF AXIAL FORCE TO THE BEHAVIOUR OF COMPOSITE STEEL CONCRETE SHEAR WALLS

Daniel Dan[a], Alexandru Fabian[a], Valeriu Stoian[a]

[a]Politehnica University of Timisoara, Department of Civil Engineering, Romania

daniel. dan@ upt. ro, alexfabian10@ yahoo. com, valeriu. stoian@ upt. ro

KEYWORDS: Composite steel-concrete walls, Axial force ratio, Strength, Ductility, Seismic behaviour.

ABSTRACT

The seismic resistant structures developed in the last decade and new solutions for structural elements, as composite steel reinforced concrete walls (CSRCW) were proposed. The solution is used in the construction of medium and high-rise buildings due to the easy technology and high structural performances. The existing theoretical and experimental studies have shown a different behaviour of elements function of the normalized axial force ratio. Using the results obtained in the experimental program performed at thePolitehnica University of Timisoara, consisting in tests on 1: 3 scale steel-concrete composite elements, the paper presents a numerical investigation on the influence of axial force to the dissipative behaviour of the elements. Finally, few issues for the design of CSRCW walls are discussed related to the strength, stiffness and deformation capacity of the walls.

CONCLUSIONS

The effective flexural stiffness of the elements is a modelling parameter that has a significant effect on the system response in structural analysis. The value of effective flexural stiffness recommended to be used in structural analysis is smaller than the gross flexural stiffness, which reflects the influence of concrete cracking and bond slippage. For RC walls, the Romanian code CR2-1-1. 1/2013 specifies the effective flexural stiffness EI_{eff} as 0. 8EI_g for axial force ratio 0. 40 and 0. 4EI_g for axial force ratio near to 0, where EI_g denotes the gross flexural stiffness of the wall section. According to Ji et al. [17], different values are used for the upper and lower bounds when evaluating the flexural stiffness, which takes into account the effect of axial force ratio. In Fig. 1 is presented the effective flexural stiffness of the CSRCW wall specimens normalized with the gross flexural stiffness. The effective flexural stiffness of the CSRCW wall specimens was calculated assuming the wall behaves as cantilever. The value of effective flexural stiffness for CSRCW walls significantly increases with the increase of the axial force ratio.

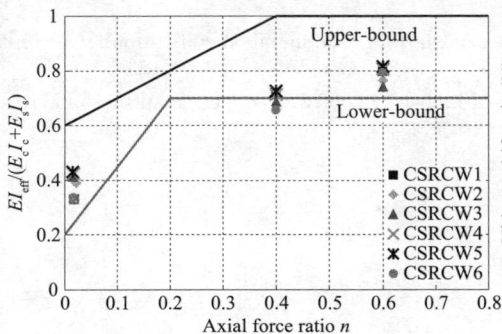

Fig. 1. Effective flexural stiffness versus axial force ratio.

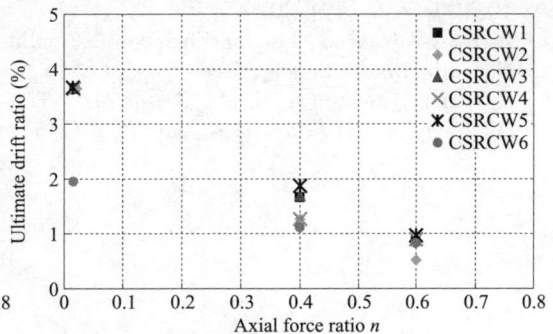

Fig. 2. Ultimate drift ratio versus axial force ratio.

Fig. 2 summarizes the analysis results for the ultimate drift ratios of CSRCW walls under various axial force ratios. The CSRCW walls under low axial force ratios had an ultimate drift ratio of around 3. 6%, while the walls under high axial force ratio had an ultimate drift ratio of approximately 0. 8%.

For moderate axial force ratios the ultimate drift values are in the range of 1.3 and 1.9. EC8 requires that the inelastic deformation capacity to be at least 1.1% for composite structural wall systems.

The paper presents a series of comparisons between the results obtained from experimental tests and numerical analysis performed in order to examine the seismic behaviour of CSRCW walls under low axial force ratios. Cyclic numerical analysis with different axial force ratios were performed to observe the influence on the seismic behaviour of CSRCW. The following conclusions and design recommendations can be drawn within the limitation of the current research:

(1) The CSRCW wall specimens shown increased ultimate strength and deformation capacity compared with the RC wall for normalized axial force smaller than 0.4;

(2) The CSRCW wall specimens shown greater ultimate lateral drift when compared with the RC wall. The CSRCW walls with axial force ratios 0.4 had an ultimate lateral drift ratio of approximately 1.3% ~ 1.9% while for higher value 0.6, the ultimate lateral drift ratio decreases under 1%.

(3) The numerical analysis results provided a relatively accurate estimation of the flexural strength of the CSRCW walls.

(4) The effective flexural stiffness of CSRCW walls is highly dependent on the applied axial force ratio.

(5) The recommendation for effective flexural stiffness of RC walls suggested by Adebar et al. [20] appeared to be appropriate for use in calculation of the effective flexural stiffness of CSRCW walls under small and high axial force ratios, being slightly different for medium normalized axial force ratio.

ACKNOWLEDGMENTS

The presented work was supported by CNCSIS-UEFISCSU project number PNII-IDEI ID_1004, Contract 621/2009, entitled, Innovative Structural Systems Using Steel-Concrete Composite Materials andFiber Reinforced Polymer Composites".

REFERENCES

[1] Dan D., Fabian A., Stoian V. "Theoretical and experimental study on composite steel-concrete shear walls with vertical steel encased profiles". *Journal of Constructional Steel Research*, 67:800-813,2011.

[2] Fabian A., Stoian V., Dan D. "Theoretical and experimental studies on composite steel-concrete structural shear walls with steel encased profiles". *Proceedings of ICSA* 2013 *Structures and architecture, Concepts, Applications and Challenges*, 445-446. 2013.

[3] Liao F. Y, Han L. H., Tao Z, "Performance of reinforced concrete shear walls with steel reinforced concrete boundary columns", *Engineering structures*, 44:186-209,2012.

[4] Ji X., Sun Y., Qian J., Lu X. "Seismic behaviour and modelling of steel reinforced concrete (SRC) walls", *The Journal of the International Association for Earthquake Engineering*, DOI: 10.1002/eqe. 2494,2014.

[5] ATENA-Advanced Tool for Engineering Nonlinear Analysis, Technical specifications, Cervenka Consulting Ltd., Prague, Czech Republic.

[6] Adebar P., Ibrahim A. M. M., Bryson M., "Test of high-rise core wall: effective stiffness for seismic analysis", *ACI Structural Journal*, 104(5):549-559,2007.

RETROFIT ANALYSIS AND DESIGN OF BUILT-UP STEEL COLUMNS

Zhichao Lai[a], Amit H. Varma[b] and Robert J. Connor[c]

[a] Ph. D. , Postdoctoral Research Engineer, School of Civil Engineering, Purdue University, USA
laiz@ purdue. edu

[b] Professor, School of Civil Engineering, Purdue University, USA
ahvarma@ purdue. edu

[c] Associate Professor, School of Civil Engineering, Purdue University, USA
rconnor@ purdue. edu

KEYWORDS: Retrofit; Strengthening; Built-up; Columns; Design.

ABSTRACT

Existing structures sometimes require retrofit or strengthening to satisfy increased strength demands, or to repair damage after extreme events such as earthquakes and explosions. There are several retrofit methods available. The retrofit of steel structures may include the replacement of existing components or the addition of new components. For example, chord members in girders or steel-arch bridges are usually retrofitted by adding cover plates to the flanges or webs of the existing section. This paper presents the analytical investigation of the behavior and strength of retrofitted built-up steel columns using two analysis approaches: (i) 3D finite element analysis, and (ii) nonlinear inelastic column buckling(NICB) analysis. The finite element analysis approach is first used to evaluate the fundamental behavior of the retrofitted columns. The NICB approach is then developed and benchmarked using the results from the finite element analysis. This paper also presents the development and verification of the design equation for estimating the axial compressive strength of built-up steel columns. The design equation is developed using the results from comprehensive parametric studies conducted using the benchmarked NICB analysis program, and verified by comparing the strengths estimating using the developed equation with that obtained from the parametric studies.

CONCLUSIONS

Built-up steel columns were strengthened in the weak axis by adding cover plates to the webs(as shown in *Fig. 1*). To investigate the fundamental behavior of these retrofitted members, comprehensive finite element models were developed. Results from the finite element analysis indicated that: (i) the failure mode of the short column is entire cross-section compression yielding, with existing section yields first, followed by the yielding of the cover plates; (ii) the failure mode of the slender column is flexural buckling in the weak axis, with existing section yielded first.

A nonlinear inelastic column buckling(NICB) analysis program was also developed and benchmarked. Comprehensive parametric analyses were conducted using the benchmarked NICB analysis program. Results from the parametric analyses indicated that the beneficial effect by adding cover plates could be represented by the effects of the following four factors: locked-in dead load ratio(P_{LD}/P_{ori}), relative strength ratio F_y^{cpl}/F_y^{ori}, relative stiffness ratio I_{cpl}/I_{ori} and column slenderness ratio KL/r.

Using the results from the parametric studies, the design equation ready for inclusion into the AASHTO Bridge Specifications to calculate the axial compressive strength(P_n) of retrofitted built-up steel columns was developed. The axial compressive strength(P_n) was estimated by modifying the axial strength predicted using the AASHTO[4] design equation(*Equation 1*) with a modification factor β_r, as shown in *Equation 2*. This modification factor(β_r) included the effects of P_{LD}/P_{ori}, F_y^{cpl}/F_y^{ori}, I_{cpl}/I_{ori}, and KL/r. The proposed design equation can conservatively evaluate the axial compressive strength of retrofitted members.

$$\text{When } \frac{P_e}{P_y^{ori}} \geqslant 0.44 \qquad P_{AASHTO} = [\,0.658^{(P_y/P_e)}\,]P_y \qquad (1)$$

When $\dfrac{P_e}{P_y^{ori}} < 0.44$ $P_{AASHTO} = 0.877 P_e$ (2)

Where, $P_e = \pi^2 EI / (KL)^2$ (3)

$$P_n = \beta_r P_{AASHTO} \tag{4}$$

$$\beta_r = 1.36 + 0.036\left(\frac{I_{cpl}}{I_{ori}} - 10\frac{P_{LD}}{P_{ori}} - 4.5\frac{F_{cpl}^y}{F_{ori}^y} - 0.02\frac{KL}{r}\right) \leqslant 1.0 \tag{5}$$

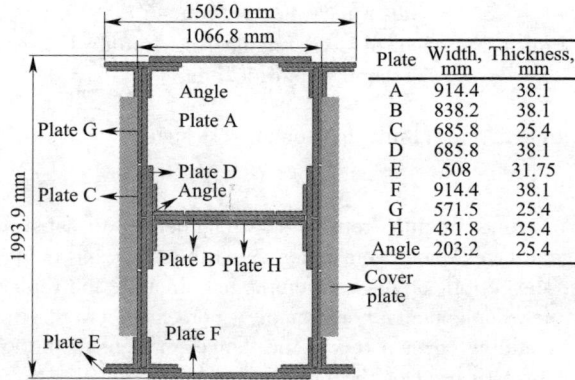

Plate	Width, mm	Thickness, mm
A	914.4	38.1
B	838.2	38.1
C	685.8	25.4
D	685.8	38.1
E	508	31.75
F	914.4	38.1
G	571.5	25.4
H	431.8	25.4
Angle	203.2	25.4

Fig. 1. Cross-section details of the prototype built-up double box member

REFERENCES

[1] Lai Z, Varma AH, Connor RJ. , "Retrofit analysis and design of built-up steel columns: modeling and fundamental behavior", ASCE *Journal of Bridge Engineering* 2014(Revision submitted).

[2] ABAQUS. *ABAQUS Version* 6.12 *Analysis User's Manuals.* Providence, RI, USA: Dassault Systemes Simulia Corporation;2012.

[3] Lai Z, Varma AH, Zhang K. , "Noncompact and slender rectangular CFT members: experimental database, analysis, and design", *Journal of Constructional Steel Research* 2014;101:455-68.

[4] AASHTO. *LRFD bridge design specifications.* Washington, DC. , USA: American Association of State Highway and Transportation Officials, AASHTO;2012.

[5] Lai Z, Varma AH, Connor RJ, Liu J. , "Retrofit analysis of steel built-up members for bottom chords of Bayonne Bridge", *Bowen Laboratory Research Report*(*BLRR*) No. 2013-02. Purdue University, 2013.

[6] Chen WF, Lui EM. *Structural stability.* Upper Saddle River, NJ. ; Prentice Hall;1987.

ANALYSIS ON SECONDARY STRESSES ON MAIN MATERIAL OF NARROW BASE TUBULAR TRANSMISSION TOWER

Bin Huang[a], Hong-Zhou Deng[a], Yun Wu[b], and Tian-You Li[c]

[a]Tongji University, College of Civil Engineering, China
89690966@ qq. com

[b] Fujian Yong Fu Project Consultant Co. , Ltd, China.
wuy@ fjyongfu. com

[c]State Grid Fujian Electric Project Consultant Co. , Ltd, China.

KEYWORDS: Narrow base tubular Transmission Tower; Secondary Stress; Slenderness Ratio of Main Member; Angle Between Main Member and Diagonal Member

ABSTRACT

Transmission towers are usually assumed that the joints are pinned. But when the section of member is large, the rigidity of joint is relatively significant that will arise secondary stresses. Many scholars have done some research on secondary stresses of transmission tower structures, and point out that, when the secondary stress is larger, it can not completely ignore. In this paper secondary stresses on main member of narrow base tubular transmission towers are calculated. Based on tubular towers with different slender ratio and angle between main member and diagonal member, the impact of which on the secondary stresses on main members of tower leg is researched. The result indicates that, the maximum secondary stress on main leg members with a lattice is mainly controlled by the slenderness ratio, and decreases with the increasing of slenderness ratio. while the maximum secondary stress on the main leg member divided into two lattices mainly depends on the angle between main member and dignonal member. and decreases with increasing of the angle. With the same angle, the secondary stesses are larger when the tower leg with two lattice.

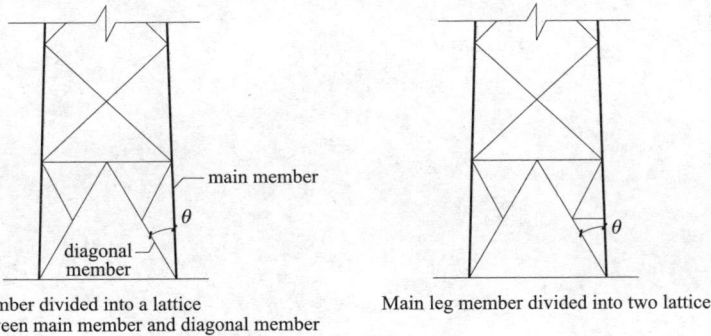

Main leg member divided into a lattice
θ:Angle between main member and diagonal member

Main leg member divided into two lattice

Fig. 1. Tower Leg Schematic Diagram

CONCLUSIONS

The following items summarize the conclusions reached as a result of the studies described in this paper.

1. The larger secondary stresses on main members mainly located at the height of the cross arm or tower diaphragm.

2. The secondary stresses of main leg members divided into two lattices are governed by the angle between main member and diagonal member, and, the secondary stresses of the leg main member increase as the angle decreasing.

3. When the main leg member with a lattice, the secondary stresses depend on the slender ratio of main leg member. the secondary stress as the slenderness ratio increasing more obviously reduces.

4. Under the same utilization rate of the cross section and the same angle between main member and diagonal member, the secondary stress of main member of tower leg divided into two lattice is lar-

ger than that of tower leg with a lattice.

REFERENCES

[1] Ministry of Construction of the People's Republic of China, "Code for Design of Steel Structures (GB50017—2003)" ,2003.

[2] Ze-Yi Dan, Xi-YuanZhao "Secondary Stresses Distribution of Steel Trusses and Design Suggestion" ,Industrial Building,14(1):30-37,1984.

[3] Jun-ke Han, Jing-boYang. "Value selection of slenderness ratio and diameter thickness ratio of steel tube for 1000KV transmission steel tubular tower leg" ,Power System Technology,33(19): 17-20,2009.

[4] Li MaoHua, YangJingbo. "Analysis on secon-dary stress for steeltube tower of 1000 kV double circuit transmission lines on same tower" ,Power System Technology,34(2):20-23,2010.

[5] Qun Shuai, "Analysis on secondary stress of primary member of UHV transmission tower structure" ,Journal of Building Structures,33(8):109-116,2012

[6] S. Roy, Shu-Jin Fang, "secondary stresses on transmission tower structures" ,J. Energy Eng,110: 157-172. 1984.

[7] G. M. S. Knight, A. R. Santhakumar, "Joint Effects on Behavior of Transmission Towers", J. Energy Eng,119:698-712. 1993.

[8] Jing Li, "Research on application of Narrow Base Tower in the City " ,SCIENCE & TECHNOLOGY INFORMATION,2:234-235,2009.

STUDIES ON AXIALLY COMPRESSED SRC COLUMN USING Q460 HIGH-STRENGTH STEEL

Su-Wen Chen[a,b], Pei Wu[a], Qing Liu[a], Zhao-Xin Hou[c], Lin-Bo Qiu[c]

[a]State Key Laboratory for Disaster Reduction in Civil Engineering, Shanghai, China
swchen@ tongji. edu. cn, 5pei@ tongji. edu. cn, 2011_liuqing@ tongji. edu. cn
[b]College of Civil Engineering, Tongji University, Shanghai, China
[c]Central Research Institute of Building and Construction Co. , Ltd. MCC Group, Beijing 100088
hzxlxl126@ vip. sina. com, nrcsc2007@ 126. com

KEYWORDS: Steel reinforced concrete column, High-strength steel, Axial compressive loading test, Numerical analysis.

ABSTRACT

High-strength steel reinforced concrete (HSRC) structure refers to Steel Reinforced Concrete (SRC) structure using high-strength steel. So far, the mechanical behaviours, ultimate bearing capacity and cyclic behaviour of SRC columns have been studied comprehensively[1]. However, research on the HSRC column is still limited.

This paper presents an experimental study on the mechanical performance of axially compressed HSRC columns with Q460 high-strength steel section. Two specimens SP1 and SP2 were made of C50 high-performance concrete and HRB400 reinforced bars containing cross-shaped steel sections with flanges(*Fig. 1.* (a)). The specimens were tested under axial compressive load. The axial and lateral displacement was monitored by LVDTs and the development of material strain was evaluated by strain gauges placed at the flanges of the steel section and the surface of the concrete.

Cracks firstly occurred at the top and bottom of the specimens and gradually extended to the entire specimens with the increasing of load. Finally, the columns failed because of the crushing of concrete(*Fig. 1.* (b))

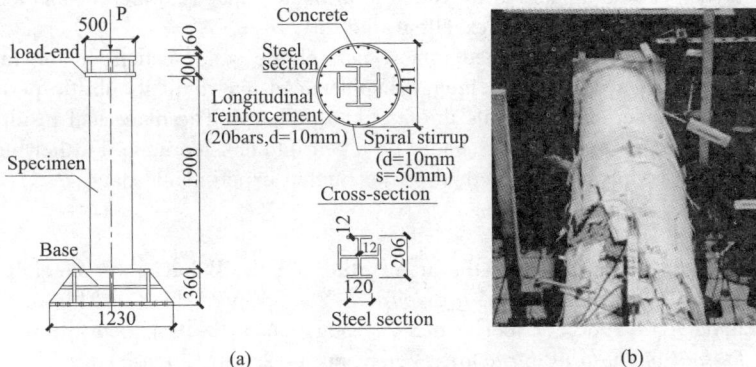

Fig. 1. (a) Dimensions and test setup; (b) Failure pattern of SP2

Fig. 2. (a) and *Fig. 2.* (b) show the axial and lateral Load-displacement curves of the specimens. The displacement was limited before the cracking of concrete. After the peak load, large deformation was observed, which indicated that the high-strength steel section enabled the columns to have excellent ductility under axial compression. The load-strain curves for steel and concrete were gained from the test. It is found that most part of the steel section yielded but the plastic deformation was not uniform in the mid-height section.

A numerical model has been established for simulatingthe experiments based on ABAQUS. An unconfined concrete model[2] is applied for the concrete cover and the Kent-Scott-Park constitutive relation[3] is used for concrete core to consider the confined effect from spiral stirrup reinforcement. The predicted peak load for the model is 11390kN, which fits the test data perfectly. The load-displacement

131

curve is gained from the numerical model and agrees well with the experimental result (see *Table 1*, *Fig. 2.* (c)).

Table 1. Test and numerical results

Specimens	First yied load P_y (kN)	Axial displacement δ_y (mm)	Ultimate load P_u (kN)	Axial displacement δ_u (mm)	Ductility factor μ ($=\delta_u/\delta_y$)
SP1	9907	5.31	10890	39.93	7.52
SP2	9323	4.94	11363	33.75	6.83
M1	10823	4.88	11391	–	–

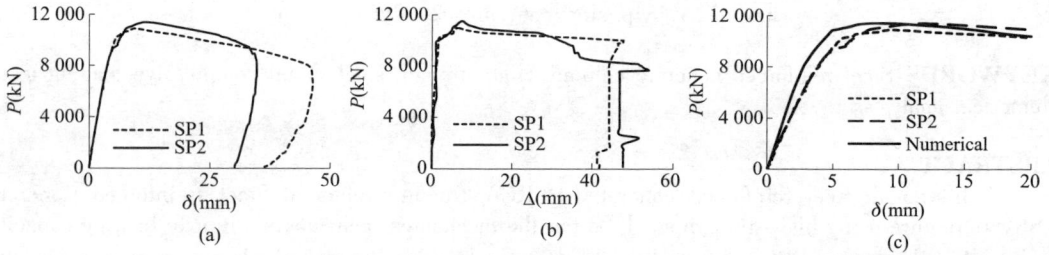

Fig. 2. (a) Axial load-displacement curves; (b) Lateral load-displacement curves; (c) Numerical result

CONCLUSIONS

In the presented study, two steel reinforced concrete columns are tested with Q460 high-strength steel and C50 high-strength concrete to analyse its mechanical behaviour under axial compressive load. A numerical model is established adopting Kent-Scott-Park constitutive relation model for confined concrete to simulate the experiments and predict the bearing capacity of the specimens. The results of the experimental and numerical studies are summarized as follows:

1. The experimental result shows that there are great increases on lateral and axial deformations but with little decrease of loading after it exceeds the ultimate capacity, which means the high-strength steel section enables the column to have excellent ductility.

2. Large plasticity has been developed inmost part of the steel section, so the high strength of Q460 steel can be well utilized. However, lateral displacement may limit its plastic performance.

3. The non-linear numerical model fits the test data perfectly. The numerical result supports that the confinement effect from spiral stirrups should be taken into consideration. Further numerical calculation can be used as a supplementary method of the limited experimental data.

REFERENCES

[1] Ye LP., Fang EH. ,2000. "State-of-the-art of Study on the Behaviors of Steel Reinforced Concrete Structure". *China Civil Engineering Journal*, Vol. 33, pp. 1-11. (in Chinese)
[2] Guo ZH. ,Zhang XQ. ,1982. "Experimental Investigation of the Complete Stress-Strain Curve of Concrete". *Journal of Building Structures*, Vol. 3, pp. 1-12. (in Chinese)
[3] Scott BD. ,Park R. ,Priestley MJN. ,1982. "Stress-strain behavior of concrete confined by overlapping hoops at low and high strain rates". *ACI Journal*, Vol. 79, pp. 13-27.

THE BUCKLING-RESTRAINED BRACE WITH HIGH FATIGUE PERFORMANCE

Kazuhisa Koyano[a], Shuichi Koide[a], Kazuaki Miyagawa[b], Mamoru Iwata[a]

[a] Department of Architecture and Building Engineering,
Kanagawa University, Yokohama, Japan
koyano@ kanagawa-u. ac. jp, r201470136ay@ kanagawa-u. ac. jp, iwata@ kanagawa-u. ac. jp
[b] JFE Civil Engineering and Construction Corporation, Tokyo, Japan
miyagawa@ jfe-civil. com

KEYWORDS: Buckling-restrained brace, Fatigue performance, Energy absorption performance, Cumulative plastic strain energy ratio, High strain amplitude domain.

ABSTRACT

A great variety of earthquakes such as the long period and long duration are predicted in the future. A high-performance buckling-restrained brace that is effective for those earthquakes is necessary. The authors have studied the buckling-restrained brace that has high-energy absorption capacities even in the high strain amplitude domain. However, not only the energy absorption capacities but also the fatigue performance is important. Therefore the authors confirm whether the buckling-restrained brace that had high-energy absorption capacities would become high-performance in the fatigue life. As a result, this buckling-restrained brace shows the high fatigue performance at the high strain amplitude domain.

TEST OVERVIEW

The shape of the specimens is illustrated in Fig. 1. Four high-performance type specimens with decreased core plate plastic zone area and a spacer(H05, H20, H30, and H40) and three basic types of specimens with non-decreased core plate plastic zone area(B05, B15, and B25) are used. The numbers assigned to the specimens indicate strain amplitude. Using the H series specimens, loading is conducted at strain amplitudes of 0. 5%, 2%, 3%, and 4%. For B series specimens, loading is conducted at strain amplitudes of 0. 5%, 1. 5%, and 2. 5%. First, preconditioning loading is conducted to fit the specimen and jigs together, after which loading is repeated at the predetermined strain amplitudes until the specimen's strength drops to 80% of its maximum strength or tensile fracture occurs. The number of loading cycles immediately before these points is regarded as the number of loading cycles that can be withstood(called "tolerable loading cycles" hereafter).

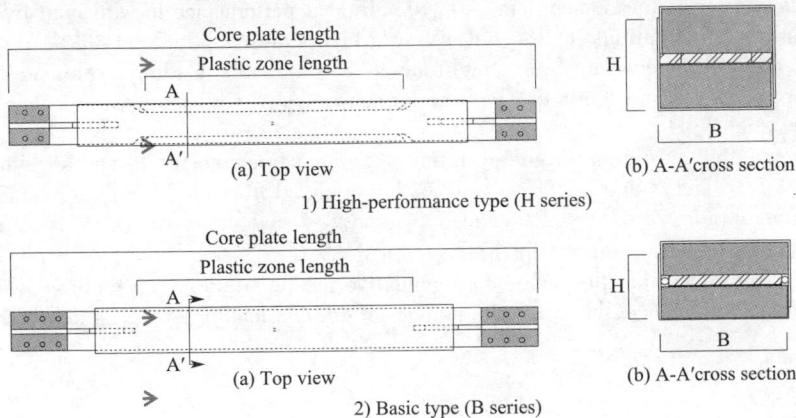

Fig. 1. Specimen shape

PERFORMANCE EVALUATION

The relationship between strain amplitude ε and tolerable loading cycles is plotted in *Fig. 2* with an approximate curve. The durable number of loading cycles of the B series specimens is 522 at a strain amplitude of 0.5% ,44 at a strain amplitude of 1.5% , and 11 at a strain amplitude of 2.5% . For the H series specimens, the tolerable loading cycles is 924 at a strain amplitude of 0.5% ,30 at a strain amplitude of 2% ,14 at a strain amplitude of 3% , and 5 at a strain amplitude of 4%. The above set of data verifies that the buckling-restrained brace developed in this study provides excellent tolerable loading cycle performance and fatigue performance even at high strain amplitudes. For both the H series and B series specimens, strain amplitude ε and tolerable loading cycles have a linear relationship on a double logarithmic graph, from which the fatigue performance of the buckling-restrained brace can be estimated.

The relationship between the cumulative plastic strain energy ratio ω and Restraining index R ($= P_E/P_y$, P_E : Euler's buckling load of restraining material, P_y : yield load of core plate) obtained from the past study results (incremental axial strain loading max. 3%) is plotted in *Fig. 3* along with the test results obtained from this study. The past study[1] proposed a performance evaluation formula ($\omega = 150 \times R (R \leqslant 6)$, $\omega = 900 (R > 6)$) obtained from the ω-R relationship. The ω of Specimens B05, B15, B25, H05, H20 and H30 is 3850,1654,1138,6442,1899, and 1541, respectively. These values greatly exceed those obtained from the performance evaluation formula. On the other hand, the ω value of Specimen H40 is 855, which is slightly below the value derived from the performance evaluation formula.

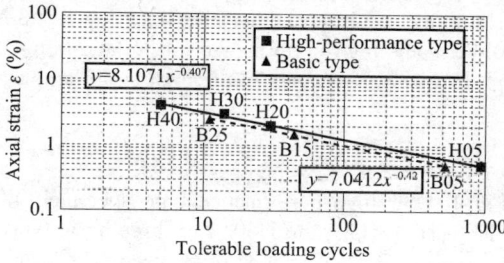

Fig. 2. Strain amplitude ε-tolerable loading cycles relationship

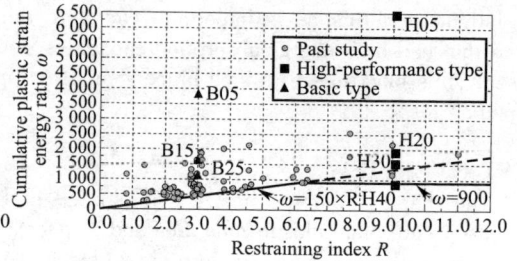

Fig. 3. Cumulative plastic strain energy ratio ω-restraining index R relationship

CONCLUSIONS

Focusing on the buckling-restrained brace using steel mortar planks, a constant strain amplitude cyclic loading test was conducted at each strain amplitude, using high-performance type and basic-type specimens. The following results were obtained:

1) The buckling-restrained brace demonstrated sufficient performance to withstand five loading cycles not only at a strain amplitude of 3% but also at a high strain amplitude of 4%.

2) The relationship between strain amplitude ε and tolerable loading cycles was plotted and showed an approximate curve. Both the high-performance type and basic-type specimens showed a nearly linear relationship.

3) The cumulative plastic strain energy ratio ω obtained from the basic-type specimens at strain amplitudes of 0.5% ,1.5% and 2.5% as well as that obtained from the high-performance type specimens at strain amplitudes of 0.5% ,2% , and 3% greatly exceeded the values derived from the performance evaluation formula proposed in the past study[1].

4) The test confirmed that the larger the cumulative plastic strain energy ratio ω, which indicates energy absorption performance, the larger the number of durable loading cycles, which indicates fatigue performance.

REFERENCES

[1] Iizuka, R. , Koyano, K. , Midorikawa, M. , and Iwata, M. : Study on Buckling-restrained Braces having Large Cumulative Plastic Strain Energy Ratio, Journal of Structural and Construction Engineering (Transaction of AIJ) , Volume 79, No. 701 , pp1015-1023, July 2014.

PLASTIC DEFORMATION CAPACITY OF RHS COLUMN WITH WELD DEFECTS

Masayuki Takakura[a], Tsuyoshi Tanaka[a], Hayato Asada[a], Ryo Ueta[b]

[a]Kobe University, Department of Architecture, Japan

t. masayuki25@ gmail. com, tanaka@ arch. kobe-u. ac. jp, asada@ people. kobe-u. ac. jp

[b] Okumura Corporation, Japan

ryo. ueta@ okumuragumi. jp

KEYWORDS: Welded Column Connection, Weld Defect, Loading Test, Finite Element Analysis, Plastic Deformation Capacity.

ABSTRACT

It is commonly understood that seismic design has provided steel structures with sufficient seismic performance. However, there are cases in which improperly applied welds can become the starting points of fractures that may lead to extensive damage to steel building structures as observed in 1995 Kobe and 1994 Northridge earthquake. Current standard[1] for inspection of weld defects in Japan is based on the experimental works without explicitly consideration for stress and strain condition of welded connection of members(such as column and beam). In the case of a rectangular hollow section (RHS) column, weld defects tend to occur in the corner part of complete joint penetration(CJP) groove weld which correspond to the strain concentration point under symmetrical bi-axial lateral loading. Thus, it should be discussed the effects of weld defects on plastic deformation capacity of welded column connections considering the stress condition and position of weld defects. This paper presents cyclic loading tests of cantilever RHS columns and Finite element analysis to assess the effects of the weld defect size such that lack of fusion, its position in CJP and loading direction on plastic deformation capacities of columns.

CONCLUSIONS

Loading tests consisted of the two series were conducted with the specimens inserted the artificial defects. In series 1, six specimens are tested to discuss the effects of loading direction and the position of weld defect on plastic deformation capacity. In series 2, four specimens are tested to discuss mainly the effect of weld defect size.

The plastic deformation capacity(η) until fracture occurrence or 10% degradation from maximum strength and the failure modes of all specimens are shown in *Fig. 1* (a).

As a result, Presence of weld defects resulted in notable reduction of the plastic deformation capacity, when subjected to lateral load in 45 degrees direction, compared to that in 0 degrees direction. Also, weld defects exist in outer layer of CJP compared to that in inner layer induce crack initiation resulting in more significant reduction of plastic deformation capacity compared to weld defects exist in inner layer of CJP. However, weld defects in inner layer having more than height of 10mm induce fracture initiated from weld defect and lead to significant reduction of plastic deformation capacity

To evaluate the reduction of plastic deformation capacity induced by weld defect in finite element analysis, Crack tip opening displacement(CTOD) criterion was used. CTOD criterion was obtained by using 45-degrees method[2]. Regardless of loading directions, largest CTOD occurred in the middle of defect length.

Fig. 1 (b). shows the relationship between normalized plastic deformation capacity η/η_0 obtained from cyclic loading tests and CTOD at normalized plastic rotation angle of 4. 0(0. 04rad). Normalized plastic deformation capacity of each specimen with weld defect was obtained by dividing by that of non-defect specimen subjected to load in same direction. It can be observed the considerable relation between cumulative plastic deformation capacity until fracture and CTODs. Whereas when CTOD value is smaller than 0. 4mm at rotation angle of 4. 0θ_p, specimens failed by local buckling or fractured by crack initiation from weld toe at column side and Significant reduction of plastic deformation capacity did not occurred, when CTOD at $4.0\theta_p$ is larger than 0.7mm, the fracture initiated from weld defect and they lead to notable reduction of the plastic deformation capacity. Taking into account this result, it is concluded that CTOD criterion can be used for evaluation of plastic deformation capacity of welded connection with weld defect.

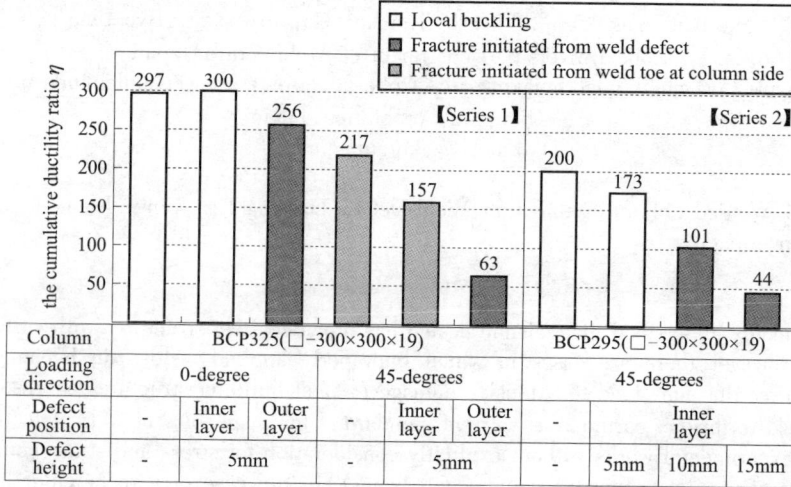

(a)

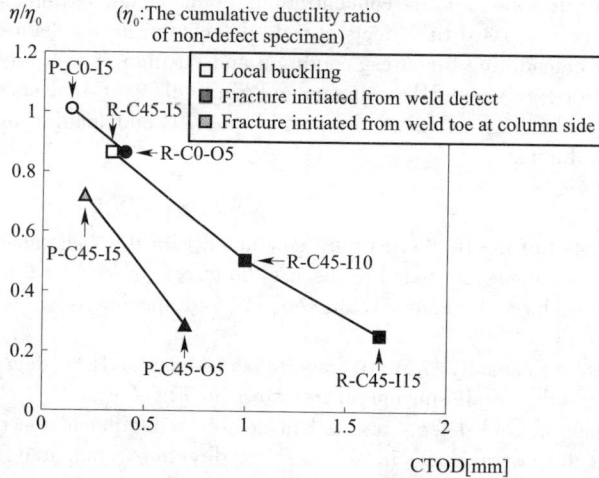

(b)

Fig. 1. (a) Plastic Deformation Capacity; (b) CTOD-Plastic Deformation Capacity

REFERENCES

[1] Architectural Institute of Japan, "Standard for the Ultrasonic Inspection of Weld Defects in Steel Structures", 2008.

[2] K. Azuma, T. Suzuki, R. Kanno and H. Simanuki, "Slit Tip Opening Behavior of Electoro-slag Weld Zone between Interior Diaphragm and Box Column on Steel Moment Frange", Journal of Structural and Construction Engineering (Transactions of Architectural Institute of Japan), vol. 573, pp. 161-168, 2003 (in Japanese)

DIFFERENT BRACING TYPES IN SEISMIC RESISTANT STRUCTURES

Marina Stoian[a], Helmuth Köber[a]

[a] Technical University of Civil Engineering Bucharest, Romania, Steel Structures
Department carinastoian@ yahoo. com,
e-mail: koberhelmuth@ yahoo. de

KEYWORDS: Centrically bracing, eccentrically bracing, buckling restrained bracing, dynamic nonlinear analyses, steel consumption.

ABSTRACT

The present paper is intended to illustrate some advantages and disadvantages of different structure types (centrically braced frames, eccentrically braced frames, buckling restrained braced frames), frequently used in the braced frames of buildings located in seismic areas.

Six different braced frames (as shown in *Fig. 1*) were sized for the same seismic action according to the provisions of EN 1998-1[1] and the Romanian seismic design code P100-1/2013[2]. The value of the q-factor was considered equal to 4 for all the analyzed bracing types. The analyzed frames had all two spans of 6. 6m and ten storeys of 3. 5m.

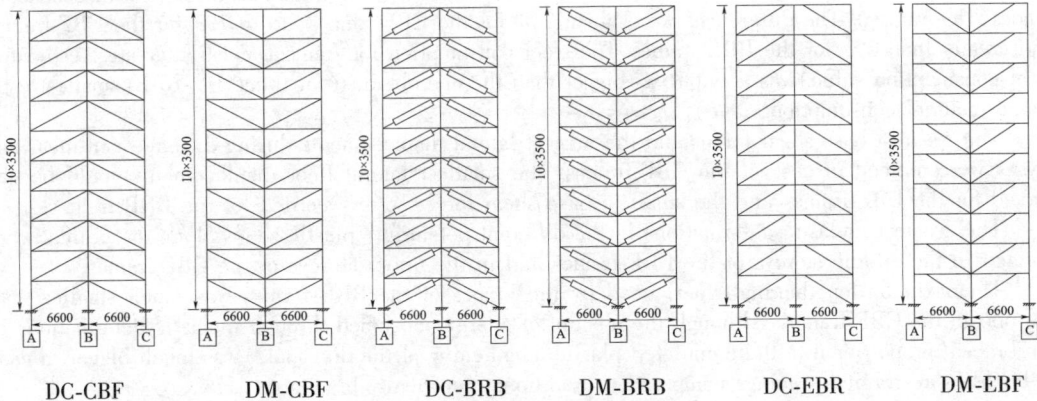

DC-CBF   DM-CBF   DC-BRB   DM-BRB   DC-EBR   DM-EBF

Fig. 1. Analyzed braced frames

A favourable global plastic failure mechanism was sized by design for each analyzed frame type. In case of the CBF-and BRB-frames, plastic deformations were accepted in the diagonals, at the bottom of first-story columns and in potentially plastic zones located near the ends of frame girders[3]. For the EBF-frames, inelastic deformations were accepted in the dissipative members and in the potentially plastic zones located near the bottom of first story columns and braces.

Dynamic nonlinear analyses were performed for each braced frame[4], considering the acceleration records of several Vrancea earthquakes (the N-S and E-W components of Vrancea earthquakes from 1977, 1986 and 1990). The acceleration records were calibrated to a peak ground acceleration value of approximately 0. 25 times the acceleration of gravity. Damping was taken into account using the Rayleigh procedure, considering mass and stiffness proportional damping factors[4].

CONCLUSIONS

In most cases during dynamic nonlinear analyses the greatest axial forces could be observed for the components of frame DM-CBF. Generally bigger axial forces werenoticed for the members of the frames with descendant braces (DM-frames) compared to one with ascendant braces (DC-frames). Only the maximum axial forces in the girders were quite in the same range for each pair of the analyzed bracing types. The values in the CBF-frames were up to 50% greater than those in the EBF-frames and about 32% bigger than the one recorded in the girders of the BRB-frames.

The smallest axial force values were recorded in the girders of frame DC-EBF(up to 55% smaller than the one in the girders of frame DM-CBF), in the lateral columns of frame DC-CBF(about 86% smaller values compared to the one in frame DM-CBF) and in the central columns of frame DC-EBF (over four time smaller values than for frame DM-CBF).

The greatest bending moments in the girdersappeared by far in frame DM-CBF, whilst the smallest bending moments values were noticed in the girders of the two EBF-frames, which are equipped with short dissipative members that work mainly in shear.

The potentially plastic zones along the girders increased the bending moment values in the columns located in their neighbourhood. The greatest bending moments appeared in the lateral columns of frame DM-EBF(for the lower storeys) and of frame DM-CBF(for the upper storeys). The smallest bending moments in the lateral columns were noticed for frame DC-EBF(about 50% smaller values compared to frame DM-EBF).

Except for the first story, the smallest bending moments in the central column were recordedfor frame DM-EBF. These values were over two times smaller than those recorded for frame DC-EBF.

Buckling of braces conducts to a sudden unloading of the compressed diagonals and leads to great values of bending moments and axial forces especially in the central columns and girders of thetwo CBF-frames.

The smallest estimated steel consumption was obtained for the BRB-frames, whilst the greatest values were noticed for the EBF-frames. For all considered bracing types greater steel consumptions were observed for the frames with descendant braces(DM-frames) compared to the one with ascendant braces(DC-frames). The differences were about 2% for the EBF-frames, up to 7% for the CBF-frames and greater than 6% for the BRB-frames. The steel consumption for frame DM-EBF(greatest estimated steel consumption value) was about 16% bigger than the one obtained for frame DC-BRB(smallest estimated material consumption value).

The greatest base shear forces and the biggest lateral displacements during dynamic nonlinear analyses were noticed in case of the CBF-frames. The smallest lateral floor displacements could be observed for the EBF-frames and the smallest base shear forces were recorded for the BRB-frames.

The greatest inelastic deformations in the different potentially plastic zones along the girders were noticed in the middle storeys of the EBF-frames and in the upper storeys of the CBF-frames.

Themaximum lengthening experienced by the braces of the BRB-frames was much smaller than the one in the CBF-frames. Although the amount of energy consumed through inelastic deformations in the braces(proportional to the cumulated plastic lengthening of the diagonals) was much bigger in case of buckling restrained bracings compared to traditional centrically bracings.

In most cases the plastic deformations in the diagonals and girders of the CBF-and BRB-frames were smaller(with about $4 \div 7$ %) for the DC-frames compared to the DM-frames. The plastic deformations in the dissipative members of frame DC-EBF and frame DM-EBF were in the same range along the height of the structures.

Taking into consideration the loading state, the maximum inelastic deformations noticed during dynamic nonlinear analyses and the estimated steel consumption, frame DC-BRB appear to be the most favourable from the six considered bracing types.

REFERENCES

[1] Eurocode 8, EN 1998-1:2004, Design of structures for earthquake resistance, Part 1: General rules, seismic actions and rules for buildings.
[2] Ministry of Transportations, Public Works and Territory Planning, 2013, Code for aseismic design-Part I-Design prescriptions for buildings, P100-1/2013, Bucharest, Romania.
[3] Köber, H. , Beţea, Ş. , 2008. An Alternative Method for the Design of Centrically Braced Frames, Proc. Eurosteel 2008-5th European Conference on Steel and Composite Structures; 03-05 September, 2008, Graz, Austria, Vol. B, pp. 1425-1430.
[4] Tsai K. C. ; Li J. W. , 1994, Drain2D + A General Purpose Computer program for Static and Dynamic Analyses of Inelastic 2D Structures.

DOG-BONE DETAILS IN SEISMIC RESISTANT STEEL STRUCTURES

Helmuth Köber[a], Bogdan Cătălin Ştefănescu[a]

[a] Technical University of Civil Engineering Bucharest, Romania, Steel Structures Department
koberhelmuth@ yahoo. de, bogdan@ utcb. ro

KEYWORDS: Dog bone details, potentially plastic zones, eccentrically braced frames, centrically braced frames, global plastic failure mechanism.

ABSTRACT

The present paper intends to illustrate some advantages of using "dog-bone" details in several seismic resistant steel structures equipped with different types of braced frames. All kind of structural elements of the analyzed centrically or eccentrically braced frames had built-up I-shaped cross-sections.

The braces cross-sections of the considered centrically braced frames had the web orientated normally to the plane of the frame, in order to avoid out of plane buckling

In a well-designed seismic resistant eccentrically braced frame, the greatest part of the energy induced in the structure by strong earthquakes is dissipated through plastic deformations concentrated in the links. To provide this, the links are sized for code specific lateral loads[1] and all the other structural members of the frames (columns, braces, beam segments outside the links) are designed for the forces generated by the fully yielded and strain hardened links[1].

Making the link to be the weakest element of the frame, the designer can force the yielding to occur in the ductile link elements while preventing inelastic deformations in members with a non-ductile behaviour. Generally this design concept conducts to smaller cross-sections for the links than for the beam segments near them. Especially in eccentrically braced frames with long links the smaller cross-sections of the links leads to smaller values for the lateral stiffness of structure.

On the other hand if the same cross-section is used for the dissipative members and for the beam segments outside the links, the lateral stiffness of the frame increases but the plastic deformations may occur in other members than links too (a favourable general failure mechanism would be difficult to obtain see *Fig. 2*). Many plastic hinges appeared in unwanted zones, mainly in the braces and beam segments outside the dissipative members, as shown in *Fig. 2*.

A configuration with reduced beam flanges sections at the member end is proposed. If the same cross-section (reduced or not) is used for the whole flexural dissipative member the positions of the plastic hinges along this element cannot be predicted exactly and so the maximum inelastic link deformations cannot be predicted exactly. The reduced cross-sections near the ends of the links ensure a better control of the positions where plastic hinges can occur along the link. A greater distance between the positions of plastic hinges along the dissipative member, reduces the link axis rotation angle.

Configurations resembling the "dog-bone" near the ends of the upper storeys diagonals in centrically braced frames ensure on one hand, the abidance of the slenderness demand for the braces[1] ($\bar{\lambda} \leqslant 2.0$) and on the other hand, small values of the overstrength ratio[1] ($\Omega_i^{(N)} = N_{pl,Rd,i}/N_{Ed,i}$).

The reduced cross-section of the diagonal in the "dog-bone" detail zones ensures an adequate tensile capacity of the braces and small $\Omega_i^{(N)}$ values. At the same time the buckling behaviour and slenderness of the braces is not significantly affected by the reduced cross-sections near the member ends and the slenderness demand of the braces will be satisfied. Compared analyses led to the conclusion that the increase of labour cost due to the supplementary cuttings of the flanges at the braces ends is fully compensated by the reduction of overall material cost.

An explication for the poor behaviour of centrically braced frames during dynamic nonlinear analyses could relay in the fact that according to most seismic design procedures a global plastic failure mechanism is not sized clearly by design for this kind of seismic resistant structures. Generally as long as inelastic deformations are concentrated only in the diagonals, the centrically braced frames have amore or less predictable behaviour. When the loading level increases and plastic deformations appear in other type of members(columns and girders) the behaviour of a centrically braced frame is difficult to control. Unfavourable distributions of plastic hinges may lead to the apparition of several local plastic mechanisms.

In order to size a favourable global plastic failure mechanism for a centrically braced frame potentially plastic zones are provided along the girders[2]. A configuration with reduced beam flanges sections(resembling the dog-bone detail) near the connections with the braces can be adopted for the girders of the centrically bracedframes. Inelastic deformations in these potentially plastic zones appear under the combined action of bending moments and axial forces.

So for centrically braced frames accepting plastic deformations in the diagonals, at the bottom of first-story columns and in the potentially plastic zones located along the frame girders near the connections with the braces can ensure a global plastic failure mechanism[3]. Also a better control of the positions of plastic hinges along the girders of the centrically braced frames is ensured. The same concept can be used for designing a favourable global plastic mechanism for frames equipped with buckling restrained braces.

Several structural details were analyzed for the potentially plastic zones located near the bottom end of the columns considering: reduced flanges cross-sections and/or transversal and longitudinal stiffeners for the columns bottom zone.

The maximum bending moment values, which can appear at the bottom of the columns during strong earthquakes, can be limited by using adequate "dog-bone" configurations near the bottom of first-story columns. This can lead to smaller anchor bolts for the columns.

Configurations with "dog-bone" details near the base of bottom-story columns appear to be safer from the point of view of assuring the general stability of first-story columns[4], in the situations when plastic deformations occur in the potentially plastic zone at the bottom of the column. Further, a better control of the inelastic deformations along the first story columns is ensured.

CONCLUSIONS

In case of eccentrically braced frames with long links, a dog-bone configuration at the link ends ensures significant lateral stiffness for the eccentrically braced frame and a better control of the inelastic deformations along the dissipative members.

Details resembling the "dog-bone" near the ends of the upper storeys diagonals in centrically braced frames conduct to an adequate slenderness for the braces and to a proper tensile capacity of the diagonals, with smalloverstrength ratio values.

A favourable global plastic hinge mechanism can be sized by design for centrically braced frames or for frames equipped with buckling restraint braces, by providing "dog-bone"-shaped configurations for potentially plastic zones located in the frame girders.

Configurations with "dog-bone" details near the base of bottom-story columns appear to be safer from the point of view of ensuring the general stability of the first-story columns in the situations when plastic deformations occur in the bottom story columns.

REFERENCES

[1] Eurocode 8, EN 1998-1:2004, Design of structures for earthquake resistance, Part 1: General rules, seismic actions and rules for buildings.
[2] Köber, H. , Beţea, Ş. , An Alternative Method for the Design of Centrically Braced Frames, Proc. Eurosteel 2008-5th European Conference on Steel and Composite Structures;03-05 September,2008, Graz, Austria, Vol. B, pp. 1425-1430,2008.
[3] Tsai K. C. ; Li J. W. ,1994, Drain2D + A General Purpose Computer program for Static and Dynamic Analyses of Inelastic 2D Structures.
[4] Eurocode 3, EN 1993-1-1:2005, Design of steel structures-Part 1-1: General rules and rules for buildings.

EXPERIMENTAL INVESTIGATION ON SEISMIC BEHAVIOR OF COLD-FORMED STEEL TRUSSING SHEAR WALLS WITH STEEL SHEET SHEATHING

Huiwen Tian[a], Yuanqi Li[a,b]

[a]Tongji University, College of Civil Engineering, China

crazythw2008@126. com

[b] Tongji University, State Key Laboratory of Disaster Reduction in Civil Engineering, China

liyq@ tongji. edu. cn

KEYWORDS: Shear wall, Cold-formed steel, Cyclic test, Seismic, Tension field.

ABSTRACT

In this paper, a new cold-formed steel(CFS) trussing shear wall with steel sheet sheathing(CFSTSW) different from the conventional one in skeleton configuration was suggested as a kind of lateral force resisting member. Four 2. 4 metres wide 3 metres high specimens were designed to investigate the failure mode and seismicbehavior of the shear walls with such new detailing requirements under lateral cyclic loading. The influences of the skeleton configurations of the frame skeleton and truss skeleton, of the sheathing configurations of the corrugated and plain steel sheet, and of the sheathing layout patterns including attaching the sheathing on one side of the wall and on both sides of the wall on the overall performance of CFSTSW under lateral loading were analyzed. The results indicated that CFSTSWs gave obviously higher shear strength, unit elastic stiffness and dissipated more energy than those obtained from conventional shear walls. Finally, the suggested response modification factor value of 4. 5 for CFSTSW is proposed for design purpose.

1. Discussion of the results

The test programincluded a total of 4 shear wall tests(Configurations 1-4), whose behavior was investigated under the reversed cyclic loading protocol(Table 1).

Table 1. Matrix of shear wall test specimens.

Configuration	Skeleton construction	Sheathing type	Chord studs	Number of tests and protocol[a]
1	Frame	Corrugated steel sheet	3 lipped channels	1C
2	Truss	Corrugated steel sheet	3 lipped channels	1C
3	Truss	Plain steel sheet	3 lipped channels	1C
4	Truss	Plain steel sheet(both sides)	3 lipped channels	1C

Table 2 provides a summary of the values of R_d, R_0 and R for CFSTSW. According to AISI standard[1], the R values are recommended to be as 6. 5 and 2. 0 for the lighted-framed walls sheathed with wood structural panels and shear panels of all other materials, respectively. The average value of response modification factor for CFSTSW obtained from this test is 4. 732. It is suggested that a value of 4. 5 may be used as the response modification factor for CFSTSW.

Table 2. Response modification factor for CFSTSW

R_0	R_d	R
1. 64	2. 89	4. 732

The lateral load vs. displacementenvelope curves from both positive and negative directions for all shear walls tested are shown in Fig. 1. Firstly, a comparison of seismic performance parameters is made between Configurations 1 and 2, to discuss the impact of the skeleton configuration on the overall performance of CFSTSW compared to conventional shear walls. The results showed that the increase in ultimate shear resistance and energy absorption were 67% and 57%, respectively, in Configuration 2 compared to Configuration 1. It concluded that CFSTSWs gave obviously higher shear strength and dissipated more energy than those obtained from conventional shear walls.

In addition, a comparison of seismic performance parameters of similarly configured CFSTSWs with different sheathing(Configurations 2 and 3) is performed to discuss the impact of the sheathing configuration on the overall performance of CFSTSW. The results showed that the ultimate strength and the energy dissipated increased by 13% and 39% in Configuration 2 compared to Configuration 3. It concluded that the corrugated panel resulted in a little increase by 13% in the ultimate strength for CFSTSW than the plain panel under the cyclic loading.

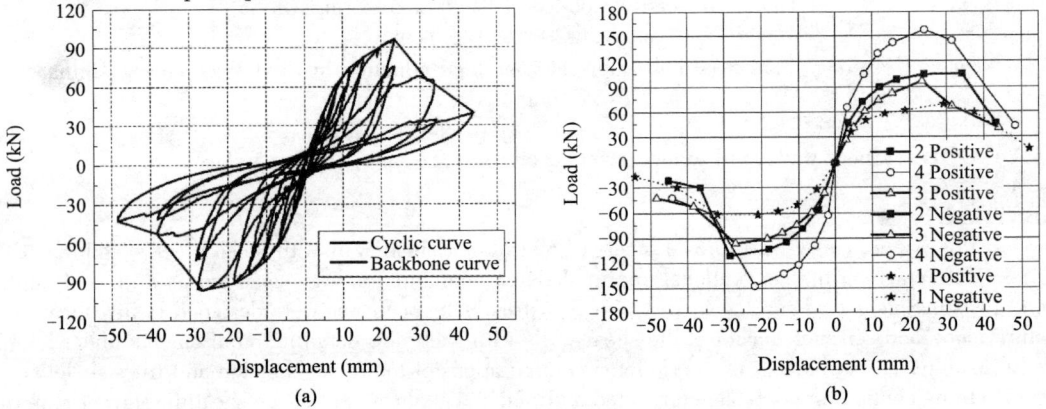

Fig. 1. (a)Cyclic curve of Configuration 2; (b)Load vs. displacement envelope curves

CONCLUSIONS

A test program was conducted to investigate the seismic behavior of steel sheathed cold-formed steel trussed shear wall(CFSTSW)under lateral cyclic loading. The influence of the skeleton configuration, sheathing configuration and sheathing layout pattern on the overall performance of shear walls is discussed in detail. The test results showed that the failure mode for CFSTSW was the sheathing-to-framing screw connection failure and brace-to-track connection buckling, while the failure mode for the conventional wall was the sheathing-to-framing screw connection failure. The sizable strength and stiffness degradation took place between the first and the second cycles of the same displacement amplitude. Also, considerable pinching behavior was observed. CFSTSWs gave obviously higher shear strength(67%) and dissipated more energy (57%) than those obtained from conventional shear walls. The corrugated panel resulted in a increase by 13% and 39% in the ultimate strength and energy dissipated for CFSTSW than the plain panel under the cyclic loading. The ultimate strength, energy absorption and ductility ratio for the CFSTSW sheathed on both sides were significantly higher by 60% ,74% and 66% than those for the CFSTSW sheathed on one side. A suggested value of 4. 5 may be used as the response modification factor for CFSTSW for design purpose. Based on the above advantages, this new cold-formed steel(CFS)trussing shear wall with steel sheet sheathing(CFSTSW)different from the conventional one in skeleton configuration was suggested as a kind of lateral force resisting member

REFERENCES

[1] American Iron and Steel Institute(AISI). North American standard for cold-formed steel framing-lateral design. AISI S213. Washington, USA. 2007.

[2] Building Seismic Safety Council. NEHRP recommended provisions for seismic regulations for new buildings and other structures—part 2 commentary. FEMA-450;2003.

[3] Newmark NM, Hall WJ. Earthquake spectra and design. California, USA: Earthquake Engineering Research Institute, Berkeley;1982.

OPENSEES MODELING OF COLD FORMED STEEL FRAMED WALL SYSTEM

Guanbo Bian[a], David A. Padilla-Llano[b], Jiazhen Leng[a],
Stephen G. Buonopane[c], Cristopher D. Moen[b], Benjamin W. Schafer[a]
[a]Johns Hopkins University, Civil Engineering Department, USA
bian@jhu.edu, jleng1@jhu.edu, schafer@jhu.edu
[b] Virginia Tech, Civil&Environmental Engineering Department, USA
dapadill@vt.edu, cmoen@vt.edu
[c]Bucknell University
stephen.buonopane@bucknell.edu

KEYWORDS: Cold-formed steel, Seismic, Gravity wall, Shear wall.

ABSTRACT

The objective of this paper is to present and explore an efficient spring-element-based finite element model of a wood sheathed cold-formed steel framed wallsystem. The model is developed in OpenSees and has the potential to be an important building block tool for modeling full structures framed with cold-formed steel. The lateral stiffness of the gravity wall is currently ignored in both the design and modeling of multi-story cold-formed steel(CFS) framed building. However, full-scale experimental work on a two-story cold-formed steel(CFS) framed buildings, as part of the CFS-NEES effort, shows that gravity walls can provide a contribution to the lateral response and potentially should be considered in the design of lateral force resisting systems. Recently an engineering model implemented in OpenSees employing fastener-based characterization as the essential nonlinearity in a CFS framed shear wall has shown that OpenSees models are capable of predicting full shear wall hysteretic performance. In the work presented here, the fastener-based shear wall model is extended to provide a model capable of capturing coupled shear wall and gravity wall behavior in a wall system. The contribution of gravity walls on lateral resistance is explored by comparing the wall system behavior with and without gravity walls. The model provides practical design advantages and also a means to model the lateral system resistance in CFS framed buildings.

CONCLUSIONS

Wood sheathed cold-formed steel framed wall systems may be efficiently modeled utilizing a fastener-based model in OpenSees. This provides engineers with an efficient solution that can predict the shear-deformation response for different shear wall and gravity wall combinations.

Fig. 1 provides the shear load-displacement curve for different wall systems. The stiffness and peak load capacity increase significantly when gravity walls, especially those with OSB and/or gypsum board, were added to the shear wall. Although the tensile stiffness for the connection between a gravity wall stud and the foundation was small, the direction of the shear deformation(left vs. right)is not observed to have a significant difference in terms of lateral stiffness. The single OSB shear wall can carry 20 kN lateral load while the combination of shear wall and gravity wall with both OSB and gypsum is 65 kN. The lateral resistance for gravity walls with OSB is significantly larger than that with gypsum board since the fastener behaviour differs significantly for these different materials.

A wall model(SW with OSB + GW with Gypsum)was selected as a representative case to provide the reaction force information. The hold down forces are always the greatest but the forces in the other stud end locations are consequential. The shear wall and gravity walls deform together, to the same direction, and with very similar magnitude. Base shear at the bottom track is provided in *Table 1*. There are two low velocity fasteners in the shear wall segment and four in the gravity walls. Although there is no hold down in the gravity wall studs and the stud ends have only 1/10th the tensile stiffness, the gravity wall does have twice the length and the large number of fastener locations make the gravity wall carry most of the base shear. The sum of the last four base shear columns, which is the

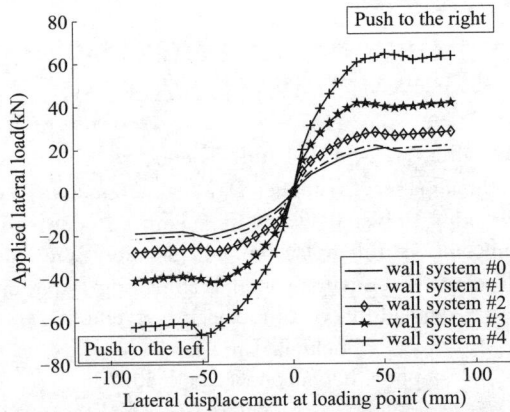

Fig. 1. Lateral load-displacement curve for wall models

base shear carried by gravity wall, is over half of the total force.

Table 1. Base shear at bottom track in wall model #3

		shear wall			gravity wall		
		LVF1	LVF2	LVF3	LVF4	LVF5	LVF6
Right lateral loading	value at peak load(kN)	−9. 68	−8. 75	−6. 38	−5. 72	−5. 63	−5. 66
	percentage	23. 1%	20. 9%	15. 3%	13. 7%	13. 5%	13. 5%
Left lateral loading	value at peak load(kN)	9. 88	9. 33	6. 43	5. 66	5. 27	5. 46
	percentage	23. 5%	22. 2%	15. 3%	13. 5%	12. 5%	13. 0%

Significant additional work remains to extend the model to more wall configurations and utilize the model more formally in seismic shear wall design, to better understand system reliability, and in full building models. Nonetheless, the model represents a significant advancement for efficient computational modeling of cold-formed steel framed walls and has wide potential application.

REFERENCES
[1] Buonopane, S. , Tun, T. , and Schafer, B. (2014). "Fastener-based computational models for prediction of seismic behavior of CFS shear walls. " *Proceedings of the 10th National Conference in Earthquake Engineering.*
[2] Bian, G. , Buonopane, S. , Ngo, H. , and Schafer, B. (2014). "Fastener-Based Computational Models with Application to Cold-Formed Steel Shear Walls. " *Proceedings of the 22nd International Specialty Conference on Cold-Formed Steel Structures.* St. Louis, Missouri, USA, 825-840.
[3] Peterman, K. D. (2014). "Behaviour of full scale cold-formed steel buildings under seismic excitations. " Ph. D. thesis. Johns Hopkins University.
[4] AISI-S213-07. (2007). "AISI North American Standard For Cold-Formed Steel Framing-Lateral Design. " American Iron and Steel Institute.
[5] Mazzoni, S. , McKenna, F. , Scott, M. H. , and Fenves, G. L. (2003). "Open System for Earthquake Engineering Simulation User Command-Language Manual. " Berkeley, California.

SEISMIC BEHAVIOR OFLARGE-SECTION RECTANGULAR CFT COLUMNS WITH DISTRIBUTIVE BEAM AND INNER DIAPHRAGMS

Yuanzhi Zhang[a], Jinhui Luo[a], Yuanqi Li[a], Zuyan Shen[a], Xueyi Fu[a,b]

[a]Tongji University, Department of Structural Engineering, China

zyzyuanzhi@ 163. com, Linuojh@ 163. com, liyq@ tongji. edu. cn,

zyshen@ tongji. edu. cn @ tongji. edu. cn, fu_xue_yi@ yahoo. com. cn

[b] CCDI(Shenzhen), China

fu_xue_yi@ yahoo. com. cn

KEYWORDS: LRCFT columns; Distributive beam; Inner diaphragms; Longitudinal stiffeners; Seismic behaviour.

ABSTRACT

Large-section rectangular concrete-filled steel tubular(called LRCFT hereinafter) columns with distributive beam and inner diaphragms had been proved that the plane strain assumption could be expected under axial compression applied on steel tube directly by previous researches. numerical simulation on 1: 5 reduced scale LRCFT column specimens under lateral cyclic loading were conducted to investigate the failure modes and seismic behavior of such columns. The influence of slenderness ratio of columns, T shaped longitudinal stiffeners on inner skin of tube, as well as axial compression ratio of columns on seismic performance and failure modes of LRCFT columns was investigated. The analytical study results indicated that, not only local buckling on the tube was delayed, but also the ultimate bearing capacity of LRCFT column was increased for specimens with longitudinal stiffeners. However ductility decreased as the axial compression ratio or slenderness ratio of specimens increased. It could be concluded that LRCFT columns with distributive beam and inner diaphragms, especially the ones with longitudinal stiffeners on steel tube could obtain the similar seismic behaviour to general rectangular CFT column.

1. Discussion of the numerical results

Figure. 1 gives the P-Δ hysteretic curves of LRCFT specimens. The results of a numerical simulation study show that LRCFT column are reliable and viable since the hysteretic curves for all connections are plump which indicated that the LRCFT column structure system has good plastic deformation capacity and energy-dissipated capacity. Hysteretic curves in figure. 1 indicates that slope of the latter cycle load is smaller than the former one, which shows that stiffness of the connection was degenerating gradually.

It can be found that in general, the degradation of the stiffness of hysteretic curves increases as the axial compression ratio increase. Especially when the axial compression ratio is 0. 6, it clearly can be seen that the hysteretic curve degrade rapidly after the peak load, and the horizontal load of such models convert to the negative direction when the horizontal displacement reached a certain values, as shown in Fig. 5(a), (c)and(e). It is plumper for specimens with small width to thickness ratio. The specimens with stiffener has a better energy dissipation performance, compared with the specimens without stiffeners(LRCFT-30-6mm series and LRCFT-50-6mm series). The ultimate bearing capacity of LRCFT column was increased for specimens with longitudinal stiffeners.

CONCLUSIONS

Numerical simulation on 1: 5 reduced scale LRCFT column specimens under lateral cyclic loading were conducted in the paper. The influence of slenderness ratio of columns, T shaped longitudinal stiffeners on inner skin of tube, as well as axial compression ratio of columns on seismic performance and failure modes of LRCFT columns is investigated. The ductility performance of numerical models considering different factors is evaluated and analyzed. The analytical study results indicates that.

(1) The ultimate bearing capacity of LRCFT column was increased for specimens with longitudi-

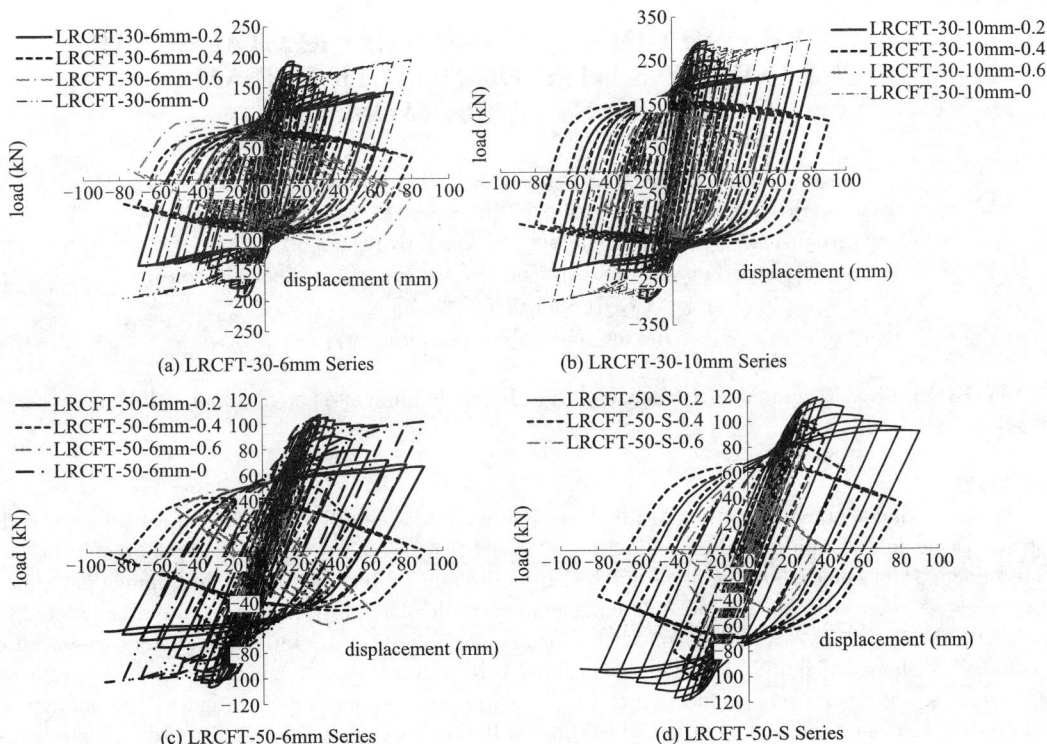

Fig. 1. P-Δ hysteretic curve

nal stiffeners, and the specimens with stiffener has a better energy dissipation performance.

(2) Axial compression ratio leads a significant influence on the stiffness degradation and the ductility of the column. The stiffness degradation became more rapid and the ductility became worse with the axial compress ratio increasing and with the slenderness ratio of specimens increasing.

(3) Axial compression ratio and the width to thickness is the key factor influencing the failure mode of steel hollow section in beam-columns cyclic loading result. The number of local bucking waves increase as the width to thickness ratio and axial compression ratio increase.

(4) LRCFT columns with distributive beam and inner diaphragms, especially the ones with longitudinal stiffeners on steel tube could obtain the similar seismic behavior to general rectangular CFT column.

ACKNOWLEDGMENT

The research reported in the paperwere funded by Natural Science Foundation of China (No : 51208375), with deep gratitude for the support by CCDI.

REFERENCES

[1] Luo Jinhui, Li Yuanqi, Zhang Yuanzhi, Fu Xueyi, Shen Zuyan, " Experimental study on axial compression load transfer of large rectangular-section CFT columns with distributive beam embedded in beam-column joints", *China Civil Engineering Journal*, 47(10):49-60, 2014

STEEL SLIDING-CONTROLLED COUPLED BEAM MODULES FOR IMPROVING SEISMIC RESILIENCE OF BUILDING SYSTEMS

Ying-Cheng Lin[a]

[a]University of Alabama in Huntsville, USA

yingcheng. lin@ uah. edu

KEYWORDS: Sliding-Controlled, Coupled Beams, Modules, Seismic.

ABSTRACT

This paper presents a steel sliding-controlled coupled beam(SC-CB) module that employs relative sliding mechanism combined with contributions of built-in-unit post-tensioning(PT) steel and supplemental energy-dissipating devices. SC-CB-to-column connection moment is derived. A prototype building was designed; nonlinear analyses were performed to study the limit states and seismic response. Statistic results indicated that SC-CB modules offer steel moment frame building systems with the potential of: (1) superior damage-free moment resistance; (2) eliminating residual drifts; (3) resolving deformation incompatibility concerns about the secondary structures of the moment frames with rocking beam-to-column connections; (4) less damage to acceleration-sensitive secondary non-structural components and equipment.

Acknowledgments are optional.

CONCLUSIONS

The conclusions that can be drawn from this study are listed as follows:

1. Static pushover analysis results showed the SC-CB-to-column connections reached IRS and PTU limit states as expected. Yield in column bases occurred before PTU limit state, but only modestly reduced the connection moment-rotation stiffness.

2. Dynamic time history analysis results showed small peak story drifts with the mean equal to 0. 6% (DBE) and 0. 9% (MCE) , which indicates using SC-CBs in steel MRFs could reduce damage to displacement-sensitive non-structural components of building systems.

3. The residual story drifts of the CB-MRF structure after the DBE and MCE ground motions were very small. SC-CBs can minimize residual drifts of steel MRFs and provide the building system with the potential of reducing repairing costs that are proportional to magnitude of residual drifts.

4. The mean value of peak floor acceleration issmall, indicating low floor accelerations. Using SC-CBs can provide MRF buildings with the potential of less damage to acceleration-sensitive secondary non-structural components and equipment.

REFERENCES

[1] Fahnestock LA, Ricles JM, Sause R. Experimental evaluation of a large-scale buckling-restrained braced frame. ASCE Journal of Structural Engineering 2007; 133(9) : 1205-1214.

[2] Tremblay R, Bolduc P, Neville R, DeVall R. Seismic testing and performance of buckling-restrained bracing systems. Canadian Journal of Civil Engineering 2006; 33: 183-198.

[3] Herrera RA, Ricles JM, Sause R. Seismic performance evaluation of a large-scale composite MRF using pseudodynamic testing. ASCE Journal of Structural Engineering 2008; 134(2) : 279-288.

[4] Erochko J, Christopoulos C, Tremblay R. Residual drift response of SMRFs and BRB frames in steel buildings designed according to ASCE7-05. ASCE Journal of Structural Engineering 2011; 137(5) : 589-599.

[5] McCormick J, Aburano H, Ikenaga M, Nakashima M. Permissible residual deformation levels for building structures considering both safety and human elements. Proceedings of the 14th World Conference on Earthquake Engineering 2008.

[6] Miranda E. Enhanced building-specific seismic performance assessment. Proceedings of ACES Workshop: Advances in Performance-Based Earthquake Engineering 2009.

[7] Ricles JM, Sause R, Garlock M, Zhao C. Post-tensioned seismic-resistant connections for steel frames. ASCE Journal of Structural Engineering 2001;127(2):113-121.

[8] Garlock M, Sause R, Ricles JM. Behavior and design of post-tensioned steel frame systems. ASCE Journal of Structural Engineering 2007;133(3):389-399.

[9] Kim JK, Christopoulos C. Seismic design procedure and seismic response of post-tensioned self-centering steel frames. Earthquake Engineering and Structural Dynamics 2008;38:355-376.

[10] Tsai KC, Chou CC, Lin CL, Chen PC, Jhang SJ. Seismic self-centering steel beam-to-column moment connections using bolted friction devices. Earthquake Engineering and Structural Dynamics 2008;37,627-645

[11] Lin YC, Sause R, Ricles JM. Seismic Performance of a large-scale steel self-centering moment-resisting frame: hybrid simulations under design basis earthquakes. ASCE Journal of Structural Engineering 2013;139(11):1823-1832.

[12] Eartherton MR, Fahnestock LA, Miller DJ. Computational study of self-centering buckling-restrained braced frame seismic performance. Earthquake Engineering and Structural Dynamics 2014;DOI:10. 1002/eqe. 2428

[13] ASCE/SEI 7-10. Minimum design loads for buildings and other structures. Published by the American Society of Civil Engineers(ASCE), Prepared by the Structural Engineering Institute of ASCE, 2010.

[14] Mazzoni S, McKenna F, Scott MH, Fenves GL. Open system for earthquake engineering simulation user command-language manual. OpenSees Version 2. 4, 2013.

[15] Federal Emergency Management Agency(FEMA). Quantification of building seismic performance factors. FEMA P695 Published by the Federal Emergency Management Agency, 2009.

[16] Herning G, Garlock M, Vanmarcke E. Reliability-based evaluation of design and performance of steel self-centering moment frames. Journal of Constructional Steel Research 2011;67(10):1495-1505.

[17] Federal Emergency Management Agency(FEMA). Design guide for improving hospital safety in earthquakes, floods, and high winds," FEMA-577 Published by the Federal Emergency Management Agency, 2007.

[18] Mayes RL, Goings C, Naguib W, Harris S, Lovejoy J, Fanucci JP, Bystricky P, Hayes JR Comparative performance of buckling-restrained braces and moment frames. Proceedings of the 14th World Conference on Earthquake Engineering 2004.

[19] Celebi M. Response of Olive View Hospital to Northridge and Whittier earthquakes. ASCE Journal of Structural Engineering 1997;123:389-39.

EXPERIMENTAL TESTS OF COMPOUND BATTENED COLUMN AND ITS BASE-PLATE CONNECTION SUBJECT TO AXIAL AND HORIZONTAL FORCES

Gaetano Della Corte[a], Raffaele Landolfo[a]

[a] Department of Structures for Engineering and Architecture,
University of Naples "Federico II", Italy
gdellaco@ unina. it, landolfo@ unina. it

KEYWORDS: Columns, Connections, Experimental tests, Steel, Seismic response.

ABSTRACT

Industrial steel buildings may comprise compound columns, where two I-shaped profiles are connected together to form a single column. Such columns are usually anchored to a concrete foundation by means of base plates and anchor bars. Commonly, response of connections (both between the two constituting profiles andat the column base) markedly affects the system response, because of significant elastic flexibility and partial strength of such connections due to large column dimensions. Therefore, evaluating seismic risk of such industrial structures requires appropriate modelling of the inelastic response of compound columns with due account for connections. Eurocode 3 (EC3) covers analysis of both compound columns and base-plate connections with anchors[1], but performance of the method for elastic stiffness calculation and/or specific geometries is still worthy of investigation.

Two specimens (*Fig. 1*) of compound battened columns, built-up from individual I-shaped profiles with different schemes of battening and different base connections, were fabricated and tested under both monotonic and cyclic loading conditions. The two specimens were samples from an existing industrial steel building in Italy, which was formerly analysed theoretically for seismic risk assessment[2]. This paper is focused on the first of the two tests, which was carried out on a typical battened column, with a single base plate anchored to a concrete foundation by means of round anchor bars.

First, monotonic loading tests were carried out, in the elastic range of response with varying values of column axial force. Aims of these elastic tests were (i) to characterize initial stiffness, (ii) to investigate the effect of the column axial force on the initial stiffness and (iii) to verify theoretical calculations of yield displacements. Results clearly illustrate that the elastic response is characterized by two phases; initially, the column base is fully compressed and the system shows relatively large lateral stiffness; subsequently, decompression occurs, with neutral axis rapidly moving towards the compressed base plate edge and generating significant loss of lateral stiffness.

Secondly, inelastic cyclic loading tests (*Fig. 2*) were carried out, by applying a cyclically varying horizontal displacement at the column top. Repetition of three cycles (for given displacement amplitudes) was adopted. Plastic deformations localized in the column base connection, while the compound column remained essentially elastic. A mixed plastic mechanism, with yielding of both the base plate in bending and anchor bars in tension, was observed. Correspondingly, the specimen showed large pinching of hysteresis response loops. Ultimate failure occurred when one of the anchor bars fractured in tension, at a total drift angle of approximately 0.075 rad. The energy dissipation capacity was also quantified by means of an equivalent viscous damping ratio, showing a peak of 8% at a ductility of 5, approximately.

CONCLUSIONS

This paper has shortly described some of the main results from tests on a compound battened column and relevant base-plate connection. The connection exhibited a mixed plastic mechanism, with yielding of both base plate and anchor bars. This mechanism originated strong pinching of hysteresis loops. Notwithstanding, the connection exhibited good ductility and energy dissipation capacity.

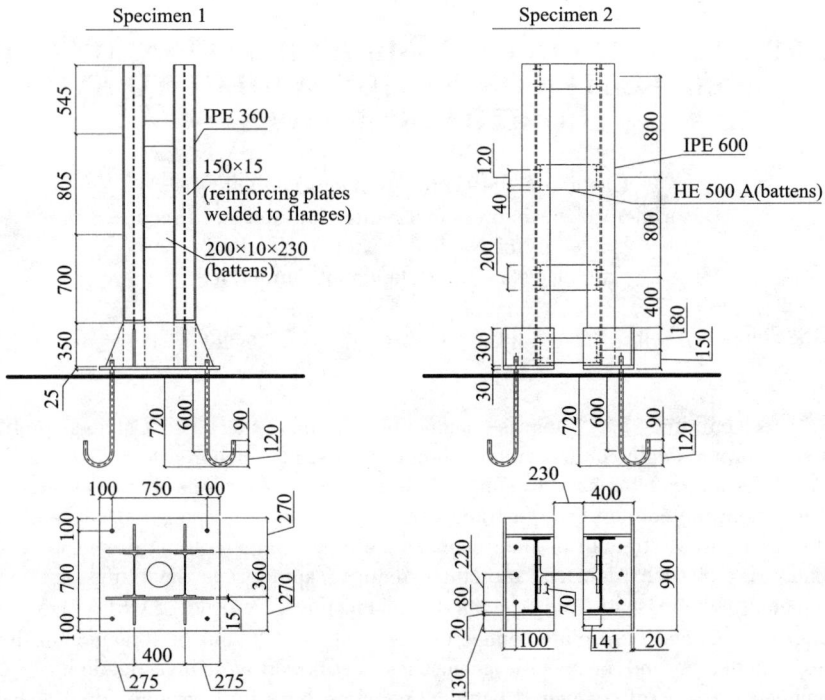

Fig. 1. Specimens

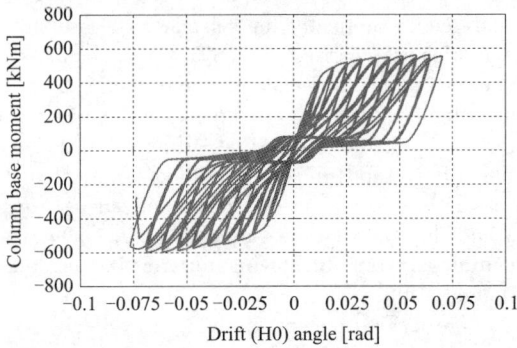

(a)   (b)

Fig. 2. Cyclic loading test of specimen 1: (a) hysteresis response; (b) plastic mechanism and ultimate failure

ACKNOWLEDGMENT

This experimental research was financially supported by theReLUIS consortium (Italian network of University laboratories for research on earthquake engineering), whose contribution is gratefully acknowledged.

REFERENCES

[1] European Committee for Standardization (CEN), "Eurocode 3: Design of steel structures-Part 1-8: Design of joints" *EN* 1993-1-8, Eurocode 3, Brussels, 2005.

[2] Della Corte G, Petruzzelli F, Iervolino I, "Structural modelling issues in seismic performance assessment of industrial steel buildings", *Proceedings of* 4th *Conference on Computational Methods in Structural Dynamics and Earthquake Engineering*, Kos (Gr), 12-14 June, Paper ID 1461, 2013.

ON THE USE OF PERFORATED METAL SHEAR PANELS FOR SEISMIC-RESISTANT APPLICATIONS

A. Formisano[a], L. Lombardi[a], F. M. Mazzolani[a]

[a]Department of Structures for Engineering and Architecture,
University of Naples "Federico II", Naples, Italy
antoform@ unina. it, luca. lombardi. 88@ gmail. com, fmm@ unina. it

KEYWORDS: Steel Plate Shear Walls, perforated plate, circular hole, FEM model, parametric analysis

ABSTRACT

Steel Plate Shear Walls are an innovative system able to confer to either new or existing structures a significant capacity to resist earthquake and wind loads. Many tests have shown that these devices may exhibit high strength, initial stiffness and ductility, as well as an excellent ability to dissipate energy. When traditional SPSWs are used as bracing devicesin buildings, they may induce excessive stresses in the surrounding structure, so to require the adoption of large cross-section profiles. For this reason, perforated steel panels, which are weakened by holes aiming at limiting the actions transmitted to the surrounding frame members, represent a valid alternative to the traditional panels. In this work, a FEM model has been calibrated on the basis of recent experimental tests. Subsequently, a parametric FEM analysis by changing the number and diameter of the holes, the plate thickness and the metal material, has been carried-out. Finally, an analytical tool to estimate the non-linear response of perforated metal shear panels has been proposed.

CONCLUSIONS

The FEM study on unstiffened perforated shear panels have shown some relevant results. The available experimental results on panels with a central opening[1] allowed to setup and calibrated an appropriate FEM model, where geometric imperfections and material non-linearity can be considered. This model can also take into account the presence of the bolted plate-to-frame connections and their imperfections. The proper calibration of the model allows to obtain a satisfactory numerical-to-experimental agreement in terms of the overall behaviour.

In this framework, a parametric FEM analysis on panels with different perforation patterns, material and thickness has been carried-out. The different perforation patterns have been considered by modifying disposition, number and diameter of the holes. Two material types have been considered: steel and aluminium, with the latter already investigated in[2]. From the results it is observed that, despite the presence of holes, the inclination of tension-field essentially remains to 45°. Comparing to traditional panels, the number of active bands decreases and it is reduced to one in the case of one centred hole. Furthermore, there is a different activation of the yielding mechanism with respect to traditional panels, where yielding is activated in corner zones penalizing the connection system. Contrary, for perforated panels, yielding activates around the holes, without stressing the system joints. Also, for perforated panels with a high percentage of holes, a considerable reduction of the stresses in the perimeter area is found. Furthermore, by adopting thicker perforated plates, very large drifts can be attained without failure around holes. In conclusion, it has been shown that aluminium panels have a better dissipative behaviour, characterized by a more negligible pinching effect than in case of steel panels.

On the basis of the obtained numerical results, the design charts reported in *Fig. 1* are derived. These diagrams can be used to evaluate the modification factors C_{m1} and C_{m2}, which appear in *Eq.* (1) and (2) proposed by Sabouri-Ghomi et al. [3] for traditional panels, to correctly predict the non-linear behaviour of perforated panels.

$$F_{wu} = bt\left(\tau_{cr} + \frac{C_{m1}}{2}\sigma_{ty}\sin2\theta\right) \tag{1}$$

$$K_w = bt(\tau_{cr} + \frac{C_{m1}}{2}\sigma_{ty}\sin2\theta)/\left[d\left(\frac{\tau_{cr}}{G} + \frac{2C_{m2}\sigma_{ty}}{E\sin2\theta}\right)\right] \tag{2}$$

In the above equations, t, b, d are the thickness, width and height of the steel plate, respectively, E and G are the normal and shear elasticity modulus of the plate materials, σ_{ty} is the tension-field stress in the plate yielding condition, θ is the diagonal tension-field angle, measured from the horizontal di-

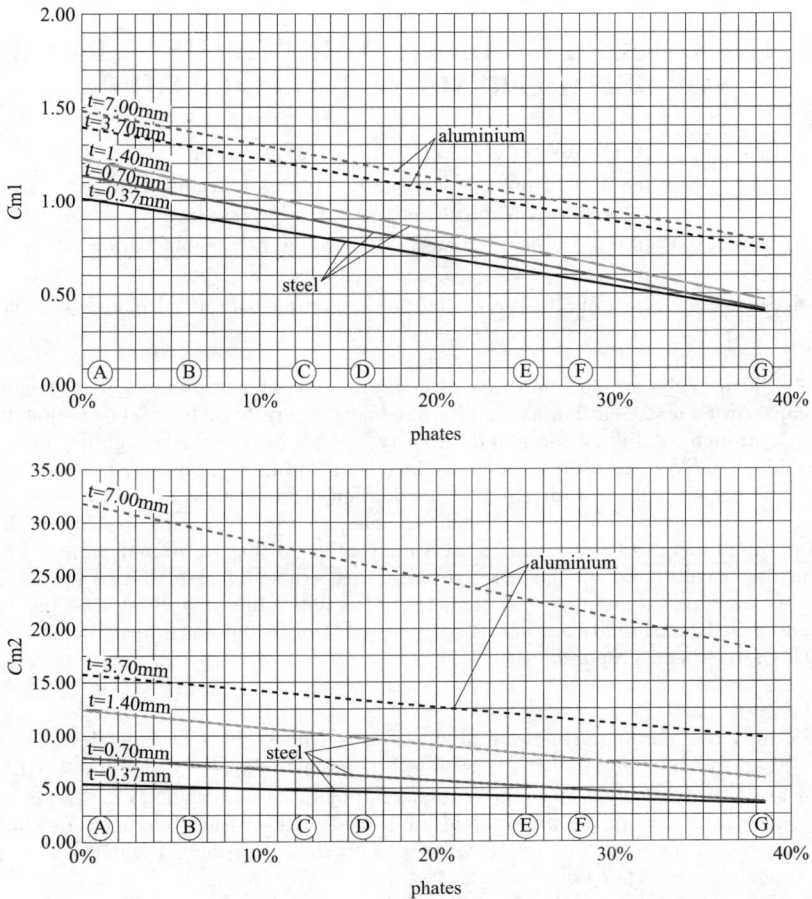

Fig. 1. Design charts for estimating the modification factors C_{m1} and C_{m2} used to predict shear strength and initial stiffness of perforated shear panels according to *Eq.* (1) and (2).

rection, and τ_{cr} is the critical buckling shear stress evaluated according to the Timoshenko's theory.

Finally, it isshown that the use of conventional steel panels with different perforation patterns can be a viable alternative to traditional panels for strengthening and stiffening both new and existing structures. In fact, if perforated panels are applied for example to an existing structure, by choosing an appropriate drilling configuration, it is possible to improve the resistance of the base building without aggravating the main structure with high stresses deriving from tension-field generated by the plates. Therefore, perforated panels appears to be also more economic than traditional plates, because the reinforcement interventions of the original building are lower in comparison to those required by traditional panels, due to weakening effect induced by the holes in the panels.

REFERENCES

[1] Valizadeh H. ,Sheidaii M. ,and Showkati H. ,"Experimental investigation on cyclic behaviour of perforated steel plate shear walls",Journal of Constructional Steel Research,70,308-316,2012.
[2] Formisano A. ,De Matteis G. ,and Mazzolani F. M. ,"Numerical and experimental behaviour of a full-scale RC structure upgraded with steel and aluminium shear panels",Computers and Structures,88,1348-1360,2010.
[3] Sabouri-Ghomi S. ,Ventura C. E. ,and Kharrazi M. H. K. ,"Shear analysis and design of ductile steel plate walls",Journal of Structural Engineering,131,878-889,2005.

BEHAVIOUR OF STEEL i-BEAMS WITH WEB OPENINGS

Luis Calado

CEris, ICIST, Instituto Superior Técnico, Universidade de Lisboa

luis. calado@ tecnico. ulisboa. pt

KEYWORDS: Steel I-beams; Web openings; Shear moment interaction curves; Parametric analysis; Finite element method.

ABSTRACT

The purpose of this paper is to present the study and analysis of the structuralbehaviour of steel beams with web openings, the influence of web openings in the load carrying capacity of steel beams and failure mechanisms. The non-linear numerical analysis performed was calibrated against the results of other similar non-linear numerical analysis[1,2,3] and experimental test data[4]. Comparison of analytical results with the available experimental results for yielding patterns, ultimate load values and load-deflection relationships show good agreement between the finite element and the experimental results thus validating the accuracy of the proposed model. The proposed finite element method was extended to carry out a parametric study taking into account some parameters, such us: opening geometry, opening size and the location of the opening throughout the span, for three different beam spans. The FE model was established with both material and geometrical non-linearity, allowing load redistribution across the web openings and formation of the "Vierendeel" mechanism. The behaviour and ultimate load of the beams with web openings can also be predicted. It is presented a contribution to the analysis and selection of the web openings best solutions. The behaviour of steel I-beams with web openings under seismic loads is presently under investigation.

CONCLUSIONS

A comprehensive parametric study using finite element method on steel beams with web openings of shapes and sizes wascarried out[5]. The study is based on a finite element model with 4-node shell elements calibrated against test results of steel beams with web openings of similar configurations. As a result of the parametric study, some qualitative recommendations can be made:

-For the entire beam spans studied and for web openings size equal or greater than 0. 50h, it is observed that the load carrying capacity of steel beams with circular web openings is higher than the load carrying capacity of steel beams with square web openings.

-For beam spans equal or greater than 6. 0m, when the opening can be located at mid-span, and for the same opening height, if it is needed a larger opening, it is possible to increase the length of the opening to a maximum of 0. 75h without a decrease ofthe beam load carrying capacity.

-For beam spans $\geqslant$ 10. 0m, and for web openings size equal or smaller than 0. 50h, the openings must be located between the support and one quarter of the span, as shown in *Fig. 1*.

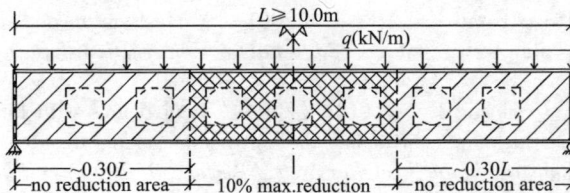

Fig. 1. Load capacity reduction vs opening position

-For beam spans < 10. 0m, and for circular and square web openings size of 0. 75h, the opening must be located near de mid-span, as shown in *Fig. 2*.

-For beam spans $\geqslant$ 10. 0m, and for web opening ratio B/H equal or greater than 0. 5 and equal or

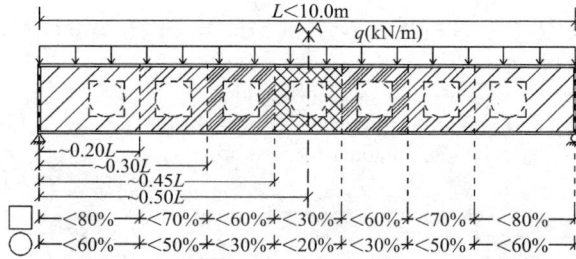

Fig. 2. Load capacity reduction vs opening position

smaller than 2. 0, the opening can be located between the supports and one quarter of the beam span, as shown in *Fig. 3*.

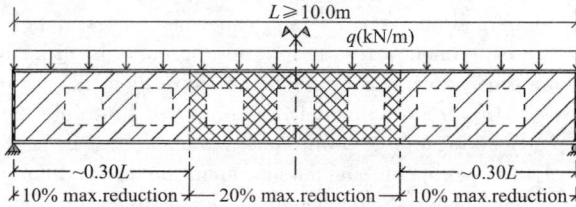

Fig. 3. Load capacity reduction vs opening position

-At mid span (neutral zone) , for a fixed height dimension H and for any opening length B ≤ 0. 75h, the reduction of the load carrying capacity is very small (less than 5%) , as shown in *Fig. 4.*

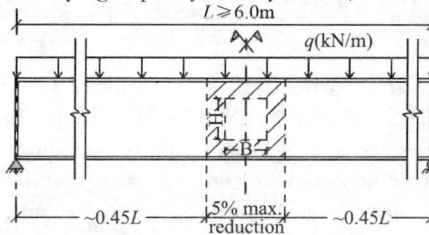

Fig. 4. Neutral zone

REFERENCES

[1] Liu, T. C. H. and Chung, K. F. (2003) , "Steel beams with large web openings of various shapes and sizes: finite element investigation", *Journal of Constructional Steel Research*, Elsevier, 59 (2003) ,1159-1176.

[2] Chung, K. F. , Liu, T. C. H and Ko, A. C. H. (2001) , "Investigation on Vierendeel mechanism in steel beams with circular web openings", *Journal of Constructional Steel Research*, Elsevier, 57 (2001) ,467-490.

[3] Chung, K. F. , Liu, T. C. H and Ko, A. C. H. (2003) "Steel beams with large web openings of various shapes and sizes: an empirical design method using a generalised moment-shear interaction curve", *Journal of Constructional Steel Research*, Elsevier,59 (2003) ,1177-1200.

[4] Redwood, R. G. , McCutcheon, J. O. (1968) , "Beam tests with un-reinforced web openings", *J. Struct. Div. Proc. ASCE* ;94 (ST1) ,1-17.

[5] Bernardino, P. J. C. (2013) , "Influência de aberturas nas almas no comportamento de vigas de aço", MSC Dissertation, Instituto Superior Técnico, Lisbon. (in Portuguese)

EXPERIMENTAL ASSESSMENT OF THE BEHAVIOR OF RUBBERIZED CONCRETE FILLED STEEL TUBE MEMBERS

Y. Jiang[a,b], A. Silva[a], J. M. Castro[a] and R. Monteiro[b]

[a] Faculty of Engineering of University of Porto, Portugal

yadong. jiang@ fe. up. pt, ajms@ fe. up. pt, miguel. castro@ fe. up. pt

[b] Istituto Universitario di Studi Superiori di Pavia, Italy

ricardo. monteiro@ iusspavia. it

KEYWORDS: concrete filled steel tube, rubberized concrete, numerical model.

ABSTRACT

The main objective of the research presented in this paper is to investigate the structural behaviour of Concrete Filled Steel Tube(CFST) columns made with Rubberized Concrete(RuC) and to identify behavioural differences between this type of composite members and typical CFST members made with standard concrete, namely in terms of the influence of the rubber aggregate replacement ratio on member strength, ductility, and energy dissipation capacity. This paper describes the preparation and development of the experimental campaign, which involves the testing of 36 specimens(28 RuCFST and 8 standard CFST). The selection of the specimens considered a number of geometrical and material parameters, namely the shape of the cross-section(circular, square and rectangular), cross-section slenderness and rubber aggregate replacement ratio. A special device was developed as part of an innovative testing setup, aimed at reducing both the cost and preparation time of the specimens. The specimens will be tested under both monotonic and cyclic lateral loading conditions and considering different levels of normalised axial load. The paper also describes the development of a finite element model in ABAQUS which was thoroughly validated against test results available in the literature. The model was used to perform a number of sensitivity analysis to both geometrical and material parameters and is used in the prediction of the test results.

CONCLUSIONS

In this paper, 5 different steel tube sizes are adopted to be part of the experimental tests, regarding 3 section shapes of both high and moderate cross-section slenderness according to Eurocode 8[1]. Normal concrete and rubberized concrete with two aggregate replacement ratios, 5% and 15% are selected. A special steel box, shown in *Fig. 1*, is designed not only to provide adequate base constraint to the specimen but also to reduce cost and preparation time of each test. A numerical model in the finite element software ABAQUS is also developed, involving a sensitivity study for the selection of element type, mesh type, material constitutive model, interaction simulation between two instances and analysis algorithm. The model is calibrated with existing experimental results from Han[2] and Varma et al[3]. Numerical models of two specimens to be tested are developed with the aim of predicting the behaviour of the CFST column under monotonic and cyclic lateral load. In *Fig. 2*, the lateral force level vs drift curves of the two specimens are plotted.

The following observations and conclusions can be drawn based on the research reported in the paper.

● The concrete core can restrain the steel tube and delay the local buckling occurrence. Thus, the in-filling of concrete can increase the ductility of steel tube.

● The concrete core can prevent the steel tube from concaving inside, therefore it can avoid or reduce the strength degradation effects caused by local buckling. Thus, the CFST column can show a ductile behaviour even after the occurrence of local buckling.

● Since the local buckling has a reduced influence on the CFST member's ductility, the member strength degradation behaviour is mainly affected by the occurrence of concrete damage. Smaller cross-section slenderness ratios can also delay the concrete damage and increase the CFST column ductility.

Fig. 1. Designed steel box

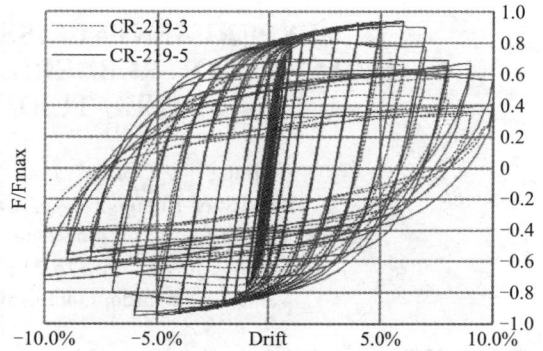

Fig. 2. Numerical results of two circular
specimens under cyclic loading

ACKNOWLEDGMENT

The research presented in this paperwas funded by the Portuguese Science Foundation through Project PTDC/ECM/117774/2010. The authors wish to express their gratitude to Ferpinta for kindly providing the steel tubes and to Presdouro for providing all the materials and conditions for the casting of the specimens.

REFERENCES

[1] Eurocode 8 BS ENV 1998-1 : 2004 , Design of structures for earthquake resistance. Part 1 , General rules , seismic actions and rules for buildings (with UK National Application Document) . *London* : *British Standards Institution.*

[2] Tao , Z. , Wang , Z. B. , Yu , Q. , 2013. "Finite element modelling of concrete-filled steel stub columns under axial compression," *Journal of Constructional Steel Research*, Vol. 67 , pp. 1719-1732.

[3] Han L. H. , 2004. "Flexural behaviour of concrete-filled steel tubes" , *Journal of Constructional Steel Research* , Vol. 60 , pp. 313-337.

SEISMIC BEHAVIOR OF SHORT-CORE BUCKLING RESTRAINED BRACES

NaderHoveidae[a], Behzad Rafezy[b]

[a]Assistant Professor, Civil Engineering Department,
Azarbaijan Shahid Madani University, Tabriz, Iran,
Hoveidae@ gmail. com.

[b] Associate Professor, Faculty of civil Engineering,
Sahand University of Technology, Tabriz, Iran

KEYWORDS: Short Core Buckling Restrained Brace, Finite element analysis, Cyclic test

ABSTRACT

The current paper investigates the seismic behavior of a new type of buckling restrained braces (BRBs) called "Short Core BRBs" in which a shorter core segment is used as an energy dissipating part and an elastic part is serially connected to the core. It seems that a short core BRB is easy to be fabricated, inspected and replaced after a severe earthquake. In addition, the energy dissipating capacity in a short core BRB is higher because of larger core strains. However, higher core strain demands result in high potential of low-cycle fatigue fracture. In this paper, a strategy is proposed to estimate the minimum core length in a short core BRBs. The seismic behavior of short core buckling restrained brace is experimentally examined. The results revealed that the short core buckling restrained brace is able to sustain large inelastic strains without any significant instability or strength degradation.

CONCLUSION

The SCBRB includes a short steel core as the brace axial load carrying member, and an outer tube as the buckling restraining mechanism. The extension of the outer tube beyond the core act as the brace core-less member which is expected to remain elastic while inelastic cyclic excursion of the core element. The current study numerically investigates the low-cycle fatigue fracture life of the SCBRBs by introducing a minimum core length for different brace orientation angles and different yielding load of the steel material used in the core member. The Coffin-Manson fatigue criterion and the AISC loading protocol for BRBs were applied to estimate and propose the minimum core length of SCBRBs. Moreover, the seismic behavior of SCBRBs was experimentally studied. Two SCBRB specimens with different core lengths were selected for the test. The AISC standard loading protocol was applied and the load-deformation relation of the brace members were obtained. The test results showed that the SCBRB system is able to sustain large plastic deformations without any significant degradation of load carrying capacity or instability. Based on the test results, none of the core length with the theoretical damage indices of 0. 9 and 1. 64 did not experienced any fracture during the cyclic loading of the brace up to the end of standard loading protocol. The known Coffin-Manson fatigue criterion and also the standard loading protocol used for evaluate the minimum core length of SCBRBs seems to be conservative. More experimental and numerical observations are required to investigate the seismic response of SCBRBs.

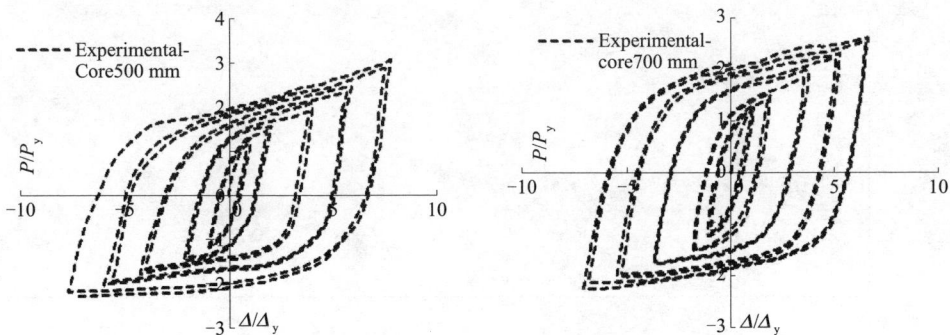

Fig 1. Hysteretic behavior of SCBRB specimens

REFERENCES

[1] Tremblay R, Bolduc P, Neville R, and DeVall R. *Seismic testing and performance of buckling restrained bracing systems.* Can. J. Civ. Engineering 2006; 33(1):183-198.

[2] Mirtaheri, M., Gheidi A., Zandi, A. P., Alanjari P., Samani H. *Experimental optimization studies on steel core lengths in buckling restrained braces.* Journal of constructional steel researches, 67 (2011) 1244-1253.

[3] Hoveidae N., Rafezy B. *Overall buckling behavior of all-steel buckling restrained braces.* Journal of constructional steel researches, 79(2012)151-158.

[4] Usami T., Wang CH., Funayama J. *Low-Cycle Fatigue Tests of a Type of Buckling Restrained Braces.* 2011, Journal of constructional steel researches; 14:956-964.

RESEARCH ON THE HYSTERETIC BEHAVIORS OF COLD-FORMED THICK-WALLED STEEL COLUMNS UNDER THE AXIAL CYCLIC LOADING

Xiao-Chao Fu[a], Yuan-Qi Li[a]

[a]Department of Structural Engineering, Tongji University,
Shanghai 1239 Siping Road, 200092
fxc-nc@ 163. com, liyq@ tongji. edu. cn

KEYWORDS: Cold-formed Thick-walled steel, Buckling mode, Hysteretic behaviour, Stiffness degradation.

ABSTRACT

In order to researchthe hysteretic behaviour of cold-formed C-section steel members, 6 cold-formed thin-walled steel columns and 9 thick-walled steel columns under the axial cyclic loading were studied by finite element method (FEM) in ANSYS. The influence of slenderness ratio about the weak axis y(λ_y) ranging from 30 to 90, width-thickness ratio(h/t) from 25 to 90 for web and buckling modes of such members was investigated. The global initial geometric imperfection, geometric and material nonlinearities and the Bauschinger effect are simultaneously included in the finite element model. Analysis results indicated that the cyclic load-deformation curve is asymmetric obvious because of buckling deformation occurred in compression. The thin-walled steel($h/t > 60$) survived local buckling and the curve is poor, but the thick-walled steel ($h/t \leqslant 30$) experienced overall buckling or strength failure and the curve is plump relatively. Members who with higher slenderness ($\lambda_y > 60$) or large width-to-thickness ratio($h/t > 60$) will have lower axial stiffness and hysteretic behaviours seriously deteriorate. The premature appearance of local buckling or overall buckling is primary reason to effect the deterioration of the hysteretic behaviour. So, the width-thickness ratio(h/t) and the slenderness ratio about the weak axis y(λ_y) are the most important factors to effect the hysteretic behaviour

CONCLUSIONS

According to the analysis above, conclusion can be obtained as follows:

(1) The cyclic load-deformation curve is asymmetric obvious because of buckling deformation occurred in compression. The cold-formed thin-walled steel survived local buckling at some stage and the curve is poor, but the cold-formed thick-walled steel experienced overall buckling or strength failure at some stage and the curve is plump, as show in *Fig. 1*.

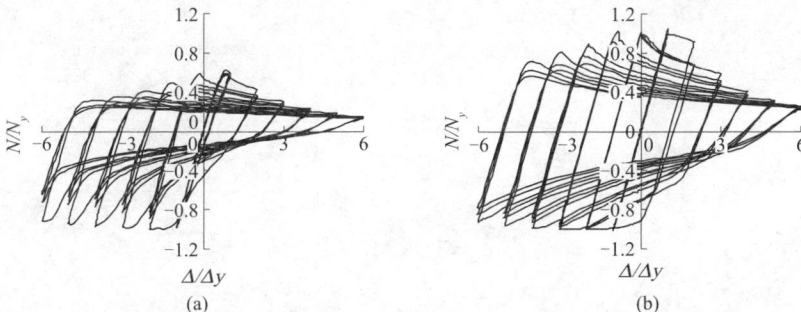

Fig 1. Cyclic load-deformation curve, (a) C1-t2-L30, (b) C5-t16-L30

(2) Members who with higher slenderness or large width-to-thickness ratio have lower axial stiffness and a gradual transition through peak load into the post-buckling range, stiffness is rapidly descendent and hysteretic behaviours seriously deteriorate.

(3) For the cold-formed C-section steel have different buckling model, such as local buckling, o-

verall buckling, and have diverse hysteretic behaviours, but meaningful comparisons of hysteretic response across limit states are challenging to make because of the different cross-sections, specimen lengths, and buckling failure modes considered in this study, this conclusion has also been proved by previous researchers.

In a word, the premature appearance of local buckling or overall buckling is primary reason to effect the deterioration of the hysteretic behaviour, further experiment needed to research in future.

REFERENCES

[1] Elchalakani M, Zhao X-L, Grzebieta R, 2003. "Tests on cold-formed circular tubular braces under cyclic axial loading". *Journal of Structural Engineering*, ASCE 2003, Vol. 129, pp. 507-514.

[2] Tremblay R, 2002. "Inelastic seismic response of steel bracing members". *Journal of Constructional Steel Research* Vol. 58, pp. 665-701.

[3] D. Padilla-Llano, M. Eatherton, C. D. Moen, 2012. "Compression-tension hysteretic response of cold-formed steel c-section framing members". 21 *International Specialty Conference on Cold-Formed Steel Structures*, America, pp. 1518-1528.

[4] DONG Jun, LU Xi, WANG Shi-qi, 2006. "Numerical analysis of the hysteretic behaviour of cold-formed thin-wall C steel members under constant compression and cyclic bending". *Journal of Disaster Prevention and Mitigation Engineering*, Vol. 26, pp. 419-424. (in Chinese).

A CLOUD SERVICE FOR AUTOMATED DESIGN OF SEISMIC BUCKLING-RESTRAINED BRACES AND CONNECTIONS

Ming-ChiehChuang[a,b], Keh-Chyuan Tsai[c], Pao-Chun Lin[d] and An-Chien Wu[a]

[a] Assistant Researcher, National Center for Research on Earthquake Engineering, Taipei, Taiwan
mcchuang@ ncree. narl. org. tw, acwu@ ncree. narl. org. tw
[b] Ph. D. Student, Dept. of Civil Engineering, National Taiwan University, Taipei, Taiwan
[c] Professor, Dept. of Civil Engineering, National Taiwan University, Taipei, Taiwan
kctsai@ ntu. edu. tw
[d] Structural Engineer, Chuang Wei Structural Engineering Inc. , Taipei, Taiwan
r97521211@ ntu. edu. tw

KEYWORDS: buckling-restrained brace, gusset plate, seismic design, cloud service, RESTful.

ABSTRACT

The welded end-slot buckling restrained brace(WES-BRB) has been developed in the Taiwan National Center for Research on Earthquake Engineering(NCREE). A steel frame equipped with the WES-BRBs can offer a cost-effective solution to meet stiffness, strength and ductility requirements for seismic steel buildings. According to the seismic design requirements of a buckling-restrained braced frame(BRBF), there are seven key checks for WES-BRB and connection designs. In order to assist the engineer with the design of the WES-BRB members and connections, an innovative cloud service named Brace on Demand(BOD) has been constructed in NCREE. This paper describes the development of the BOD service. The BOD server provides an up-to-date, on-demand service for users to effectively and automatically obtain a complete design of the WES-BRBs and connections meeting all seismic design requirements. Users are only required to enter the frame geometry and the BRB yield strength in the BOD browser. The BOD server can immediately compute and return a suitable design configuration, including all the detailed dimensions of the WES-BRB, gussets, and weld sizes. The final design results, including structural calculations in. txt and. xls format, can be downloaded and conveniently incorporated into the construction documents. This paper demonstrates that the BOD is a very efficient tool for structural engineers engaged in the seismic design of buildings using WES-BRBs to obtain suitable designs. In this study, the effectiveness of the proposed strategies for improving the initial design to meet all design checks is verified by a series of diagonal-configuration BRBF design cases using BOD.

CONCLUSIONS

As shown in the *Fig. 1*, for using cloud computing, the BOD service consists of the client-server architecture including the BOD server in the NCREE and the BOD browser in the client side. When the server receives the request sent from the client-side BOD browser, the RESTful web service of performing the WES-BRB design and connections will be invoked automatically before returning the results to the client-side browser. The BOD browser provides a friendly user interface to display the results. It allows the designers to evaluate all the WES-BRB and connection detail dimensions[1 ,2] and all DCRs[3 ,4]. In this paper, the effectiveness and efficiency of the BOD has been demonstrated by using an example(*Fig. 2*). Therefore, the BOD has been shown to be a useful tool for structural engineers to effectively reduce the iteration work inherent in BRBF design, and enhance the engineers' productivity. The cloud computing service possesses many useful characteristics, including ease of maintenance. The RESTful web service can be conveniently incorporated into other platforms, such as hand-held devices, if desired. Since the BOD was released, a number of applications have confirmed that the BOD is effective for users engaging in research and practice of seismic resistant building design. Currently, the BOD is in service and the BOD browser can be downloaded for free (http://bod. ncree. org. tw).

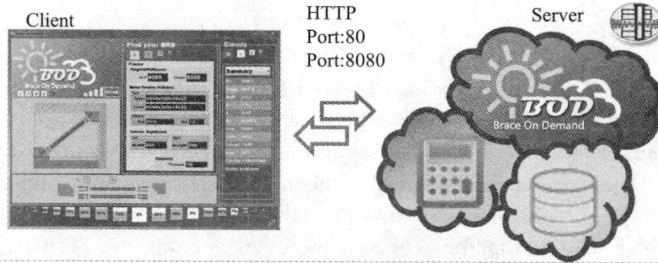

Fig 1. Client-server architecture of BOD

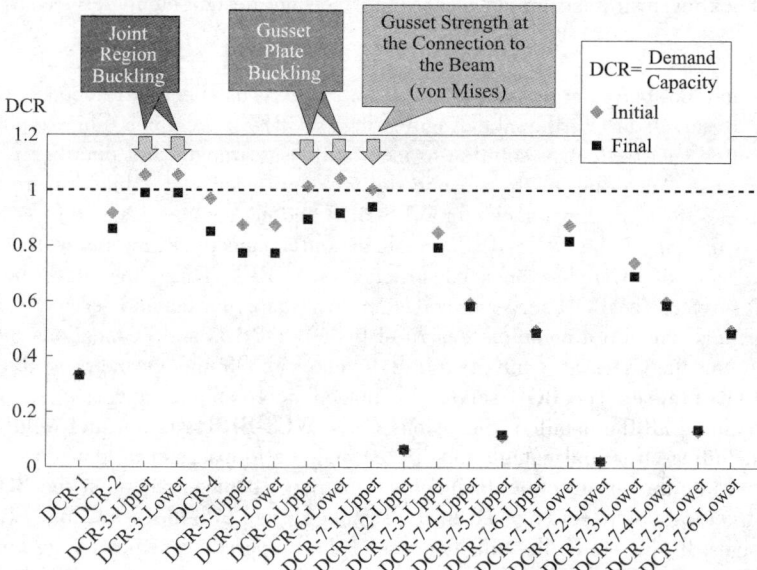

Fig 2. The results of an improved design using BOD automatic adjustment

REFERENCES

[1] Tsai KC, Wu AC, Wei CY, Lin PC, Chuang MC and Yu YJ, "Welded end-slot connection and un-bonding layers for buckling-restrained braces", *Earthquake Engineering and Structural Dynamics*, DOI: 10. 1002/eqe. 2423, 2014.

[2] Lin PC, Tsai KC, Wu AC and Chuang MC, "Seismic design and test of gusset connections for buckling-restrained brace frames", *Earthquake Engineering and Structural Dynamics*, 43 (4): 565-587, 2014.

[3] Lin PC, Tsai KC, Wu AC and Chuang MC, "User Guide for BOD: Buckling-Restrained Brace and Connection Design Procedures", http://bod. ncree. org. tw, NCREE, 2014.

[4] Chuang MC, Tsai KC, Lin PC and Wu AC, "Critical limit states in seismic buckling-restrained brace and connection designs", *Earthquake Engineering and Structural Dynamics*, DOI: 10. 1002/eqe. 2535, 2014.

SHS STUB COLUMNS UNDER CYCLIC LARGE STRAIN LOADING: AN EXPERIMENTAL AND NUMERICAL STUDY

Liang-Jiu Jia[a,b], Tsuyoshi Koyama[c], Hitoshi Kuwamura[c]

[a]Tongji University, Research Institute of Structural Engineering and Disaster Reduction, China
[b]Meijo University, Advanced Research Center for Natural Disaster Risk Reduction, Japan
[c]The University of Tokyo, Department of Architecture, Japan
ljjia@ tongji. edu. cn, koyama@ arch. t. u-tokyo. ac. jp
kuwamura@ arch. t. u-tokyo. ac. jp

KEYWORDS: Ductile fracture, Post-buckling, Cyclic loading, SHScolumn, Steel.

ABSTRACT

Brittle fracture was found to be induced by ductilefracture under cyclic loading during the 1994 Northridge earthquake and the 1995 Kobe earthquake[1], where a number of research efforts were devoted to the micro-mechanism of the ductile fracture recently, e. g. , [2]. A cyclic ductile fracture model was newly proposed by the authors[3,4], where only a monotonic tensile coupon test is required to calibrate parameters of the fracture and plasticity models. This paper aims to simulate post-buckling cracking behaviors of stub columns using the proposed cyclic fracture model and a validated plasticity model. Experiments on 12 SHS (square hollow sections) stub columns illustrated in *Fig. 1* with the test setup shown in *Fig. 2* were conducted under different loading histories listed in Table 1. Numerical simulations were also carried out to predict cracking initiation of the specimens, and investigate the capacity of the newly proposed cyclic ductile fracture model.

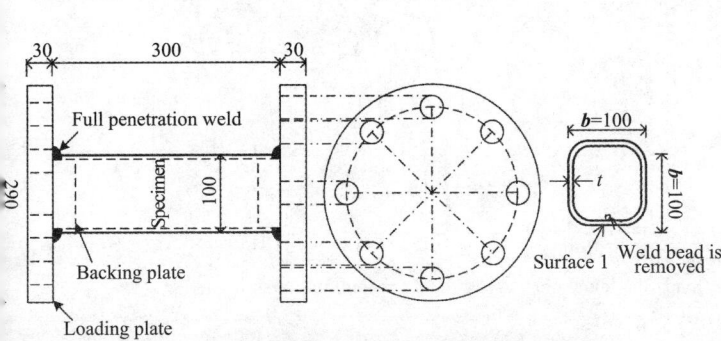

Fig 1. Configuration of specimen

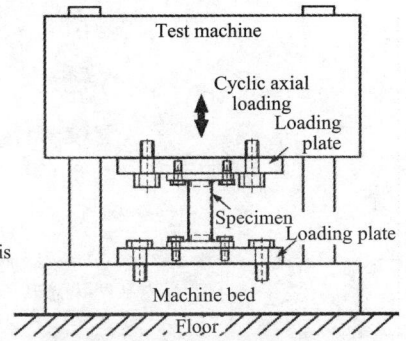

Fig 2. Test setup

Table 1. Geometrical and mechanical properties of specimens

Specimens	t(mm)	Width-to-thickness ratio	Average initial yield stress(N/mm^2)	Loading history
RH1-1			259	Monotonic tension
RH1-2 *	2. 1	47. 6	—	Two cycles *
RH1-3			260	Incremental
RH1-4			—	Constant amplitude *
RH2-1			271	Monotonic tension
RH2-2	4. 2	23. 8	227	A single full cycle
RH2-3			242	Incremental
RH2-4			217	Constant amplitude
RH3-1			242	Monotonic tension
RH3-2	8. 4	11. 9	226	A single full cycle
RH3-3			210	Incremental
RH3-4			244	Constant amplitude

CONCLUSIONS

Numerical simulations were conducted to simulate post-buckling ductile crack initiation and propagation of the specimens. The cyclic plasticity model and the ductile fracture model can well predict ductile crack initiation and the load-displacement relationships of the experiments as illustrated in *Fig. 3* and *5*. Two cracking modes, i. e. , single-crack mode shown in *Fig. 4* and multiple-crack mode shown in *Fig. 5*, are captured by numerical simulation, while the instants at the final rupture cannot be captured accurately for some specimens by the element deletion method. The accuracy of the method for the crack propagation requires further investigated. The instants at crack initiation predicted by the numerical simulations of the specimens with a large width-to-thickness ratio under a single full cycle loading are slightly overestimated probably due to that the cyclic fracture model underestimates the damage under negative stress triaxialities in extremely large plastic strain ranges, where further study is necessary at the extremely large plastic strain ranges.

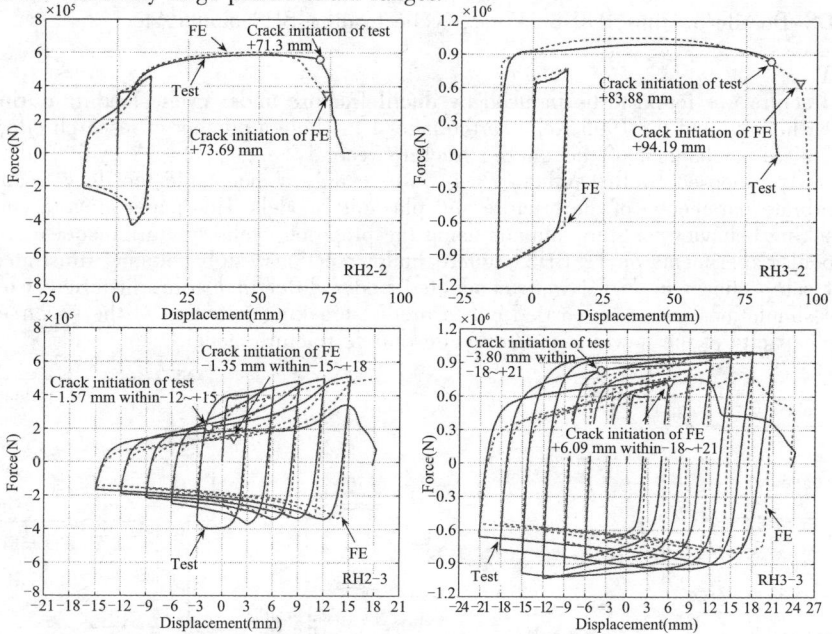

Fig. 3. Comparison of load-displacement curves for representative specimens

Fig. 4. Single-crack mode

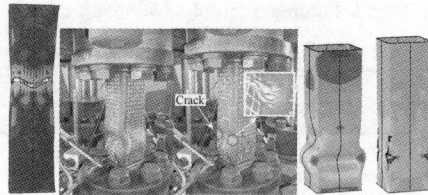

Fig. 5. Multiple-crack mode

REFERENCES

[1] Kuwamura, H. and Yamamoto, K. , 1997. "Ductile crack as trigger of brittle fracture in steel". *J Struct Eng(ASCE)* , Vol. 123 , No. 6 , pp. 729-735.

[2] Kanvinde, A. M. and Deierlein, G. G. , 2007. "Cyclic void growth model to assess ductile fracture initiation in structural steels due to ultra low cycle fatigue". *J Eng Mech(ASCE)* , Vol. 133 , No. 6 , pp. 701-712.

[3] Jia, L. J. and Kuwamura, H. , 2014. "Ductile fracture simulation of structural steels under monotonic tension". *J Struct Eng(ASCE)* , Vol. 140 , No. 2.

[4] Jia, L. J. , 2013. "Ductile fracture of structural steels under cyclic large strain loading". *Doctoral dissertation* , The University of Tokyo , Tokyo.

EXPERIMENTAL INVESTIGATION ONSTABILITY BEHAVIOR OF Q420 HIGH STRENGTH STEEL Y-SECTION COLUMNS

Hong-Zhou Deng, Xiang-Lin Yu, Ming-Yu Wei

Tongji University, Department of Structural Engineering, China

denghz@ tongji. edu. cn, 6yux@ tongji. edu. cn, nextinnovation90@ gmail. com

KEYWORDS: Experiment, Buckling, High strength, Y-section, Residual stresses.

ABSTRACT

A new type of Y-section columns consisting of a steel plate welded at the back of steel angles with equal legs was put forward and subject to experimental investigation for the purpose of sectional optimization and engineering design reference. The experiments include: Material properties tests for Q420 high strength steel; Longitudinal residual stresses tests; and structural buckling tests under axial forces. Through experimental results and numerical analysis, it was concluded that: The axial buckling strength of Y-column is generally higher than that corresponds to b-category specified in code GB50017-2003. Therefore, it is safe enough to design the columns on the basis of b-category column curves; Slenderness ratio not only affects buckling strength but also has an impact on failure modes. As the slenderness increases, the failure modes generally take on local buckling, flexural buckling accompanied by torsional buckling, pure flexural buckling, respectively.

CONCLUSIONS

Through experimental investigation of buckling behaviour of Q420 high strength steel Y-shaped columns, the following conclusions are drawn,

(1) The design force can be calculated according tothe maximum flexural slenderness ratio around symmetry axis to meet practical engineering application and do not need to consider flexural-torsional effects around non-symmetry axis.

(2) Slenderness ratios affect both buckling strength and failure modes. The failure modes take on local buckling ($\lambda \leqslant 50$), flexural-torsional buckling ($50 < \lambda \leqslant 70$), and pure flexural buckling ($\lambda > 70$), respectively.

(3) The axial buckling strength of Y-column is higher than that of b-category. Thus, it is abundant to design the columns in accordance with b-category column curves;

Comparison of test buckling strength and code results for M11XX and M88XX is listed in *Table 1*.

Table 1. Comparison of buckling strength from tests and code[1] results for M11XX and M88XX

Specimen	Test buckling strength(kN) (1)	Category *b* (neglecting torsion) (2)	Category *b* (considering torsion) (3)	[(1) − (2)]/(2) (%)	[(1) − (3)]/(3) (%)
M1130	1596	1513	1053	5. 49	51. 57
M1135	1551	1460	1047	6. 23	48. 14
M1140	1505	1401	1042	7. 40	44. 38
M1145	1469	1336	1032	9. 96	42. 32
M1150	1445	1264	1019	14. 32	41. 86
M1155	1419	1186	998	19. 61	42. 16
M1160	1417	1105	968	28. 24	46. 45
M8830	1388	1340	973	3. 59	42. 60
M8840	1285	1235	949	4. 04	35. 41
M8850	1190	1106	908	7. 56	31. 08
M8860	1165	960	839	21. 37	38. 80
M8870	1130	813	743	38. 98	52. 01
M8880	1023	682	641	50. 05	59. 52

According to GB 50017-2003[1], the design compression strength of members shall be taken as:

$$N = \frac{\varphi A f_y}{\gamma_R} \qquad (1)$$

where φ is the stability factor of axially loaded compression members (smaller value of factors around two principal axes), γ_R is the partial safety factor.

$$\varphi = 1 - \alpha_1 \lambda_n^2, \text{for } \lambda_n = \frac{\lambda}{\pi} \sqrt{f_y/E} \leqslant 0.215 \qquad (2)$$

$$\varphi = \frac{1}{2\lambda_n^2}[(\alpha_2 + \alpha_3 \lambda_n + \lambda_n^2) - \sqrt{(\alpha_2 + \alpha_3 \lambda_n + \lambda_n^2)^2 - 4\lambda_n^2}], \text{for } \lambda_n > 0.215 \qquad (3)$$

where λ_n is the normalized slenderness, α_1, α_2, α_3 are imperfection coefficients taken from Table C-5 in accordance with the cross-section classification (a, b, c and d) of Table 5.1.2 of this code. Comparison of experimental results and design column curves of GB 50017-2003 is shown in *Fig. 1*. It is recommended that category b column curve in the code be adopted for engineering design.

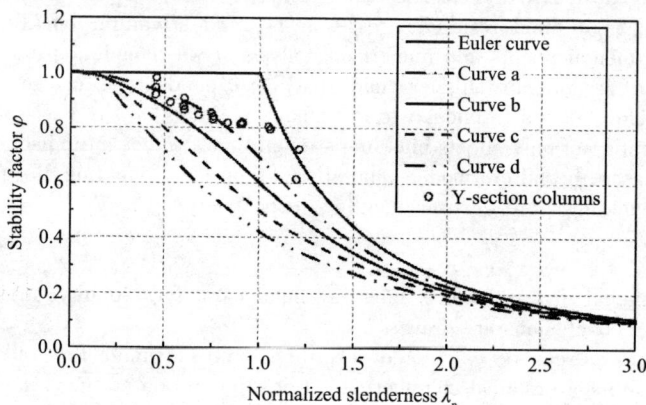

Fig 1. Comparison of experimental results and design column curves of GB 50017-2003

ACKNOWLEDGEMENT

Acknowledgments are extended to the following organizations: China Energy Engineering Group Co., Ltd. (Energy China); Energy China GEDI; Building Structure Laboratory of Tongji University; Hefei Hongyuan Transmission Tower Co., Ltd.

REFERENCES

[1] GB50017-2003, "Code for design of steel structures", Beijing: China Planning Press, 2003. (in Chinese)

SEISMIC BEHAVIOR OF CONCRETE FILLED STEEL TUBES SUBJECTED TO CYCLIC TORSION

Yu-Hang Wang[a], Jian-Guo Nie[b], Jian-Sheng Fan[b]

[a]Chongqing University, School of Civil Engineering, China

wangyuhang@ cqu. edu. cn

[b] Tsinghua University, Department of Civil Engineering, China

nejianguo@ tsinghua. edu. cn, fanjiansheng@ tsinghua. edu. cn

KEYWORDS: Concrete filled steel tube, Seismic, Torsion.

ABSTRACT

In order to study the hysteretic response and failure modes of concrete filled steel tube columns under cyclic torsion load, the quasi-static test on four concrete filled steel tube columns subjected to pure torsion was carried out. The various section types, steel ratios were chosen, and a self-design load device was used for apply the cyclic torsion load. From test results, it could be observed that the cracking and buckling of the steel tube were the main failure modes of concrete filled steel tube columns subjected to pure torsion. The vertical crack appeared at the surface of the steel tube with circular section and extended in the diagonal direction, and then the concrete near the crack of the steel tube crushed. The cross diagonal waves appeared on the long side of the steel tube with rectangular section due to the buckling of the steel tube, and the specimen with thinner steel tube had more waves than that with thicker steel tube. The hysteretic relations of concrete filled steel tube columns under cyclic pure torsion were very plump, indicating that concrete filled steel tube columns had good seismic capacity. The unloading stiffness of concrete filled steel tube columns was nearly equivalent to the initial elastic stiffness, and the ductility was also very good. When concrete filled steel tube columns beard pure torsion load, spiral diagonal compressive struts would be created in the in-filled concrete, and the axial components of the diagonal compressive force produced in the in-filled concrete was equal to the axial tensile force produced in the steel tube due to the section force equilibrium condition, corresponding to the tensile stress observed in the steel tube of specimens. The steel tube of concrete filled steel tube columns subjected to pure torsion with high steel ratio has more contribution to the torsion capacity than those with low steel ratio.

CONCLUSIONS

An experimental investigation on concrete filled steel columns under pure cyclic torsion load has been carried out in this study. The following observations and conclusions are made based on the study in the paper: The CFST columns under cyclic torsion have high energy dissipation capacity according to the torsion moment-rotation angle hysteretic curves. Furthermore, the degradation of the strength and rigidity of specimens is not obvious. The pure torsion state of CFST columns could be regarded as the combination of tension-torsion state of the steel tube and compression-torsion state of the in-filled concrete, and the total force of the in-filled concrete is equal to the total force of the steel tube.

REFERENCES

[1] Chua LO, Stromsmoe KA. , "Mathematical model for dynamic hysteresis loops", International Journal of Engineering Science, 9:433-450, 1971.

[2] ASCCS. Concrete filled steel tubes-a comparison of international codes and practices. In: ASCCS seminar, Innsbruck, Austria, 1997.

[3] Shanmugam N. E. , Lakshmi B. . State of the art report on steel-concrete composite columns. J Construct Steel Res 2001;57(10):1041-80.

[4] Gourley B. C. , Tort C. , Hajjar J. F. , Schiller P. H. . A synopsis of studies of the monotonic and cyclic behavior of concrete-filled steel tube beam-columns. Structural engineering report no. ST-01-4. Minnesota, USA: Department of Civil Engineering, University of Minnesota,

December 2001.

[5] Han L. H. Tests on stub columns of concrete-filled RHS sections. J Construct Steel Res 2002;58 (3):353-72.

[6] Schneider S. P. . Axially loaded concrete-filled steel tubes. J Struct Eng 1998;124(10):1125-38.

[7] Beck J. , Kiyomiya O. . Fundamental pure torsional properties of concrete filled circular steel tubes. J Mater Conc Struct Pavements,JSCE no. 739/V-60,2003. pp. 85-96.

[8] Xu Jishan, Gong An. Experimental study on short concrete filled steel tube column under compression and torsion. The third annual meeting of China Steel Construction Society Association for Steel-Concrete Composite Structures. 1991.

[9] Gong A. . Experimental study on concrete filled steel tubular short column under combined torsion and compression. MSc thesis,Beijing Institute of Civil Engineering and Architecture,1989.

[10] Lee G. , Jishan X. , An G. , Zhang K. C. . Experimental studies on concrete-filled tubular short columns subjected under compression and torsion. In:Wakabayashi Proceedings 3rd International Conference on Steel-Concrete Composite Structures, Fukuoka, Japan, 26-29 September 1991. Tokyo:Association for International Cooperation and Research in Steel-Concrete Composite Structures. 1991. 143-148.

[11] Zhou J. The experimental research of concrete filled steel tubular slender column under combined compression and torsion. MSc. thesis,Beijing Institute of Civil Engineering and Architecture,1990

[12] Han L. H. , Yao G. H. , Tao Z. . Performance of concrete filled thin-walled steel tubes under pure torsion. Thin-Walled Struct. ,2007,45,24-36.

[13] Han L. H. , Yao G. H. , and Tao Z. . Behaviors of concrete filled steel tubular members subjected to combined loading. Thin-Walled Struct. 2007,45:600-619.

RESEARCH AND APPLICATION OF STEEL PLATE COMPOSITE SHEAR WALLS

Fan Zhong[a], Wang Jinjin[a], Zhang Lili[b]

[a] China Architecture Design Institute CO. LTD, China

fanz@ cadg. cn, wjj. ncut@ 163. com

[b] Beijing Liujian Construction Group Co. LTD, China

zhlili@ sohu. com

KEYWORDS: Steel plate composite shear walls; Structural details; Test; Analysis; Construction technique.

ABSTRACT

Steel plate composite shear walls combine the advantages of steel and concrete members, which mean that steel plate composite shear walls are able to meet the requirement of axial compression ratio with small wall thickness. As a result, the weight of the structure will be much lighter, the spaces will be cut down and the seismic response will be alleviated. In addition, their abilities of fire protection, thermal insulation, sound insulation effect and durability are good.

According to the structure details, steel plate composite shear walls can be divided into embedded steel plate composite shear walls and steel encased composite shear walls. Besides, there are some new types of SPCSWs emerged in recent years.

After years of research and practice, SPCSWs are widely applied in high-rise building, especially super high-rise building in high earthquake intensity regions. More than half of the domestic super-tall adopts steel plate composite shear walls for the reason that they can improve the structural seismic performance, increase the shear walls' bearing capacity and ductility, and reduce the bottom walls' thickness, axial compression ratio and the structure space occupancy.

Many researches have been done on SPCSWs and calculation methods and relevant construction measures have been put forward. But the emphases are mainly on the component compression-bending performances. Studies on morphology under tension-bending force are few and the relevant calculation methods have not been established. Usually, finite element analysis software XTRACT, which is based on the assumption of plane section, is used to check the design of components. Besides, general finite element software, like Msc. Marc, Ansys and Abaqus can also be used to calculate and analyze.

Most kinds of shear wall formulas are only applicable to in-line wall, but in the practical construction, there are a large number of complicated L shaped section walls. On this point, the XTRACT software can conveniently be used to analyze bearing capacity of the normal section of SPCSWs, as shown in *Fig. 1*.

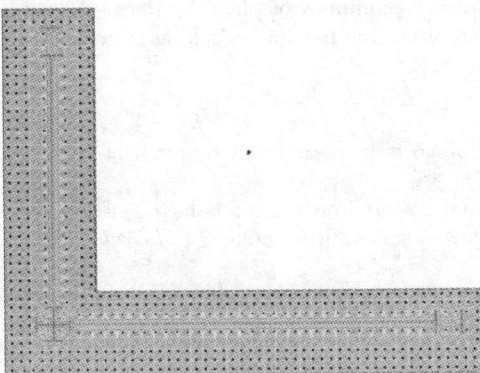

(a) L shape composite wall (b) P-M curves

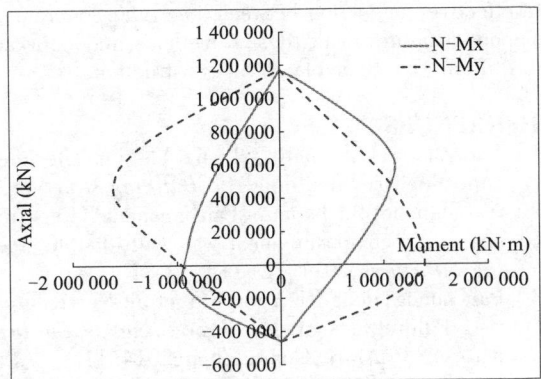

Fig. 1. esults of XTRACT

Based on the Yinchuan Greenland Center project, experiments on the performance of SPCSWs under tension-bending are taken at Tsinghua University. Preliminary test results show that the SPCSWs have a good performance under tension-bending, test phenomena are shown in *Fig. 2*, *Fig. 3*:

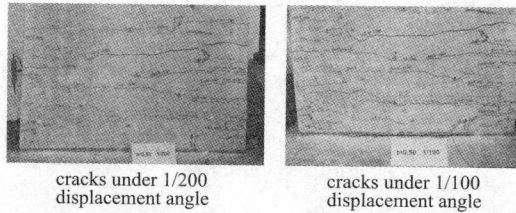

cracks under 1/200
displacement angle

cracks under 1/100
displacement angle

Fig. 2. Cracks on the component root under 0. 5 axial tension ratio

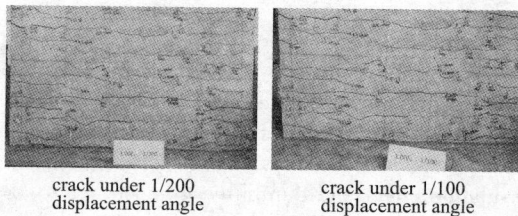

crack under 1/200
displacement angle

crack under 1/100
displacement angle

Fig. 3. Cracks on the component root under 0. 3 axial tension ratio

SPCSWs are mix force shoulder components which combined of steel and concrete. Solving the problem of bond-slip and ensuring the collaborative work between the two materials is a prerequisite to achieve their stress performances. However, too many rebar passing through the steel plate will weaken the steel plate and closed stirrups with 135 degree hooks, which passing through steel plate, are hardly to construct. Therefore, reasonable optimize of wearing plate rebar can largely improve construction efficiency and quality.

At present, more and more cracking problems in SPCSWs are emerged to the different extent. Wall cracks bring large hidden structural danger to high-rise and have a detrimental influence on structural durability and applicability. Therefore, in-depth researches on cracks in SPCSWs are carried out combining Greenland of Wangjing Center project, and have obtained very good effects in practical application.

CONCLUSIONS

Compared with ordinary reinforced concrete shear wall, the composite of steel plate and concrete can effectively improve the bearing capacity, ductility and hysteretic energy dissipation of the composite shear walls. The bearing capacity and hysteretic energy dissipation are enhanced with the increase of steel ratio of SPCWs. But there are still some problems remain to be solved and improved, such as the effective connection between steel and concrete, applying conditions of plane section assumption, component failure mode, design methods under the action of tension-bending, cracking of concrete and the difficulty and complexity in construction.

REFERENCES

[1] Fan Zhong, Liu Xuelin, Huang Yanjun. The latest progress in design and research of shear wall super high-rise building[J], *Building structure*, 2011. 04.
[2] Nie Jianguo, Bu Fanmin, FanJiansheng. Experimental research on seismic behavior of low shear-span ratio composite shear wall with double steel plates and infill concrete[J]. *Journal of Building Structures*, 2011, 32(11): 74-81.
[3] Fan Zhong, Peng Yi, Yang Kai, et al. Key technology of Greenland of Wangjing center super high-rise building design[M]. *Proceedings of the twenty-third national academic meeting of high-rise building structure*, Guang Zhou, 2014. 11.
[4] Fan Zhong, Kong Xiangli, Liu Xianming, Yang Su. Bundled tower structure and application in Olympic Sightseeing Tower[M]. *Proceeding of IStructE Conference on Structural Engineering in Hazard Mitigation*, October 2013, Beijing China.

DETECTION OF NONLINEAR BEHAVIOR IN EXPOSED COLUMN BASES USING THE SECOND TIME DERIVATIVE OF ABSOLUTE ACCELERATION

Masaki Wakui[a], Jun Iyama[a], Tsuyoshi Koyama[a]

[a]the University of Tokyo, Department of Architecture, Japan

wakui@stahl.arch.t.u-tokyo.ac.jp, iyama@arch.t.u-tokyo.ac.jp,

koyama@arch.t.u-tokyo.ac.jp

KEYWORDS: Snap, Structural health monitoring, Nonlinearity, Exposed-type column base, Shaking test

ABSTRACT

This paper proposes a method to detect damage due to change in the stiffness of dynamically loaded exposed-type steel column bases utilizing the second derivative of the absolute acceleration, called "snap". The column bases are designed so that either the anchor-bolt or the base-plate initially yields. An exposed-type column base typically shows a slip-type behavior with a load-displacement hysterisis shown in *Fig. 1*. A change in the stiffness in the load-displacement relationship should occur at the following 4 states: A(yielding), B(unloading), C(onset of slip), D(end of slip)

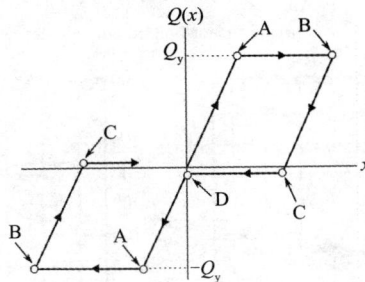

Fig 1. A slip-type of restoring force characteristic

In order to observe the detection of nonlinearity in detail, the snap time histories and the load-displacement relationship of relevant cycles (for example the cycle shown in *Fig. 2*), are selected. Two types of thresholds are defined to detect nonlinearity in the load-displacement. One is shown as the two dotted line. This threshold value varies with time, such that it depends on the ground acceleration. The other is shown as the two dashed line. This threshold value is a constant. The circled numbers denote the extremum points exceeding the range of the constant threshold. Moreover, the alphabets in the framed box designate the type of change in the stiffness observed, either State A, B, C, or D as shown in *Fig. 1*. This method succeeded the detect the anchor-bolt yielding at a high rate of 86%, however, the rate of the detection of the base-plate yielding was only 64%. This is because the stiffness change when the base-plate yielding occurs is more gradual than the case of anchor-bolt yielding, which makes the detection by "snap" difficult.

CONCLUSIONS

This paper investigates the capability of a method to detect damage due to the change in the stiffness through the second derivative of the response absolute acceleration, called "snap", on actual structures. The method is applied to data obtained from two full scale exposed-type steel column bases subjected to dynamic loading in shaking table tests. One specimen was designed so that the anchor-bolt initially yields(ABY), the other so that the base-plate initially yields(BPY). The specimens showed nonlinear behavior at the yielding, unloading, opening and closing of the gap between the footing and the base-plate in the shaking table test. In this paper, it is investigated whether this method using "snap" of the response acceleration could detect these nonlinear behavior properly or not.

From the comparison between the analysis by snap and the load-displacement relationship of the shaking table test, it is shown that this method succeed in detecting the nonlinear behavior at a high rate of 86% for ABY, however, the success rate for BPY was only 64%. Detailed observation clarified that the success rate of detecting yielding is 100% for both specimens, but the success rate of detection of the slip behavior is 33% to 71% for the BPY specimen. This reason is because in the BPY specimen the slip behavior occurs more gradually compared to the ABY specimen, which means that the stiffness change is slower and the value of snap smaller. This observation implies that it may be necessary to modify the threshold bound shown in this paper according to the type of nonlinear behavior expected.

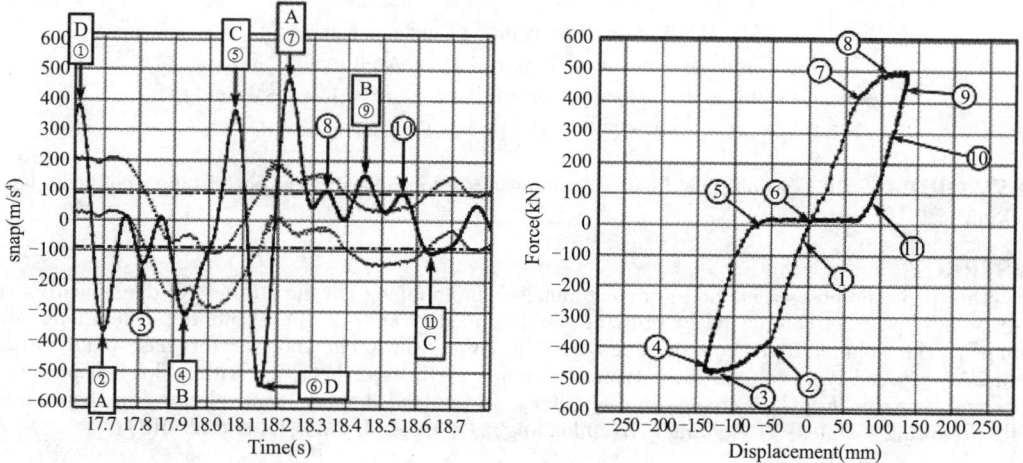

(a) the anchor-bolt initially yields

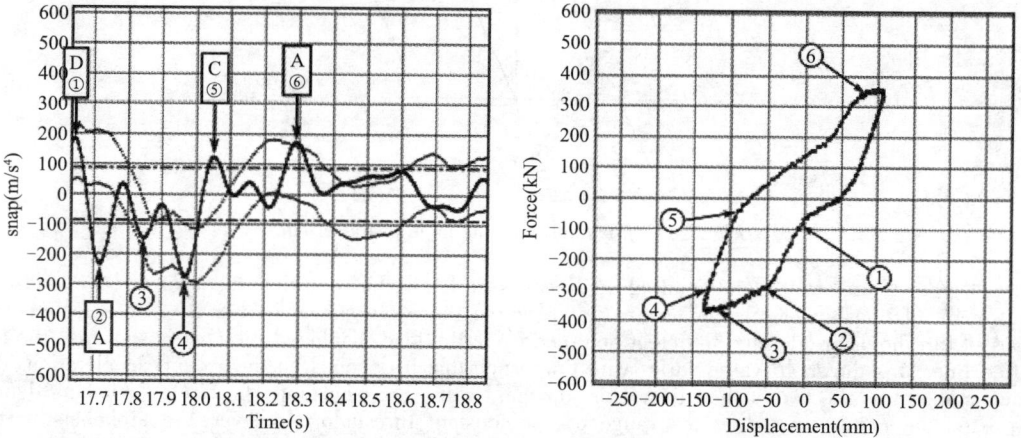

(b) the base-plate initially yields

Fig 2. The snap time history and the load-displacement relationship

ACKNOWLEDGMENT

The authors would like to thank Prof. S. Yamada of Tokyo Institute of Technology for providing the data used in this paper. This work was supported by TAKENAKA IKUEIKAI Architecture Grant.

REFERENCES

[1] Wakui M. , Iyama J. , Koyama T. , "Detection of nonlinearity in vibrational systems using second time derivative of absolute acceleration", *Second European Conference on Earthquake Engineering and Seismology*, Istanbul, 2014.

[2] Wakui M. , Iyama J. , "Detection of nonlinearity in vibrational systems using the high-order derivative of absolute response acceleration containing noise", *Summaries of technical papers of annual meeting AIJ*, :89-90, 2014 (in Japanese).

[3] Akiyama H. , Yamada S. , Takahashi M. , Katsura D. , Kimura K. , Yahata S. , "Full scale table test of the exposed-type column bases", *Journal of Structural and Construction Engineering*, AIJ, 514: 185-192, 1998 (in Japanese).

SEISMIC BEHAVIOUR OF X BRACINGS: ANALYSIS OF MODELS AND DESIGN CRITERIA

Antonio Formisano, Beatrice Faggiano, Giuseppe Marino, Federico M. Mazzolani
Department of Structures for Engineering and Architecture
Piazzale Tecchio 80, 80125 Naples
e-mails: antoform@ unina. it, faggiano@ unina. it, gmarino1986@ libero. it, fmm@ unina. it

KEYWORDS: Buckling, Design Criteria, Numerical Model, X-Bracings, Pushover Analysis.

ABSTRACT

The design rules for X bracing[1,2] generally do not make any difference when the compression diagonal is considered or neglected. Actually, the tensile and compression diagonals interact each other more or less depending on their connection type. For this reason, in the current study the critical analysis of the design provisions for CBF-X structures is done with the aim to detect situations where the calculation models neglecting the compression braces are reliable or not. First, the calibration of appropriate numerical models of CBF-X systems on the basis of available experimental tests[3,4] is set-up by means of the "SeismoStruct" software[5]. Subsequently, a parametric analysis on one-storey CBF-X structure is performed by varying several geometrical parameters. Finally, some provisions for the braces slenderness limits are given aiming at identifying the ranges where the calculation models based on single tensile diagonals only can be adopted for sake of simplicity or not.

CONCLUSIONS

On the basisof the results obtained from the performed analysis, the main concluding remarks on the design provisions for CBF-X systems, whose numerical test setup and geometrically imperfect configuration are respectively depicted in *Fig. 1* a and *1*b, are summarized in the following.

The design chart shown in *Fig. 2* is based on the results of the parametric analysis. It can be useful for selecting the design criteria for X-CBF systems. In particular, three behavioural fields are identified:

-for stocky diagonals (normalised slenderness values less than 0. 8), the CBF-X systems can be designed considering only the strength contribution and neglecting buckling in both diagonals;

-for intermediate slenderness diagonals (normalised slenderness values between 0. 8 and 1. 5), the influence of both diagonals should be taken into account. Therefore, the buckling behaviour of the compressed diagonal must be considered.

-for slender diagonals (normalised slenderness values greater than 1. 5), the influence of the compressed diagonal can be neglected and the CBF-X system design is made by considering the tensile brace only.

Given the complex nature of the behaviour of CBF-X, these results based on simplified assumptions represent a first attempt for identify the main physical aspects. They must be considered as the prelude to a series of future developments, aiming at achieving a better knowledge on the global performances of X-bracing systems against earthquake. As a first important result, the parametric analysis has shown that the Italian code has serious shortcomings and does not interpret correctly the actual mechanical behaviour of these structural configurations. This is the main reason why it is necessary to undertake a wide experimental and numerical tests campaign on single bracing members and on CBF-X systems, in order to deepen not only the knowledge on their actual mechanical behaviour, but also to better calibrate numerical models to be implemented in subsequent parametric analysis, also on multi-story systems, with the goal to further improve the seismic code provisions.

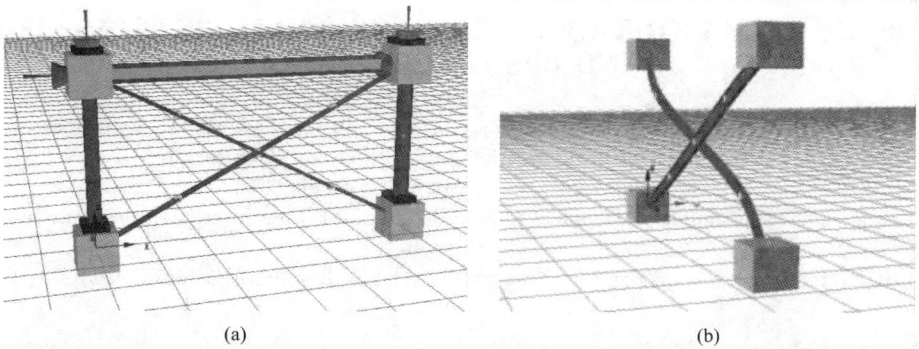

Fig 1. The Seismo Struct numerical model : (a) test setup ; (b) initial geometrical imperfections

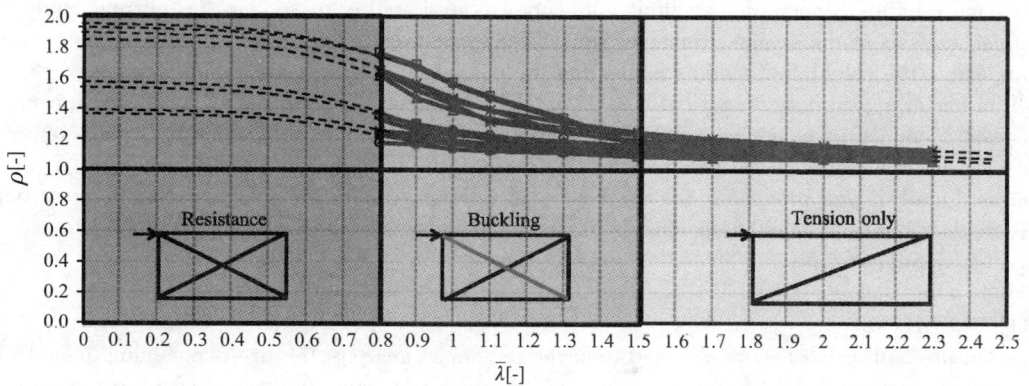

Fig 2. Comparison curves ρ-$\bar{\lambda}$: final design chart for CBF-X systems

REFERENCES

[1] Ministero delle infrastrutture , D. M. 14/01/2008 , 2008. *Technical codes for construc- tions* (in Italian).

[2] CEN , 2005. "EN 1998-1 : 2005. Eurocode 8 : Design of structures for earthquake resistance-Part 1 : General rules , seismic actions and rules for buildings". *European Committee for Standardization*, *Bruxelles.*

[3] Black G. R. , Wenger B. A. and Popov E. P. , 1980. "Inelastic Buckling of Steel Struts Under Cyclic Load Reversals". *UCB/EERC*-80/40 , Earthquake Engineering Research Center , Berkeley , CA , USA.

[4] Wakabayashi M. , Matsui C. , Minami K. and Mitani I. , 1970. "Inelastic Behaviour of Full Scale Steel Frames". *Kyoto University Research Information Repository* , *Disaster Prevention Research Institute annuals.*

[5] Seismosoft , 2011. "*SeismoStruct-A computer program for static and dynamic nonlinear analysis of framed structures*". Available from URL : www. seismosoft. com.

FINITE ELEMENT ANALYSIS OF COLUMN BASE WEAK AXIS ALIGNED ASYMMETRIC FRICTION(WAFC)

M. Hatami[a], J. Borzouie[b], G. A. MacRae[b], M. Yekrangnia[c], S. Abubakar[a]

[a] Department of Civil Engineering, University of Technology Malaysia, Johor, Malaysia, School of Civil and Environmental Engineering,

[b] University of Canterbury, Christchurch, New Zealand.

e-mails: hmahdi3@ live. utm. my, jamaledin. borzouie@ pg. canterbury. ac. nz

[c] Department of Civil engineering, Sharif University, Tehran, Iran

e-mails: Gregory. macrae@ canterbury. ac. nz, yekrangnia@ mehr. sharif. edu, suhaimi@ utm. my

KEYWORDS: Low damage construction, Column Base Weak Axis Aligned Asymmetric Friction Connection, Base connection, Finite element, Asymmetric friction connection, Column shortening

ABSTRACT

This paper presents finite element modeling of a base connection with column base weak axis aligned asymmetric friction connection(WAFC) considering 3D solid elements, dynamic implicit analysis, and non-linear geometry. The hysteretic curve of the developed model is compared with results from five experimental tests covering a wide range of design variables. According to the in-plane, and out of plane cyclic loading with and without axial force, it is shown that the presented finite element model can replicate the overall behaviour of connection. The peak moment resistance at 4% drift was predicted to an accuracy of 90%. Also, while column shortening of this type of base connection was small (less than 10mm) under different magnitude of axial forces a parameter study shows that axial shortening becomes more significant for greater axial loads, especially when out-of-plane loading occurs.

SUMMARY AND CONCLUSIONS

Observations from past earthquakes have shown that current structural design codes can provide life safety. However, significant damage and economic loss has resulted. Low damage construction techniques have been recently developed to minimize damage in severe earthquakesMost of the work carried out on low damage systems has been focused on beam to column joint connection, or braces. Relatively little research has been conducted on column base. If the structure remains undamaged after a strong earthquake, but the earthquake caused damage to the base of structure, that building is not a low damage building. Recently, weak axis aligned asymmetric friction connection(WAFC) was developed and experimentally tested at the University of Canterbury(Borzouie, et al. [1], Borzouie, et al. [2], Borzouie[3]).

Steel columns are subjected to axial compression load and inelastic cyclic displacement during earthquakes that cause column axial shortening. This can result in column shortening as a consequence of material inelastic deformation without occurring any buckling. If buckling occurs, the buckling can enhance the rate of axial shortening. The magnitude of column shortening is affected by the cumulative inelastic deformation demand as well as the axial force level according to studies by MacRae et al. [4],[5]. Finally some methods to estimate column shortening and the way to mitigate that were developed.

In this paper, numerical analysis of WAFC was conducted in Abaqus to answer the following questions:

1. What is the performance of weak axis aligned asymmetric friction connection due to cyclic and axial compression loads?

2. What is the effect of out-of-plane and in-plane loading considering different amount of axial forces on quantity of column shortening?

To evaluate the reliability of the FE model, force-displacement curves of experimental tests and FE model were compared. In *Fig. 1* (Tests #1), lateral displacement was applied causing strong axis bending without axial load. It can be observed the FE model indicates more elastic rotational base flexibility than the experimental test that could be justified because of some differences in exact material properties, initial stiffness, friction between surfaces, mesh sizes and bolt tightening load. The strengths seem to be similar between the two models.

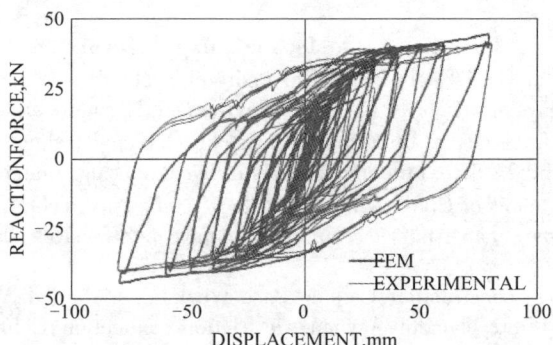

Fig. 1. Force-displacements curve of model without axial load (Test #1)

This paper described the finite element simulation of column base weak axis aligned asymmetric friction connections (WAFC) subjected to cyclic displacement causing strong and weak axis bending, with and without axial force. It was shown that:

1. The configuration behaved well with almost no damage. However, some localized yielding at the end of the column reduced the lateral stiffness.

2. The axial shortening observed was less than 0. 5% of the column length for columns with different levels of axial force and lateral displacement. It was not affected significantly by the number of loading cycles. This is different to fixed base tests where large axial shortening displacements have been observed.

REFERENCES

[1] J. Borzouie, G. W. Rodgers, G. A. MacRae, G. J. Chase, and C. G. Clifton, " Base Connections Seismic Sustainability and Base Flexibility Effects," presented at the Steel Innovations conference, Christchurch, New Zealand, 2013.

[2] J. Borzouie, G. A. MacRae, G. J. Chase, and C. G. Clifton, "Experimental studies on cyclic behaviour of steel base plate connections considering anchor bolts post tensioning," presented at the NZSEE, Auckland, New Zealand, 2014.

[3] J. Borzouie "Low Damage Steel Base Connection," PhD, Civil and Natural Resources Engineering, University of Canterbury, University of Canterbury, 2015, to be submitted.

[4] G. A. MacRae, "The seismic response of steel frames", PhD Thesis, University of Canterbury, 1989.

[5] G. A. MacRae, C. R. Urmson, W. R. Walpole, P. Moss, K. Hyde, and C. Clifton, "Axial Shortening of Steel Columns in Buildings Subjected to Earthquakes," *Bulletin of the New Zealand Society for Earthquake Engineering*, vol. 42, p. 275, 2009.

COLUMN BASE WEAK AXIS ALIGNED ASYMMETRIC FRICTION CONNECTION CYCLIC PERFORMANCE

J. Borzouie[a], G. A. MacRae[a], J. G. Chase[b], G. W. Rodgers[b], G. C. Clifton[c]

[a] University of Canterbury, Department of Civil and Natural Resources Engineering, New Zealand.

jamaledin. borzouie@ pg. canterbury. ac. nz, gregory. macrae@ canterbury. ac. nz

[b] University of Canterbury, Department of Mechanical Engineering, New Zealand.

geoff. chase@ canterbury. ac. nz, geoff. rodgers@ canterbury. ac. nz

[c] University of Auckland, Department of Civil and Environmental Engineering, New Zealand.

c. clifton@ auckland. ac. nz

KEYWORDS: Base connection; Asymmetric Friction Connection; Low Damage Design; Steel Connection, Wafc

ABSTRACT

Observations from past earthquakes and experimental tests show that base plate connections can be damaged under severe loading. Since replacing any structural element on this level is difficult and costly, any damage to base connections can lead to demolition and rebuilding. Low damage designs are a new concept to increase the economic performance while maintaining or increasing life safety. In this concept, damage is mitigated by replaceable dissipativeelements rather than sacrificing structural yielding. Research on low-damage construction in steel frame structures has concentrated on beam-to-column moment resisting joints(Clifton[1], Rodgers, et al. [2], MacRae, et al. [3], Rodgers, et al. [4]) and braces(Chanchí, et al. [5]). However, no study has focused specifically on low-damage connections at the base of the columns.

This research presents the experimental results of a low damage design column base weak axis aligned asymmetric friction connection(WAFC). In this connection, the column is connected to the base plate by four asymmetric friction connections(AFC) rather than direct welding of the column to the base plate. Cyclic tests were carried out in-plane about the strong axis, out of plane about the weak axis, both with and without applied axial force. Displacement inputs ranged from 0. 2% drift(4 mm) ratio to 4% (80mm). The increase in each new drift is 1. 25 to 1. 5 time that of the previous step. Three full cycles and one cycle of half of the drift level in the studied direction were applied. Also, a simple analytical model is developed to represent the behaviour of this base connection. This method considers the friction resistance, prying and axial force effects for calculation of moment resistance.

CONCLUSIONS

The results show this type of rocking base connection can tolerate high levels of drift without any significant damage and strength degradation at the base. However, some stiffness degradation due to flange yielding of the column was observed. The bolts and shims can be reused after an earthquake and level of shims degradation was minor. Also, the boundary plates remained elastic. If it is required any repair, it may be undertaken by retightening and replacing of the bolts and shims. Overall, experimental results illustrate that this base connection has potential to perform as a low damage connection, even for large drifts, thus offering significant potential economic savings while still maintaining the same life safety.

A simple analytical model developed which can calculate the neutral axis of the column, column uplift, and the moment resistance at 4% drift. Frictional resistance, axial force and prying effects were considered in the analytical model, and it can provide a reasonable estimate for strength. Also, for untested column it can predict the stiffness with acceptable accuracy. However, during cyclic loading for further tests some yielding happened on the column that reduced the stiffness from the predicted value during subsequent test.

ACKNOWLEDGMENT

The authors would like to acknowledge the MBIE Natural Hazards Research Platform(NHRP)for its support of this study as part of the Composite Solutions project. All opinions expressed remain those of the authors.

REFERENCES

[1] C. G. Clifton, "SEMI-RIGID JOINTS FOR MOMENT. RESISTING STEEL FRAMED SEISMIC-RESISTING SYSTEMS," PhD, University of Auckland, 2005.

[2] G. W. Rodgers, J. G. Chase, J. B. Mander, N. C. Leach, and C. S. Denmead, "Experimental Development, Tradeoff Analysis and Design Implementation of High Force-To-Volume Damping Technology," *NZSEE Bulletin*, vol. 40, pp. 35-48, 2007.

[3] G. A. MacRae, G. C. Clifton, H. Mackinven, N. Mago, J. Butterworth, and S. Pampanin, "The Sliding Hinge Joint Moment Connection," *Bulletin of the New Zealand Society for Earthquake Engineering*, vol. 43, p. 10, 2010.

[4] G. Rodgers, K. Solberg, J. Mander, J. Chase, B. Bradley, and R. Dhakal, "High-Force-to-Volume Seismic Dissipators Embedded in a Jointed Precast Concrete Frame," *Journal of Structural Engineering*, vol. 138, pp. 375-386, 2012.

[5] J. C. Chanchí, R. Xie, G. MacRae, G. Chase, G. Rodgers, and C. Clifton, "Low-damage braces using Asymmetrical Friction Connections(AFC)," presented at the NZSEE, Auckland, New Zelaand 2014.

BEHAVIOR OF EXTERNAL DIAPHRAGM CONNECTIONS FOR SQUARE CFST COLUMNS UNDER BIDIRECTIONAL LOADING

Helmy Tjahjanto[a], Gregory MacRae[a], Anthony Abu[a], Charles Clifton[b], Tessa Beetham[c], Nandor Mago[d]

[a] Department of Civil and Natural Resources Engineering, University of Canterbury, Christchurch, New Zealand
helmy. tjahjanto@ pg. canterbury. ac. nz

[b] Department of Civil and Environmental Engineering, University of Auckland, Auckland, New Zealand

[c] Aurecon, Christchurch, New Zealand

[d] Heavy Engineering Research Association, Auckland, New Zealand

KEYWORDS: concrete-filled steel tubular column, connection, external diaphragm, bidirectional loading

ABSTRACT

External diaphragm connections are suitable for two-way frames with concrete-filled steel tubular (CFST) columns. The connections have efficient force transfer mechanisms and require simple construction methods. However, studies on the behavior of such connections under bidirectional loading are still very limited. This paper presents a numerical study on external diaphragm connections under one-way and two-way loading. Initial yielding and residual deformation of steel tubes are considered as the design limit states for external diaphragm capacity. It is shown the perpendicular loading does not significantly affect the capacity demand. A simple design method is proposed based on the basic characteristics from numerical analysis results.

CONCLUSIONS

Numerical analyses have been carried out to understand the behavior of external diaphragm connections for CFST columns. The application of perpendicular loads does not significantly affect the maximum stress and the nonlinear deformation of steel tubes(*Fig. 1*). In all cases, maximum stresses occur on the tension sides near CFST column corners. A design limit state based on nonlinear deformation under tension on the middle part of CFST column is recommended. A simple design method is proposed to estimate the capacity of external diaphragm connections(*Fig. 2*). The capacity is determined by the critical size of the external diaphragm without considering the effect of perpendicular loading.

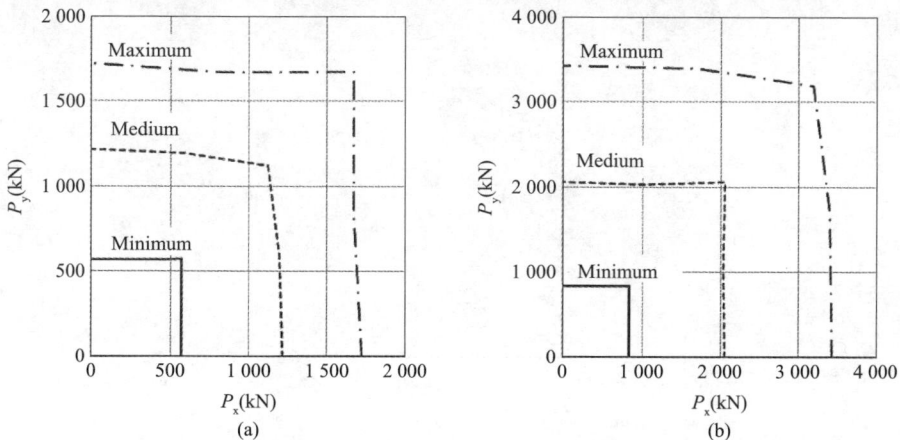

Fig. 1. Perpendicular load interaction.
(a) Based on initial yielding; (b) Based on 0. 5 mm residual deformation

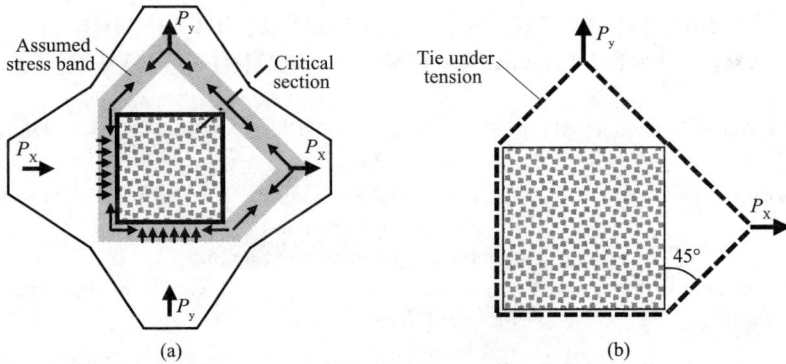

Fig. 2. The tie method concept. (a) Simplified stress band; (b) Tie element approach

REFERENCES

[1] Morino S , Uchikoshi M , and Yamaguchi I , "Concrete-filled steel tube column system-its advantages" , *Steel Structures* , vol. 1 , pp. 33-44 , 2001.

[2] Guo L , Zhang S , Kim WJ , and Ranzi G , "Behavior of square hollow steel tubes and steel tubes
filled with concrete" , *Thin-Walled Structures* , vol. 45 , pp. 961-973 , 2007.

[3] De Oliveira W , De Nardin S & de Cresce El Debs A , "Evaluation of passive confinement in CFT
columns" , *Journal of Constructional Steel Research* 66 , pp. 487-495 , 2010.

[4] Inai E , Mukai A , Kai M , Tokinoya H , Fukumoto T , and Mori K. , "Behavior of Concrete-Filled
Steel Tube Beam Columns" , *Journal of Structural Engineering* , vol. 130 , pp. 189-202 , 2004.

[5] Alostaz Y , and Schneider S , "Analytical behavior of connections to concrete-filled steel tubes" ,
Journal of Constructional Steel Research , vol. 40 , no. 2 , pp. 95-127 , 1996.

[6] Schneider S , and Alostaz Y , "Experimental behavior of connections to concrete-filled steel
tubes" , *Journal of Constructional Steel Research* , vol. 45 , no. 3 , pp. 321-352 , 1998.

[7] Chiew SP , Lie ST , and Dai CW , "Moment Resistance of Steel I-Beam to CFT Column Connections" , *Journal of Structural Engineering* , vol. 127 , pp. 1164-1172 , 2001.

[8] Tjahjanto HH , MacRae GA , Abu AK , Clifton GC , Beetham T , and Mago N , "Design capacity of
external diaphragm connections for square CFST columns" , *Steel and Composite Structures* , (submitted for publications).

A FINITE ELEMENT INVESTIGATION OF SKEWED AND SLOPED MOMENT CONNECTIONS IN STEEL CONSTRUCTION

Kevin E. Wilson[a], Gian A. Rassati[a], and James A. Swanson[a]

[a]University of Cincinnati, College of Engineering and Applied Science, Ohio, USA

wilso2ke@ mail. uc. edu, rassatga@ ucmail. uc. edu, swansojs@ ucmail. uc. edu

KEYWORDS: Non-Orthogonal, Special Moment Frame, Reduced Beam Section (RBS), Welded Unreinforced Flange-Welded Web (WUF-W).

ABSTRACT

Modern architectural designs for buildings require unconventional structural solutions such as non-orthogonally framed connections. Moment frames provide architecturalfreedom as building space can be utilized more efficiently than braced frames or structural walls. The ANSI/AISC 358 document provides engineers with a selection of prequalified connections that can be directly implemented in seismic applications. [1] The connections listed, however, do not consider non-orthogonal framing. Because of this, an engineer required to design a non-orthogonal moment frame would be limited to engineering judgement. Furthermore, if the engineer decides to implement an intermediate or special moment frame, ANSI/AISC 341 would require that the connection be qualified by cyclic testing evaluation. [2] Recent strong-motion earthquakes, such as the Northridge earthquake of 1994, have resulted in undesirable failure modes of steel moment frames, particularly, those of welded beam-to-column connections. The associated failures with these connections were mainly due to poor detailing and execution of field welds. In the aftermath of the Northridge earthquake, several welded connection configurations were developed to ensure the desired performance, among which are the welded unreinforced flange-welded web (WUF-W) and the reduced beam section (RBS) connections. With non-orthogonal framing, additional forces and different states of strain are present in comparison with traditional perpendicular framing due to the skewed or sloped geometry of the connection. The research presented provides a finite element investigation of the various skewed and sloped configurations of two prequalified connections presented in ANSI/AISC 358. The welded unreinforced flange-welded web (WUF-W) and the reduced beam section (RBS) moment connections were selected, as they are the most typical solution for moment frames. The definitions for skewed and sloped configurations were taken from the ANSI/AISC 360 specification. [3] Previous experimental testing of orthogonally framed connections validates the modelling techniques. [4,5] The results are then compared with the non-orthogonally framed connections. The investigation essentially examined two response quantities, namely the moment-rotation relationship and the equivalent plastic strain. The moment versus rotation relationship of the connection was documented from the analytical models to investigate the effects on connection stiffness and moment capacity. Also, the equivalent plastic strains (PEEQ) were investigated based on a Von Mises yield criterion. A visual comparison of the change in equivalent plastic strain demands was attained by scaling the plastic strains to the maximum observed value in the orthogonal model and contrasting the various configurations.

CONCLUSIONS

For both the WUF-W and RBS connections, the skewed configurations had minimal impact on the plastic moment capacity. The results indicated that with proper bracing as required, the full plastic moment capacity could be obtained for skew angles ranging from 0 to 30 degrees. The sloped configurations displayed an increase in plastic moment capacity for each increasing slope angle. This was attributed to the increase in effective depth of the cross section as a direct result of the slope modification. The distributions of PEEQs provided interesting results. The WUF-W connections with skew configurations observed localized increases in plastic strain in both flanges near the acute angle of the connection. The WUF-W connections with sloped configurations showed an increase in the plastic strain demand in the bottom flange near the acute angle formed (see *Fig. 1*). This was a result of the hinging

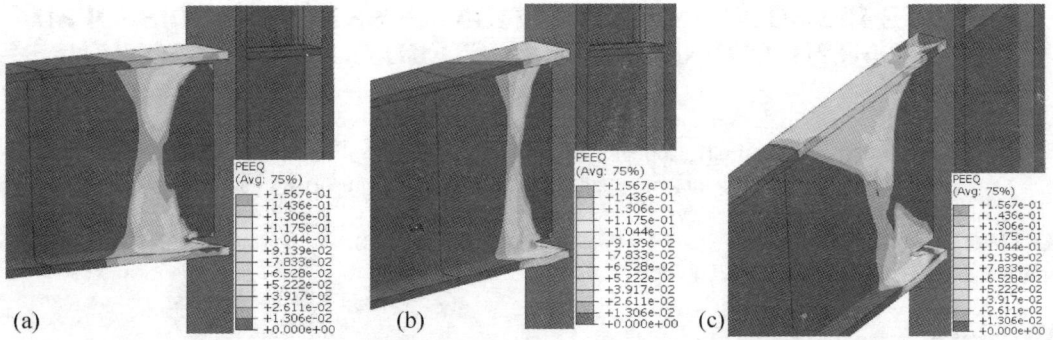

Fig. 1. Plastic accumulated equivalent strain(PEEQ) for the WUF-W connections with sloped configurations. (a) Orthogonal; (b) 30-degree skew; (c) 30-degree slope

mechanism remaining perpendicular to the beam cross section and therefore forcing strain demands near the beam/column interface at the bottom flange. The RBS connections with skewed configurations were successful in focusing the strain demand within the radius cut portion of the beam. With skew angles exceeding 10 degrees, there were some localized strain demands in the top and bottom flanges. For the RBS connections with slope configurations, the radius cut was successful in controlling plastic strains for angles less than 10 degrees. Once the slope angle exceeded 10 degrees, the strain demand focused in the bottom flange near the acute angle of the connection, with values exceeding 60% of the maximum observed plastic strain. The results indicate that for WUF-W connections, skewed or sloped configurations exceeding 10 degrees may not behave as intended, and need careful consideration. Similarly, for RBS connections, skew angles exceeding 30 degrees, and slope angles exceeding 10 degrees may give rise to unexpected behaviours. If a designing engineer is considering larger angles in their configurations, the authors advise that the increased localized strain demands be accounted for in the design of the connection and welds. More research is recommended on this topic.

REFERENCES

[1] AISC, "Prequalified Connections of Special and Intermediate Steel Moment Frames for Seismic Applications", *American Institute of Steel Construction*, 2010.
[2] AISC, "Specification for Structural Steel Buildings", *American Institute of Steel Construction*, 2010.
[3] AISC, "Seismic Provisions for Structural Steel Buildings", *American Institute of Steel Construction*, 2010.
[4] Hassan, T. , "SAC Task 7. 05-Inelastic Cyclic Analysis and Testing of Full Scale Welded Unreinforced Flange Connections. "*Advanced Technology for Large Structural System*, Paper 4, 2012.
[5] Engelhardt, M. D. , "Experimental Investigation of Dogbone Moment Connections", *Engineering Journal-American Institute of Steel Construction*, Vol. 35, No. 4, pp. 128-139, 1998.

LATERAL STIFFNESS AND STRENGTH OF STEEL COLUMN-TO-FOOTONG CONNECTIONS

Paul W. Richards[a], and Nicholas Barnwell[a]

[a]Brigham Young University, Department of Civil and Environmental Engineering, USA
prichards@ byu. edu, nbarnwel@ gmail. com

KEYWORDS: Steel columns, column-to-footing connection, anchor bolts.

ABSTRACT

Most steel columns in buildings are embedded in some concrete, but the effects of shallow embedment are usually neglected in design calculations. In U. S. practice, steel buildings with a slab on grade are constructed with block-outs in the slab located over the footings to permit the installation of the steel columns(*Fig. 1*). Once the columns are in place and bolted, the block-outs are filled with unreinforced concrete. Design calculations for the lateral strength and stiffness of these column-to-footing connections neglect any contribution from the block-out concrete or surrounding slab. Often such connections are considered pinned unless the base plate and anchor rods are specially designed to transmit moments, and a grade beam is integrated with the footing.

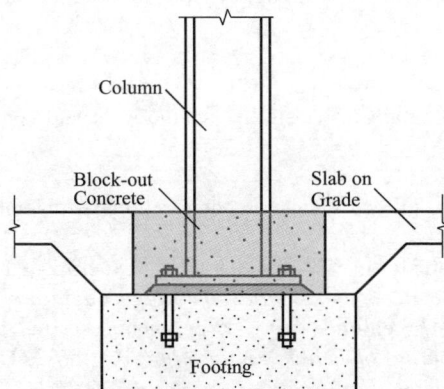

Fig 1. Typical steel column-to-concrete foundation with slab block-out detail

Limited studies indicate that shallow embedded steel shapes have much more rotational stiffness and strength than typically assumed. A better understanding of the behaviour of typical steel column-to-footing connections may lead to reduced column sizes, thinner base plates, and fewer anchor rods. In addition, quantifying the stiffness and strength of column-to-footing connections will lead to more accurate representation of the participation of gravity columns in the seismic response of steel buildings.

Experimental and analytical work is underway to quantify the stiffness and strength of typical steel column-to-footing connections. This paper presents preliminary results from experimental testing of eight specimens that represent typical connections between gravity columns and footings.

Each two-thirds scale specimen consisted of a steel column attached to a footing with four anchor bolts. The footing was anchored to the laboratory floor. A slab was poured on top of the footing, with a block-out for the column. The block-out was later filled with concrete. The parameters that were varied for the specimens were: column shape(W8 × 35 or W8 × 48), block-out depth[20 or 41 cm(8 or 16 in.)], and column orientation. Columns were subjected to cyclic lateral loading by an actuator mounted on a reaction frame.

The specimens exhibited essentially no damage in the block-concrete for drifts up to 3 percent. The strength of the specimens was limited by yielding of the anchor bolts; the ductility of the specimens was limited by anchor bolt rupture. The slab uplifted at the larger deformations after the anchor bolts had yielded(*Fig. 2*).

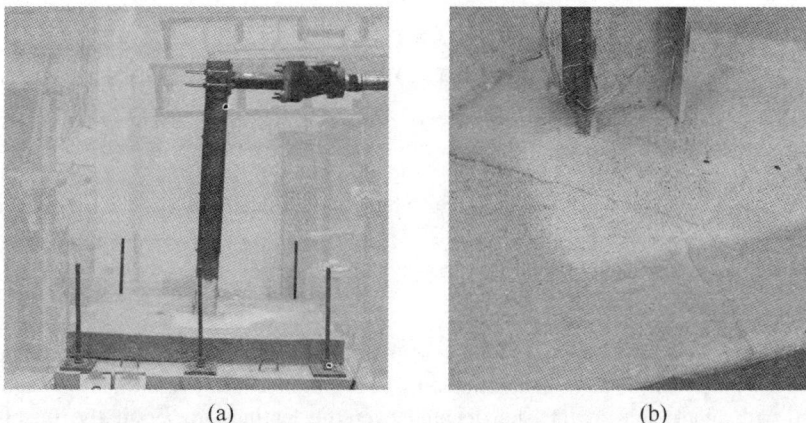

(a)   (b)

Fig. 2. (a) Slab uplift at 4 percent drift, (b) Minor cracking in block-out concrete at 4 percent drift

CONCLUSIONS

Preliminary results from the experiments support the following conclusions : 1) steel columns with block-out concrete have significantconnection stiffness, similar to fully embedded connections ; and 2) block-out concrete enables higher flexural strength by effectively increasing the size of the base plate.

ACKNOWLEDGMENT

This work was sponsored by the American Institute of Steel Construction(AISC) , with Tom Schlafly as research director ; the authors retain responsibility for the work and conclusions.

REFERENCES

[1] Marcakis K , Mitchell D. , " Precast concrete connections with embedded steel members " , *PCI Journal* , 25 : 88-116 , 1980.

[2] Xiao Y , Wu H , Yaprak TT , Martin GR , Mander JB. , " Experimental studies on seismic behaviour of steel pile-to-pile cap connections " , *Journal of Bridge Engineering* , 11 : 151-159 , 2006.

[3] Richards PW , Rollins KM , Stenlund TE. , " Experimental testing of pile-to-cap connections for embedded pipe piles " , *Journal of Bridge Engineering* , 16 : 286-294 , 2011.

[4] Eastman RS. , *Experimental Investigation of Steel Pipe Pile to Concrete Cap Connections* (*M. S. Thesis*) , Brigham Young University , 2011.

[5] AISC. , *Seismic Provision for Structural Steel Buildings* , American Institute of Steel Construction (AISC) , Chicago , 2010.

[6] Fisher JM , Kloiber LA. , *Base Plate and Anchor Rod Design* , American Institute of Steel Construction , Chicago , 2006.

EXPERIMENTAL INVESTIGATION ON BEHAVIOR OF CAST STEEL CONNECTORS FOR BEAM-TO-COLUMN CONNECTIONS UNDER CYCLIC LOADING

Ying-Zhi Chen[a], Le-Wei Tong[a], Yi-Yi Chen[a]

[a]Department of Structural Engineering, College of Civil Engineering, Tongji University, China
State Key Laboratory of Disaster Reduction in Civil Engineering, Tongji University, China
09yzchen@ tongji. edu. cn, tonglw@ tongji. edu. cn, yiyichen@ tongji. edu. cn

KEYWORDS: T-stub, Cast modular connector, Bolted connection, Experimental study.

ABSTRACT

Under strong earthquakes, energy is usually dissipated by the development of plastic hinge at the beam end in beam-to-column connections of steel structures, which always leads to large plastic deformation and makes it difficult to repair or replace. For bolted beam-to-column connections, the concentration of plastic strain and bolt slippage exist in stressed process of T-stubs, which limits its energy dissipation capacity and leads to the deterioration of stiffness and strength. In order to improve energy dissipation capacity of T-stubs, This paper presents two kinds of cast modular connectors used as an energy dissipating component in beam-to-column connections, one is set on both beam top and bottom flanges, and the other is only set on beam bottom flange, hoping that these modular prefabricated connectors possess highly efficient and stable energy dissipation capacity and be convenient to be replaced after earthquakes. Three specimens were tested under cyclic loading to simulate their mechanical behavior in a beam-to-column connection. Base on test results, several interesting conclusions were got.

1. Cast modular connector concept

To release the strain concentration on the ends of the T-stub flange, a simple way of reducing the section to match the bending moment diagram was adopted, as shown in *Fig. 1*(a). The finite element analysis software ABAQUS was used to obtain the optimal geometry dimensions, as shown in *Fig. 1* (b). To eliminate the influence of catenary action, two balancing plates were designed and added to both sides of the modified T-sub flange, which constituted a self-balancing frame as illustrated in *Fig. 2*. Considering the existence of concrete slab on the beam top flange, the replacement of cast modular connector C1 would not be easy, so another connector C2 was proposed, which was thickened at both ends to provide the compression space and could be set only on beam bottom flange, as given in *Fig. 3* and *Fig. 4*.

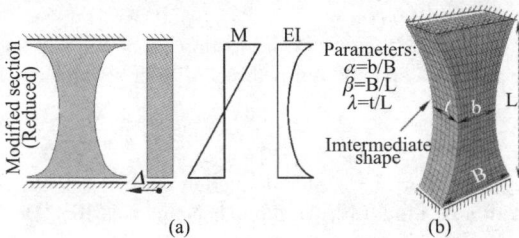

Fig. 1. (a) Modified section;
(b) Finite element analysis model

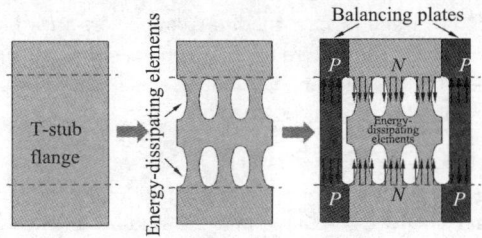

Fig. 2. Function of balancing plates

2. Properties of the cast steel material

Based on the steel casting engineer's advice, the material G20Mn5QT was chose for the cast connectors C1 and C2. The result of coupon tests is shown in *Table 1*.

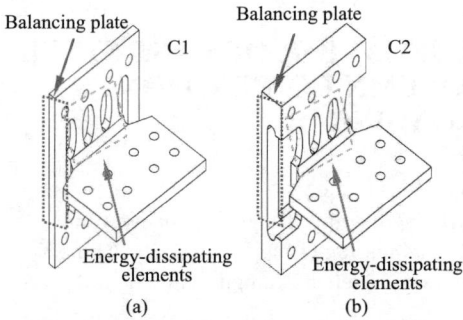

Fig. 3. Cast modular connector C1 and C2

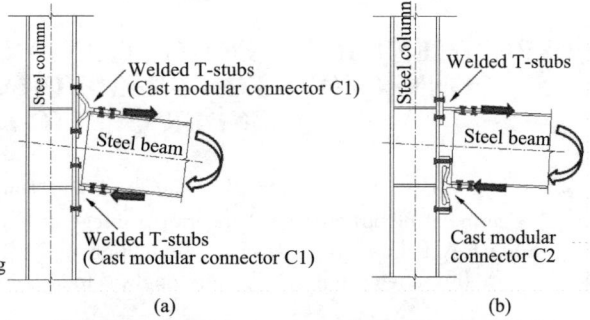

Fig. 4. Beam-to-column connections with C1 and C2

Table 1. Mechanical properties

Components	Yield strength f_y (MPa)	Tensile strength f_u (MPa)	Elongation(%)
Cast modular connectors(G20Mn5QT)	384	570	25

3. Full-scale tests

Three specimens including a welded T-stub WT, cast modular connector C1 and C2 were tested under cyclic loading to simulate their behaviour in a beam-to-column connection. Hysteretic curves and energy dissipation analysis are depicted in *Fig. 5* and *Fig. 6*, respectively.

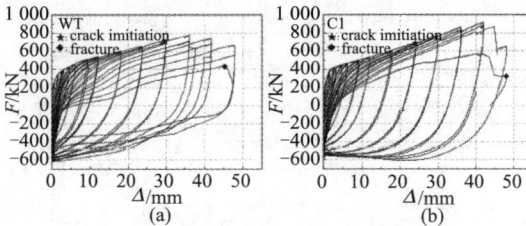

Fig. 5. Hysteretic curves: (a) WT; (b) C1; (c) C2.　　　Fig. 6. Equivalent viscous damping coefficients

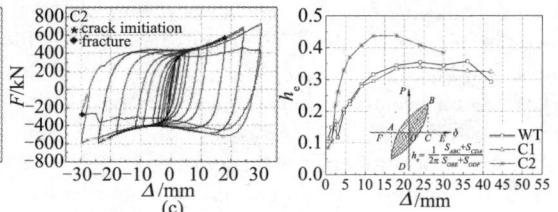

CONCLUSIONS

Based on the test results, several conclusions could be got as follows:

(1) The plastic strain was spread all over the energy-dissipating elements instead of focusing in two areas near the flange-to-web weld and the bolt axis, which shows the effectiveness of the proposed special shape of the energy-dissipating element.

(2) Energy dissipation capacities of the new cast modular connectors C1 and C2 were both better than that of specimen WT for plumper and more stable hysteretic curves, which verifies the possibility of the usage of cast modular connectors in a beam-to-column connection.

(3) The difference between specimen C1 and C2 shows that the energy dissipation capacity of specimen C2 was higher than that of specimen C1 while the bearing capacity and deformability of specimen C1 was better.

ACKNOWLEDGMENT

The presented work was supported by the State Key Laboratoryof Disaster Reduction in Civil Engineering(SLDRCE)of China and the Natural Science Foundation of China(NSFC)through Grant no. 51038008.

REFERENCES

[1] Latour M. , Rizzano G, 2011. "Experimental behavior and mechanical modeling of dissipative T-stub connections". *Journal of Structural Engineering*, ISSN 0733-9445, Vol. 138, pp. 170-182.

[2] Fleischman RB. , Sumer A, 2006. "Optimum arm geometry for ductile modular connectors". *Journal of Structural Engineering*, ISSN 0733-9445, Vol. 132, pp. 705-716.

[3] De Oliveira J. , Packer JA. , Christopoulos C, 2008. "Cast steel connectors for circular hollow section braces under inelastic cyclic loading". *Journal of Structural Engineering*, ISSN 0733-9445, Vol. 134, pp. 374-383.

IMPROVING THE SEISMIC BEHAVIOUR OF THE SLIDING HINGE JOINT USING BELLEVILLE SPRINGS

Shahab Ramhormozian[a], G. Charles Clifton[a], Gregory A. MacRae[b], Hsen-Han Khoo[c]

[a]Department of Civil and Environmental Engineering, The University of Auckland, Auckland, New Zealand.

sram732@ aucklanduni. ac. nz, c. clifton@ auckland. ac. nz

[b] School of Civil and Environmental Engineering, The University of Canterbury, Christchurch, New Zealand.

gregory. macrae@ canterbury. ac. nz

[c]KTA(Sarawak)Sdn. Bhd. , Selangor, Malaysia.

hkhoo@ unitec. ac. nz

KEYWORDS: Sliding hinge joint; Asymmetric friction connection, Belleville spring; Low damage; Self-centring.

ABSTRACT

The Sliding Hinge Joint(SHJ)is a low damage beam-column connection developed for seismic Moment-Resisting Steel Frames(MRSFs). The SHJ is intended to be rigid under Serviceability Limit State(SLS) conditions both before and after being subjected to an ultimate limit state(ULS) earthquake. For shaking up to and beyond the ULS levels, rotation of the column relative to the beam is developed through frictional sliding in the Asymmetric Friction Connection(AFC), the key energy dissipating component in the SHJ. At the end of the shaking, the SHJ is expected to seize up and become rigid again.

The seismic behaviour of the SHJ isclosely related to the sliding behaviour of the AFC. Previous research has shown once significant sliding has occurred, there is a considerable reduction in the elastic strength and stiffness of the SHJ, such that re-tightening or replacement of the bolts is likely to be required. This is because the AFC bolts lose part of their installed tension during sliding. The SHJ therefore does not fully satisfy the performance requirement of a low damage structural system, which ideally should not require reinstatement of structural components after a severe earthquake. This shortcoming can be significantly reduced through the optimum use of Belleville springs.

This paper reports the displacement controlled dynamic testing of the bottom flange AFC assemblage in the SHJ with and without Belleville springs. The bolts were installed in their elastic range as is proposed in this research. A loading regime was designed and applied to simulate the SLS-ULS-SLS condition to investigate the pre and post-earthquake behaviour of the AFC assemblage in the SHJ too. The seismic hysteretic behaviour of the system as well as the coefficient of friction value of the AFC sliding surfaces is also discussed.

RESULTS AND DISCUSSIONS

BeSs can significantly reduce the bolt tension loss if they are used in the optimum way. The more BeS maximum deflection the more reduction in bolt tension loss, however the greater the cost and complexity of installation so an optimum point where these two factors intersect needs to be found. The average bolt tension loss for the three NS tests "and still not having fully tensioned bolts" was 60%. This was 35% for the three S4 tests.

Fig. 1 demonstrates the AFC sliding surfaces coefficient of friction, μ, calculated for the cases with no BeSs "NS" and with four BeSs "S4" respectively plotted with the same scale, each one for three test repeats.

Using BeSs can provide more stable behaviour for the AFC and the bolt group. The hysteresis curve is better from the self-centring point of view for the cases that use the BeSs. The coefficient of friction of the AFC found in this research is higher than the values proposed so far and is significantly increased by using the BeSs. Since the bolt tension is monitored for all of the bolts, the coefficient of

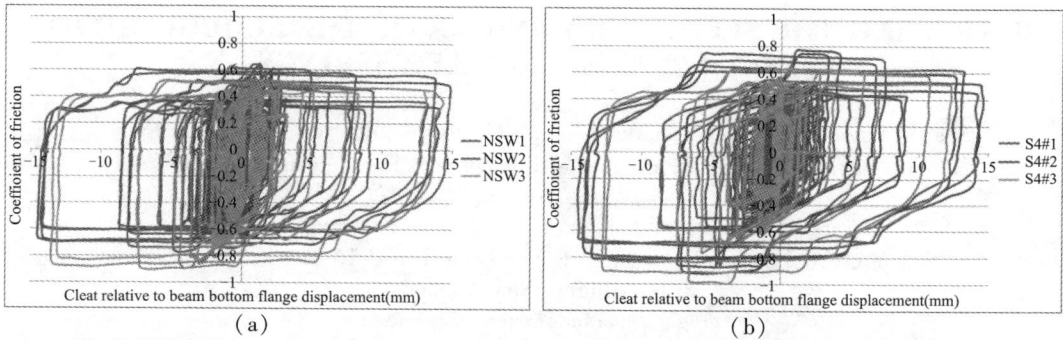

Fig 1. AFC sliding surfaces coefficient of friction, μ (a) without BeSs "NS", (b) with BeSs "S4"

friction value calculated in this research can be considered as a reliable value in practice.

The prying effect observed during the tests in absence of the BeSs was large, causing the bolts to stretch plastically. BeSs could strongly eliminate the prying effect if they are not installed fully squashed. However, consideration of the boundary conditions on the "beam flange part" of the component test setup has shown that the boundary conditions are an order of magnitude at least stiffer than they are for a beam flange in practice. This will significantly increase the prying effect measured off the test rig and means the test rig must be redesigned to bring the boundary condition stiffness on the beam flange part into line with what it is in practice. That is to be done by the authors.

All of the bolts were longer after removing compared with the initial length, consistent with the plastic elongation. This can also be because of the plastic torsion that had happened at installation. This issue of plastifying the bolts in torsion during tightening has been researched and addressed by the authors and can be significantly reduced by pre-installation physical inspections and lubricating the bolts.

The temperature was found to be an un-important parameter with a bolt temperature rise of not more than 4 degrees Centigrade; not sufficient to influence the behaviour. Usinga depth micrometre to measure the BeSs height found to be not accurate. No turn of the nut was observed during the tests.

ACKNOWLEDGMENT

This research was financially supported by Earthquake Commission Research Foundation (Project 14/U687 "Sliding Hinge Joint Connection with Belleville Springs"). The authors are grateful for this support. The first author PhD studies at the UoA is financially supported by a departmental scholarship through the support and kindness of the PhD supervisor, Associate Professor G. Charles Clifton. This support is much appreciated. The authors would also like to acknowledge the efforts and help of the P4P students as well as technical and laboratory staff at the Department of Civil and Environmental Engineering, the University of Auckland involved in this research in conducting experimental testing.

REFERENCES

[1] Clifton, G. C. , *Semi-rigid joints for moment-resisting steel framed seismic-resisting systems*, in *Department of Civil and Environmental Engineering* 2005, University of Auckland: Auckland, New Zealand.

[2] MacRae, G. A. , et al. , *The Sliding Hinge Joint Moment Connection*. Bulletin of the New Zealand Society for Earthquake Engineering, 2010. 43(3) :p. 202.

THE OPTIMUM USE OF BELLEVILLE SPRINGS
IN THE ASYMMETRIC FRICTION CONNECTION

Shahab Ramhormozian[a], G. Charles Clifton[a], Gregory A. MacRae[b], Hsen-Han Khoo[c]

[a] Department of Civil and Environmental Engineering,
The University of Auckland, Auckland, New Zealand.
sram732@ aucklanduni. ac. nz, c. clifton@ auckland. ac. nz

[b] School of Civil and Environmental Engineering, The University of Canterbury,
Christchurch, New Zealand.
gregory. macrae@ canterbury. ac. nz

[c] KTA (Sarawak) Sdn. Bhd. , Selangor, Malaysia.
hkhoo@ unitec. ac. nz

KEYWORDS: Asymmetric friction connection, Sliding hinge joint, Belleville spring, Low damage seismic systems, Prying.

ABSTRACT

The Asymmetric Friction Connection (AFC) is a key energy dissipating component used in the Sliding Hinge Joint connection (SHJ) , a low damage beam-column connection developed for seismic Moment-Resisting Steel Frames (MRSFs). The moment-rotational behaviour of the SHJ during sliding depends on the AFC sliding characteristics. In current practice, the AFC bolts are defined to be fully tensioned at installation (i. e. yielded) based on the design procedure that has been developed primarily for the SHJ.

The AFC bolts are subjected to a moment, shear, and axial force (MVP) interaction resulting from bending in double curvature during joint sliding. It has beenshown by previous experiments that the AFC stable sliding bolt tension is considerably lower than the AFC installed bolt tension. In this paper, the reasons of AFC bolt tension loss are discussed. These include the MVP interaction, prying effects, and thinning of the AFC plies. The use of Belleville springs in the AFC can have considerable improving effects on the retention of installed bolt tension during stable sliding and consequently the AFC and SHJ seismic behaviour.

An analytical discussion and recommendation of the way that Belleville springs can maintain very close to the installed level of AFC bolt tension along with presentation of a real example is presented in this paper. This involves considering the spring-type behaviour of the AFC. A way of installing Belleville springs to reduce the negative effect of prying actions is also proposed along with simulating the prying effect on the bolt with and without Belleville springs. The optimum level of installed bolt tension and the way of using Belleville springs in the AFC is also recommended.

CONCLUSIONS

It is shown in this paper that Belleville springs (BeSs) can considerably reduce the post sliding bolt tension loss in the AFC or any other friction-type connection. An overloaded BeS acts similar to a hardened washer until the bolt tension drops to the BeS flat load. From that point, the BeSs pushes out to compensate for the most of the bolt tension loss. Hence, it would be optimum to choose the BeSs with the flat load not less than the installed bolt tension.

Using BeSs considerably reduces the sensitivity of the AFC bolts to the factors causing the bolt tension loss. Hence the post sliding bolt tension variability of different bolts of a given AFC incorporating the BeSs will be much less than the AFC with no BeS. This provides the AFC with more consistent and predictable seismic behaviour. *Fig. 1* shows the AFC post sliding bolt tension vs combination of AFC bolt plastic stretch and AFC plies thinning for the cases with and without BeSs.

Installing the BeSs in the not flattened state provides a degree of flexibility under the bolts head and or nut causing to considerably reduce the possibility of the bolt to be plastically stretched specially during the first cycle of sliding, as otherwise the BeS is supposed to be still flat. *Fig. 2* shows the pry-

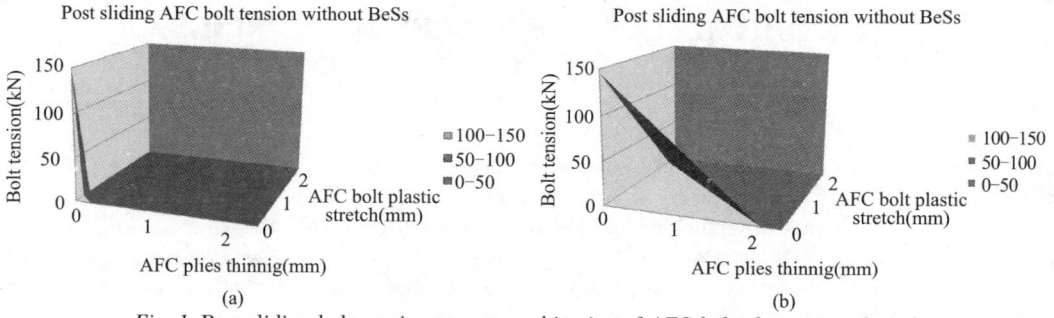

Fig. 1. Post sliding bolt tension versus combination of AFC bolt plastic stretch and plies thinning for the cases (a) without BeSs and (b) with BeSs

ing bolt tension versus prying expansion of the system.

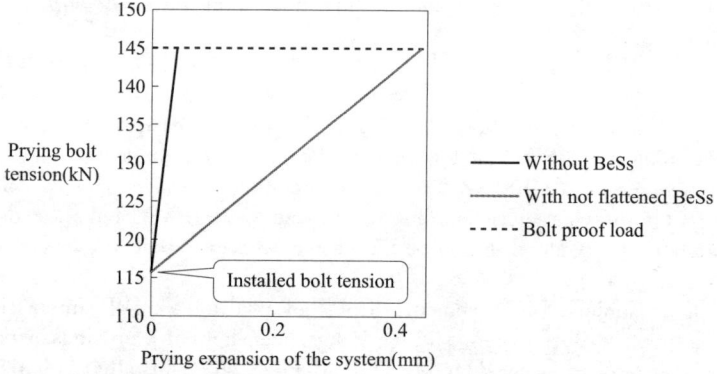

Fig. 2. AFC prying bolt tension versus prying expansion of the system

It is also recommended to install the bolts with BeSs in the bolt elastic range. This decreases the probability of the bolt tension loss too. The most feasible way of installing the bolts with BeSs up to a certain level of bolt tension within the elastic range is developed as a key part of the ongoing research.

ACKNOWLEDGMENT

This research was financially supported by Earthquake Commission Research Foundation(Project 14/U687 "Sliding Hinge Joint Connection with Belleville Springs"). The authors are grateful for this support. The first author PhD studies at the UoA is financially supported by a departmental scholarship through the support and kindness of the PhD supervisor, Associate Professor G. Charles Clifton. This support is much appreciated.

NUMERICAL SIMULATION OF Q690 GRADE STEEL EXTENDED END PLATE CONNECTIONS

SUN Feifei[a], RAN Mingming[b] and SUN Mi[b]

[a]State Key Laboratory for Disaster Mitigation in Civil Engineering,
Tongji University, Shanghai 200092, China;
ffsun@ tongji. edu. cn

[b] School of Civil Engineering, Tongji University, Shanghai 200092, China
ranfanchuang@ 126. com

KEYWORDS: High Strength Steel, Extended End Plate, Cyclic Loading, Numerical Simulation

ABSTRACT

A large number of researches on the behaviour of extended end plate connection have been done in the last several years [1-3]. There were many researches on HHS used in beams and columns, but very limited experimental work to study HHS used in end plate has been performed. To solve this problem, an experimental investigation was undertaken of moment connections with extended end plates made up of the high strength steel Q690. The experimental program involved the testing of two extended end plate moment connections under cyclic loading. The difference between them is the end plate thickness, which specimen S-1 has a 12mm thickness end plate and specimen S-2 has a 10mm thickness end plate(*Fig. 1*).

The M-Φ response for the two specimens using 10mm and 12mm end plate are reported in *Fig. 2*, respectively. The skeleton curves of them are showed in *Fig. 3*.

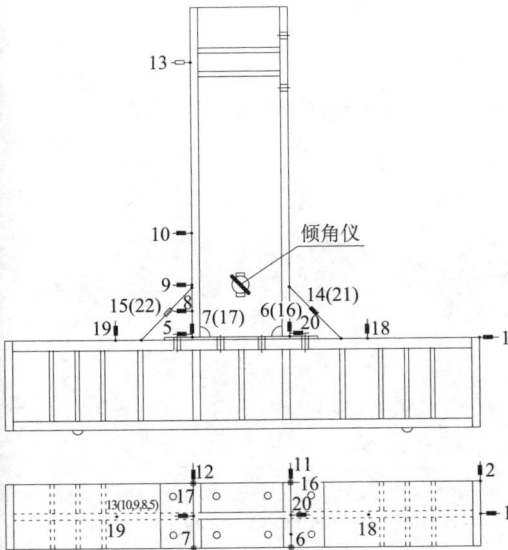

Fig. 1. Location of the displacement transducers

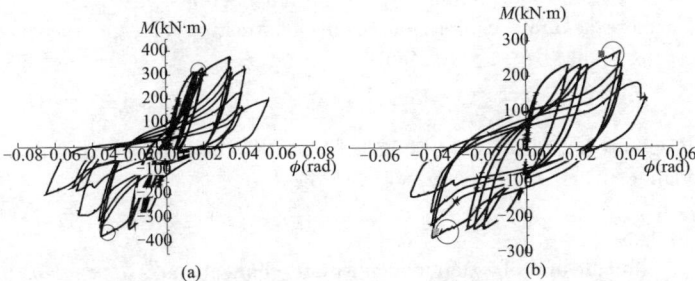

Fig. 2. Moment versus end plate rotation
(a) S-1 with 12mm end plate; (b) S-2 with 10mm end plate

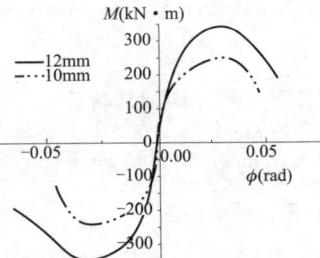

Fig. 3. Skeleton curves

EC3-1-8 gives quantitative rules for the prediction of joint flexural resistance, initial stiffness. These structural properties are evaluated below using the actual geometrical characteristics and the actual material properties of the steel. *Table 1* summarizes the code predictions for the full moment resistance, initial stiffness and rotational capacity of the two specimens. The ratio between EC3 calculations and experimental results was also computed.

Table 1. Comparisons of EC3 predictions and test results (Ratio = Theoretical value/Test value)

Specimens	M_y (kN $\cdot$ m)			$S_{j,\text{ini}}$ / (kN $\cdot$ m $\cdot$ rad^{-1})			φ_m / rad	φ_{cra} / rad	δ
	Theoretical value	Test value	Ratio	Theoretical value	Test value	Ratio			
S-1(12mm)	309.433	301.351	1.027	64413	32339	1.992	0.044	0.018	4.930
S-2(10mm)	229.263	214.480	1.069	42257	35278	1.198	0.041	0.029	7.573

ABAQUS 6.12-3 is selected for equivalent three dimensional(3D) analysis. *Fig. 4* displays both the numerical and experimental results for ABAQUS/Standard, summarized in hysteresis loops. *Fig. 5* displays the results of experimental and numerical hysteresis loops for ABAQUS/Explicit.

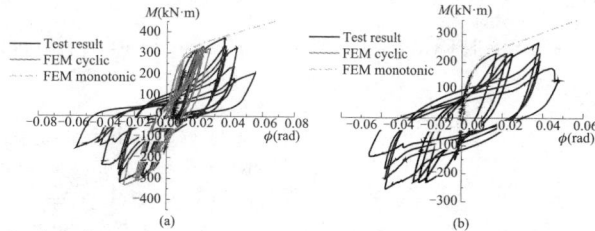

Fig. 4. Experimental and numerical hysteresis loops for ABAQUS/Standard   (a)S-1;(b)S-2

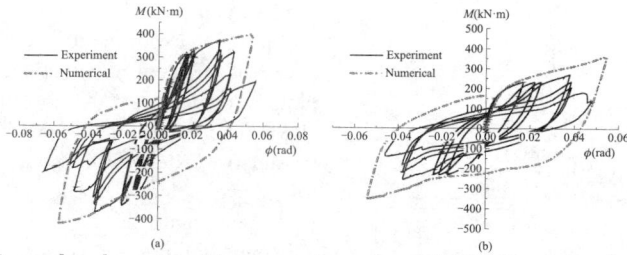

Fig. 5. Experimental and numerical hysteresis loops for ABAQUS/Explicit   (a)S-1;(b)S-2

CONCLUSIONS

1. With the increase of the thickness of the end plate, plastic flexural resistance and the maximum bending moment had obvious increase, and displayed hysteresis loops with less pinching and it was able to undergo a larger number of inelastic cycles and dissipated significantly more energy.

2. EC3 gives accurate results in terms of prediction of the plastic flexural resistance, even when Q690 grade steels are employed. However, the prediction of initial stiffness is overestimated. Also, EC3 should give a method to quantitatively calculate the rotation capacity of connections.

3. Implicit model can provide a more accurate calculation results contrasting on experiment, but model is difficult to converge comparing with ABAQUS/Explicit.

ACKNOWLEDGMENT

Financial support bySupport Program(No. SLDRCE14-B-12)of State Key Laboratory for Disaster Mitigation in Civil Engineering, Tongji University, is greatly acknowledged.

REFERENCES

[1] Mann, A. P., and L. J. Morris, "Limit design of extended end-plate connections", *Journal of the Structural Division*, Vol. 105, No. 3, 1979, pp. 511-526.

[2] Girão Coelho, A. M., F. S. K. Bijlaard, and L. Simões Da Silva, "Experimental assessment of the ductility of extended end plate connections", *Engineering Structures*, Vol. 26, No. 9, 2004, pp. 1185-1206.

[3] Adey, B. T., G. Y. Grondin, and J. J. R. Cheng, "Cyclic loading of end plate moment connections", *Canadian Journal of Civil Engineering*, Vol. 27, No. 4, 2000, pp. 683-701.

STUDY ON PERFORMANCE OF FLANGE COVER PLATE IN WEB-CLAMPED BEAM-TO-COLUMN CONNECTION

Tong Su[a], Keita Araki[a], Jun Iyama[a]

[a]the University of Tokyo, Department of Architecture, Faculty of Engineering, Japan
673038710st@ gmail. com, araki@ stahl. arch. t. u-tokyo. ac. jp,
iyama@ arch. t. u-tokyo. ac. jp

KEYWORDS: Beam-to-column Connection, Web-clamped connection, Flange Cover Plate, Finite Element Analysis, Stress Concentration.

ABSTRACT

A new type of beam-to-column connection, named the web-clamped connection has been proposed by the authors as an alternative to conventional welded connections. This connection utilizes prefabricated attachments, which connect the beam flange and column web through slits opened on the column flange. [1] It is presumed that there may be stress concentration on the column flange between two slits, which may cause local fracture or local buckling. To prevent that, flange cover plates are supposed to be welded on the column flange. In this paper, finite element analysis results are presented to clarify the stress distribution around slits of H-shaped column connected by the web-clamped connection with and without flange cover plate, and compared to those of a conventional welded connection.

Three types of finite element models shown in *Fig 1* are made for analysis. Finite element analysis software ABAQUS6. 12 is used. The panel zone is designed to be weaker than the column so that the panel zone shear failure dominates the overall strength of the model. *Fig 2* shows the loading setup of models. The columns are simply supported. For web-clamped type model, horizontal displacement is applied to the inner surfaces of the bolt holes to simulate the column behavior under anti-symmetric loading. For conventional welded type model, the beam is also modeled as well as the column, and the vertical displacement is applied to the both end of the beam.

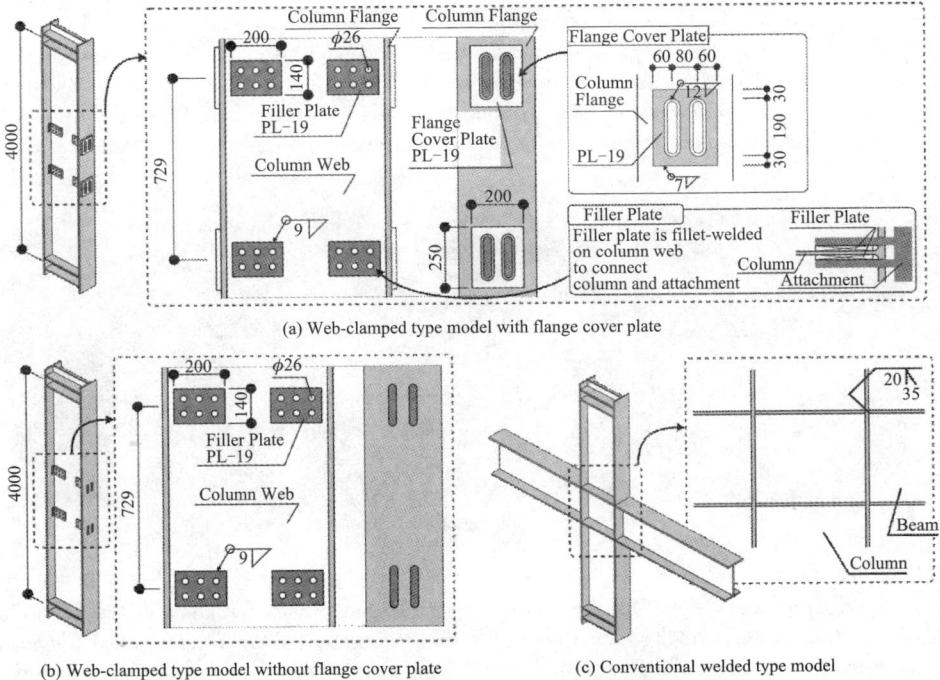

(a) Web-clamped type model with flange cover plate

(b) Web-clamped type model without flange cover plate   (c) Conventional welded type model

Fig 1. Shapes of finite element models for analysis

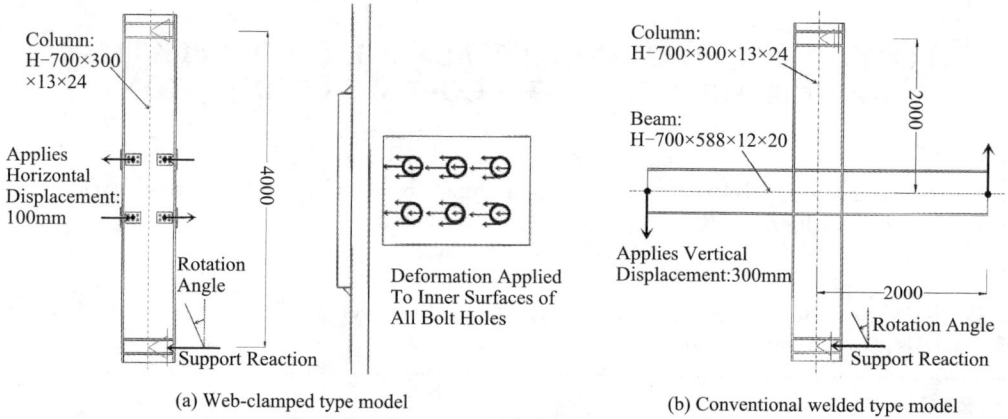

(a) Web-clamped type model   (b) Conventional welded type model

Fig 2. Loading setup of finite element analysis

When the analysis was completed, the values of support reaction force and rotation angle at the both ends of the column were measured to compare the stiffness, strength, and deformation capacity of the connection. The axial stress distribution curves at the column flange with two slits were also compared at the ultimate state for the sake of evaluation of the effect of flange cover plate.

CONCLUSIONS

According to the result of finite element analysis, in all three models, ultimate state is determined by the panel zone shear failure and column flange did not yield. However, it was shown that the web-clamped type model with flange cover plate can bear a higher bending moment than conventional welded type model.

As shown in *Fig 3*, axial stress concentration in the area between the two slits on the column flange and local bending deformation of the column flange around the slits are observed in Web-clamped type model without flange cover plate as expected. However, the analysis of the model with the flange cover plate showed some reduction of stress concentration on the column flange after panel zone yields. The out-of-plane deformation of the column flange is also much smaller compared to the results without the flange cover plate. The conventional welded type model also showed some stress concentration on the center of the column flange. The peak stress on the column flange of the conventional welded type model is similar to that of the web-clamped type model with the flange cover plate. This result shows that the flange cover plate reduces the stress concentration and effectively prevents local bending deformation.

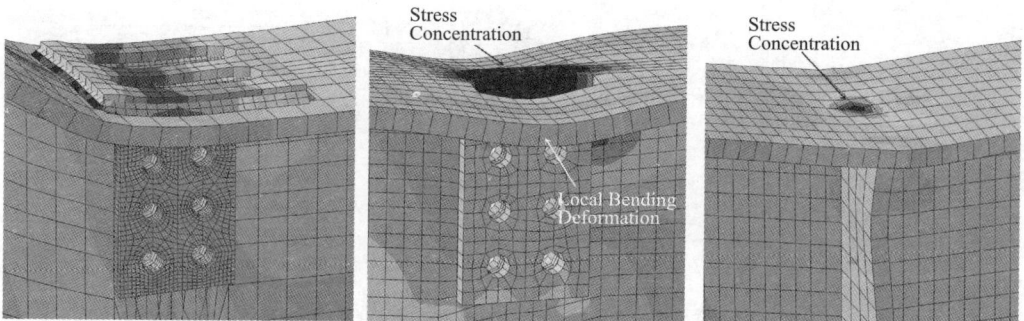

Fig 3. Axial stress contour of finite element models

REFERENCES

[1] Keita Araki, Jun Iyama, Piao Shiwan. , "Experimental Study on Performance of the Web-clamped Beam-to-column Connection", *Joint Conference Proceedings of the 9th International Conference on Urban Earthquake Engineering and 4th Asia Conference on Earthquake Engineering*, pp. 1081-1085, 2012.

NUMERICAL STUDY ON MECHANICAL BEHAVIOR OF SHEAR PLATE IN WEB-CLAMPED TYPE BEAM-TO-COLUMN CONNECTION

Keita Araki[a], Jun Iyama[a] and Shiwan Piao[b]

[a] The University of Tokyo, Tokyo, Japan

araki@ stahl. arch. t. u-tokyo. ac. jp, iyama@ arch. t. u-tokyo. ac. jp

[b] I-TEC Corporation, Tokyo, Japan

piao_shiwan@ itec-c. co. jp

KEYWORDS: Beam-to-column connection, Web-clamped connection, Shear plate, Finite element analysis

ABSTRACT

The web-clamped type connection is proposed as a new connection which has stiffness comparable to conventional welded connections and high workability comparable to conventional high strength bolted connections. As shown in *Fig. 1*(a), in the structural design, it is essential to properly estimate the internal force of the attachments and the shear plate at the minimum joint section.

However, according to previous studies, there is little information available on the mechanicalbehavior of the shear plate. A previous study by finite element analysis focused on the shear deformation of the panel zone and the axial deformation of the attachments and investigated internal force transfer mechanism at the minimum joint section[1]. In the study, yielding and local buckling of the beam were strictly prevented by adjusting the beams strength and flange width-to-thickness ratio. Therefore, the effect of the shear plate is not investigated in the case plastic deformation of the beam occurs.

In this paper, the numerical analysis of a cantilever beam with the web-clamped type connection is conducted, which is shown in *Fig. 1*(b). As shown in *Table 1*, parameters of the models are beam lengths l_b and the existence of the shear plate, in order to control the stress share between the attachments and the shear plate. All models are designed so that the ultimate strength of the connection is determined by the full plastic strength of the beam. Through this study, the effect of the shear plate on the moment-rotation relationship and deformed shape of the connection is investigated, in the condition the local deformation of the beam occurs.

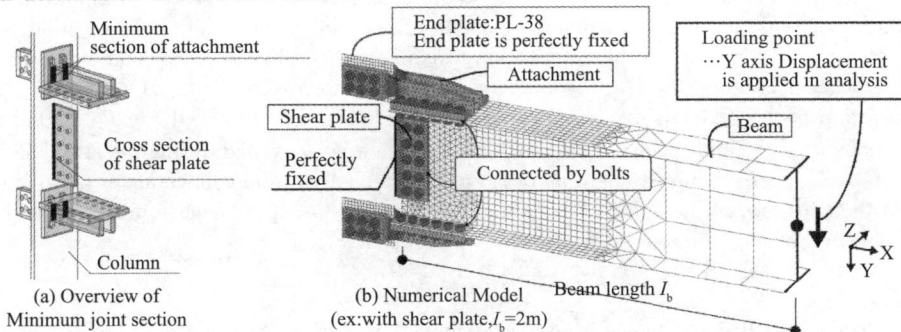

(a) Overview of Minimum joint section

(b) Numerical Model (ex:with shear plate, l_b=2m)

Fig. 1. FEA model

Table 1. Main specifications of the models conducted in this paper

No	Beam Length l_b (m)	Existence of shear plate	Beam size
1	2	○	H-582 × 300 × 12 × 17
2		×	
3	6	○	
4		×	

※ ○: model with shear plate, ×: model without shear plate

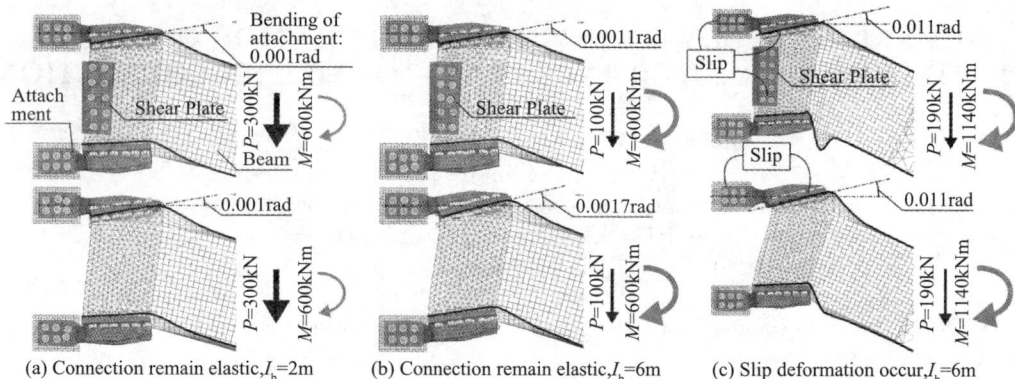

(a) Connection remain elastic,I_b=2m   (b) Connection remain elastic,I_b=6m   (c) Slip deformation occur,I_b=6m

Fig. 2. Comparison of deformed shapes of the connection

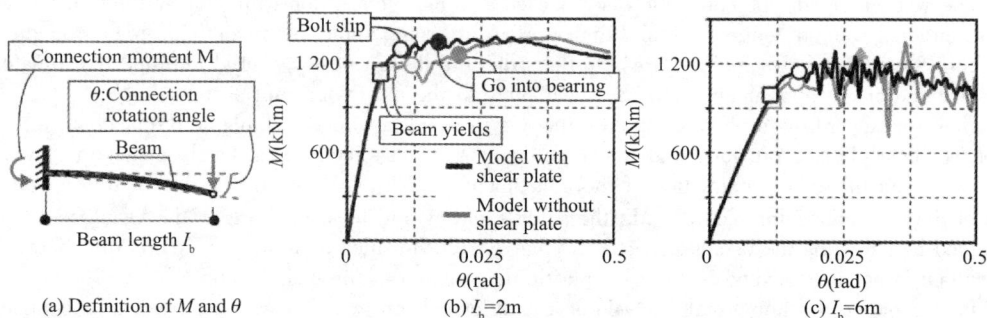

(a) Definition of M and θ (b) I_b=2m   (c) I_b=6m

Fig. 3. Comparison of M-θ relationships

CONCLUSIONS

Comparison of the deformed shapes is presented in *Fig. 2*. *Fig. 2*(a) and (b) shows the deformed shape when the connection remains elastic and (c) shows the one after the connection yields and the slip deformation occurs. When the deformed shapes are compared with the case of with and without the shear plate, it suggests that the shear plate is capable of restraining the bending deformation of the attachment. However, the amount of the deformation which is reduced by the shear plate may vary depending on the beam length and the slip deformation. From the comparison of *Fig. 2*(a) and (b), the bending deformation of the attachments is apparently smaller when l_b = 6m (see *Fig. 2*(b)). From the comparison of *Fig. 2*(b) and (c), the effect of the shear plate is weakened during the slip deformation.

As shown in in *Fig. 3*, the strength and deformation capacity of the connection are also similar regardless of the existence of the shear plate in the M-θ relationships. It is because the effect of the shear plate is weakened by the slip deformation which is shown in *Fig. 2*.

REFERENCES

[1] Keita Araki, Jun Iyama, Piao Shiwan: An analytical study on the force distribution between the shear plate and attachments in web-clamped beam-to-column connections, Study on web-clamped beam-to-column connection Part 1, J. Struct. Constr., AIJ, Vol. 79, No. 706, 1931-1940, Dec., 2014(In Japanese)

EXPERIMENTAL PROGRAM AND NUMERICAL SIMULATIONS OF BOLTED BEAM TO COLUMN JOINTS WITH HAUNCHES

Cosmin Maris[a], Cristian Vulcu[a], Aurel Stratan[a], Dan Dubina[a,b]

[a]Politehnica University of Timisoara, Dept. of Steel Structures and
Structural Mechanics, Romania

[b]CCTFA, Romanian Academy, Timisoara Branch

cosmin. marsi@ upt. ro; cristian. vulcu@ upt. ro; aurel. stratan@ upt. ro; dan. dubina@ upt. ro

KEYWORDS: bolted connections with haunches, numerical simulations, component method

ABSTRACT

Modern seismic design codes require that the seismic performance of beam-to-column connections in steel moment resisting frames to be demonstrated through experimental investigations, which often prove to be expensive and time-consuming. A solution to this problem is the pre-qualification of typical connections for the design practice, which is common in U. S. and Japan. Nevertheless, the standard joints pre-qualified according to codified procedures in U. S. and Japan cannot be extended to Europe, due to differences in materials and section shapes. Moreover, the beam-to-column joint types are not commonly used in Europe. As a result, the existing scientific and technical background on pre-qualification may not be directly extended to European context. For this reason, a European research project entitled EQUALJOINTS(European pre-QUALified steel JOINTS)[1], is currently underway and aims at seismic pre-qualification of several beam-to-column connection typologies common in the European practice(see *Fig. 1* abc).

The current paper outlines the experimental program on bolted extended end-plate connections with haunches(see *Fig. 1*a) that will be investigated at the Politehnica University of Timisoara in the framework of EQUALJOINTS project[1]. In addition, results are presented from a set of pre-test numerical simulations carried out for the evaluation of the designed beam-to-column joint assemblies, and the influence of several parameters: member size, haunch geometry, column web panel strength, joint configuration(internal/external), and loading conditions(monotonic/cyclic). The pre-test numerical simulationcomprised the calibration of a T-stub model(see *Fig. 1*d), including the material model for bolts. Based on the outcomes of the FE simulations, the design procedure of the joints was adjusted considering the actual position of the compression centre and the active bolt rows(see *Fig. 1*e). The cyclic response of a joint was evaluated as well(see *Fig. 1*e).

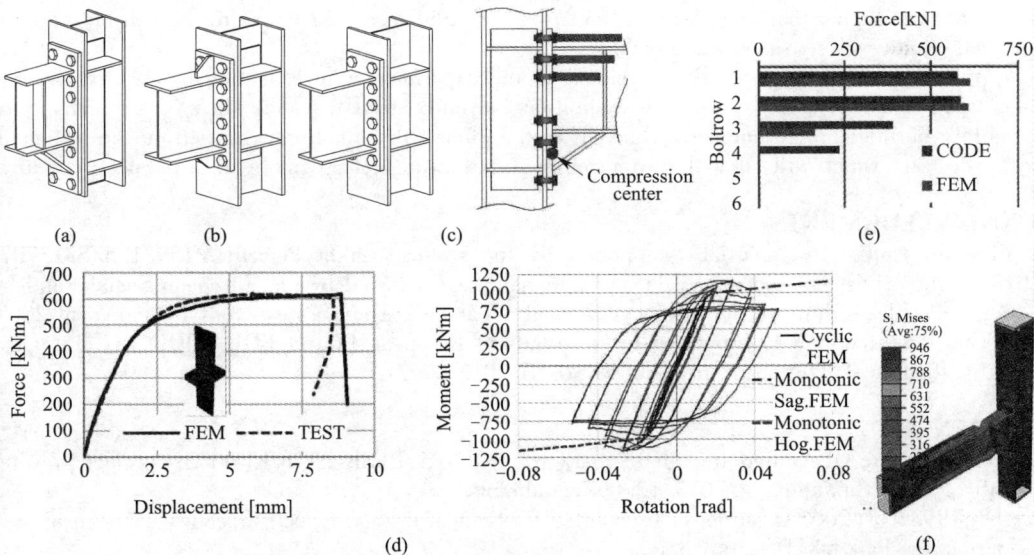

Fig. 1. Connection typologies investigated within EQUALJOINTS project[1]: (a) with haunches, (b) with stiffened end-plate, (c) with un-stiffened end-plate; (d) Calibration of T-stub model(including material model for bolts); (e) Active bolt rows and actual position of the compression centre; (f) Cyclic response of joint assembly

CONCLUSIONS

The current paper presented the framework of the investigations carried out at thePolitehnica University of Timisoara, for the pre-qualification of bolted beam-to-column joints with haunches. In addition, the experimental program on joint assemblies was presented, with the emphasis on joint configurations, and parameters to be investigated in the program. Further, the outcomes of the pre-test numerical investigations were shown in relation to: (i) calibration of a numerical model of T-stub, (ii) distribution of forces in the connection components, (iii) parametric study.

The calibrated T-stub model was able to reproduce with good accuracy the response from the test in terms of force-displacement curve, particularly: initial stiffness, ultimate capacity and failure mode (bolts). Consequently, a material model was established for the behaviour of the bolts, which was further used within the numerical models of the bolted beam-to-column joint assemblies.

The pre-test numerical simulations carried out for the joint assemblies designed based on EN 1993-1-8[2] confirmed the intended failure mode, i. e. formation of the plastic hinge in the beam close to the haunch ending. However, other assumptions considered in design were not confirmed, in particular: force distribution in the bolt rows, and position of the compression centre (middle of the compressed haunch flange). In contrast, the active bolt rows were those situated near the flange in tension, and the compression centre was located at a distance equal to 60% of the haunch depth measured from the bottom flange of the beam. As a result, the design procedure was adjusted and the beam-to-column joint configurations were re-designed and analysed.

The parametric study allowed investigating the influence of: member size, haunch geometry, web panel strength, and cyclic loading. Consequently, the study evidenced higher strain in members (beams) of larger cross-section, for the same joint rotation. In addition, the capacity and the initial stiffness of the connection increased for higher values of haunch angle and depth, but for economic and architectural reasons, haunches with minimum depth and a smooth slope are recommended.

Finally, the joint response was evaluated under cyclic loading conditions, which lead to an increase of capacity corresponding to the cycles of 20 and 30mrad amplitudes, while for higher amplitudes, the capacity decreased compared to the monotonic curve due to local buckling of flanges and web under compression and bending. In addition, the seismic performance was evaluated constructing an envelope curve through the peak values of the first cycle for each amplitude, and further the rotation of the joint assembly was obtained from the intersection of the envelope curve with the horizontal line representing 20% reduction of the maximum capacity. Consequently, a rotation of 41 mrad was obtained, which is higher than the 35 mrad limit from EN 1998-1[3], and therefore the seismic performance of the joint was considered acceptable.

Future research activities will be related to the experimental investigation of material samples (standard tensile tests, and in addition cyclic tests) and beam-to-column joint assemblies. Furthermore, the numerical models of the joints will be calibrated based on the monotonic and cyclic tests, which will allow the extension of the experimental program with a parametric study.

ACKNOWLEDGMENTS

The first author was partially supported by the strategic grant POSDRU/159/1. 5/S/137070 (2014) of the Ministry of National Education, Romania, co-financed by the European Social Fund-Investing in People, within the Sectorial Operational Programme Human Resources Development 2007-2013. The present work was supported by the funds of European Project EQUALJOINTS: "European pre-QUALified steel JOINTS", Grant No. RFSR-CT-2013-00021.

REFERENCES

[1] European pre-QUALified steel JOINTS-EQUALJOINTS, RFSR-CT-2013-00021, project website: http://dist. dip. unina. it/2013/12/09/equaljoints/
[2] EN 1993-1-8 (2005). European Committee for Standardization-CEN. Eurocode 3: Design of steel structures-Part 1-8: Design of Joints.
[3] EN 1998-1 (2004). European Committee for Standardization-CEN. Eurocode 8: Design of structures for earthquake resistance-Part 1, General rules, seismic actions and rules for buildings.

FULL SCALE TESTING OF EXTENDED BEAM-TO-COLUMN AND BEAM-TO-GIRDER SHEAR TAB CONNECTIONS SUBJECTED TO SHEAR

Jacob Hertz[a], Dimitrios G. Lignos[a], Colin A. Rogers[a]

[a]McGill University, Department of Civil Engineering and Applied Mechanics, Canada

Jacob. hertz@ mail. mcgill. ca, dimitrios. lignos@ mcgill. ca, colin. rogers@ mcgill. ca

KEYWORDS: Extended shear tab connections, beam-to-girder connections, girder mechanism, plate buckling.

ABSTRACT

A design method is not available in Canada for extended shear tab connections, which are typically detailed having an 'a' distance greater than 76 mm and up to 200 mm; measured from the centerline of the first column of bolts to the face of the supporting column or web of the supporting girder. In order to make recommendations on a generally applicable shear tab design method further investigation of extended shear tabs is necessary.

The objective of this paper is to first document the laboratory phase of a 3-year research project comprising beam-to-column and beam-to-girder connections with extended shear tab connections. Included in the scope of testing are four beam-to-column connection specimens(W310 beams-2 columns of bolts, W690 beams-3 columns of bolts) with extensions of 150 mm and 200 mm, as well as eight beam-to-girder connections(W310 to W610 beams & W610 to W760 girders with various configurations and stiffening details of connections with 2 columns of bolts). The following steel grades were used; ASTM A992 345 MPa beams, girders and columns, ASTM A325 19 mm to 25 mm diameter bolts, E49 electrodes(490 MPa) and 350 MPa W Grade plate. Variables to be examined included; the distance from the shear plate weld to the first column of bolts('a' distance), detailing of the shear plate(full height vs. partial height) and detailing of stiffeners on the back side of the girder web. The paper describes the testing program, observed and measured performance and failure modes, as well as compares the results with predictions from a modified version of the AISC extended shear tab method.

The laboratory results showed that extended beam-to-column shear tab connections could reach resistance levels consistent with typical predicted failure modes and still maintain an adequate level of plastic rotation capacity. The beam-to-girder connection tests revealed that(a) plate local buckling for full-height connections; and(b) localized deformation of the supporting girder web and flange for partial-height connections should be explicitly considered as part of the design process of such connections. *Fig. 1* shows an example of full-height beam-to-girder shear tab connections in which plate buckling occurred in all cases.

CONCLUSIONS

This paper summarizes an extensive experimental program that was conducted at full-scale to characterize the behaviour of extended beam-to-column and beam-to-girder shear tab connections subjected to direct shear. The connections were designed according to the extended shear tab design method per[1,2]. The main findings of the experimental program are summarized as follows:

● The beam-to-column tests showed good agreement between the predicted and measured resistances.

● The AISC[1] design method specifies sizing the welds 5/8ths of the plate thickness. This ratio is based on experimental observation[3] as well as theory[4]. It is recommended that this ratio be replaced by the Muir and Hewitt design equation[4] with the plate yield stress taken as 110% of the nominal.

● The full-height beam-to-girder tests were characterized by plastic buckling of the stiffener portion of the shear tab. A design check is proposed taking into account the vertical stresses due to the connection shear and the horizontal stresses due to flexural action of the beam.

- The partial-height beam-to-girder shear tab tests revealed that girder web and flange deformation is significant when the top flange of the supporting girder is unrestrained. In these cases, including a stiffener opposite to the shear tab for flexible connections can reduce this deformation.

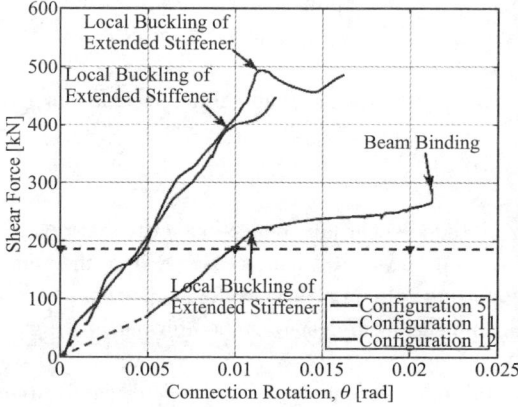

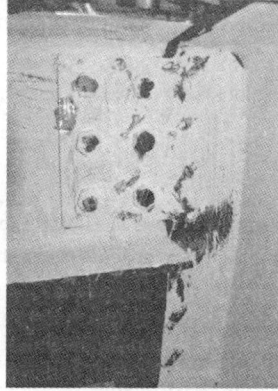

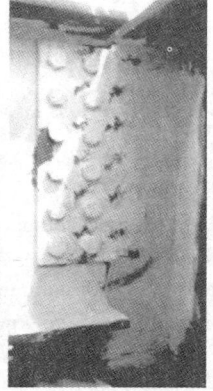

(a) Connection shear force-connection rotation   (b) Configuration 5   (c) Configuration 11

Fig. 1. Performance of full-height beam-to-girder shear tab connections

ACKNOWLEDGEMENTS

This paper is based on research supported bythe ADF Group Inc. and DPHV Structural Consultants, as well as the Natural Sciences and Engineering Research Council of Canada(NSERC). This financial and technical support is gratefully acknowledged. The assistance of the undergraduate student researchers Farbod Pakpour, Milad Moradi, and Harrison Moir for the execution of the laboratory experiments is greatly appreciated. Any opinions, findings, and conclusions or recommendations expressed in this paper are those of the authors and do not necessarily reflect the views of sponsors.

REFERENCES

[1] AISC "Steel construction manual, 14th Edition", American Institute of Steel Construction. Chicago, IL, 2010.

[2] CSA S16-09. , "Design of steel structures", *Canadian Standards Association*, Mississauga, ON, 2009.

[3] Astaneh A, McMullin KM, Call SM. , "Behaviour and design of steel single plate shear connections", ASCE, *Journal of Structural Engineering*, 119(8):2421-2440, 1989.

[4] Muir LS, Hewitt CM. , "Design of unstiffened extended single-plate shear connections", *Engineering Journal*, 46(2):67-79. , 2009.

ASSESSMENT OF RWS BEAM-COLUMN CONNECTIONS USING CELLULAR BEAMS WITH MULTIPLE CLOSELY SPACED WEB OPENINGS

Konstantinos Daniel Tsavdaridis[a] and Theodore Papadopoulos[b]

[a] Assistant Professor of Structural Engineering, School of Civil Engineering,
University of Leeds, LS2 9JT, Leeds, UK
k. tsavdaridis@ leeds. ac. uk

[b] Design Engineer, Pell Frischmann Consultants, 5 Manchester Square,
W1U 3PD, London, UK
TPapadopoulos@ pellfrischman. com

KEYWORDS: RWS, perforated beams, web openings, cellular, seismic design, cyclic loading

ABSTRACT

Steel moment resisting frames (MRFs) have traditionally been built in areas susceptible to high seismicity and are greatly dependent on their beam-to-column connection behaviours. In the past, fully weldedconnections were considered to provide the optimum combination of strength, stiffness and ductility which according to codes are the major factors in the seismic design of joints[1,2]. A more recent trend of earthquake resistant design comes by the name of performance-based design. Reduced Beam Section (RBS) or "Dogbone" connections with different ways of reducing locally the cross-sectional area of the beams are embracing this concept by achieving material efficiency. Recently, a pioneer concept of merging the purposely weakened beams/connections with the use of perforated beams was introduced, as a trend of achieving material efficiency and considering material reduction through beam web cuts, also known as Reduced Web Section (RWS)[3]. From another perspective, perforated beams offer the advantages of a deeper web (i. e. increased second moment of area) depending on the manufacturing procedure, but without the increase in weight; hence reduced material volume without a decrease of the bending capacity. They can therefore span longer distances as well as incorporate services within the floor-to-ceiling structural zone. There has been a lot of research on perforated beam webs with the geometry of the perforation ranging from circular, hexagonal, to even elliptically-based shapes[4,5]. However, very limited research has been conducted up to date regarding the design limitations of connections when such standard and non-standard perforated beams are used.

Research undertaken in RWS connections is related to the fabrication process (increased cost due to web cutting and welding of the section), buckling issues (i. e. stability issues due to the increased depth in certain cases), number of web perforations as well as their use in MRFs. RWS connections provide a higher rotational capacity of the order of 0. 05 radian, whilst 0. 035 radian is suggested by EC8 and FEMA-350 to be acceptable in seismic design[2,6]. On the other hand, the local shear capacity of the beams is decreased because of the opening existence. Nowadays' trend is to increase the rotational capacity of connections from what was suggested by post-Northridge codes. Recent research was conducted for the seismic-resistant design of MRFs using beams with isolated web openings by Yang et al. in 2009[7]. A comparison with typical un-perforated (solid webbed) beam-to-column connections as well as with connections consisting one of the three different typical sizes of circular web openings was established. All results revealed a ductile failure of the frame with adequate stiffness which was not restricted significantly by the presence of the web openings. The focus on the beam-to-column connections promoted the creation of the Vierendeel mechanism with no brittle web fracture. It is worth to note that the Vierendeel mechanism is a ductile failure mode with the formation of four plastic hinges and the redistribution of the load in the vicinity of the opening.

A semi-rigid steel beam-to-column connection with circular beam web openings is examined while subjected to cyclic loading. Only conventional circular web openings are used, while the web-post buckling behaviour is extensively analysed by having multiple closely spaced web openings along the length of the beam. The structural performance of the RWS connection with the use of perforated beams, instead of one local perforation (i. e. local weakening) at the cross-sectional area of the beamand with the connection being partially restrained, has not yet been examined under cyclic loading. In order to fully understand the response of RWS connections with perforated beams and evaluate their ability to dissipate seismic energy, numerous FE models including a solid beam model with no perforations is also developed. A total of 12 FE models were developed and categorised into four Sets. The same connection configuration from Díaz et al. (2011) (specimen T101. 010) is used in this study, whilst the length from the end-plate to the load application point, Lload, is different[8]. The beam length was increased in order to be able to accommodate periodical multiple web openings and simulate a typical span of 6m. Consequently, Lload was extended from 1. 25m in the literature to 3m in the parametric Sets. The main geometrical parameter examined in this paper is the distance from the face of the column to the centreline of the web opening, S.

CONCLUSIONS

The study indicates that RWS connections using cellular beams behave in a satisfactory manner and provide an enhanced performance in terms of stress distribution when subjected to cyclic loading, especially when the web openings are located at a particular distance from the face of the column. The desirable weak beam-strong column mechanism was accomplished by the introduction of web openings, proving the performance of RWS connections to be used in seismic resistant designs. Models with uniform material properties were also examined (S355 for all components except of the bolts with Class 10. 9), as the initial difference in material properties taken from the specific experiment hindered the desired mechanism. The addition of column stiffeners, similar to practice, was investigated as the flanges restrain the out-of-plane buckling. Specimens with closely spaced web openings demonstrated positive results, with the latter ones mobilising the critical stresses in the vicinity of the first web opening. Column face distance, S, of 520mm decreases the ultimate rotational capacity from 0. 05radian to 0. 04radian, while distance of 200mm or less leads to critical stress concentration very close to the columns' face and the connections' components. The ideal distance, S, was therefore identified as being 350mm (i. e. 1. 2do or 0. 96h). Limitations of this study suggest future work to be conducted regarding the use of perforated beams in the aseismic design of steel MRFs. This study is part of an extensive research project investigating different types of connections and the use of novel patented elliptically-based web openings, aiming to establish comparisons of seismic and progressive collapse codes and design guidelines.

REFERENCES

[1] Eurocode 3 [2005]: Design of Steel Structures-Part1: General Rules and Rules for Buildings, EN 1993-1-1, Brussels, Belgium.
[2] EC8. Part 3: Design of Structures for Earthquake Resistance. Assessment and Retrofitting of Buildings. EN 1998-3, June 2005.
[3] Tsavdaridis, K. D. , Faghih, F. , and Nikitas, N. 2014. Assessment of Perforated Steel Beam-to-Column Connections Subjected to Cyclic Loading. Journal of Earthquake Engineering, vol. 18, issue 8, pp. 1302-1325.
[4] Tsavdaridis, K. D. , and D'Mello, C. 2011. Web Buckling Study of the Behaviour and Strength of Perforated Steel Beams with Different Novel Web Opening Shapes. Journal of Constructional Steel Research, vol. 67, pp. 1605-1620.
[5] Tsavdaridis, K. D. , and D'Mello, C. 2012. Optimisation of Novel Elliptically-Based Web Opening Shapes of Perforated Steel Beams. Journal of Constructional Steel Research, vol. 76, pp. 39-53.
[6] FEMA 350, "Recommended seismic design criteria for the new steel moment frame Buildings", Report no. FEMA 350, SAC Joint Venture, CA, 2000.
[7] Yang, Q. , Li, B. , and Yang, N. 2009. AseismicBehaviors of Steel Moment Resisting Frames with Opening in Beam Web. Journal of Constructional Steel Research, vol. 65, pp. 1323-1336.

SUBASSEMBLAGE TESTS OF THE IN-PLANE STRUCTURAL BEHAVIOR OF BUCKLING RESTAINED BRACE WELDED END CONNECTIONS

Junxian Zhao[a,b], Zhan Wang[a,b], Fuxiong Lin[b]

[a] State Key Laboratory of Subtropical Building Science, South China University of Technology, Guangzhou 510641, China
ctjxzhao@ scut. edu. cn, wangzhan@ scut. edu. cn

[b] School of Civil Engineering and Transportation, South China University of Technology, Guangzhou 510641, China
ctjxzhao@ scut. edu. cn, wangzhan@ scut. edu. cn, linfuxiong2012@ 163. com

KEYWORDS: Buckling-restrained brace, Subassemblage test, Frame action, End rotation, End bending moment.

INTRODUCTION

Buckling-restrained braces(BRBs), serving as both lateral loading-resisting members and energy dissipation fuses(EDF) to engineering structures, are now gaining wide acceptance in seismic-prone areas. It is known that bolted or welded end connections(hereafter called fix-ended condition) are generally used to connect BRBs to frame structures for easy construction. Damage of the BRB end connections in subassemblage and frame tests were often reported in previous studies, leading to unexpected failure of the EDF systems. However, much attention was paid to the out-of-plane stability of BRB end connections and few discussions on their in-plane structuralbehaviour were reported.

On the basis of the first author's prior work[1-3], this paper aims to further investigate the effect of in-plane frame action on the BRB fix-ended connections. Subassemblage tests, designed to simulate the frame action mode with single diagonal bracing configuration, were conducted on 3 BRB specimens with welded end connections and varying flexural rigidities of end zone. The cyclic behaviour of BRBs and the rotational and flexural behavior of BRB end zones are first presented. The predictions for the rotational and flexural demand on BRB ends are also determined based on rigid body motion analysis and validated experimentally. The negative effect of frame action on BRB end zone is discussed finally.

TEST PROGRAM

Fig. 1 shows the loading setup for the specimens, including the left and right columns and the top and bottom beams connected by pins. Two MTS actuators with total capacity of 500kN were connected to the loading beams welded to the upper end of the right column to impose horizontal cyclic loading. With the presence of the top beam, the two columns could rotate around their bottom pins with identical rotational angles. Therefore, both horizontal and rotational deformations could be imposed on the upper gusset with only the rotational one on the lower gusset, simulating the frame action of BRBF with single-diagonal bracing configuration. The BRB subassemblage, i. e. the BRB specimen and two gussets, were connected to the loading frame by high-strength bolts for easy replacement.

TEST RESULTS AND ANALYSIS

Fig. 2 presents the normalized bending moment responses M_{end}/M_y (M_y is the yield moment of the end zone) versus story drift angle α relationship based on the flexural strain responses and elastic beam theory. The magnitudes of bending moment ranged from 0. 5 to 1. 2M_y, showing a significant influence on the structural behaviour of BRB end zones. Significant inelastic responses could be easily expected when both axial force and bending moment were imposedon BRB end zones. This is an undesirable case in design and may lead the end zone to buckling. Based on the above observation, it is recommended that the negative effect of bending moment induced by frame action should be carefully considered in design of BRB ends.

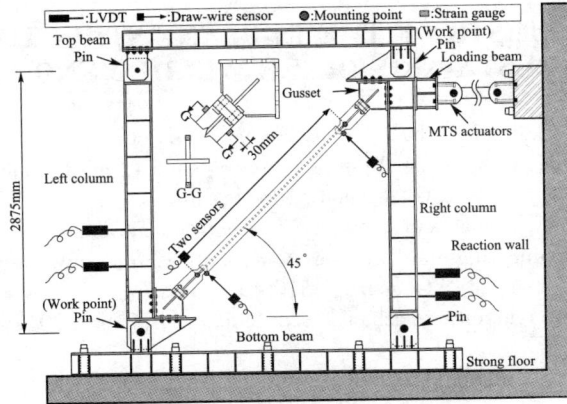

Fig. 1. Loading setup

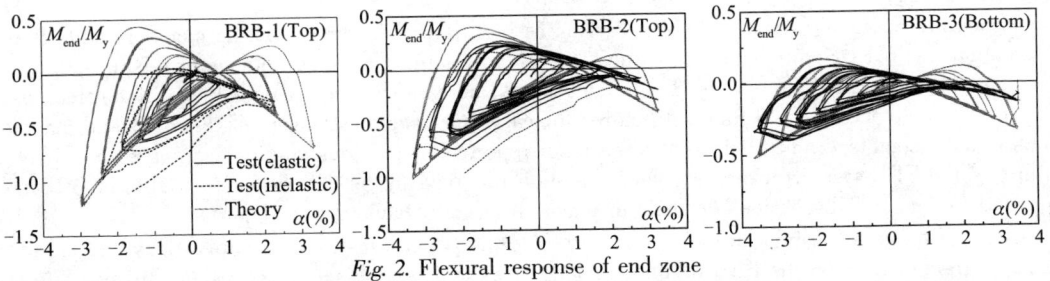

Fig. 2. Flexural response of end zone

CONCLUSIONS

The effect of seismic frame action on the structural behaviour of BRB welded end connections was investigated by subassemblage tests on 3 BRB specimens with single-diagonal bracing configuration. Main conclusions can be drawn as follows:

1) The proposed BRB end rotational mechanism induced by the expected frame action is validated experimentally. Test results show that the relative rotational response between end zone and plastic zone is comparable with the story drift angle.

2) The initial imperfections induced by the construction error between gusset zone and end zone may have significant influence on the rotational demand on BRB ends, whichshould be properly considered in design.

3) Test confirmed that the presence of frame action subjects the BRB end zone to additional bending moment up to its yield moment, which cannot be ignored in design. The rotational and flexural demands on BRB ends can be simply estimated by the proposed rigid-body motion analysis method. However, future study is still needed to consider the effect of flexural deformation of end zone on the bending moment.

REFERENCES

[1] Zhao JX, Wu B, Ou JP, 2011. "A novel type of angle steel buckling-restrained brace: Cyclic behavior and failure mechanism". *Earthquake Engineering and Structural Dynamics*, Vol. 40 pp. 1083-1102.

[2] Zhao JX, Wu B, Ou JP, 2012. "Flexural demand on pin-connected buckling-restrained braces and design recommendations". *Journal of Structural Engineering*, ASCE, 138: 1398-1415.

[3] Zhao JX, 2012. "Seismic behavior and stability design methods of all-steel buckling-restrained braces", Dissertation: Harbin Institute of Technology. [in Chinese]

AXIAL STRENGTH AND DEFORMATION DEMANDS FOR T-STUB CONNECTION COMPONENTSAT CATENARY STAGE IN THE BEAMS

Florea Dinu[a,b], Dan Dubina[a,b], Ioan Marginean[a], Calin Neagu[a], Ioan Petran[c]

[a] Politehnica University Timisoara, Department of Steel Structures and Structural Mechanics, Romania

florea. dinu@ upt. ro, dan. dubina@ upt. ro, ioan. marginean@ upt. ro, calin. neagu@ upt. ro

[b] Laboratory of Steel Structures, Romanian Academy, Timisoara Branch, Romania

[c] Technical University of Cluj Napoca, Department of Structures, Romania

ioan. petran@ bmt. utcluj. ro

KEYWORDS: steel connection, bolted T-stub, catenary action, experimental, numerical.

ABSTRACT

Capacity of multi-storey steel frame buildings to resist extreme loading may depend on the performance of beam-to-column joints to provide continuity across the damaged area, and thus to allow the development of alternate loads paths(AP). The AP method, with its emphasis on continuity and ductility, is similar to current seismic design practice. However, there are specific problems which need to be considered when localized failures, particularly of columns, occur, i. e. development of the catenary forces in the beams and admissibility criteria to be considered in the design, considering the interaction between axial loads and bending moments. It is therefore of interest to study the capacity of actual design procedures to provide enough robustness for connections under extreme loading conditions.

A large experimental program on connection componentswas developed within the framework of the CODEC research program at PU Timisoara[1]. The scope of the investigations was to evaluate the influence of initial design conditions of bolted T-stub components(i. e. resistance, failure mode) on the tensile response under different strain rates. Different T-stub components, designed to fail in mode 1 and 2, were tested in tension until failure, in order to simulate their behaviour in catenary action phase (axial strength, deformation capacity). Numerical models were validated against test results.

RESULTS

The following typologies were selected for the experimental program: T-10-16-100; T-10-16-120; T-10-16-140; T-12-16-100; T-12-16-120; T-12-16-140. First letter is the T-stub, second term is the thickness of the end plate, third is the diameter of the bolt and then the distance between the bolts, all distances in mm. The steel grades were S235 for flanges(end-plate) and S355 for webs, and bolts were class 10. 9. Specimens were tested under two levels of strain rate, i. e. low strain rate(C)-imposed displacement of 0. 05mm/sec, and high strain rate(CS)-imposed displacement of 10 mm/sec. Experimental results(Fig. 1)showed that more ductile configurations can be obtained by reducing the end plate thickness or by increasing bolt row distance, due to flexural yielding of the end-plate. The loading rate has a reduced impact on the ultimate capacity and failure mode.

Numerical modelswere validated against the test data. It can be seen the response follows with high accuracy the actual behaviour of the specimens(Fig. 2). A bending moment-axial force interaction curve was obtained through numerical tests on the bolts(Fig. 2). The evolution of bending moment and axial force in bolts showed that, after approaching the interaction curve, the bending moment started to reduce, allowing the bolt to resist larger axial forces. The fracture points for several configurations of T-stubs were close, although the capacity of the T-stubs varied.

Four bolts T-stubs with the same material properties, end plate thickness and bolt distance were constructed in the following configurations: T-stubs with no stiffeners(I), T-stubs with one stiffener (T), and T-stubs with two stiffeners(X)(Fig. 3. a). X-configurations showed a 40% ~ 50 % greater capacity than I configurations, while the ductility decreases by approximately 25% to 30 %. The ductility and ultimate force of T configurations is between the values obtained for X and I configurations (Fig. 3. b). If the sum of forces in the bolts is normalised to the loading force, F_{bolt}/F_{T-stub}, it is possible to observe the axial force demands on the T-stub due to prying effects. These prying forces may lead to additional axial forces(Fig. 3. c).

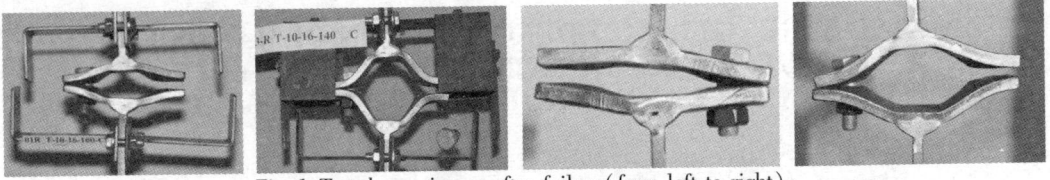

Fig. 1. T-stub specimens after failure(from left to right):
T-10-16-100;T-10-16-140;T-12-16-100;T-12-16-140

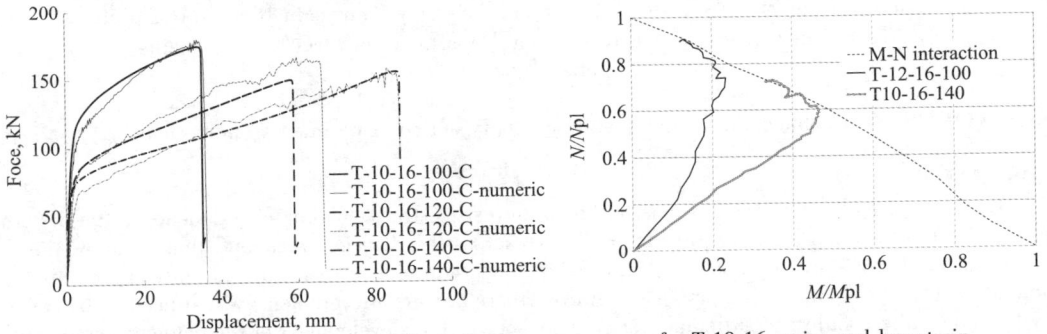

Fig. 2. Experimental vs numerical force-displacement curves for T-10-16 series and low strain rate(left);axial force-bending moment for bolts in T-stubs and bolt interaction curve(right)

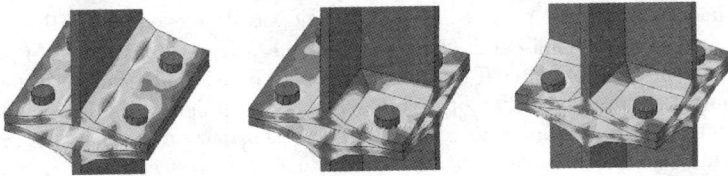

(a)

(b)

(c)

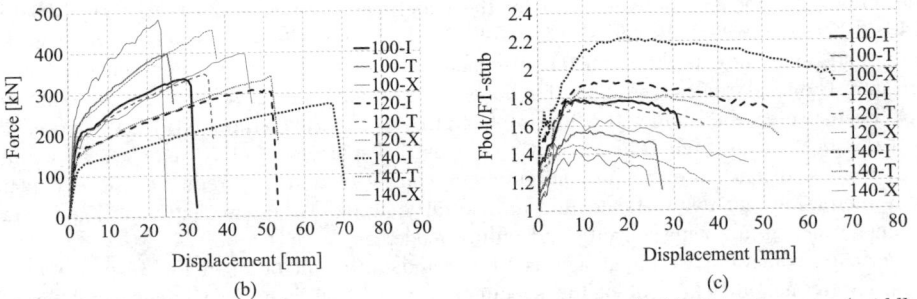

Fig. 3. (a)Deformed configuration of 4 bolts T-stubs:I configuration(left),T configuration(middle) and X configuration(right);(b)force-displacement curves;(c)$F_{bolt}/F_{T\text{-stub}}$ vs. T-stub displacement

ACKNOWLEDGMENT

Funding for this research was provided by UEFISCDI,grant PCCA 55/2012(2012-2016)and by th strategic grant POSDRU/159/1.5/S/137070(2014)of the Ministry of National Education,Romania,co financed by the European Social Funds-Investing in People,within the SOP HRD 2007-2013.

REFERENCES

[1] CODEC. Structural conception and collapse control performance based design of multistory struc tures under accidental actions,2012-2016. Executive Agency for Higher Education,Research,De velopment and Innovation Funding,Romania,grant PNII PCCA 55/2012,the Partnerships Pro gram Joint Applied Research Projects,2012.

VELOCITY EFFECTS ON THE BEHAVIOUR OF ASYMMETRICAL FRICTION CONNECTIONS (AFC)

Jose C. Chanchi Golondrino[a], Gregory A. MacRae[b], James G. Chase[b],
Geoffrey W. Rodgers[b] and George C. Clifton[c]

[a]University of Canterbury-New Zealand, National University of Colombia-Colombia
jcchanchigo@ unal. edu. co
[b]University of Canterbury-New Zealand
gregory. macrae@ canterbury. ac. nz, geoff. chase@ canterbury. ac. nz,
geoffrey. rodgers@ canterbury. ac. nz
[c]University of Auckland-New Zealand
c. clifton@ auckland. ac. nz

KEYWORDS: Asymmetrical Friction Connection, Velocity dependence, Low damage dissipaters, Energy dissipation, Shim materials.

ABSTRACT

Asymmetrical Friction Connections (AFC) are friction connections assembled with a slotted plate sandwiched by two plates with standard holes. One of these plates is termed fixed plate and it is attached to a support that restricts its displacement, and the other is a floating plate that is attached to the fixed and slotted plate by means of high strength bolts. Thin plates termed shims can be inserted at the interface between the slotted plate and the fixed plate, and at the interface between the slotted plate and the cap plate. By inserting shims with hardness greater than the hardness of the slotted plate a stable hysteretic behaviour with low degradation can be achieved on this type of friction connection. AFCs can be used to dissipate seismic energy in different structural systems by attaching the slotted plate to a member or joint that forces the slotted plate to slide across the two interfaces generated with the fixed and cap plates. In doing that, the seismic force on the member or on the joint is dissipated via friction by overcoming the clamping force induced by the high strength bolts on the AFC and producing the sliding of the slotted plate. The behaviour of AFCs using different shim materials under quasi-static conditions has been characterized recently by Clifton 2005, Khoo et al. 2011, and Chanchi et al. 2012. However, (i) the behaviour of AFCs at high velocities has not been described, (ii) a model for describing the velocity dependence of AFCs has not been available, and (iii) ranges of variation of the effective friction for AFCs using different shim materials at high velocities are not known. In order to address the issues stated above testing of AFCs using different shim materials was carried out at low and high velocities of 10mm/s and 190mm/s respectively. This paper describes the effects of the high velocity on the hysteretic behaviour of AFCs, presents a model for representing the velocity dependence of AFCs, and quantifies a range of variation of the effective friction coefficient for the low and high velocity cases. Results show that at the high velocity regardless of the shim material the hysteresis loop is less stable, and greater forces are required to activate the sliding mechanism of the slotted plate when compared with the hysteresis loop at the low velocity. Results also show that the velocity dependence of AFCs can be quantified through the effective friction coefficient at quasi-static conditions and a velocity exponent that ranges between 0. 0020 and 0. 0045.

CONCLUSIONS

This paper describes the velocity effects on the AFC using different shim materials. It is shown that:

1. At high velocities the hysteresis loop of AFCs loses stability, and greater forces are required to activate the sliding mechanism when compared with the forces required in quasi-static conditions.

2. The effective friction coefficient of AFCs increases with increments on the velocity. Magnitude of the increments on the effective friction coefficient depends on the ratio between the hardness of the shim material and the hardness of the slotted plate.

3. For ratios between the hardness of the shim material and the hardness of the slotted plate (hardness ratio) varying on the range 0.58-3.85, the effective friction at low and high velocities vary in the ranges 0.22-0.27 and 0.24-031 respectively.

4. The effective friction coefficient at high velocities can be quantified as the product between the effective friction coefficient at low velocities and the ratio between the high and low velocity raised to the power of an exponent that varies in the range of 0.020-0.045.

ACKNOWLEDGMENT
The authors acknowledge the financial support given by Natural Hazard Research Platform (NHRP)-New Zealand for undertaking the research associated with this publication.

REFERENCES
[1] Chanchi, J. C., MacRae, G. A., Chase, G., Rodgers, G, and Clifton, C. (2012). Behaviour of a-symmetrical friction connections using different shim materials. *New Zealand Society for Earthquake Engineering Conference.*

[2] Clifton, G. C. (2005). Semi-Rigid Joints for Moments Resisting Steel Framed Seismic Resisting Systems. *Published PhD Thesis, Department of Civil and Environmental Engineering.* University of Auckland-New Zealand.

[3] Chanchi, J. C., Xie R., MacRae, G. A., Chase, G., Rodgers, G, and Clifton, C. (2014). Low damage brace using Asymmetrical Friction Connections. *New Zealand Society for Earthquake Engineering Conference.*

[4] MacRae, G. A., Clifton, C. G., MacKinven, H., Mago, N., Butterworth, J., Pampanin, S. (2010). The Sliding Hinge Joint Moment Connection. *Bulletin of the New Zealand Society for Earthquake Engineering.*

[5] Khoo, H. H., Clifton, C., Butterworth, J., MacRae, G. and Ferguson, G. (2011) Influence of steel shimhardness on the Sliding Hinge Joint. *Journal of Constructional Steel Research.*

[6] Pekcan, G., Mander, J. B., M. Eeri, and Chen, S. S. (1995). The Seismic Response of a 1:3 Scale Model R. C. Structure with Elastomeric Spring Dampers. *Earthquake Spectra*, Vol. 11, No. 2, pp 249-267.

A STEP FORWARD IN THE CYCLIC ASSESSMENT OF THE *F-Δ* COMPONENTS USING COMPLETE FINITE ELEMENTS MODELS OF BEAM-TO-COLUMN STEEL END PLATE BOLTED JOINTS

Hugo Augusto[a], José Miguel Castro[b], Carlos Rebelo[a], Luís Simões da Silva[a]

[a] ISISE, University of Coimbra, Department of Civil Engineering, Portugal

hugo. augusto@ dec. uc. pt, crebelo@ dec. uc. pt, luisss@ dec. uc. pt

[b] University of Porto, Department of Civil Engineering, Portugal

miguel. castro@ fe. up. pt

KEYWORDS: Cyclic, joints, components, Finite Elements, end-plate.

ABSTRACT

In recent years, the component method (CM) used in assessing beam-column joints behaviour became highly disseminated through the scientific community, due to simplicity of use and proved accuracy in the behaviour characterization of beam-to-column and column base connections. The method was also adopted in the engineering practice through its implementation in codes of practice such as Eurocode 3[1], and also in commercial software. Despite the many advantages of the method, a key limitation is that it is only applicable to static monotonically increasing loading.

Experimental tests proved to be so far the most accurate way to calibrate the response obtained by the CM for the joints behaviour. However there are some limitations in obtaining all the data necessary for the proper characterization of the individual components that contribute to the joints behaviour. Finite elements models provide the opportunity to complement the experimental test data, due to the discretised nature of the FE models, providing all the relevant information in any element, and even providing unavailable data for the experimental tests like the contact or friction forces. The objective of the research work under development is to contribute to the extension of the CM to account for the cyclic behaviour of the various components. In this paper is presented the procedure to obtain the *F-Δ* components response from the FE models results. This is needed to characterize the spring mechanical models, hence allowing its application in the characterization of end plate beam-to-column joints subjected to cyclic loading conditions. To that end, a set of detailed numerical models of beam-to-column end plate joints have been developed using the ABAQUS FE package, capable of representing the various connections components and the several sources of nonlinearity associated to them. The numerical models were previously validated against monotonic and cyclic experimental test data available, obtained by the research team, and were additionally validated in this paper for the strains developed in the column web.

The results presented in this paper allow setting the base for the implementation of the dissipative behaviour in the context of the component method and also provide a methodology to extract the needed force-displacement relationships of the column web components of the joint.

CONCLUSIONS

The finite elements models developed for this work followed the experimental tests performed previously[2]. The excellent agreement allowed to conclude that the model is capable of simulating accurately the end-plate beam-to-column joints, and that it can be used in detailed analyses of the components behaviour.

A procedure is proposed, based on the considerations presented in[3] for welded joints. The forces are obtained by numerical integration of the stress fields, obtained in the FE models in predefined paths (P1 to P3) along the web mesh. For the components column web in transverse tension or compression the *Eq.* (1) is used, and for the column web panel in shear the *Eq.* (2) is used, see *Fig. 1*a) and c). The deformations are obtained by the nodal displacements obtained in the FE models, see *Fig. 1*b) and c). The accuracy of the model depends on the element size and the number of stress fields analysed from the available load increments.

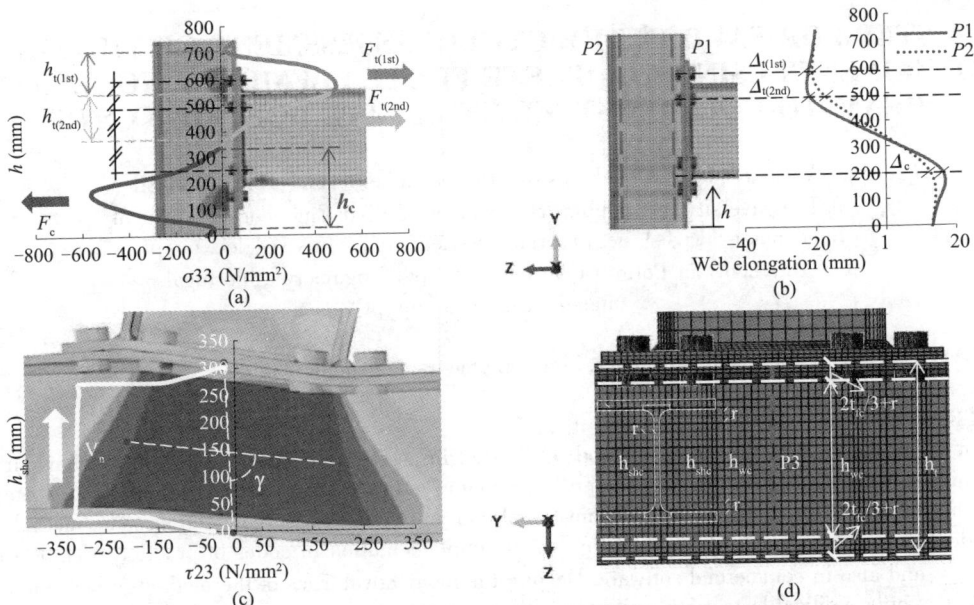

Fig. 1. F-Δ/V-γ determination by stress field integration and nodal displacements

$$F_c = \left(\int_{}^{h_c} \sigma_{33} dy \right) \cdot t_{wc} \ or \ F_{t,i} = \left(\int_{}^{h_{t,i}} \sigma_{33} dy \right) \cdot t_{wc} \tag{1}$$

$$V_n = \left(\int_{(2t_{fc}/3+r)_{back}}^{} \tau_{23} dz \right) \cdot t_1 + \left(\int_{}^{h_{wc}} \tau_{23} dz \right) \cdot t_{wc} + \left(\int_{(2t_{fc}/3+r)_{front}}^{} \tau_{23} dz \right) \cdot t_1 \tag{2}$$

The procedure was applied to the developed FE models, loaded monotonically, and compared with the Atamaz-Jaspart[4] model and also to the Eurocode 3[1] procedure, revealing the good accuracy of the models and the procedure developed. The promising results will allow the extension of the procedure to other components and to models loaded cyclically.

ACKNOWLEDGMENT

Financial support from the Portuguese agency *Fundação para a Ciência e a Tecnologia* (FCT) under contract grant PTDC/ECM/116904/2010 is gratefully acknowledged. Also the financial support from the Portuguese agency FCT within QREN-POPH-Typology 4. 1-Advanced Education, reimbursed by the European Social Fund and Portuguese national funds MEC, under contract grants SFRH/BD/91167/2012, for Hugo Augusto, is gratefully acknowledged.

REFERENCES

[1] Eurocode 3, 2005. *Design of steel structures — Part* 1-8 ; *design of joints*, EN 1993-1-8, European Committee for Standardization, CEN, Belgium, Brussels.

[2] Nogueiro, P., 2009. "Dynamic Behaviour of Steel Connections", PhD Thesis, Civil Engineering Department, University of Coimbra, Portugal, (in Portuguese).

[3] Jordão S., Simões da Silva L., Simões R., 2013. "Behaviour of welded beam-to-column joints with beams of unequal depth", *Journal of Constructional Steel Research*, ISSN 0143-974X, Elsevier Science Limited, Vol. 91, pp. 42-59.

[4] Jaspart J. P., 1990. "Shear and load-introduction deformability and strength of column web panel in strong axis beam-to-column joints", *Internal Report MSM*, $n°$ 202, University of Liége, April Belgium.

DEFORMATION LIMIT FOR DUCTILE FRACTURE IN WELDED TUBULAR JOINTS

Xudong Qian, Aziz Ahmed

Centre for Offshore Research and Engineering, Department of Civil and Environmental
Engineering, National University of Singapore, Singapore
qianxudong@ nus. edu. sg, ceeaziza@ nus. edu. sg

KEYWORDS: Circular hollow section, deformation limit, fracture resistance, ductile fracture.

ABSTRACT

Circular hollow sections have evolved over the years as a primary structural member for both on-shore and offshore infrastructures. The welded connections between the circular hollow sections often entail sharp changes in the local geometry, which leads to high concentration of stresses. Such high stress concentrations frequently nucleate and grow fatigue cracks near the weld toes along the profiled connection between adjacent circular hollow section members. The unstable fracture of these fatigue cracks under an extreme environmental action, e. g. , an earthquake, creates critical threats for the safety of the steel tubular frames.

The prevailing engineering design guidelines do not provide an explicit treatment of the ductile fracture failure for welded connections. The ductility requirement on the structure remains as a qualitative material requirement, prescribed either as the material elongation level or theCharpy energy values. On the other hand, the fracture-mechanics type of analysis on the welded connections often requires significant efforts in both the model preparation and the calibration of the material damage parameters. The latter prohibits the scalable application of fracture mechanics to welded connections fabricated potentially from an extensively wide range of steel grades. Engineering assessment of ductile fracture and the subsequent unstable fracture failure requires a simple physical criterion, readily implementable by practicing engineers.

This study aims to propose a simple deformation limit at the onset of ductile fracture failure for welded tubular X-joints. The deformation criterion derives from an extensive numerical study covering a comprehensive matrix of material and geometric parameters for CHS X-joints with an initial fatigue crack. The critical deformation in this study equals the deformation level at which the elastic-plastic crack driving force reaches the material fracture toughness. The proposed deformation limit follows a non-dimensional form, which addresses its scalability to a practical range of steel materials with different fracture toughness levels. The deformation limit for intact joints follows conservatively the extrapolated deformation level for joints with a zero crack size. The comparison of the experimental results for circular hollow section X-joints confirms the validity of the proposed deformation limit.

VALIDATION

The comparison of the between experimentally measured deformation limit corresponding to the fracture failure and the computed deformation level at the critical fracture toughness confirms the feasibility of a fracture toughness-based deformation level, as illustrated in *Fig. 1*. The experimental specimen reported by *Qian et al.* [1] entails a fatigue crack at the hot-spot position in a circular hollow section X-joint generated by cyclic loading prior to the monotonic load test. The large-scale X-joint specimen has an outer diameter of 750 mm, with the brace-to-chord diameter ratio, $\beta = 0.54$, the chord radius to thickness ratio, $\gamma = 15$, and thebrace-to-chord wall thickness ratio, $\tau = 0.5$. The fatigue crack locates at the weld toe in the brace member, which utilizes the S355 steel and has a measured toughness of $J_{\text{Ic}} = 90 \text{ kJ/m}^2$.

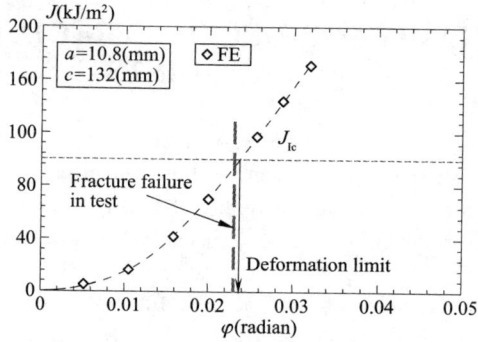

Fig. 1. Geometric configuration of an X-joint

DEFORMATION LIMIT

The critical deformation corresponding to the material toughness values exhibit significant scatters. The proposed deformation limit thus follows a normalized procedure to minimize the scatter against different geometric parameters and fracture toughness levels. The normalized deformation limit follows a linear relationship with respect to the crack depth ratio,

$$\varphi_n = \frac{\varphi_{cr}}{\left(\frac{J_{Ic} \times \beta \times \tau^3}{E \times t_0}\right)^{\frac{1}{2}}} = 22 - 19.8\left(\frac{a}{t_1}\right) \tag{1}$$

where J_{Ic} refers to the fracture toughness, E defines the Young's modulus, t_0 denotes the thickness of the chord, t_1 represents the thickness of the brace and a indicates the crack depth.

CONCLUSIONS

The proposed deformation limit represents a lower-bound deformation level at the onset of ductile tearing in fatigue-cracked(or otherwise cracked)CHS X-joints. The deformation level extrapolated at a zero crack depth ratio provides a conservative deformation limit for the intact X-joint. The proposed deformation limit agrees closely with the deformation limit derived from the failure assessment diagram.

REFERENCES

[1] Qian X,Jitpairod K,Marshall PW,Swaddiwudhipong S,Ou Z,Zhang Y,Pradana MR. ,"Fatigue and Residual Strength of Concrete-Filled Tubular X-Joints with Full Capacity Welds",Journal of Constructional Steel Research,100:21-35,2014.

NUMERICAL STUDY ON THE LOCAL BUCKLING BEHAVIOR OF END-PLATE CONNECTION IN STEEL GABLED FRAMES

Yun-Dong Shi[a], Yi-Yi Chen[b]

[a] Tianjin University, Key Laboratory of Coast Civil Structure Safety of Ministry of
Education, China
yundong@ tju. edu. cn
[b] Tongji University, State Key Laboratory of Disaster Reduction in Civil Engineering, China
yiyichen@ tongji. edu. cn

KEYWORDS: End-plate connection, Non-compact members, Local buckling, Panel zone.

ABSTRACT

Recently, tests on end-plate connection in the light-weight steel gabled frame were conducted in Tongji University by the authors[1]. Local buckling of beam and column members was observed, and it caused strength deterioration of the connection and degraded the seismic performance of the structure. On the other hand, the thin-walled panel zone was found to have stable behavior during the test. This indicated that the panel zone might be used to improve the seismic behavior of the steel gabled frame. In order to further investigate the behavior of the end-plate connection, finite element (FE) models using the ANSYS 9.0 package[8] were developed and parametrical study was conducted.

Totally 10 specimens are designed in the parametric study as shown in *Table 1*.

Table 1. Specimens' properties.

Specimen ID	EW/T of		Yield Strength of Column	Diagonal Stiffener	Deformation Modes[a]
	Column Web	Panel Zone			
C6B6E18HO	91	118	345	No	LBC
C6B6E18HW	91	118	345	Yes	LBC
C6B4E18HO_1	91	177	345	No	LBP
C6B4E18HO_2	91	177	322	No	LBP
C6B4E18HO_3	91	177	310	No	LBC& LBP
C6B4E18HW	91	177	345	Yes	LBC
C4B4E18HO_1	136	177	345	No	LBC& LBP
C4B4E18HO_2	136	177	390	No	LBP
C4B4E18HO_3	136	177	375	No	LBC& LBP
C4B4E18HW	136	177	345	Yes	LBC

Note: [a] LBC, LBP, SDP represent local buckling of column, local buckling of panel zone, and shear deformation of panel zone, respectively.

Fig. 1 shows three typical deformation modes observed in the parametric study: local buckling of panel zone (LBP), local buckling of column (LBC), and the combination of LBC and LBP. The differences of the three deformation modes are discussed with respect to the strength deterioration, deformation capacity, ductility behavior and energy dissipation capacity.

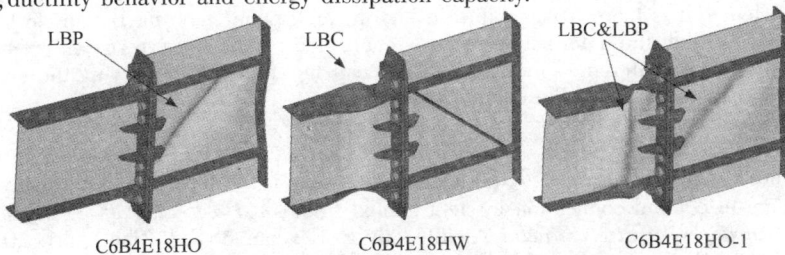

LBP   LBC   LBC&LBP

C6B4E18HO   C6B4E18HW   C6B4E18HO-1

Fig. 1. Three typical deformation modes

Fig. 2 gives the moment ($M = PL_{1p}$) vs. panel zone rotation curve M-θ_{pz}, moment vs. column rotation curve M-θ_{co} and moment vs. specimen rotation M-θ_s ($\theta_s = \Delta_p / L_{1p}$) of the three typical deformation modes, where P is the applied load; Δ_p is the column end displacement excluding the elastic deformation; L_{1p} is the distance from column end to the beam flange.

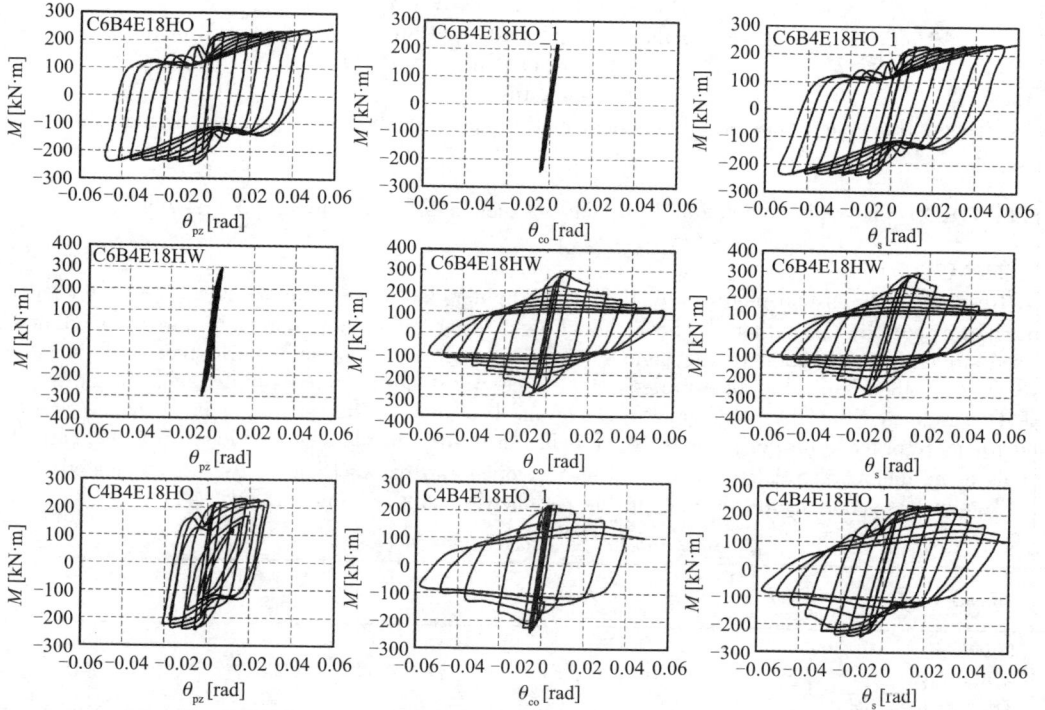

Fig. 2. Moment-rotation relationship

Different deformation modes are achieved by varying the configuration of the panel zone and the strength of the column. LBP occurs for those specimens with large effective width to thickness ratio in panel zone (EW/T = 177) and without diagonal stiffeners. Large deformation capacity up to 0.05 rad and ductility ratio up to 7 are achieved for the LBP mode. Also it could effectively dissipate energy. On the other hand, for those specimens with small EW/T (= 118) or with diagonal stiffener, LBC occurs. Decreasing the strength of column can also initiate the LBC mode for the specimens without a diagonal stiffener. Compared with the LBP mode, the deformation capacity is only about 0.02 rad and the ductility ratio is only about 2. The energy dissipated by the specimen is significantly less for the LBC mode (about 20 kN. m) compared with the LBP mode (about 120 kN. m).

CONCLUSIONS

A parametric study is conducted with the emphasis on the hysteretic behavior of end-plate connection in single layered gabled frames. The deformation modes of LBC and LBP are discussed. The LBP mode is found to have a relatively higher ductility behavior and energy dissipation capacity than the LBC mode. Hence, it is more reasonable to design a weak panel zone for the single layered gabled frame. It is also found that the deformation mode is controlled by the relative strength between the column and the panel zone. Different deformation modes can be achieved by varying the configuration of the panel zone and the strength of the column.

REFERENCES

[1] Shi Y, Chen Y, Xu Y, Zhao X., "Experimental analysis on the local buckling behaviour at panel zone of beam-to-column connection in steel gabled frames", *Proceeding of 7th International Conference on Advances in Steel Structures*, edited by S. L. Chan & S. P. Shu, April 2012, Nanjing, China, Southeast University Press, Vol. 1, pp. 443-451.
[2] ANSYS Element Reference. ANSYS 9.0 documentation. Online help.

SEISMIC BEHAVIOR OF BRACED FRAME COLUMN BASE CONNECTIONS

Yao Cui[a], Shoichi Kishiki[b], Satoshi Yamada[c]

[a] Dalian University of Technology, State Key Laboratory of Coastal and
Offshore Engineering, China
cuiyao@ dlut. edu. cn

[b] Osaka Institute of Technology, Department of Architecture, Japan
kishiki@ arch. oit. ac. jp

[c] Tokyo Institute of Technology, Structural Engineering Research Center, Japan
yamada. s. ad@ m. titech. ac. jp

KEYWORDS: Exposed Column Base, BRB Frames, Cyclic Loading, Anchor Rod, Slip

ABSTRACT

In Buckling Restrained Braced (BRB) frames, the axial and shear force on the column base is more critical compared with conventional moment resisting frames. Quasi-static tests of exposed column bases with BRB were conducted. It was found that the column base was uplifted when the BRB force increased to maximum tension strength. The friction force between base plate and mortar will be absent and then slip of base plate will occur. In this paper, numerical study was conducted to investigate the effect of the slip of base plate on the BRB behavior. And the effect of the number and size of anchor rods on the slip were briefly discussed.

CONCLUSIONS

Buckling restrained braces (BRBs) have an extremely stable hysteretic behavior under both tensile and compressive force and are commonly used as a type of energy dissipation device both in new construction and seismic retrofit projects. Exposed column bases are commonly used in low-to-medium rise building because of the easy construction and reliable seismic behavior. Such column bases are generally designed to consider the bending moment and axial force. Previous experimental studies on the seismic behavior of exposed column base in conventional braced frames[1,2] indicated that the anchor rods of exposed column base would deform severely in horizontal. Different from the conventional brace, BRB showed near equal tension and compression capacity. The different directions of the shear force caused by the BRB and the frame story drift are the unique behaviours for exposed column base with BRB. The effect of such different directions of the story drift and base plate shear on the hysteretic behavior of exposed column base is unclear. It is highly required to clarify the hysteretic behavior of exposed column base in BRB frames in order to develop the corresponding design procedures.

Three specimens were designed following AIJ recommendation[3,4] and fabricated at approximately 2/3 scale. From the test results, it is notable that the base plate rotation contribute mostly to the story drift angle in the positive loading (when BRB in tension) and the column deformed mainly in the negative loading (when BRB in tension) due to the increased compressive axial force from BRB vertical component.

The slip of base plate was suddenly increased when the shear force at the base plate is around 200kN for all specimens. The slip of base plate occurred when the friction force between base plate and mortar layer is broken, point e in *Fig. 1*, and recovered when the base plate touched with the mortar ($\theta_{bp} > 0$), point D' in *Fig. 1*. In this region, the resisting mechanism of column base is described in *Fig. 2*.

In moment frames, the shear is often transferred through friction developed in the bearing potion of the base plate. However, when such a bearing zone is not present (e. g. , in braced frames or in moment frames where the columns have tensile axial loads), shear transfer mechanisms must be carefully considered and designed. The main findings of this paper are summarized as follows:

1) The shearbehavior is critical for the exposed column base in BRB frames. The friction force

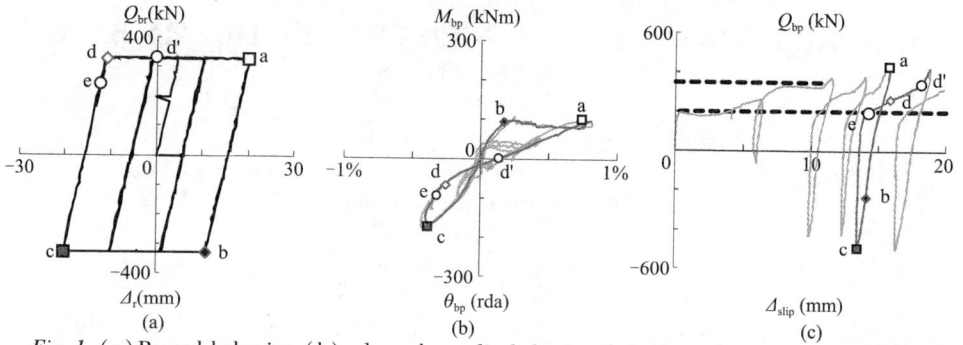

Fig. 1. (a) Braced behavior; (b) column base slip behavior; (c) column base rotation behavior

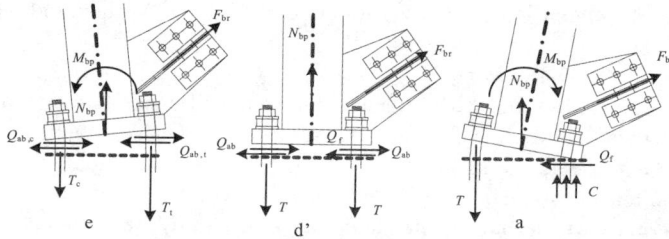

Fig. 2. Resisting mechanism of column base with BRB

between the base plate and foundation beam was broken when the BRB force was in tension. Significant slip occurred then.

2) The slip of base plate was cumulated in cyclic loadings. And such slip would reduce the effect of BRB behavior.

3) The slip was reduced by increasing thenumber of anchor rods. However, the arrangement of the anchor rods had slightly contribution to increase the seismic behavior of exposed column base.

4) Other methods to reduce the slip of base plate should be studied, such as stopper in front of base plate or shear key under base plate.

5) In the design, not only the shear resistance should be checked, but also the deformations at serviceability.

ACKNOWLEDGMENT

The authors acknowledge support fromthe Japan Iron and Steel Federation, KAKENHI (21686051) and NSFC(No. 51208076).

REFERENCES

[1] Hatanaka R. ,Tabuchi M. ,Tanaka T,and Masui A. ,"Study on Seismic Retrofit for Exposed Column Bases of Steel Frames with Braces",Summaries of Technical Papers of Annual Meeting Kinki Branch AIJ. No. 42 ,221-224 ,2002.

[2] Fukuhara A. ,Yamanishi T. ,et al. ,"Anchor-bolt-yield-type Exposed Column-Base Subjected to Tensile and Shearing Forces : Part 2 Cyclic Loading Test" ,Summaries of Technical Papers of Annual Meeting AIJ. C-1 (III) ,843-844 ,2010.

[3] Architectural Institute of Japan(AIJ). Recommendations forDesign of Connections in Steel Structures. Tokyo ,2006.

[4] Japanese Society of Steel Construction(JSSC). Recommendation and Commentary for Construction of Exposed Column Base with Structural Anchor Bolts. Tokyo ,2009.

[5] Cui Y. ,Nagae T. ,Nakashima M. ,"Hysteretic Behavior and Strength Capacity of Shallowly Embedded Steel Column Bases" , *Journal of Structural Engineering, ASCE*, Vol. 135, No. 10, pp. 1231-1238 ,2009.

FINITE ELEMENT ANALYSIS OF STEEL FRAME BEAM-COLUMN JOINTS UNDER LOW-CYCLIC LOADING BASED ON OPENSEES

Sui Weining[a], Shi Qingze[a]

[a]Shenyang Jianzhu University, College of Civil Engineering, China

swnwzf19790627@ hotmail. com, shiqingze@ 126. com

KEYWORDS: Steel frame, Finite element analysis, Low-cyclic loading.

ABSTRACT

In order to study the elastic-plastic hysteretic properties of the steel frame beam-column joints under low-cyclic loading, OpenSees, an object-oriented framework for finite element analysis, is used to investigate the seismic performance of the steel frame beam-column joints. A nonlinearBeamColumn element object in OpenSees which is based on the non-iterative(or iterative)force formulation, and considers the spread of plasticity along the element, is used to simulate the steel beam and the column. The Joint 2D in OpenSees is used to simulate the seismic performance of the connection of steel column and two H-shaped beams. The comparison between the computed results and the experimental results shows that finite element analysis can simulate the elastic-plastic hysteretic properties of the steel frame beam-column joints under low-cyclic loading well, it provides a reliable simulation method and reference for the seismic design of steel structure joints.

CONCLUSIONS

OpenSees is used to investigate the seismic performance of the steel frame beam-column joints in this paper, The joint elemen Joint-2D is used to simulate the joints of circular steel column and box steel column in OpenSees. The yield moment, capping moment strength, pre-capping plastic rotation for monotonic loadingand post-capping plastic rotatio are obtained according to the test, the results of simulation have good agreement with the observed one, the following conclusions can be drawn:

(1) The joint elemen Joint-2D is used to simulate the joints of circular steel column and box steel column in OpenSees. The seismic performance of joints would be taken into consideration in finite element analysis, the analysis results are well agreement with the test.

(2) For the joints of circular steel column, when the same displacement of beam end is loaded, the bearing capacity of the test is a little bigger than the bearing capacity of simulation, but in general the analysis results are well agreement with the test.

(3) For the joints of box steel colum, with the anchorage force increased, the trend of simulation results are agreement with the test, but the bearing capacity of the test is a little bigger than the bearing capacity of simulation with the same loaded-displacement of beam end.

ACKNOWLEDGMENT

Acknowledgments are optional. Numbered reference list shall be given as follows.

REFERENCES

[1] Mazzoni S, McKenna F, Scott MH, Fenves GL. "Open system for earthquake engineering simulation(OpenSEES)command language manual";2007.

[2] Lowes L, Mitra N, Altoontash A. A beam-column joint model for simulating the earthquake response of reinforced concrete frames. Berkeley, CA: Pacific EarthquakeEngineering Research Center, University of California at Berkeley;2003.

[3] Dimitrios G. Lignos, A. M. ASCE1, Helmut Krawinkler, M. ASCE2. "Deterioration modeling of steel components in support of collapse predictionof steel moment frames under earthquake load-

ing. Journal of Structural Engineering", Journal of Structural Engineering, February 16, 2010.

[4] Filippou, F. C. , Popov, E. P. , Bertero, V. V. (1983). "Effects of Bond Deterioration on Hysteretic Behavior of Reinforced Concrete Joints". Report EERC 83-19, Earthquake Engineering Research Center, University of California, Berkeley.

[5] Kishin N, Chen W. Moment-Rotation Relations of Semi-Rigid Connections with Angles[J]. Journal of Structural Engineer-ing, 1990, 116(7): 1813-1834

[6] Sims J. "Flange Angle Behavior in Semi-Rigid Connections for Steel PR Frames"[D]. Indiana: University of Notre Dame, 2000

[7] SuiWeining, Shi Qingze, Wang Zhanfei, Bai Xue. "Finite Element Analysis of Steel Frame Beam-Column Irregular Joints with A Strengthened External Diaphragm", Mechanics and Materials, 2014.

[8] Arash Altoontash, 2004, "Simulation and damage models for performance assessment of reinforced concrete beam-column joints", PhD Dissertation, Stanford University, California, USA

[9] Eads L. Pushover and dynamic analyses of 2-story moment frame with panel zonesand RBS. http://opensees. berkeley. edu/wiki/index. php/Pushover_and_Dynamic_Analyses_of_2-Story_Moment_Frame_with_Panel_Zones_and_RBS.

[10] Chen JiaJia Meng Shaoping Cai Xiaoning, "Numerical Analysis of Seismic Behavior for Bolted-angle Connection Based on OpenSees"

ULTRA-LOW CYCLE FATIGUE DEMAND ON COPED BEAM CONNECTIONS UNDER VERTICAL EXCITATIONS

Huajie Wen[a] and Hussam Mahmoud[a]

[a]Colorado State University, Department of Civil and Environmental Engineering, USA
huajie. wen@ colostate. edu, hussam. mahmoud@ colostate. edu

KEYWORDS: fracture cyclic loading, damage, ultra-low cycle fatigue

ABSTRACT

Studies on the behaviour of coped beams under cyclic loadings arenonexistent. Logically, under vertical excitations, resulting from vertical ground motions, the corresponding behaviour may show instability and therefore should be explored. This study evaluates, through numerical analysis, the behaviour of coped beams with double bolted clip angle connections under reverse vertical excitation protocols. First, a new ductile fracture model is introduced and its features highlighted including its dependency on stress triaxiality and lode angle parameters. The numerical simulation method, with the integrated fracture model is then verified by comparing the results to laboratory tests under monotonic loadings. Following the verification under monotonic loading, simulations are conducted under reverse cyclic loading to assess the ultra-low cycle fatigue behaviour and the susceptibility to fracture while accounting for nonlinear damage evolution. Specifically the effects of connection geometry and beam end rotations on the fundamental fracture characteristics of the connections are explored. Recommendations on connection design and detailing are put forward based on the numerical results. The results also highlight the importance and advantages of employing an accurate fracture model for predicting the response of steel structures under complex loading conditions.

CONCLUSIONS

In this paper, numerical simulations of block shear in coped beams with double angle bolted connections are conducted, through the application of a newly developed ductile fracture criterion that accounts for stresstriaxiality and lode angle parameter and includes a nonlinear damage accumulation law. The laboratory test results from Franchuk et. al. [1] serves as experimental comparison in the verification analysis. Following the verifications, the connection behaviour up to failure under cyclic loading conditions was also evaluated in order to assess the ultra-low cycle fatigue and the susceptibility to fracture while accounting for nonlinear damage evolution. Based on these analysis results, the effects of connection geometry and beam end rotations on the fundamental dynamic characteristics of the connections are explored.

Wen and Mahmoud[2] proposed a new fracture criterion, with dependency on both the stress triaxiality and Lode angle parameter. Only three parameters are included in the criterion, making the calibration process significantly less challenging. The mathematical representation of the fracture model is shown in Eq. 1,

$$\bar{\varepsilon}_f(\eta, \bar{\theta}) = c_1 \exp(c_2 \eta) \left[\cos \left(\frac{\pi}{6} \bar{\theta} \right) \right]^{c_3} \tag{1}$$

A nonlinear damage accumulation rule, expressed in Eq. 2, is also used to predict damage where fracture is presumed to occur when the damage D exceeds the value one.

$$dD = \exp(c_4 \kappa) m \left(\frac{\bar{\varepsilon}_{pt}}{\bar{\varepsilon}_f} \right)^{m-1} \frac{d \bar{\varepsilon}_{pt}}{\bar{\varepsilon}_f(\eta, \bar{\theta})} \tag{2}$$

Simulation on the cyclicbehavior of the coped beam has been conducted to verify the new model's ability in predicting the fracture behavior of structural details under cyclic loadings, and obtain a qualiative behavior of ultra-low cycle fatigue for the coped beam bolted connections. To ensure the progression of damage in each cycle, in every cycle of the loading procedure the elastic limits is exceeded. The loading protocol is shown in *Fig. 1* (a) and the resulting load versus displacement curve is shown in *Fig. 1* (b). The reduction in strength at approximately 9 mm of displacement is an indication of the effect of cyclic loading on the response.

The connection fractured after 3 cycles, and it is a typical ultra-low cycle fatigue type failure. The fracture first occurred at 12. 9 s, and was located at the top right corner of top bolt hole. At 13. 3 s, the shear fracture was fully developed, while around 13. 7 s another tension fracture occurs on the left inner side of the top bolt hole. Finally at 14. 1 s, the main tensile fracture on the left of bottom bolt hole occurred. As the displacement was moving downward at the same time, the load decreased to zero immediately, as shown in *Fig. 3*. The damage contour with fracture progression is shown in *Fig. 8*, where the fracture profiles are also depicted.

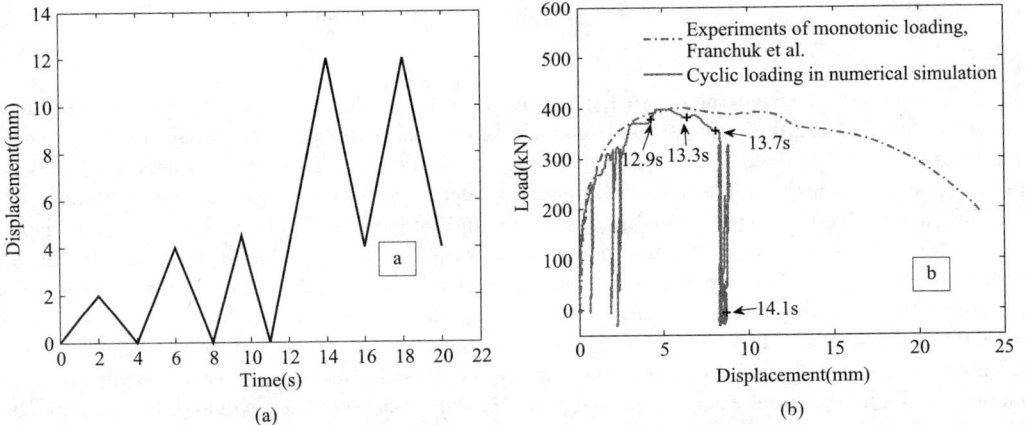

(a)

(b)

Fig. 1. (a) Cyclic loading protocol and (b) resulting load versus displacement

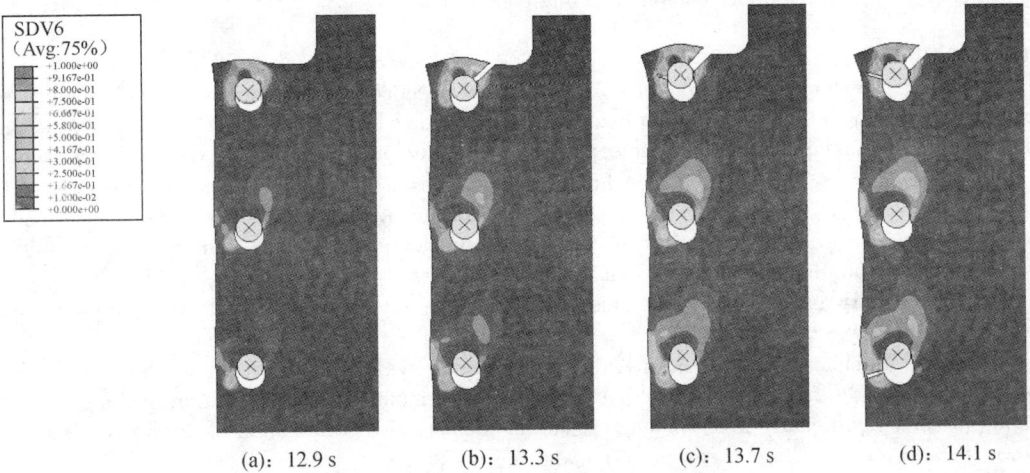

(a): 12.9 s   (b): 13.3 s   (c): 13.7 s   (d): 14.1 s

Fig. 2. Fracture progression of coped beam under cyclic loading

REFERENCES

[1] Franchuk, C. P. , Driver, R. G. , and Grondin, G. Y. , " Block shear behavior of coped steel beams" , Structural Engineering Report, No. 244, 2002. Department of Civil and Environmental Engineering, University of Alberta, Edmonton, AB, Canada.

[2] Wen, H. and Mahmoud, H. " A New Model for Ductile Fracture of Metal alloys: Part II-Reverse Loading" , ASCE Journal of Engineering Mechanics, submitted for review, 2014.

THREE-DIMENSIONAL NUMERICAL SIMULATIONS OF STEEL CONCRETE COMPOSITE BEAM-TO-COLUMN WELDED AND BOLTED JOINTS

Claudio Amadio[a], Nader Akkad[a], Marco Fasan[a]

[a]University of Trieste, Dept. Structures of Engineering and Architecture, Italy

amadio@ units. it, nader. akkad@ phd. units. it, marco. fasan@ phd. units. it

KEYWORDS: 3D simulation, composite structure, welded and bolted joints, Abaqus.

ABSTRACT

The behaviour of composite joints continues to be an issue of interest in the area of steel and composite structures. The progress achieved in the programming and computer technology made possible the application of advanced 3D Finite Element Method for the analysis of composite steel concrete frame joints and simulate their real behaviour. The goal of this 3D modelling is to obtain "Load-deflection" curves as close as possible to the ones experimentally obtained, in a practical way and with reduced cost and time. The study presents in detail two illustrative examples created by using 3D FEM for a welded joint(WJ) and a joint with extended end plate(FJ), designed on the base of Eurocode 8[1]. For both joints, a nonlinear static analysis was applied using ABAQUS commercial code, ver. 6. 11[2] and the obtained results were compared with experimental full-scale composite joints conducted by the authors. The numerical models were created by using 3D finite elements of solid type C3D4R, considering the contact between components and the pre-tensioned bolts. A Concrete damaged plasticity "CDP" model[3] was used for concrete material and a steel damage option based on Gurson's porous metal plasticity theory[4] was used for the steel structure. The analysis focused on the state of stress at each loading step, the failure mode in comparison with the one experimentally observed, and the comparison of the Load-Deflection predictions curves with the ones experimentally obtained.

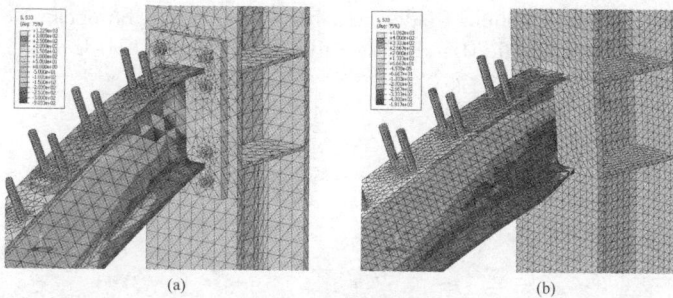

Fig. 1. Stress distribution and local buckling effects on steel under negative moment:
(a) FJ specimen; (b) WJ specimen

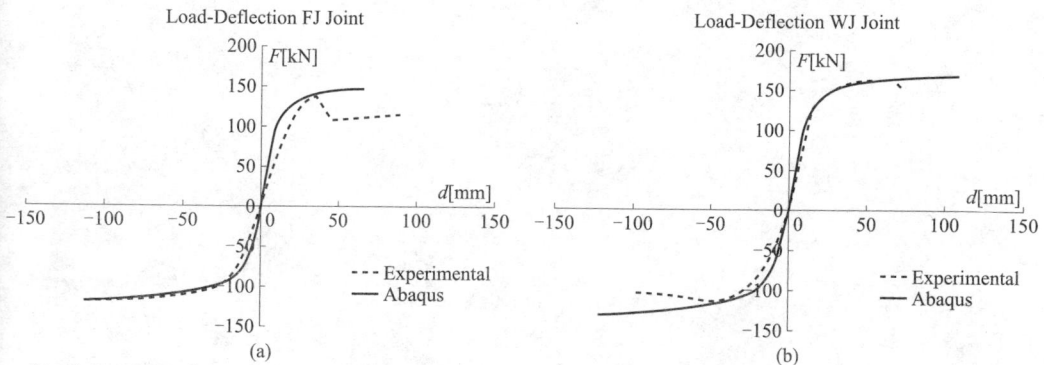

Fig. 2. Load-Displacement Experimental and FEM curve: (a) FJ composite joint; (b) WJ composite joint

CONCLUSIONS

Based on experimental test results conducted in theSannio and Naples Federico II Universities[5], a 3D solid FEM model was developed in Abaqus. The modelling was described in detail. The FEM analysis, under monotonic loading was demonstrated very adequate in the case of local and global response. The FEM overall and global response in term of Load-Deflection curve was very close to those obtained through experimental tests and a similar fail collapse was obtained. The local comparison response of FEM and Experimental results captured very well all the significant elements of the test. A similar strain gauges force-deformation response of the longitudinal and seismic rebars was obtained, together with the relationship of moment curvature of concrete slab and steel beam element. The proposed solid 3D model was then able to predict the local buckling and the overall mode of failure of both the tests. The good agreement of the FEM results with the experimental ones will give the possibility of a parametric analysis on any alternative structural detail for this type of structural joint.

ACKNOWLEDGMENT

This work was supported by research funding DPC-RELUIS 2014-15.

REFERENCES

[1] Eurocode 8, 2004. "Design provisions for earthquake resistance of structures. Part 1: General rules. Specific rules", *European Committee for Standardisation, Brussels, Belgium*

[2] ABAQUS CAE, Theory Manual & Analysis Manual version 6.11, Dassault Systems 2011.

[3] T. Jankowiak and T. Lodygowski, 2005. "Identification of parameters of concrete damage plasticity constitutive model", Foundation of civil and environmental engineering, No. 6, Poznan university of technology, Poland, 2005, pp. 53-69.

[4] G. Cricrì, 2009. "Consistent use of the Gurson-Tvergaard damage model for the R-curve calculation", National Conference IGF XX, Torino 24-26 June 2009; 138-150.

[5] C. Amadio, M. Pecce, F. Rossi, N. Akkad. Experimental study and numerical simulations of steel-concrete composite beam-to-column welded and bolted joints. 7th European Conference on steel and composite structures, Napoli 2014, p. 253-ISBN 978-92-9147-121-8.

THE NONLINEAR BEAVIOUR OF EXTERNAL STEEL-CONCRETE COMPOSITE JOINTS UNDER CYCLIC LOADING

M. Pecce and F. Rossi

Department of Engineering-University of Sannio

e-mails: pecce@ unisannio. it, fernando. rossi@ unisannio. it

KEYWORDS: steel-concrete, nonlinear response, joints.

ABSTRACT

This paper reports the experimental results of two full-scale steel-concrete composite joints (one welded and the other bolted) tested under cyclic loading, that were designed to localize the damage in the composite beam end in respect to the MRF capacity design, therefore the joint contributes to the deformability of the system essentially in the elastic field. The specimens were widely instrumented so that many experimental results are reported in terms of global and local measures. For the global response, the main role of the composite beam is highlighted. The performance of the joints was described in terms of resistance, ductility, dissipated energy and stiffness degradation. The contribution of the composite beam to the global ductility is evaluated in terms of rotational capacity and an "equivalent" plastic hinge length is identified. Two full-scale composite exterior joints were tested. This type of joint was selected because external joints have exhibited some pinched hysteresis loops, whereas the cyclic behaviors of internal nodes were more stable. The specimens were designed according to the European and national code[1-2]; in particular, the design focuses on the damage in the composite beam end with respect to the design capacity of a MRF. The tested joints are denoted throughout this paper as flanged joint (FJ) and welded joint (WJ). The specimen FJ was characterized by an extended end plate connection (Fig. 1b), whereas the specimen WJ was characterized by the steel beam that was directly welded to the column (Fig. 1a).

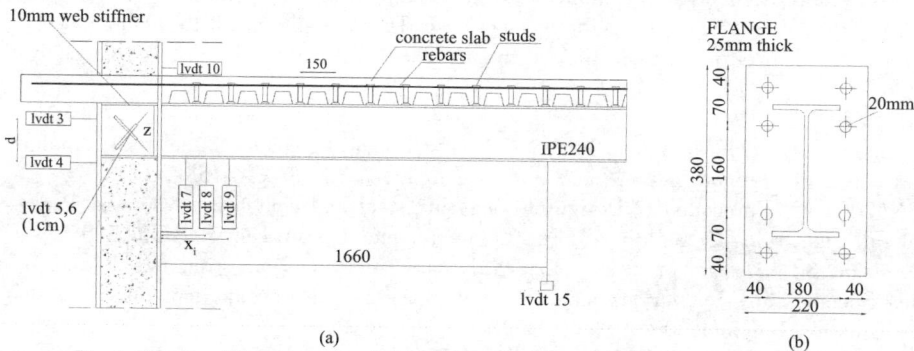

Fig. 1. Details of the composite joints: (a) welded joint (WJ) and instrumentation arrangement (LVDTs) for both specimens; (b) flange for the specimen FJ

The experimental program was previously detailed in[3] to discuss the main behavioral aspects and develop a comparison with a detailed FEM, therefore in this paper, a brief description of the experimental tests is provided (Fig. 2).

CONCLUSION

The following features have been highlighted:

-the type of joint (flanged or welded) has a certain influence on the elastic deformability of the entire system because only the initiation of plasticity of the joint has been observed for the welded joint;

-the plastic field of the sub-assemblage is significantly influenced by the plastic response of the

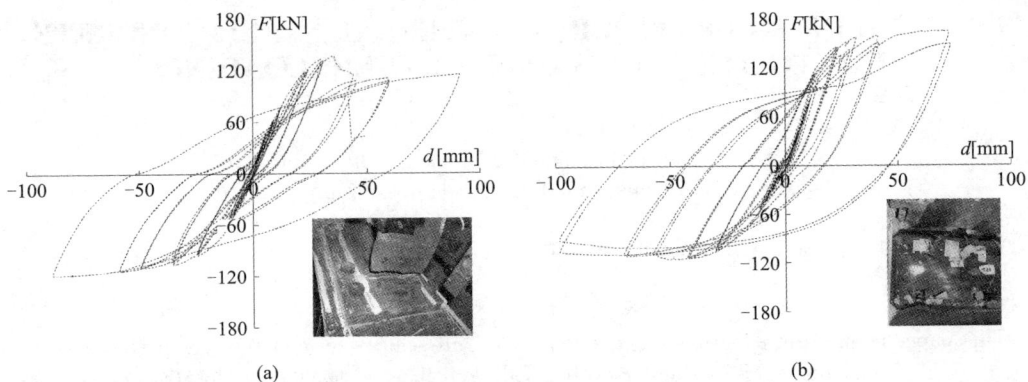

Fig. 2. Total cyclic response: (a) specimen
FJ (fracture and local buckling) and (b) specimen WJ (local bucling).

composite beam, as established by the design approach. The ductility of the composite beams exceeds 3; in one case, this limit is reduced because the test was halted. The values of the rotation capacity overcame the acceptance criteria of EC8 for a section of class 2 and FEMA356 at ultimate limit states, for both life safety and collapse;

-the ECCS procedure for evaluating the energy dissipation of the composite sub-assemblage yields an absorbed energy ratio of approximately 0. 4 for the two specimens, which is stable to collapse under a sagging moment but increases under a hogging moment, especially for the welded joint. This response is acceptable in terms of energy dissipation despite the pinching phenomena that affect the two specimens, even if more evident for the flanged joint;

-The cyclic degradation of the stiffness ratio according to ECCS is significantly higher for the flanged joint that showed a more distinct pinching.

The equivalent plastic hinge length L_{pe} of the composite beam, which is evaluated from the experimental tests, is approximately 0. 6h, where h is the height of the steel profile for the hogging moment case and approximately h for the sagging moment case. These results about the rotational capacity and plastic hinge length need to be more widely investigated to generalize for various composite beams.

REFERENCES

[1] D'Aniello M, Landolfo R, Piluso V, Rizzano G, "Ultimate behavior of steel beams under non-uniform bending", *Journal of Constructional Steel Research* 78, pp. 144-158, 2012.

[2] CEN (2008-a), Eurocode 4, "Design of composite steel and composite structures. Part 1. 1: General rules and rules for buildings", Eur. Comm. for Stand., Brussels, Belgium, 2008.

[3] Eurocode 8, "Design provisions for earthquake resistance of structures. Part 1. 3: General rules. Specific rules for various materials and elements", European Committee for Standardisation, Brussels, Belgium, 2004.

[4] Amadio C, Pecce M, Rossi F, Akkad N, Fasan M, "Experimental study and numerical simulations of composite joints", *EUROSTEEL* 2014, September 10-12, Naples, 2014.

ULTIMATE STRENGTH EVALUATION OF INCLINED FILLET WELDS BASED ON LIMIT ANALYSIS

Misaki Tanaka[a], Hayato Asada[a], Tsuyoshi Tanaka[a]

[a] Department of Architecture, Kobe University, Japan

misaki. tanaka. 1214@ gmail. com, asada@ people. kobe-ac. jp, tanaka@ arch. kobe-u. ac. jp

KEYWORDS: Inclined fillet weld, Ultimate strength, Limit analysis, Loading test

ABSTRACT

The purpose of this study is to evaluate the ultimate strength of inclined fillet welds. Inclined fillet welds are commonly used in steel building structures connection; interface welds of gusset plate connection, welded moment web connection etc. However, few reports are available on theoretical evaluation of ultimate strength of inclined fillet weld revealing relation with inclined angle (loading angle). In this study, first, limit analysis was performed to derive ultimate strength formula assuming incompressibility and von-Mises criterion. Validity of proposed evaluation formula was discussed in comparison with tensile loading test results.

CONCLUSIONS

In this paper, first, theoretical solution was derived based on limit analysis assuming incompressibility and the extended von-Mises criterion for weld metal. Fracture mechanism is shown in *Fig. 1*. Ultimate strength of inclined fillet weld per unit length is given by *Eq.*(1).

$$_w q_u = \frac{\sqrt{(\cos\theta - X\sin\theta)^2 + (\sin\theta + X\cos\theta)^2 \cos^2\beta} + (\cos\theta - X\sin\theta)\sin\beta}{(\sin\beta + \cos\beta)\cos\beta} \cdot S \cdot \frac{\sigma_u}{\sqrt{3}} \quad (1)$$

$$X = u/v \quad (2)$$

where $_w q_u$ is ultimate strength per unit weld length,
     q is inclined angle,
     b is fracture angle, (see *Fig. 1*)
     S is the size of fillet weld,
     u is the displacement rate in the loading direction
     v is the displacement rate in perpendicular to loading direction

Loading tests were performed to discuss validity of proposed evaluation formula. Comparison of experimental and calculated ultimate strength are shown in *Fig. 1*. Although theoretical solution generally underestimates the experimental results, a good correlation was observed, regardless of the inclined angle of fillet weld. Accuracy of theoretical value was improved by considering weld penetration in determination of fracture angle. The effect of weld penetration on strength became significant when inclined angle is small.

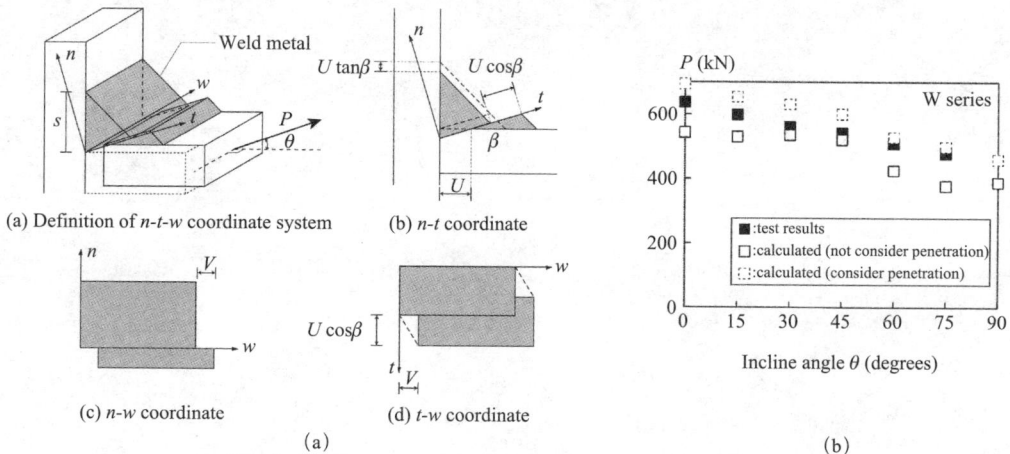

(a) Definition of n-t-w coordinate system   (b) n-t coordinate

(c) n-w coordinate   (d) t-w coordinate

(a)   (b)

Fig. 1. (a) Fracture mechanism assumed in limit analysis;
(b) Comparison of experimental and calculated ultimate strength

REFERENCES

[1] Kato B. ,Morita,K. ,"Strength of transverse fillet welded joints" ,*Welding Research Supplement*, 59-64,1974.

[2] Kamtekar A. G. ,"A New Analysis of the Strength of Some Simple Fillet Welded Connections", *Journal of Constructional Steel Research*,33-45,1982

[3] Butler L. J. ,Kulak. G. L. ,"Strength of fillet welds as a function of direction of load" ,*Welding Journal Research Supplement*,231s-234s,1971

[4] Kamtekar A. G. ,"The Strength of Inclined Fillet Welds" ,*Journal of Constructional Steel Research*,43-54,1987.

[5] Miazga G. S,Kennedy D. J. L. ,"Behaviour of fillet welds as a function of the angle of loading", *Canadian Journal of Civil Engineering*,583-599,1989.

[6] Jensen A. P,"Limit Analysis of Fillet Welds Loaded in Shear and Tension" ,*International Institute of Welding*,XV-93,1993.

[7] Yasui N. ,Suita K. ,and Inoue K. ,"Fracture mechanism and ultimate strength of inclined fillet welds" ,*Journal of Constructional and Structural Engineering*,*AIJ*, 579:111-118,2004.

EXPERIMENTAL RESEARCH OF SCREW AND RIVETED CONNECTIONS IN THE STEEL THIN-WALLED STRUCTURES UNDER STATIC AND CYCLIC LOADING

Eduard Ayrumyan[a], Ivan Katranov[b], Nikolay Kamenshchikov[a]

[a]Melnikov Central Reseach and Design Institute of Steel Structures,
Dept. Light Gauge Cold-formed Structures, Moscow, Russia
edward. ayrumyan@ gmail. com, nikolay. kamenshchikov@ gmail. com

[b]Department of Test Facilities, National Research University
Moscow State University of Civil Engineering, Moscow, Russia
katranoff@ bk. ru

KEYWORDS: Connections, rivets, screws, cyclic loading, tests.

ABSTRACT

This report presents the research results of screw and riveted connections in the light steel thin-walled structures under static and cyclic loading. The validity of calculation procedureEurocode(EN) was confirmed and the different connections were introduced depending on the types of fasteners and possible types of connection failure, the material resistance factors, as well as additional design factors.

At the present time application of the light steel thin-walled structures(LSTS), a significant difference of which is the use of the special types of fasteners for connecting profiles, mainly pop-rivets and self-drilling, self-tapping screws, has increased significantly in the world construction practice.

Studies ofthe thin-walled structure connections were carried out by such authors as: La Boube RA, Rogers Colin A., Hancock Gregory J., Makelainen P., Ayrumyan E. L., Beliy G. I. [1],[2],[3]

The aim ofconducted research was to study the actual behavior of screw and riveted connections and adaptation of the European engineering calculation procedure to the Russian conditions(preparation of National Application), as well as to improve the calculation procedure.

The test results confirmed the basic types ofconnection failure. At the same time a differentiated end-of-life criterion of connection was accepted. Thus, the load-bearing capacity of connection under shear of fastener, its rupture and tear-out, as well as detachment of sheet through the press-washer is estimated by a limit state 1a-full depletion of the load-bearing capacity. The load-bearing capacity of connection when crumbling and rupture of sheet along the cross section is estimated by a limit state 1f-unacceptable development of strains on the basis of preset limits of these strains. As a limit criterion of strain its value was taken equal to 0. 5 mm.

Thus, the tests of connections carried out with single-sided and many-sided fastener arrangement shown their identical load-bearing capacity.

It was found that theload-bearing capacity of a double-shear screw connection is more than that of one-shear connection by 20%.

During the tests a reduction in the load-bearing capacity of multiscrew connection was also revealed due to redistribution of forces between the screws under loading. Herein, the more intensive inclusion of the first row screws in behavior was established.

As a result of the carried out research the accuracy of calculation procedure Eurocode(EN) was confirmed and the different connections were introduced depending on the types of fasteners and possible types of failure and the material resistance factors γ_{M2}, which are recommended for use in the National Application[4].

On the basis of cyclic tests of connections under continuous loading to the fixed values of load with a frequency of 1 Hz and ratio $N_{min}/N_{max} = 0$, the following conclusions were made up to threshold of 10,000 cycles: 33% of the total strains under loading up to 10,000 cycles occured during the first 25 cycles regardless of load acting on the connection(*Fig. 1*). Analysis of results obtained during the cyclic tests of screw connections confirmed validity of the calculated strength values of connection upon reaching the ultimate strains of 0. 5 mm in shear-behavior of connection under cyclic loading; according to seismic forces intensity did not exceed 9 on MSK-64 scale[5].

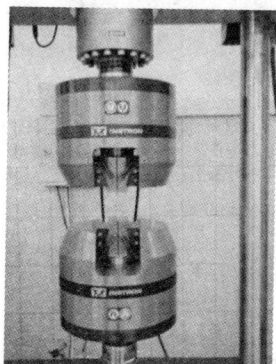

(a)

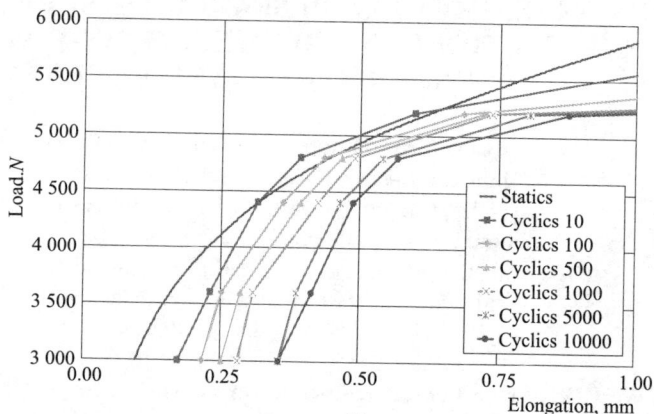

(b)

Fig. 1. (a) Test facility; (b) General diagram of cyclic tests

CONCLUSIONS

The experimental and theoretical research of the screw and rivet connection behavior in the light steel thin-walled structures allowed the researchers to formulate the following conclusions:

-the limit load-bearing capacity of connections was fixed at the different loading condition and types of connections; area of the effective application was demarcated for each type of connections.

-the accuracy of calculation procedure Eurocode (EN) was confirmed; the different material resistance factors γ_{M2} were introduced depending on the type of fasteners and possible modes of connection failure;

-a reduction factor $\beta = 0.8$ was introduced for multiscrew connections due to non-uniformity of inclusion of fasteners in their behavior;

-a mark-up factor $k = 1.2$ was introduced for double-shear connections.

REFERENCES

[1] LaBoube R. A. & Sokol, M. A. , 2002, "Behavior of screw connections in residential construction", *ASCE Journal of Structural Engineering*, 128 :1 , pp. 115-118.

[2] Rogers C. A. & Hancock G. J. , 1999, "Screwed connection tests of thin G550 and G300 sheet steels", *ASCE Journal of Structural Engineering*, 125 :2 , pp. 128-136. .

[3] Ayrumyan E. L. , Ganichev S. V. , Kamynin S. V. , 2009, "Pop-rivets or self-tapping screws?", *Journal of Erection and Special Works in Construction*, Vol. 3 , pp. 10-12.

[4] Katranov I. G. , 2011, "The issue of calculating LSTC screw connections under tensile testing", *Journal of Industrial and Civil Construction*, Vol. 3 , pp. 9-11.

[5] MSK-64 Seismic intensity scale MSK , 1964.

BEHAVIOUR OF JOINT COMPONENTS OF I BEAM TO TUBULAR COLUMNS CONNECTIONS WITH WELDED REVERSE CHANNEL

Luís Magalhães[a], Carlos Rebelo[b], Sandra Jordão[b]

[a] Polytechnic Institute of Castelo Branco, Technical-Scientific Unit of
Civil Engineering, Portugal
lmmbmagalhaes@ ipcb. pt

[b] University of Coimbra, Department of Civil Engineering, Portugal
crebelo@ dec. uc. pt, sjordao@ dec. uc. pt

KEYWORDS: Beam to column joints, tubular columns, reverse channel, nonlinear cyclic behaviour, components tests.

ABSTRACT

When compared with other steel shapes, tubular profiles show a privileged structural behaviour due to their ability to withstand axial loads, bending in several directions and torsion, be-sides considerable advantages in terms of maintenance and aesthetics. The welded reverse channel connection is a good solution since it allows for a bolted joint between I beam to hollow column. Furthermore, this type of joint detail has a reasonable construction cost, is easy to implement and possesses large ductility through the deformation of the web panel.

In this paper a comparison is made between experimental and numerical finite element model results. The main focus will be on the results concerning the main components of the reverse channel, i. e. web face in bending, flanges panels in shear, compression and tension. The main objective of these tests is determining the strength, stiffness and rotation capacity of the main components of the reverse channel joining detail.

In the experimental program the characteristics of the nonlinear behaviour of the principal reverse channel components is assessed by means of bending tests (monotonic and cyclic). The configurations selected for this tests correspond to a parametric variation on the factors with major influence on the structural behaviour of the reverse channel, the geometry of the loaded area and the dimensions of the flanges and web of the U shaped element. The prototypes are formed by the reverse channel (two flange plates welded to one web plate with holes for bolted to end plate beam). The experimental lay out corresponds to a beam to column connection between an I-beam and a vertical steel rigid structure using a reverse channel element.

The numerical modelsrepresent the real prototypes used in experimental tests. This models are developed in the software LUSAS and calibrated with the results from the experimental tests.

CONCLUSIONS

The deformationsof the channels are caused by the tension of the bolts at the top, and by end plate compression at the bottom. In both cases significative plastic deformation are observed in *Fig. 1*.

The deformation in the numerical models is identical to the one observed in the homologous experimental tests (*Fig. 2*a).

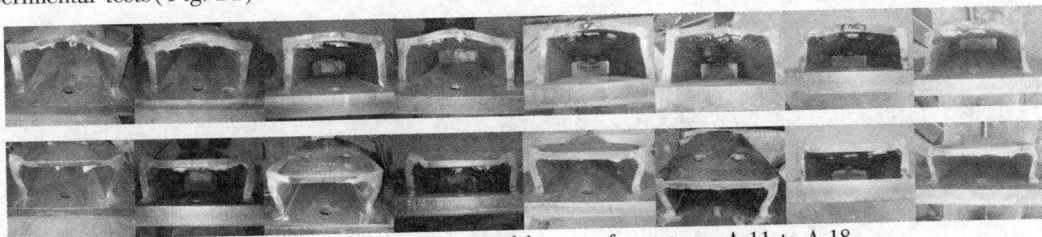

Fig. 1. Deformation on the top and bottom of prototypes A-11 to A-18.

The prototypes with smaller web thicknesses (A11, *Fig. 1*) and wider webs (A14, *Fig. 1*) show the higher out of plane bending deformation, as expected. Prototype A-18, the one with the lowest value for flange thickness (10 mm), was the only one who reached rupture at the flange plates.

Both the results obtained from the experimental tests "EC" and the numerical models "MN" can be expressed in terms of moment-rotation curves. The graphics of the *Fig.* 2b show in simultaneous the comparison of the "EC" and "MN" results, and the results of parametric study, in terms of strength.

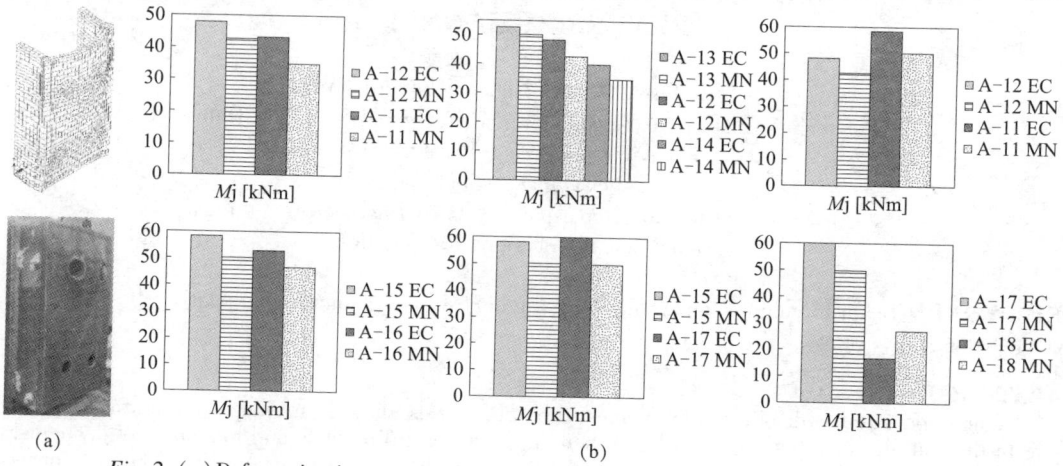

(a)   (b)

Fig. 2. (a) Deformation in numerical models; (b) Graphics of comparison of the results.

The results of the parametric variation are summarized in *Table 1*. The comparisons established in *Table 1* shows that the tendencies in parametric variations are practically identical in both types of test, finding only one case of conflicting results.

Table 1. Parametric study synthesis.

Parameters	variation	Prototype Model	Experimental [EC]			Numerical [MN]		
			M_j	$S_{j,ini}$	$S_{j,pl}$	M_j	$S_{j,ini}$	$S_{j,pl}$
Web thickness [t_{wc}]	Decrease	A-12 → A-11	<	<	<	<	<	<
Web width [h_c]	Decrease	A-12 → A-13	>	> >	> >	>	> >	> >
	Increase	A-12 → A-14	<	<	<	<	< <	<
End plate width [b_p]	Increase	A-12 → A-15	>	>	>	>	> >	>
Bolts distance [p_2]	Decrease	A-15 → A-16	<	>	<	<	<	<
Flange width [b_c]	Decrease	A-15 → A-17	=	=	>	=	=	>
Flange thickness [t_{fc}]	Decrease	A-17 → A-18	< < <	<	< <	< <	< <	< <

The loaded area, the web and flanges thicknesses and widths are parameters with important influence in the structural behavior of reverse channel beam-to-column connections. Significant changes of the structural behavior of the joint were found for the parametric variation considered in the study, confirming it was well tailored for the purpose.

From the comparisons between the numerical model and experimental results, it is concluded that the approximation between the results is good.

REFERENCES

[1] CEN, Eurocode 3, Part 1. 8: "Design of Joints", EN 1993-1-8, 2010.
[2] Jaspat, J, Pietrapertosa, C, Weynand, K, Busse, E, Klinkhammer, R, "Development a Full Consistent Design Approach for Bolted and Welded Joints in Building Frames and Trusses between Steel Members Made of Hollow and/or Open Sections", Application of the Component Method, Volume 1-Practical Guidelines, CIDECT Report 5BP-4/05, 2005.
[3] RILEM, "Tension Testing of Metallic Structural Materials for Determining Stress-Strain Relations under Monotonic and Uniaxial Tensile Loading", Draft Recommendations, 23 35-46, 1990.
[4] LUSAS, FEA Ltd, "Element Reference Manual", Version 13, United Kingdom.

INVESTIGATION OF HOLLOW STRUCTURAL SECTION BASED COLLAR CONNECTIONS UNDER SEISMIC LOADS

Dan Wei[a] and Jason P. McCormick[a]

[a]University of Michigan, Dept. of Civil & Environmental Engineering, USA

danwei@umich.edu, jpmccorm@umich.edu

KEYWORDS: Connection, Hollow Structural Sections, Moment Frame, Steel, Seismic

ABSTRACT

Traditional steel seismic moment frames typically rely on either wide flange column-to-wide flange beam connections or hollow structural section(HSS) column-to-wide flange beam connections to resist the lateral loads associated with an earthquake. However, other steel sections, such as HSS, used as beam members have the potential to perform well under seismic loads, particularly in low-rise moment frame systems. HSS provide a high strength-to-weight ratio that limits the seismic mass of a structure; good compression and bending properties; high torsional stiffness that decreases the need for beam lateral bracing; and applicability toward modular/rapid construction if the correct connections are designed. Recent studies of HSS in pure bending have shown that when used as beam members they can form stable plastic hinges under cyclic bending[1], which meets current seismic moment connection requirements where the majority of the inelastic behaviour must occur in the beam member. Further, experimental tests have shown that the use of either internal through plates or external diaphragm plates with HSS-to-HSS moment connections can ensure that beam plastic hinging occurs prior to local buckling and failure of the connection due to weld or member fracture[2]. However, limited studies have considered further ways to take advantage of the inherent properties of HSS members in seismic moment frame systems.

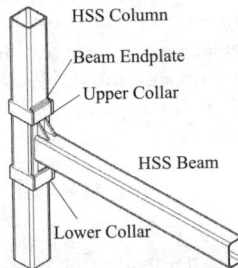

Fig. 1. Welded collar connection

To investigate optimal HSS-to-HSS moment connection configurations and to minimize the need for field welding these connections, an innovative HSS-to-HSS moment connection configuration utilizing a beam endplate and collars is devised and evaluated through a detailed finite element study. *Fig. 1* provides a schematic of the considered connection. The connection configuration utilizes HSS beams and columns to take advantage of their properties. To address field welding concerns and construction speed, the HSS beam member has a stiffened endplate that is shop welded. The beam is attached to the column using two collars that slip over the column and the beam endplate. The lower collar is shop-welded to the column with a gap left between the column face and collar to provide a location for the bottom of the beam endplate to slip into in the field. Once the beam is in place in the field, the upper collar can be slipped down the column and over the top of the beam endplate. Only fillet welds are needed in the field to attach the upper collar to the column and beam endplate and the beam endplate to the column.

A parametric study that considers 33 different HSS-to-HSS welded collar connectionswas conducted with the beam width-thickness ratio, beam endplate thickness, collar thickness, and collar depth taken as the varied parameters. A single column section size, HSS 254 × 254 × 15.9 mm, was considered and three different beam section sizes were utilized, HSS 305 × 203 × 7. 94 mm, HSS 305 × 203 × 9. 53 mm and HSS 305 × 203 × 12. 7 mm. Two stiffeners were used at the top and bottom beam flange to stiffen the beam endplate. The stiffeners had equal leg lengths of 152 mm, a thickness of 19. 1 mm, and were located 38. 1 mm on either side of the centreline of the beam flange. The finite element models of the connection were constructed in Abaqus CAE 6. 12 with S4R shell elements used for the HSS beam and column and C3D8R solid elements used for the beam endplate, collars, and stiffeners. Each connection was cycled up to 0. 06 rad. Fig. 2 shows normalized moment-rotation plots for the connections with HSS 305 × 203 × 7. 94 and HSS 305 × 203 × 12. 7 beam members where the beam endplate and collar thickness is the same.

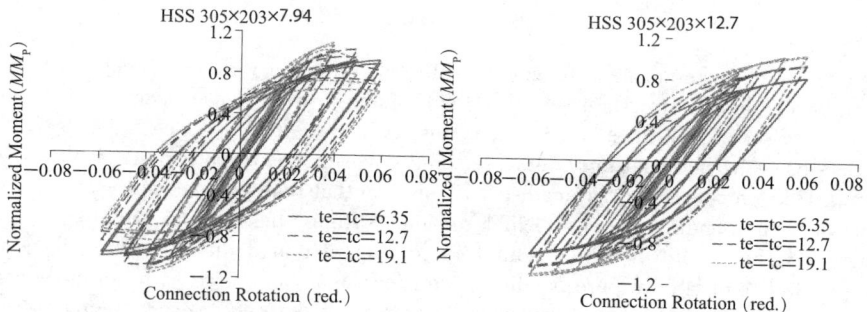

Fig. 2. Normalized moment-connection rotation results (units in mm)

CONCLUSIONS

In total, the performance of 33 different connection configurations were considered under cyclic loading associated with a seismic event. Many of the welded collar connections were able to develop stable plastic hinging of the beam member and reach the beam plastic moment capacity. The beams that had larger width-thickness ratios experienced the most plastic beam rotation. The beam endplate thickness had more influence on the beam moment capacity than the other parameters. The maximum normalized moment can be improved by increasing the thickness of the beam endplate. Collar thickness had little effect on the cyclic behavior of the connections, while the optimal collar depth was 152 mm for the three beam section sizes considered. Overall, the welded collar HSS-to-HSS moment connection showed promise for use in low-rise systems.

ACKNOWLEDGMENT

This work is supported by the American Institute of Steel Construction through the Milek Faculty Fellowship. The views expressed herein are solely those of the authors and do not necessarily represent the views of the supporting agency.

REFERENCES

[1] Fadden, M. , McCormick, J. , "Finite element model of the cyclic bending behavior of hollow structural sections", J. Constr. Steel Res. , 94:64-75, 2014.
[2] Fadden, M. , Wei, D, McCormick, J. , "Cyclic Testing of Welded HSS-to-HSS Moment Connections for Seismic Applications", J. Struct. Eng. -ASCE, 141(2):04014109-1-14, 2015.

CYCLIC BEHAVIOR OF EXPOSED COLUMN BASE JOINTS: EXPERIMENTAL ANALYSIS AND MECHANICAL MODELING

M. Latour[a] and G. Rizzano[a]

[a] University of Salerno, Department of Civil Engineering

g. rizzano@ unisa. it

KEYWORDS: Joints, Experimental, Base plate, Cyclic Modelling

ABSTRACT

Dealing with base-plate joints the number of theoretical studies and experimental programs is much lower than those regarding beam-to-column connections and, as a result, the knowledge of their monotonic and cyclic behaviour is still far from a complete understanding. Nevertheless, concerning their seismic behaviour, under the assumption of failure mode of global type and adoption of partial strength joints, base plate connections are also a dissipative zone and, therefore, their cyclic behaviour should be characterized in order to accurately predict the behaviour of the whole frame.

To this scope, in the present paper a method for predicting the cyclic behaviour of base plate joints is proposed by extending the approach already proposed in[1] for end-plate joints. In order to reach this goal, the dissipation capacity and the hysteretic laws of the main sources of dissipation of the base plate joint are modelled by properly accounting for the existing literature. Afterwards a mechanical model able to predict the behaviour of the joint starting from the mathematical laws of the single components is assembled. In order to verify the accuracy of the proposed approach, two experimental tests executed at the laboratory on materials and structures of the University of Salerno are presented and used to verify the model.

In particular, following the same approach already proposed by EC3[2], the cyclic behaviour of exposed base plate joints is modelled by characterizing the force-displacement behaviour of four non-linear springs, two representing the behaviour of the "concrete in compression including grout" and two modelling the T-stubs of the end-plate in the extended part. As an approximation the springs modelling the concrete behaviour are assumed to have an elastic-brittle behaviour, without any energy dissipation capacity, with stiffness and resistance characterized according to EC3[2]. Conversely, the springs modelling the plate, which are the main sources of energy dissipation, are characterized by means of the monotonic and cyclic models of Francavilla et al. ,2015[3] and Piluso et al ,2008[4] respectively.

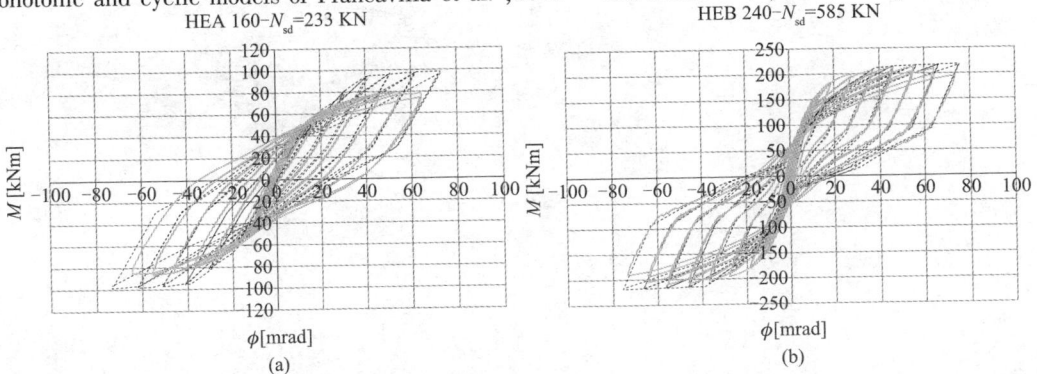

HEA 160-N_{sd}=233 KN   HEB 240-N_{sd}=585 KN

M [kNm] ϕ[mrad]   (a)

M [kNm] ϕ[mrad]   (b)

Fig. 1. Application of the model to the experimental tests

Therefore, starting from the modelling of the single components a general model for the prediction of the cyclic behaviour of the whole base plate joint has been developed. The model, which is based on a step by step algorithm, has been codified in a computer program which provides the joint moment-rotation curve starting from an assigned joint rotation history. In particular, for each assigned value of the joint rotation the corresponding displacements and forces of the joint components are determined on the base of equilibria equations and on the base of their cyclic laws.

The accuracy of the proposed model has been verified against two experimental tests carried out at the University of Salerno. As reported in the paper, the comparison between experimental results and that deriving from application of the mechanical model shows a good correlation both in terms of moment-rotation curves and in terms of energy dissipation.

CONCLUSIONS

In the present paper, an experimental campaign on column bases and a model for predicting the cyclic response of column base joints have been presented. The main feature of the model is that it provides the prediction of the behaviour of column bases by applying the component approach in the same fashion of the approach already codified in EC3. The accuracy of the model has been verified against two experimental tests in terms of energy dissipation supply and peak moment at each cycle showing that the model provides a good accuracy even though some approximations could be related the degradation laws for the T-stubs and to the hypothesis made for modelling the behaviour of the concrete in compression. The obtained results are very encouraging about the possibility of extending the component approach to cyclic loading conditions and, even though further investigations are needed, they represent a very important step towards the application of the component method to base plate joints under cyclic loading conditions.

REFERENCES

[1] Latour M, Piluso V, Rizzano G "Cyclic Modeling of Bolted Beam-to-Column Connections: Component Approach", *Journal of Earthquake Engineering*, 4 537-563, 2011.
[2] CEN, "Eurocode 3: Design of steel structures-Part 1-8: Design of joints", 2005b.
[3] Francavilla A, Latour M, Piluso V, Rizzano G "Monotonic behaviour of bolted T-stubs: a refined theoretical model for flange yielding and bolt fracture failure mode", *Nordic Steel 2015*, *Tampere*.
[4] *Piluso V, Rizzano G "Experimental Analysis and modelling of bolted T-stubs under cyclic loads". Journal of Constructional Steel Research, Volume 64, pp. 655-669. 2008*

NUMERICAL INVESTIGATION ON THE SEISMIC RESPONSE OF BOLTED EXTENDED STIFFENED END-PLATE JOINTS

Roberto Tartaglia[a], Mariana Zimbru[a], Mario D'Aniello[a], Silvia Costanzo[a],
Raffaele Landolfo[a] and Attilio De Martino[a]

[a]University of Naples Federico II, Department of Structures for
Engineering and Architecture, Italy

e-mails: roberto. tartaglia@ unina. it, zimbru. mariana@ gmail. com,
mdaniel@ unina. it, silvia. costanzo@ unina. it, landolfo@ unina. it,
attilio. demartino@ unina. it

KEYWORDS: Seismic qualification, Steel bolted Joints, moment-rotation response.

ABSTRACT

Beam-to-column joints significantly influence the seismic performance of moment resisting frames (MRFs). As it is well known, joints can be designed either full-strength or partial-strength as respect to the connected beams. These two different performance objectives may significantly modify the dissipative behaviour of seismic resistant MRFs. Indeed, in case of full strength joints plastic hinge should form in beams, while in case of partial strength joints the plastic deformation should concentrate in the connections. The former design strategy needs to guarantee that the joints have flexural overstrength larger than the beams which are connected. Unfortunately, owing to the variability of steel strength and to the actual post-yield flexural overstrength of steel beams, these connections could not have enough o-verstrength. In EN 1998-1[1], the minimum required joint extra-strength is equal to $1.1 \times 1.25 M_{b,pl,Rd}$ (being $M_{b,pl,Rd}$ the beam plastic moment) could be largely overcome in many cases. In addition, it is necessary to give effective rules to control the column web panel.

Concerning partial strength joints, this design criterion requires that joints have sufficient rotation capacity to guarantee the formation of global mechanism. At the present time, EN 1993-1-8[2] provides models to compute the strength and the stiffness of connections but no reliable analytical tools are available to predict the rotation capacity and the cyclic performance in relation to the connection typology. EN 1998-1[1] requires design supported by specific experimental testing, resulting in impractical solutions within the typical time and budget constraints of real-life projects. As an alternative to design supported by testing, the code prescribes to find existing data on experimental tests performed on similar connections in the literature. It is clear that this procedure is unfeasible from the designer's point of view.

The above considerations are general for all types of both welded and bolted joints. However, each joint typology is characterized by specific criticism which needs further investigation. In Europe, Extended stiffened (ES) bolted end-plate beam-to-column joints are widely used in moment resisting frames (MRFs), especially in seismic countries. However, this type of joint is not specifically codified for seismic applications. On the contrary, in USA specific qualification codes (e. g. AISC 341-10[3] and AISC 358-10[4]) have been developed to provide seismic design rules for ES joints.

In EN 1993:1-8[2] the influence of the rib stiffener on both strength and stiffness of this type of joints is not accounted for. On the other hand, in US codes the rib is considered but the influence of the bolt rows below the first below the beam flange in tension is neglected.

Therefore, in order to investigate on these aspects a parametric numerical study based on finite element analyses has been carried out and the main results are described and discussed hereinafter.

CONCLUSIONS

On the basis of the numerical results, the following concluding remarks can be drawn:

-The response of the joint is affected by the rib thickness. For Full strength Joints: the rib thickness influences the beam ductility; the thicker the rib is, the smaller the beam rotation capacity. For Equal and Partial strength joints both strength and stiffness increase with the rib thickness. The presence of the rib is extremely beneficial for the definition of the joint failure mode.

-The middle bolt row does not affect the global behaviour both in terms of stiffness and resistance. However, it affects the ductility of the connection providing an extra capacity that is beneficial to improve the joint robustness, as well.

ACKNOWLEDGMENT

The research leading to these results has received funding from the European Union's Research Fund for Coal and Steel (RFCS) research programme under grant agreement n° RFSR-CT-2013-00021.

The Authors thanks also the financial support of the RELUIS project "Linea diRicerca Acciaio & Composte Acciaio-Calcestruzzo" 2014-2015.

REFERENCES

[1] EN 1998-1, Design of structures for earthquake resistance-Part 1: General rules, seismic actions and rules for buildings.

[2] EN 1993-1-8, Design of steel structures-Part 1-8: Design of joints

[3] ANSI/AISC341-10(2010). "Seismic Provisions for Structural Steel Buildings". American Institute of Steel Construction.

[4] ANSI/AISC 358-10 (2010). "Prequalified Connections for Special and Intermediate Steel Moment Frames for Seismic Applications"

[5] Lee. C-H, "Seismic design of rib-reinforced steel moment connections based on equivalent strut model", *Journal of Structural Engineering*, 128 : 1121-1129, 2002.

[6] Lee. C-H, Jung J-H, Oh M-H, Koo E-S, "Experimental study of cyclic seismic behavior of steel moment connections reinforced with ribs", *Journal of Structural Engineering*, 131: 108-118, 2005.

[7] Shi Y., Shi G., Wang Y, "Experimental and theoretical analysis of the moment-rotation behaviour of stiffened extended end-plate connections", *Journal of Constructional Steel Research*, 63:1279 - 1293, 2007.

[8] Iannone F, Latour M, Piluso V, Rizzano G, "Experimental Analysis of bolted steel beam to column connections: Component identification", *Journal of Earthquake Engineering*, 15:214-244, 2011;

[9] Abidelah A., Bouchaïr A., Kerdal D. E. "Experimental and analytical behaviour of bolted end-plate connections with or without stiffeners", *Journal of Constructional Steel Research*, 76:13-27, 2012.

[10] Maquoi R, Chabrolin B, Frame design including joints, Volume I, 1998;

[11] Wang M, Shi Y, Wang Y, Shi G, "Numerical Study on seismic behaviour of steel frame end plate connections", *Journal of Constructional Steel Research*, 90:140-152, 2013.

[12] Swanson J. A., Leon R. "Stiffnessmodeling of bolted t-stub connection components". *Journal of Structural Engineering*, 127(5), 2001.

SEISMIC PERFORMANCE OF MULTISTOREY FRAMES WITH BOLTED EXTENDED END-PLATE JOINTS: THE INFLUENCE OF JOINT MODELLING ASSUMPTIONS

Silvia Costanzo[a], Mariana Zimbru[a], Mario D'Aniello[a], Roberto Tartaglia[a],
Raffaele Landolfo[a], and Attilio De Martino[a]

[a]University of Naples Federico II, Department of Structures for Engineering and Architecture

e-mails: silvia. costanzo@ unina. it, zimbru. mariana@ gmail. com, mdaniel@ unina. it, roberto. tartaglia@ unina. it,   landolfo@ unina. it, attilio. demartino@ unina. it

KEYWORDS: numerical modelling bolted joints, cyclic behaviour, moment – rotation response.

ABSTRACT

The cyclic behaviour of beam-to-column joints has a crucial role on the overall seismic response of both steel MR and dual frames. Recent studies[1] highlighted the influence of joint rotation capacity on the seismic response of mid-rise MR frames designed according to EN 1998-1 (2005)[2]. As for dissipative zones, EN 1998-1[2] allows the formation of plastic hinges in the connections in case of partial-strength and/or semi-rigid joints, provided that the following requirements are verified: i) the connections have a rotation capacity consistent with the global deformations; ii) members framing into the connections are demonstrated to be stable at the ultimate limit state (ULS); iii) the effect of connection deformation on global drift is taken into account using nonlinear static global analysis or non-linear time history analysis. At the present time, EN 1993-1-8 (2005)[3] provides models to compute the strength and the stiffness of connections but no reliable and effective analytical tools are available to predict the rotation capacity and the cyclic performance in relation to the connection typology. On the other hand, in order to carry out the seismic assessment of frames with either partial strength or dissipative bolted joints, it is necessary to account for the joint behaviour by using refined models. This consideration motivated the present study which is devoted to propose some modelling criteria for a set of bolted beam-to-column joint typologies which can be easily implemented for any beam-column assemblies. With this regard, modelling assumptions to simulate the nonlinear behaviour of both the connection and the web panel for full and partial strength bolted joints have been validated against some experimental results given by literature and finite element analyses. Finally, the effectiveness of the proposed refined modelling hypotheses has been compared with the results obtained using simplified assumptions. *Fig. 1* shows the comparison between numerical and experimental (tests by Dubina *et al.* [12]) results in terms of overall response (namely applied force vs. top displacement), column web panel response and connection response. As it can be observed, both the overall response (*Fig. 1*a) and column web panel response (*Fig. 1*b) match very well the experimental curves. Indeed, the failure mechanism experienced in the experimental test is successfully reproduced involving the yielding of both, column web panel and connection. However, the connection response (*Fig. 1*c) highlights the limit of the examined modelling criteria. Indeed, during the test, fracture occurred in the beam, while the numerical model cannot account for this event. This limitation implies that at the same imposed overall displacement the amount of rotation numerically applied to the connection results to be larger than the experimental case.

CONCLUSIONS

Refined models of bolted beam-to-column joints have been developed to investigate the influence of joint moment-rotation behaviour on the overall seismic performance of steel frames. The examined models are made of three macro-components, namely (i) the beam at intersection zone (ii) the column web panel (ii) the connection zone. Specific modelling assumptions were proposed for each zone; finally the three macro-components were combined in a comprehensive model describing the overall moment-rotation response of the joint. Three different connection configurations were considered, classified on the basis of simply capacity design criteria and thus of the expected failure mechanism: (a) haunched joint with full strength partial-rigid connection; (b) rib-stiffened extended endplate joint with equal-strength partial-rigid connection; (c) unstiffened extended endplate joint with partial-strength partial-rigid connection. The refined models developed according to the proposed assumptions generally match quite well the experimental results. In order to verify the effectiveness of the proposed modelling strategy, the refined model was compared with the results obtained using simplified models, showing the better accuracy than the simplified ones, whose response is inadequate to reproduce the hysteretic behaviour of the joint.

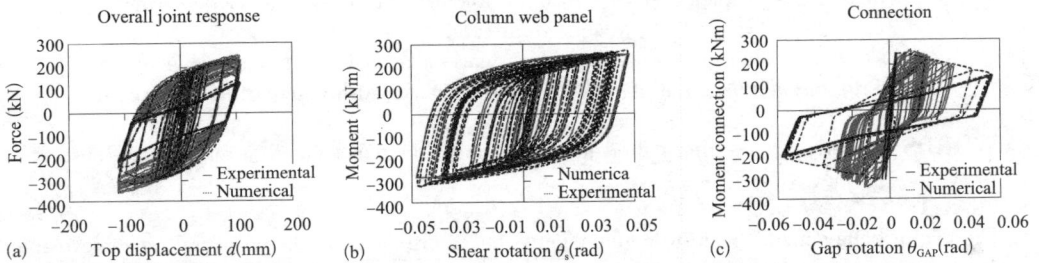

Fig. 1. Model calibration: (a) overall joint response(b) web panel response(c) connection response.

ACKNOWLEDGMENT

The research leading to these results has received funding from the European Union's Research Fund for Coal and Steel (RFCS) research programme under grant agreement n° RFSR-CT-2013-00021. The Authors thanks also the financial support of the RELUIS project "Linea diRicerca Acciaio & Composte Acciaio-Calcestruzzo" 2014-2015.

REFERENCES

[1] Kazantzi A. K. , Righiniotis T. D. , Chryssanthopoulos M. K. (2008). "The effect of joint ductility on the seismic fragility of a regular moment resisting steel frame designed to EC8 provisions". *Journal of Constructional Steel Research* 64: 987-996,2008.

[2] EN 1998-1, Design of structures for earthquake resistance-Part 1: General rules, seismic actions and rules for buildings.

[3] EN 1993-1-8, Design of steel structures-Part 1-8: Design of joints

[4] Seismosoft(2011). SeismoStruct-"A computer program for static and dynamicnonlinear analysis of framed structures". *Available from URL: www. seismosoft. com.*

[5] D'Aniello M, Landolfo R, Piluso V, Rizzano G. "Ultimate Behaviour of Steel Beams under Non-Uniform Bending". *Journal of Constructional Steel Research* 78:144-158,2012.

[6] Cermelj B. , Beg D. "Cyclic behaviour of welded stiffened beam-to-column joints-experimental tests". Steel Construction 7(4): 221-229,2014.

[7] American Institute of Steel Construction, Inc. (AISC) (2010). Seismic provisions for structural steel buildings. Standard ANSI/AISC 341-10, Chicago(IL, USA).

[8] Krawinkler H. , Bertero V. V. , Popov E. P. (1971). "Inelastic behavior of steel beam-to-column sub-assemblages" *Report No. UCB/EERC-71/07*, *Earthquake Engineering Research Center (EE-RC)*, *University of California at Berkeley*

[9] Gupta A. , Krawinkler H. (1999), "Influence of column web stiffening on the seismic behaviour of beam-to-column joints". Stanford University, Stanford, CA,1999.

[10] Kim, K. , andEngelhardt, M. D. (1995). "Development of analytical models for earthquake analysis of steel moment frames," *Report No. PMFSEL 95-2*, *Dept. of Civil Eng. , Univ. of Texas at Austin.*

[11] Ciutina A. , Dubina D. "Column Web Stiffening of Steel Beam-to-Column Joints Subjected to Seismic Actions". *Journal of Structural Engineering*,134(3):505-510,2008.

[12] Dubina D. , Ciutina A. , Stratan A. "Cyclic tests of double-sided beam-to-column joints". *Journal of Structural Engineering*,127(2):129-136,2001.

SIMPLIFIED STRUT MODELING FOR BEAM-TO-COLUMN CONNECTION RETROFITTED WITH SUPPLEMENTAL H-SECTION HAUNCHES

Takuma Uehara[a], Hayato Asada[a], Tsuyoshi Tnaka[a]

[a]Department of Architecture, Kobe University, JAPAN

takuma. u3306@ gmail. com, asada@ people. kobe-u. ac. jp, tanaka@ arch. kobe-u. ac. jp

KEYWORDS: Seismic retrofit, Beam-to-column connection, H-section haunch, Force transfer mechanism, Simplified strut modelling

ABSTRACT

During the Northridge Earthquake of 1994 and the Great Hanshin Earthquake of 1995, many moment resisting frame buildings suffered premature failure of beam-to-column connections, which has led to recent concern about the poor plastic deformation capacity of the beam-to-column connections in existing buildings. In japan, as well, earthquakes in theNankai-Trough or Tokyo metropolitan area are obvious concerns. Preparations for future large earthquakes include active continuation of the seismic upgrading of existing buildings and research into the seismic retrofitting of beam-to-column connections in steel structures[1,2,3]. A prerequisite of seismic retrofitting construction method that uses high-strength bolted connection and welding for stabilization and ensures the construction quality on-site. Authors had proposed a method to improve plastic deformation capacity of the welded flange beam-to-column connections by using supplemental H-section haunches jointed by high strength bolts and welding which expected to secure construction quality as shown in *Fig. 1*. Previous research verified retrofit effect of proposed method on reducing the plastic demands for original beam connection which led to improvement of plastic deformation capacity[4]. And also, it was indicated that force transfer of H-section haunch cannot be predicted based on classical beam theory as pointed out by other researches[3,5]. In this paper, first, parametric study using verified finite element analysis models were conducted to discuss the effect of the beam size, geometry and haunch length on force transfer mechanism of beam-to-column connection retrofitted with proposed method. From numerical results, simplified strut modelling for retrofitted beam was developed to predict the contribution of the H-section haunches that are necessary for establishing of the seismic retrofit design of proposed methods.

CONCLUSIONS

This paper investigates force transfer mechanism of beam-to-column connection retrofitted with H-section haunch(See *Fig. 1*) and identifies forces acting on haunches at beam-to-column connection using verified finite element analysis models. FEA results show that force acts on haunch cannot be predicted by classical beam theory as previous studies mentioned. Force acting on haunch in tension side is relatively smaller than that in compression side due to the force interaction in interface of beam-to-haunch bolted connection, whereas shear force acts on both haunches are comparable. Based on this observation in FEA results, simplified strut modelling(See *Fig. 2a*)) was developed to simulate the unsymmetrical force transfer mechanism of supplemental haunches. Accuracy of proposed model was verified by comparison to FEA results. SSM provides reliable design procedure of proposed retrofit method.

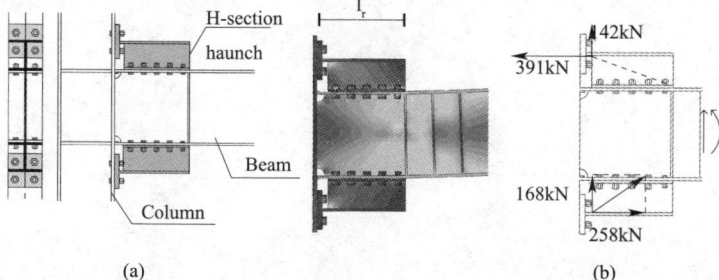

Fig. 1. (a) Beam-to-column subassembly retrofitted with H-section haunch;
(b) Force transfer mechanism of retrofitted connection

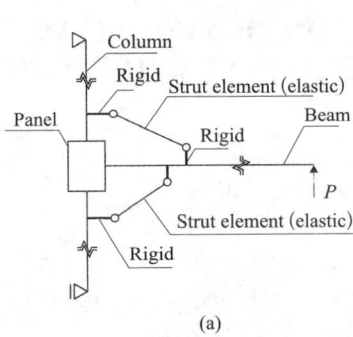

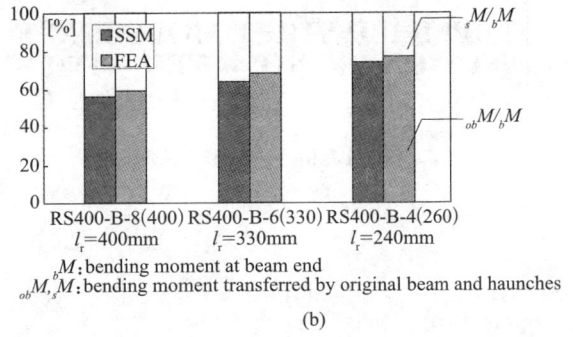

Fig. 2. (a) Simplified strut modelling ; (b) bending moment prediction: the ratio of haunch to original beam bending moment at beam-to-column connection

ACKNOWLEDGMENT

This study has proceeded with a grant from the "Steel Structures Education-Research Grant Program" of the Japan Iron and Steel Federation, 2013, for which we would like to express our appreciation.

REFERENCES

[1] Chung Y-L, Nagae T, Matsumiya T, and Nakashima M. , "Seismic capacity of retrofitted beam-column connections in high-rise steel frames when subjected to long-period ground motions", *Earthquake Engng. Strut. Dyn.* , 41: 735-753, 2012.

[2] Kim Y. J, Oh S. H, and Moon T. S. , "Seismic behavior and retrofit of steel moment connections considering slab effects", *Engineering Structures*, 26: 1993-2005, 2004.

[3] Yu, Q. , Uang, C-M. , and Gross, J. , "Seismic rehabilitation design of steel moment connection with welded haunch", *J. Struct. Eng.* ASCE, 126(1): 69-78, 2000.

[4] Asada, H. , Tanaka, T. , Yamada, S. , and Matoba, H. , "Proposal for seismic retrofit of beam-to-column connection by the addition of H-section haunches to beams using bolt connection", *Int. J. of steel structures*, 14(4): 865-871, 2014.

[5] Lee, C. H. , "Seismic design of rib-reinforced steel moment connections based on equivalent strut model", *J. struct. Eng.* ASCE, 128(9): 1121-1129, 2002.

SEISMIC BEHAVIOR ON JOINT OF PEC COLUMNS-STEEL BEAM CONNECTION WITH END-PLATE

Zhao Gentian[a], Di Hao[a]

[a] School of Architecture and Civil Engineering, Inner Mongolia University of
Science & Technology, Inner Mongolia Baotou 014010, China
zhaogentian93110@ sina. com
teamhao@ 126. com

KEYWORDS: mechanical properties of PEC columns; end plates connection; hysteretic behavior; ductility and energy-dissipation

ABSTRACT

To obtain the mechanical properties of partially encased composite column-beam joints connected by end-plate under seismic action, the seismicbehaviour of 6 specimens which steel beam connect with end plate under reversed cyclic loads were researched. The influence of end plate thickness, thickness of column flange, backing plate on the failure characteristics and seismic performance of PEC column joint were discussed. The failure pattern of each specimen in the testing process was observed. The moment rotation hysteretic curves, energy dissipation capacity and failure mode of PEC column, beam, end-plate and them connection joints were obtained by analyzing the measured data. Experiments show that, the column flange thickness and provided backing plate make a great influence on the bearing capacity of the joints. Node failure modes are characterized by plastic hinge of the beam flange and web, the end plate rupturing and tensile failure of the core area of the column flange. The tests demonstrate that the application of wide leg thin-walled component which width-to-thickness ratio is less than critical value have its possibility. The thickness of end plate and column flange width-thickness ratio is the key factor to affect the node bearing capacity, seismic performance. The bending bearing capacity of nodes increases with the increase of the thickness of theend plate, and the decrease of the column flange width-thickness ratio. The column flange thickness increased from 6mm to 12mm, the energy dissipation capability of node increase 36. 9%, the column flange thickness increased from 12mm to 16mm, the energy dissipation capability of node increase 28. 5% ; The average effective limit angle of all the nodes in test is larger than or close to 0. 03rad. The ductility factor of the displacement angle is from 3. 38 to 6. 50. The tests provide also the theoretical basis for semi rigid structure design specification and engineering application.

CONCLUSIONS

Through 6 weldingH-type partially encased composite column-beam joints connected by end-plate test under seismic action, the following conclusions can be drawn.

1. Increasing end-plate thickness hysteretic behaviour of the node can be improved, but joint behaviour rising is limited. When the end plate thickness is equated to column flange thickness, stiffness of end-plate is relatively weak and end-plate failure is prior to node other parts. As end plate thickness is 1. 5 times or twice to column flange thickness, joint bearing capacity and limit displacements are much the same, beam end has a clear plastic hinge, the node has good energy dissipation mechanism.

2. When column with broad-limb and thin-web is used simplify, its bearing capacity and hysteretic behaviour is not ideal. But setting back pad on the column with broad-limb and thin-web components, it has good seismic performance and ultimate flexural strength. By setting the back plate on the column with broad-limb and thin-web components, it can be applied to earthquake-resistant structures and the constraints of the steel flange plate thickness ratio can be magnified. Through the addition of the back plate, the local stiffness of column flange with broad-limb can be improved. And the influence on the bending capacity of nodes is more obvious, the joints bearing capacity can be up to 49%, but no significant deformation decreasing occurs on the node.

3. The influence of end-plate thickness on flexural bearing capacity and initial stiffness is not ob-

vious, but significant impact on extreme corners is obvious.

4. Increasing with column flange thickness, ultimate bearing capacity of node is gradually improving, hysteresis loop sizes are gradually increasing, the influence of column flange thickness in node area to bearing capacity and hysteretic capacity of joint is important.

5. End plate thickness has a great impact on structural elements' energy dissipation and equivalent viscous damping coefficient. The ultimate equivalent viscous damping coefficient of JD1, JD2, JD5 and JD6 is 0.39, 0.35, 0.38 and 0.29 respectively, which is close to energy dissipation capacity of steel reinforced concrete joints.

6. The column flange thickness increased from 6mm to 12mm, the energy dissipation capability of node increase 36.9%, the column flange thickness increased from 12mm to 16mm, the energy dissipation capability of node increase 28.5%.

7. The back pad make a very clearly impact on the energy dissipation capacity of nodes, the nodes energy dissipation capacity increase 18.5% added back pad.

ACKNOWLEDGMENT

This research was financially supported by the National Natural Science Foundation, China (Grant No. 51268042).

REFERENCES

[1] Prickett B S, Driver R G. 2006. "Behavior of partially encased composite columns made with high performance concrete". *Structure Engineering Report.* No. 262. Canada, Alberta, Edmonton: University of Alberta, Department of Civil & Environmental Engineering.

[2] Spacone E, EI-Tawil H. 2004. "Nonlinear analysis of steel concrete composite structures: state of the art ". *Journal of Structural Engineering*, ASCE, 130(2): pp. 159 -167.

[3] Lakshmi B, Shanmugam N E. 2000. "Behavior of steel-concrete composite columns" *Pro 6th ASCCS Int Conf on Steel and Concrete Composite Structures.* Los Angeles: Association for International Cooperation and Research in Steel-concrete Composite Structures (ASCCS), pp. 449-456.

[4] Shi Gang. 2004. "Steel frame with semi rigid end-plate connections of static and aseismic performance research". PhD thesis at Qinghua University: pp. 52-78.

[5] Zhao Gentian, CaoFubo, Wang Shan, Wan Xin. 2009. "Behavior of partially encased composite columns subjected to eccentric compression". *Proceeding of Shanghai International Conference on Technology of Architecture and Structure*, ICTAS 2009. Shanghai: Tongji University Press: pp. 602-607.

[6] Gentian Zhao, Chao Feng. 2012. "Axial ultimate capacity of partially encased composite columns". *Progress in Structures. 2nd International Conference on Civil Engineering, Architecture and Building Materials*, CEABM 2012 Yantai, China. Applied Mechanics and Materials, Vols. 166-169, pp. 292-295.

[7] Guo Bing, Gu Qiang. 2002. "The study on hysteretic behavior of beam-column end-plate connections". *Architecture Structure Journal*, 23(3): pp. 8-13.

[8] JBJ101 – 96. 1996. *Specification for seismic test method of building.* Beijing: China architecture and building press.

AN EXPERIMENTAL STUDY OF HIGH-STRENGTH BOLTED T-STUB CONNECTIONS TO SHS COLUMNS UNDER CYCLIC LOADING

Zhi-Yu Wang[a], Hui Xue[a], Xiao-Kai Liu[a], Bei-Lei Lv[a]

[a]Department of Civil Engineering & Mechanics, Institute of Architecture and
Environment, Sichuan University, Chengdu, PR China

zywang@ scu. edu. cn

KEYWORDS: Blind bolts, Cyclic loading, Hollow section columns, Life prediction.

ABSTRACT

The bolted endplate connections to open section columns have been significantly concerned for their ductility and dissipative capacity in moment resisting steel frames. In contrast, the cyclic behaviour of these connections to SHS(square hollow section) members has still not been well documented in the literature which limited the evaluation of their application in seismic area. Regarding this, this study provides some behavioural evidences of such connections in the tension region of the joint under cyclic loading condition.

This paper presents an experimental research on the cyclic behaviour of high-strength bolted T-stub connections to SHS columns. The design of the specimens aims at the investigation of the dissipation performance of the SHS column combined with the bolt in tension. Accordingly, the endplate was designed of adequate strength and rigidity which is regarded as noncritical in the performance of the connection. Study in this paper includes the examination of the cyclic load induced failure modes of the connection as well as its related components. The crack propagation on the SHS columns is presented as an evidence for representation. The cyclic loading behaviour of the connection is examined with varying variables of slenderness of SHS column and bolt size. The hysteretic force-displacement response as a basis for evaluation of the cyclic characteristics, including degradations of strength and energy absorption ability, is compared and discussed. Based on the statistical analysis of fatigue test data, the relations of the low cycle fatigue life of the connection are given with the maximum initial strength range and applied displacement range. To represent the low-cycle fatigue due to large stress condition of test connections, the Coffin-Manson relation is modified for the life estimation. The accuracy of the prediction is evaluated with the test data and the corresponding discussion is also presented.

CONCLUSIONS

In this study, the cyclic behaviour of high-strength bolted T-stub connections to square hollow section columns has been studied experimentally. The test specimens were designed to isolate the SHS column and bolt as the primary deformation sources which are not sufficiently covered by recent research work. It has been shown that the tube wall thickness and diameter of the bolt shank have significant influences on the performance of the connection subjected to cyclic loading. The connection with relatively thin tube wall and small bolt diameter tends to suffer local crack in the middle of the tube face between the centre of the bolt while the others have wider crack propagation across the tube connecting face. As shown in Fig. 1, the increase of the tube wall thickness and bolt diameter increases the strength, initial stiffness and energy dissipation of the connections. It has been shown from the comparison of the test strength degradation that the increase of the tube wall thickness and bolt diameter decreases the initial slope of the reduction of strength and energy dissipation which in turn may avoid the premature failure of the connection undergoing large displacement. The low-cycle fatigue behavior of the connection due to large stress condition can be characterized by modified Coffin-Manson relation incorporating displacement range. As listed in *Table 1*, the modified Coffin-Manson relation can be confirmed in a good correlation with test data for the life estimation of the test connections.

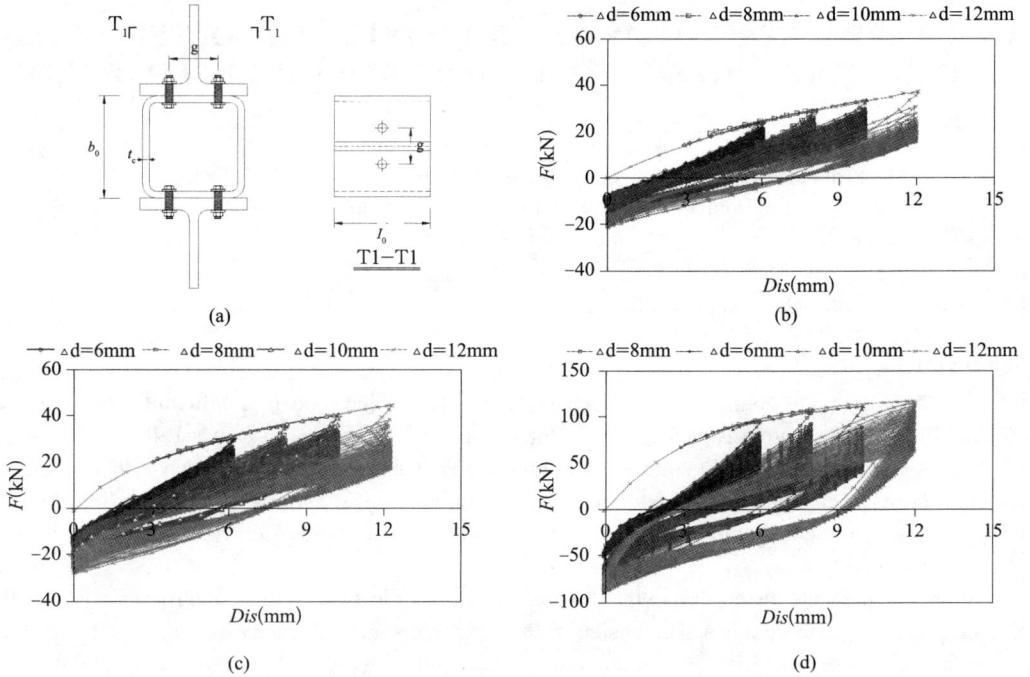

Fig. 1. (a) Illustration of test specimen; (b) Hysteretic loops for BSHS150-5-60-M12;
(c) Hysteretic loops for BSHS150-5-60-M16; (d) Hysteretic loops for BSHS150-8-60-M16

Table 1. Summary of life estimate expressions

Test ID	Expression	Correlation coefficient
BSHS150-5-60-M12	$\Delta d = 150\,(N_r)^{-0.315}$	0.998
BSHS150-5-60-M16	$\Delta d = 125\,(N_r)^{-0.315}$	0.990
BSHS150-8-60-M16	$\Delta d = 102\,(N_r)^{-0.315}$	0.991

ACKNOWLEDGMENT

The financial supports from the National Natural Science Foundation of PR China (No. 51308363) and the Scientific Research Foundation for the Returned Overseas Chinese Scholars (No. 2013-1792-9-4) are greatly appreciated.

REFERENCES

[1] Packer JA, Sherman DR, Lecce M, 2010. *Hollow structural section connections*, Steel Design Guide No. 24, American Institute of Steel Construction(AISC), Chicago.

[2] Bernuzzi C, Zandonini R, Zanon P, 1996. "Experimental analysis and modelling of semi-rigid steel joints under cyclic reversal loading", *Journal of Constructional Steel Research*, Vol. 38 (2), pp. 95-123.

[3] Mourad S, Ghobarah A, Korol RM, 1995. "Dynamic response of hollow section frames with bolted moment connections", *Engineering Structures*, Vol. 10(17), pp. 737-748.

[4] Harada Y, Arakaki T, Morita K., 2002. "Structural behaviour of RHS column-to-H beam connection with high strength bolts", *International Journal of Steel Structures*, Vol. 2: 111-121.

[5] Piluso V, Rizzano G, 2008. "Experimental analysis and modelling of bolted T-stubs under cyclic loads", *Journal of Constructional Steel Research*, Vol. 64, pp. 655-669.

Global Behaviour

COLLAPSE ASSESSMENT OF A 4-STORY BUCKLING RESTRAINED KNEE BRACED TRUSS MOMENT FRAME SYSTEM

T. Y. Yang[a], Yuanjie Li[a]

[a]University of British Columbia, Department of Civil Engineering, Canada
yang@ civil. ubc. ca, yuanjieli01@ gmail. com

KEYWORDS: Buckling restrained brace, innovative system, performance-based design, collapse assessment.

ABSTRACT

Buckling restrained knee braced truss moment frame (BRKBTMF) is a novel and innovative steel structural system that utilizes the advantages of long span trusses and dedicated structural fuses for seismic applications. Truss girders are very economical and effective in spanning large distance, however, steel trusses are typically not suitable for seismic application due to the lack of ductility and poor energy dissipation capacity. BRKBTMF utilizes buckling restrained knee braces as the designated structural fuses to dissipate earthquake energy, this allows the BRKBTMF to span long distance and yet very efficient in resisting the seismic loads. Since there is no code defined design procedure for the BRKBTMF system, this paper utilizesthe performance-based plastic design procedure proposed by Goel and Chao[1] to design the system. Unlike the conventional equivalent lateral force based design procedure, where the structures are designed for prescribed lateral forces then checked for drift, the PBPD procedure designs the element sizes to satisfy the forces and drift limits without iterations. More importantly, PBPD takes the plastic mechanism into the design consideration, which leads to controlled failure mechanism during the strong earthquake shaking. *Fig. 1* shows the plastic mechanism and capacity design procedure used to design the BRKBTMF.

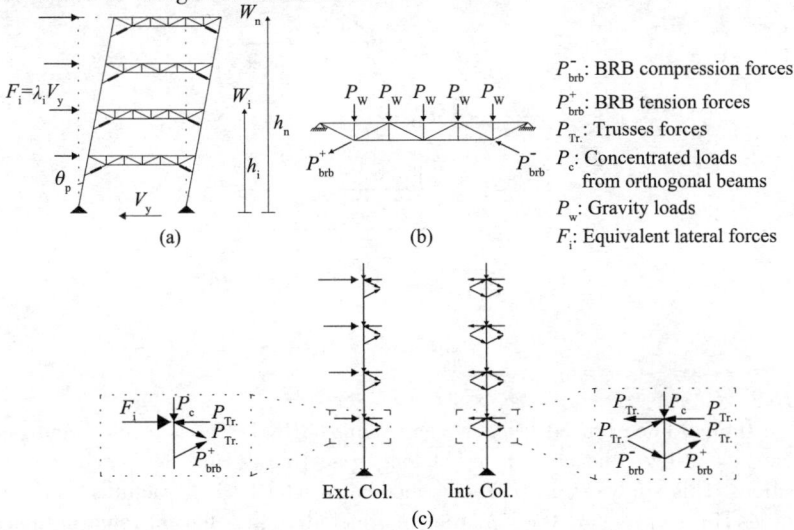

Fig. 1. (a) Plastic mechanism; (b) truss capacity design and (c) column capacity design

In order to accurately model the impact when the BRB fractured. Advanced direct element removal modelling technique was implemented in this study. This method was first proposed by Talaat and Mosalam[2] to simulate the collapse behaviour of reinforced concrete and unreinforced masonry buildings. The algorithm has been modified to simulate the fracture behaviour of BRBs in this study. *Fig. 1.* and *3.* show the element removal diagrammatic procedure and calibrated numerical and experimental result of BRB. *Fig. 4.* shows the structural performance comparison of two different modelling approaches. Significant difference can be observed in the simulation of the structural response. Time history analysis under different hazard levels was conducted. The result shows that BRKBTMF has controlled drift and acceleration. Detailed incremental dynamic analysis was conducted to evaluate the structural safety. *Fig. 5* shows the collapse fragility curve for the system.

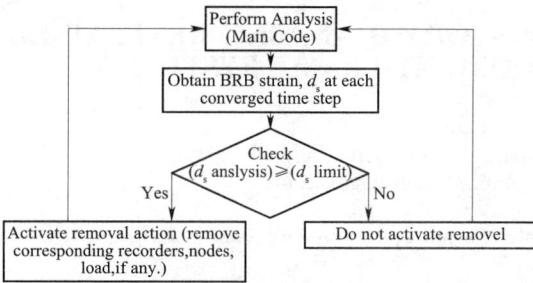

Fig. 2. Element removal procedure

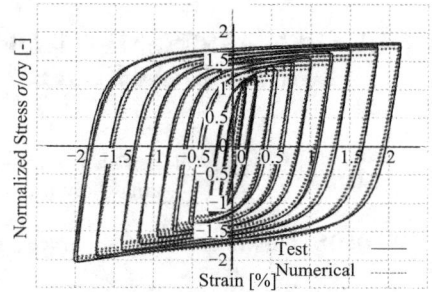

Fig. 3. BRB calibration

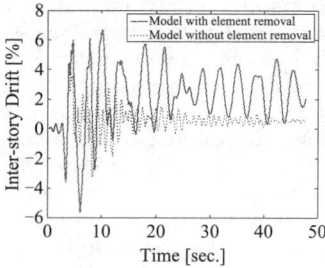

(a) Inter-story drift

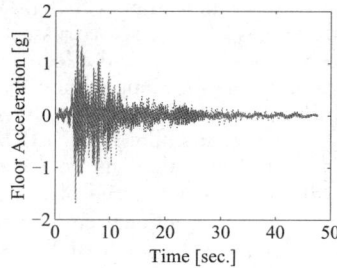

(b) Floor acceleration

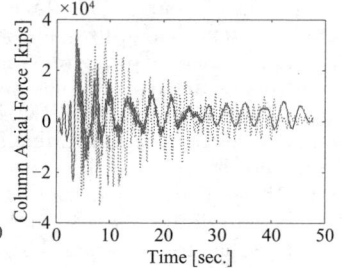

(c) Column axial demand

Fig. 4. Seismic behavior comparison on element removal modelling

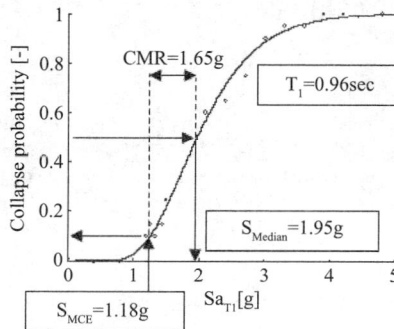

Fig. 5. Building fragility curve

CONCLUSIONS

Buckling restrained knee braced truss moment frame(BRKBTMF) is a novel and innovative steel structural system that utilizes the advantages of long span trusses and dedicated structural fuses for seismic applications. This study proposed a new method to model BRB fractures using direct element removal technique. The study shows the importance of including the element removal to model the BRB facture. A prototype 4-story office building located in Berkley, California, USA was designed and modelled using the proposed technique. The seismic performance of the prototype building was assessed using both time history analysis and incremental dynamic analysis. The result shows that BRKBTMF has controlled drift and acceleration as well as, excellence of collapse resistance. This shows the proposed BRKBTMF is an effective and viable seismic force resisting system.

REFERENCES

[1] Goel, S. C., & Chao, S. H. (2008). Performance-Based Plastic Design: Earthquake-Resistant Steel Structures. International Code Council, USA.

[2] Talaat, M., & Mosalam, K. M. (2009). Modeling progressive collapse in reinforced concrete buildings using direct element removal. *Earthquake Engineering & Structural Dynamics*, 38, 609-634.

MODELING ASPECTS FOR COLLAPSE ANALYSIS OF STEEL MOMENT-FRAME BUILDINGS

Johnn Judd[a], Andrew B. Hardyniec[b], Finley Charney[a]

[a]Department of Civil and Environmental Engineering,
Virginia Tech, Blacksburg, Virginia, USA
jjohnn11@ vt. edu, arp12@ vt. edu, meather@ vt. edu, fcharney@ vt. edu
[b]Exponent, Failure Analysis Associates, Warrenville, Illinois, USA
ahardyniec@ exponent. com

KEYWORDS: Incremental dynamic analysis, Earthquake engineering, Second-order effects.

ABSTRACT

This paper discusses the effects of the beam-column joint (panel zone) model and the method used to incorporate second-order effects on the collapse analysis of steel moment frame buildings. The FEMA P-695 methodology is used to compare the collapse capacity of ductile and non-ductile moment frames using centerline, scissors, and Krawinkler panel zone models, and accounting for second-order effects using the P-Δapproximation and corotational method.

INTRODUCTION

A common approach to model steel moment-frame buildings for earthquake collapse analysis is to idealize the structural frame using a phenomenological (concentrated plasticity) approach [e. g. [1,2,3]]. In this approach, beams and columns are represented using an assembly of rotational springs and elastic beam-column elements in order to simulate the potential for inelastic behavior at or near the end of the member. Aside from component behavior, key aspects of modeling are the beam-column joint (panel zone) representation [4] and the method used to take into account second-order (P-Δ) effects.

DISCUSSION

The differences in response from the ductile and non-ductile frames originate in the details for the design of each frame. The non-ductile frame was designed with components that cannot sustain large ductility demands without failure, as opposed to the detailing for the ductile frame. As is observed from comparing the IDA curves, the ductile frames were capable of sustaining greater interstory drift ratios than the non-ductile frames. Increasing the drift in the stories of the ductile frame increased the ductility demand on the hinges in the beams and columns, which caused collapse at lower ground motion intensity values. Therefore, modeling approaches that increased interstory drift ratios had a larger effect on the non-ductile frames. In addition, the inability for the connections in the non-ductile frame to sustain large ductility demands increased the sensitivity of the model to collapse when considering variations in ground motion records. The higher sensitivity is demonstrated by higher β_{Total} values from the non-ductile frame compared to the values from the ductile frame. Comparing the collapse response of the baseline non-ductile model to the response of the models with the centerline and scissors approaches for the panel zone indicates the effect of stiffening the panel zone joint.

The centerline approach stiffens the structure compared to the Krawinkler and scissors approaches, as is shown by comparing the first mode periods. Stiffening the structure increased the spectral acceleration associated with the median collapse value, but the collapse margin ratio is similar to the CMR from the models with the Krawinkler and scissors joints because of the increase in spectral acceleration at the MCE ground motion. However, as anticipated, changing from a second-order analysis to a first-order analysis in the non-ductile frame had a large effect on the collapse response. The first-order approach is incapable of modeling the destabilizing effects of the gravity loads, which reduces the interstory drift ratios in the model and increases the collapse resistance. The inability of the non-ductile frame to sustain large repeated interstory drift ratios prevented it from reaching large interstory drift ratios where the differences between the corotational and P-Δ approximation are amplified before the frame collapses.

As with the non-ductile frames, the largest differences in response for the ductile frame among changes inmodeling approaches occurred for the change from a first-order to second-order approach. However, the ability for the ductile frame to reach higher interstory drift ratios before collapse amplified the differences in modeling approaches, as demonstrated by the conditional collapse probability values at MCE. The larger interstory drift ratios also enabled the frame to demonstrate differences in the collapse resistances between employing the corotational and P-Δ approximation. The ability for the corotational approach to model large displacements enabled the ductile frame to better distribute ductility demand throughout the structure, increasing the collapse resistance of the frame.

Changing from the Krawinkler or scissors panel zone joint model to a centerline model stiffened the ductile frame, as was observed for the non-ductile frame. The increased collapse resistance from the centerline model also did not have a large effect on the CMR because of the larger spectral acceleration associated with the MCE ground motion. However, the frame with the Krawinkler joint representation had a greater CMR than the frame using the scissors approach. The differences in response between the models with the Krawinkler and scissors panel zone models is due to the larger deformations in the joints [5].

CONCLUSIONS

The effects of varying the modeling approaches for second-order effects and panel zone joints have been presented for a non-ductile and ductile steel moment frame. A comparison of the responses indicates that predicted collapse resistance can change significantly with a change in modeling approach. As anticipated, the largest changes occurred for a change from a first-order approach to a second-order approach. Using a centerline approach for the panel zone models also resulted in a higher median collapse spectral acceleration for both frames, though the effect on the collapse margin ratio was minimal. Interestingly, the responses from the models with the Krawinkler and scissors panel zone joints varied for the ductile frame because of higher interstory drift ratios compared to the non-ductile frame where the differences in collapse resistance were minimal. The similarity in response suggests that employing the scissors model may be a valid way to reduce analysis time. Finally, this study demonstrated that a thorough understanding of the modeling aspects is needed when analyzing frames near collapse.

REFERENCES

[1] Deierlein GG, Reinhorn AM, Willford MR. Nonlinear structural analysis for seismic design. *NIST GCR* 10-917-5, NEHRP Seismic Design Technical Brief No. 4, National Institute of Standards and Technology, Gaithersburg, Maryland; 2010.

[2] National Institute of Standards and Technology(NIST). Evaluation of the FEMA P-695 methodology for quantification of building seismic performance factors. *NIST GCR* 10-917-8, National Institute of Standards and Technology, Gaithersburg, Maryland; 2010.

[3] Lignos DG, Krawinkler H. Deterioration modeling of steel components in support of collapse prediction of steel moment frames under earthquake loading. *Journal of Structural Engineering*, 137 (11): 1291-1302; 2011.

[4] Charney FA, Marshall J. Comparison of the Krawinkler and scissors models for including beam column joint deformations in the analysis of moment resisting steel frames. *Engineering Journal*, 43(1): 31-48; 2006.

[5] Charney FA, Marshall J. Comparison of the Krawinkler and scissors models for including beam column joint deformations in the analysis of moment resisting steel frames. Engineering Journal, 43(1): 31-48; 2006

EFFECT OF COLUMN SPLICE LOCATION ON SEISMIC DEMANDS IN STEEL MOMENT FRAMES CONSIDERING SPLICE FLEXIBILITY

Fahimeh Tork Ladani[a], Gregory MacRae[a], J. Geoffery Chase[b]

[a] University of Canterbury, Department of Civil and Natural
Resources Engineering, New Zealand
fahimeh. torkladani@ pg. canterbury. ac. nz, gregory. macrae@ canterbury. ac. nz
[b] University of Canterbury, Department of Mechanical Engineering, New Zealand
geoff. chase@ canterbury. ac. nz

KEYWORDS: Column Splices, Steel Moment Frame, Splice Location, Seismic Demands.

ABSTRACT

During an earthquake, connections play a great role in preserving the integrity of structures. Column splices, which are column to column connections, are of great importance in steel structures and they should be designed for a minimum strength and stiffness. If splices are not strong enough to withstand the induced demands, it may lead to partial or total collapse of the structure. If splices are strong enough to carry the demand but not stiff enough, they may exhibit large deformations leading to further damage elsewhere. Since splices are placed at the same level of the structure, the probability of instability and a developed mechanism in the frame increases with large deformations. Thus, strength and stiffness of the connection affect the performance of the structure and should be paid attention.

This study aims to quantify the effect of splice connections in columns of steel frames on the responses for DBE and MCE level of ground motions. Inelastic dynamic time history analysis was conducted to assess splice and frame demands. Splices were characterized by strength and stiffness and were explicitly considered in the analyses. It was shown that (1) the presence of flexible splices increased the frame first and second mode periods by about 2% and 4%, respectively; (2) non-zero flexible splices increased storey drift ratios by up to 11%; (3) splice stiffnesses had almost no effect on frame displacements; (4) the splice moment demand increased with increasing splice stiffness on the frame and was as high as 47% of the column flexural capacity; (5) sheardemand on splices can reach 48% of the nominal shear capacity; and (6) splice location can affect the demands of the frame and splices.

CONCLUSIONS

Nonlinear time history analyses for a mid-rise steel moment frame have been conducted using DBE and MCE levels of ground motion. Column splices have been modelled as flexural springs characterised by strength and stiffness. Nonlinear numerical analysis of the effect of these properties and splice location on the responses of the frame clearly show that:

1. Flexible splices increased the frame first and second mode period by about 2 and 4 percent respectively when splices are located at either one or two third of column. Splice flexibility had no effect on the frame period when splices are in the middle of columns.

2. Peak frame displacements were not significantly affected by splice flexibility other than zero. But it increased the peak frame drift ratios by up to about 7% and 11% for non-zero splices when splices are at one third and two third of the column respectively.

3. It was observed that splices located in the middle of column have the least impact on the drift ratio and moment demand while splices located at two third of columns experienced larger demands.

4. Splice shear demands were up to about 41% and 48% of the nominal shear capacity under DBE and MCE shakings which is above the minimum requirement for splices according to NZS3404. The median of splice moment demands did not exceed 30% and 47% of the capacity for DBE and MCE shakings. However, there were splices with demands above 50% of their capacity in

MCE level ground motions. Future study is needed to assess the likelihood of any collapse or mechanism when splices are designed for shear and bending according to NZS3404.

REFERENCES

[1] FEMA,2000. Recommended seismic Design Criteria for New Steel Moment-Frame Building,FEMA 350,Federal Emergency Management Agency,Washington,DC.

[2] AISC,2010. Seismic Provisions for Steel Structural Buildings,AISC 341-10,American Institute of Steel Construction,Chicago,IL.

[3] NZS3404,1997. Steel structures standard. Part1. New Zealand.

[4] Popov,E. P. ,Tsai,K. ,& Engelhardt,M. D. ,1989. "On Seismic Steel Joints and Connections". *Engineering Structures*,Vol. 11(1),pp. 148-162.

[5] Bruneau,M. ,& Mahin,S. A. ,1990. "Ultimate Behavior of Heavy Steel Section Welded Splices and Design Implications". *Journal of Structural Engineering*,Vol. 116(8),pp. 2214-2235.

[6] Shen,J. A. Y. ,Sabol,T. A. ,Akbas,B. ,Sutchiewcharn,N. ,Cai,W. ,2008. "Seismic Demand on Column Splices in Steel Moment Frames",The 14th World Conference On Earthquake Engineering,Beijing,China.

[7] Shen,J. A. Y. ,Sabol,T. A. ,Akbas,B. ,& Sutchiewcharn,N. ,2010. "Seismic Demand on Column Splices in Steel Moment Frames". *Engineering Journal*,pp. 223-240.

[8] Akbas,B. ,Shen,J. ,& Sabol,T. a. ,2011. "Estimation of seismic-induced demands on column splices with a neural network model". *Applied Soft Computing*, Vol. 11(8),pp. 4820-4829.

[9] Somerville,P. ,Smith,N. ,Punyamurthula,S. ,and Sun,J. ,"Development of Ground Motion Time Histories For Phase II of the FEMA/SAC Steel Project",SAC Background Document Report SAC/BD-97/04.

[10] Open system for earthquake engineering simulation(OpenSees),Pacific Earthquake Engineering Research Center,University of California,Berkeley,CA.

[11] Gupta,A. ,Krawinkler,H. ,1999. "Seismic Demands for Performance Evaluation of Steel Moment Resisting Frame Structure".

COMPOSITE SLAB EFFECTS ON BEAM-COLUMN SUBASSEMBLY SEISMIC PERFORMANCE

Tushar D. Chaudhari[a], Gregory A. MacRae[a], Desmond Bull[a],
Geoffrey Chase[b], Stephen Hicks[c], George C. Clifton[d], Michael Hobbs[e]
[a] University of Canterbury, Department of Civil and Natural Resources
Engineering, New Zealand tushar. chaudhari@ pg. canterbury. ac. nz,
gregory. macrae@ canterbury. ac. nz, des. bull@ canterbury. ac. nz
[b] University of Canterbury, Department of Mechanical Engineering, New Zealand
geoff. chase@ canterbury. ac. nz
[c] NZ Heavy Engineering Research Association (HERA), Manukau, Auckland,
New Zealand stephen. hicks@ hera. org. nz
[d] University of Auckland, Department of Civil and Environmental
Engineering, Auckland, New Zealand c. clifton@ auckland. ac. nz
[e] Batchelor McDougall Consulting, Christchurch, New Zealand
michael. hobbs146@ gmail. com

KEYWORDS: Earthquake engineering, Moment frames, Overstrength, Panel Zones, Slabs.

ABSTRACT

The experimental behaviour of the beam-column joint subassemblies with (i) no slab, (ii) a slab fully isolated from the column, and (iii) isolation on the outside of the column face with a slab shear key between the column flanges to activate Eurocode Mechanism 2, were investigated. It is shown that with full isolation of the slab from the column, the peak strength was similar to that with no slab, but the strength degradation during cyclic loading was significantly less. Also, for the assembly with the slab shear key, the lateral strength was 23% higher than that of the other specimens with no slab touching the column faces, but the strength was not maintained through large displacements because re-bar placed to prevent shear key failure caused delamination/spalling of the concrete slab both inside and outside the shear key.

CONCLUSIONS

Experiments of three composite beam-column sub-assemblages were conducted. The first had no slab. The second fully isolated unit had a slab but no slab contact with the column, end-plate, haunches, and nuts and bolts. The third used a shear key to transfer force to the column between column flanges. It was shown that:

i) The sub-assembly with full isolation showed no sign of spalling around the column. Beam yielding occurred primarily at the beam bottom flange due to the presence of the slab. The peak strength was similar to that of the sub-assembly with no slab, but less strength degradation occurred.

ii) For the assembly with the slab shear key, the lateral strength was 23% higher than that of the other specimens with no slab touching the column faces. However, the strength was not maintained through large displacements because after an initial crack occurred beside the shear key, the rebar placed to prevent shear key failure acted in dowel action pushing against the concrete and causing slab delamination/spalling of the concrete slab both inside and outside the shear key. The behaviour seen in this economical composite assembly did not provide the reliable strength desired, and further studies are continuing.

ACKNOWLEDGMENT

The authors would like to acknowledge the MBIE Natural Hazards Research Platform for its support to conduct the proposed research study as a part of the Composite Solution Research Project. All opinions expressed remain those of the authors.

REFERENCES

[1] Standards New Zealand, "NZS3404. Steel Structures Standard-Part 1, Incorporating Amendment No. 1 and Amendment No. 2", *Wellington: Standards New Zealand*, 1997.

[2] Lee, S. J. and Lu, L. W. , "Cyclic Tests of Full-Scale Composite Joint Subassemblages", *Journal of Structural Engineering(ASCE)*, Vol. 115, No. 8: 1977-1998, 1989.

[3] Leon, R. T. , Hajjar, J. F. , and Gustafson, M. A. , "Seismic Response of Composite Moment-Resisting Connections. I: Performance", *Journal of Structural Engineering (ASCE)*, Vol. 124, No. 8: 868-876, 1998.

[4] Hobbs, M. , MacRae, G. A. , Bull, D. , Gunasekaran, U. , Clifton, G. C. , and Leon, R. T. , "Slab Column Interaction-Significant or Not?", *Steel Innovations Conference*, Steel Construction New Zealand, Wigram, Christchurch, Paper 14: 21-22, 2013.

[5] Leon, R. T. , Green, T. P. , and Rassati, G. A. , "Bidirectional Tests on Partially Restrained, Composite Beam-to-Column Connections", *Journal of Structural Engineering (ASCE)*, Vol. 130, No. 2: 320-327, 2004.

[6] Eurocode 8-Part 1, "General rules, seismic actions and rules for buildings. EN 1998-1: 2004 (E)", *CEN: Brussels*, 2004

[7] Braconi, A. , Elamary, A. , and Salvatore, W. , "Seismic behaviour of beam-to-column partial-strength joints for steel-concrete composite frames", *Journal of Constructional Steel Research*, 66: 1431-1444, 2010.

[8] ComFlor, New Zealand. , "Comflor 80 Product Guide. (2008)", Available from: "www. comflor. co. nz/ product-guides/ comflor-80-product-guide/"

[9] Steel & Tube. , "SE82 Mesh", Available from: http://steelandtube. co. nz/ product/ rei/ hrc-mesh/ seismic-mesh/ se82.

[10] Beele Engineering, Netherlands. , "Actifoam fire stop sealing system (2009)", Available from: http://www. beele. eu/ web/ images/ stories/ File/ Installation%   20instruction/ installation% 20instructions% 20ACTIFOAM% 20construction. pdf"

[11] MacRae G. A. , Chanchi J. and Yeow T. , "What Structure is Best? Auckland", *Australasian Structural Engineering Conference(ASEC)*, 9 July to Friday 11 July 2014. PN: 177.

STRUCTURAL BEHAVIOR OF STEEL FRAME WITH LOW JOINT EFFICIENCY OF BEAM WEB

Miki Norihito[a], Nohsho Masahiro[a], Yamada Satoshi[b],
Kishiki Shoichi[c], and Hasegawa Takashi[d]

[a] Graduate Student, Tokyo Institute of Technology
e-mails: miki. n. ac@ m. titech. ac. jp, nohsho. m. ab@ m. titech. ac. jp
[b] Professor, Structural Engineering Research Center, Tokyo Institute of Technology
e-mail: yamada. s. ad@ m. titech. ac. jp
[c] Lecturer, Dept. of Architecture, Osaka Institute of Technology
e-mail: kishiki@ archi. oit. ac. jp
[d] Senior researcher, Building Research Institute
e-mail: hase@ kenken. go. jp

KEYWORDS: Beam-to-column connection, Plastic deformation capacity.

ABSTRACT

Generally in Japan, the beam to column connection consists of wide flange beam and rectangular hollow section (RHS) columns. The moment of the beam web is transmitted to column through the column skin plate. When the column thickness becomes small, it easily acquires out-of-plane deformation, and the joint efficiency of the beam web decreases. As the joint efficiency of the beam web decreases, the maximum strength of the connection and plastic range of beam also decrease. As a result, as plastic deformation capacity of beam-to-column connection decreases, the beam may fracture at early stage[1,2].

Until now, many studies on the relation between plastic deformation capacity of beam-to-column connection and the joint efficiency of the beam web were conducted. However, the experiments of beam-to-column connection in which connection coefficient is small due to joint efficiency of beam web decrease were seldom conducted. Systematic experimental research focused on the region in which the connection coefficient is small is necessary in order to investigate the relation between plastic deformation capacity of beam-to-column connection and the connection coefficient.

In this study, in order to evaluate the plastic deformation capacity of the beam-to-column connection decided by ductile fracture, tests on beam-to-column connections with small connection coefficient were conducted. Here, connection coefficient is defined as the ratio of the maximum strength of beam connection to the full plastic strength of beam.

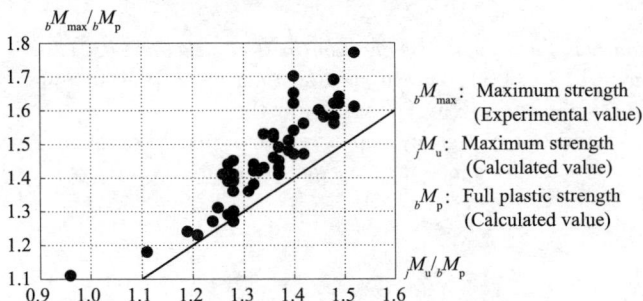

Fig. 1. Current experimental study

CONCLUSIONS

In this study, in order to evaluate the plastic deformation capacity of the beam-to-column connection decided by ductile fracture, tests on beam-to-column connections with small connection coefficient were conducted. 18 specimens fractured at beam flange, and a specimen collapsed by local buckling.

In this test result, as α decreased, η also decreased. When α was about 1.2, η was about 20. When α was between about 1.0 and 1.1, η was between about 15 and 20.

As the calculated value of maximum strength of beam connection increased, the experimental one increased. The experimental value of maximum strength of beam connection was 10% larger than that of calculated one, thus the equation in the design code is evaluated on the conservative side.

The experimental value of joint efficiency of the beam web was larger than the calculated one, and the design equation is evaluated on the conservative side.

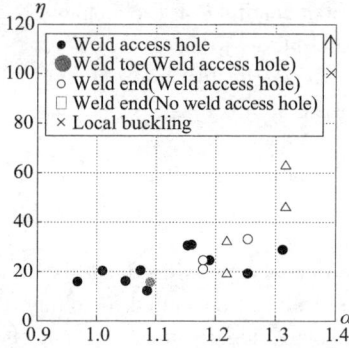

Fig. 2. Relation between η and α

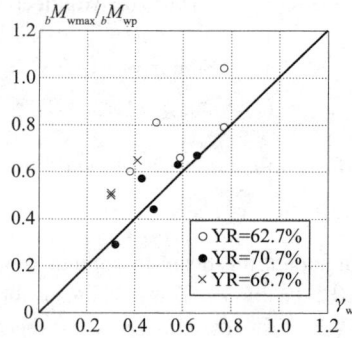

Fig. 3. Joint efficiency of beam web

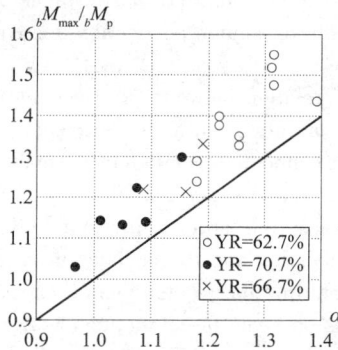

Fig. 4. Maximum strength of beam connection

REFERENCES

[1] AIJ. "Recommendations for design of connection in steel structure", 2012.
[2] Suita K. and Tanaka T. , (2000) , "Flexural Strength of Beam Web to Square Tube Column Joint" Steel Construction Engineering, JSSC, 26, 51-58, 2000.
[3] An Independent Administrative Agency, Japan Iron and Steel Federation, "Testing Methods of the Evaluation of Structural Performance for the Steel Structures", 2002.

ANALYTICAL STUDY COMPARING THE SEISMIC BEHAVIOR OF PARTIALLY RESTRAINED STEEL MOMENT FRAMES TO FULLY RESTRAINED STEEL MOMENT FRAMES

Derek A. Marucci[a], James A. Swanson[a], Gian A. Rassati[a]

[a] University of Cincinnati, College of Engineering and Applied Science, USA

maruccda@ mail. uc. edu, james. swanson@ uc. edu, gian. rassati@ uc. edu

KEYWORDS: Seismic, Partially-Restrained, Fully-Restrained, Dynamic, Stiffness

ABSTRACT

A key component in the stiffness and ductility of steel moment-resisting frames (MRFs) is the connection between the beams and the columns. Previous studies have demonstrated that partially-restrained (PR) connections can beeffective in seismic applications. The types of connections prequalified for use in special moment frames (SMFs) and intermediate moment frames (IMFs), which are provided in ANSI/ AISC 358-10, are only considered fully-restrained (FR)[1]. This paper addresses the seismic behavior of SMFs incorporating PR connections.

A series of frameswere analytically modelled and evaluated using incremental dynamic analyses (IDAs). The frame configuration utilized the 9-Story Los Angeles building design employed by the SAC Joint Venture[2]. A control frame was modelled to represent the original building design utilizing conventional FR perimeter moment frames as the seismic force resisting system (SFRS). A series of PR frames were modelled as modified versions of the original building's SFRS where all of the parallel frames in each principle direction act as the SFRS. The control frame was modelled with the same member sizes that were used in the original building design while the PR frames were modelled with members that were proportioned to account for the effect of having all six parallel frames involved in the SFRS.

The beam-to-column connections utilized in the PR frames were modelled using the procedure employed by Maison et al. [3]. Three different PR connection stiffnesses, K_c, were used with values of 5EI/L, 11EI/L, and 17EI/L where E, I, and L are the modulus of elasticity, moment of inertia, and length of the connected beam. Each connection stiffness had three connection strengths, M_{cy}, of 1.0M_p, 0.66M_p, and 0.33M_p where M_p is the plastic moment strength of the connected beam. The connections were modelled with a bi-linear moment-rotation behavior. The elastic stiffness of the connections was modelled with the abovementioned stiffness values at each of the established connection strengths. The post yield stiffness was modelled with a linear slope that extends from the yield point through a point correlating to a connection strength of 1.4M_{cy} at a rotation of 0.03 radians. A total of nine PR frames and a "base case" FR control frame were constructed as two dimensional models and evaluated using OpenSees.

A suite of ground motion records comprising of one horizontal ground motion record from each pair of the 28 earthquake records presented in the FEMA P695 "Near-Field" record set were used to conduct multi-record IDAs[4]. Each frame was subjected to an incremental progression of nonlinear dynamic analyses with increasing intensities of each horizontal ground motion record up to the maximum intensity that the analysis program could administer. The demands on each structure were recorded and measured with respect to the spectral acceleration at each structure's fundamental period and displayed in an IDA curve. The damage measures (DMs) recorded during the analyses of each frame were the maximum base shears and the maximum roof displacements. The 2% damped spectral accelerations at each structure's first-mode period were defined as the intensity measures (IMs) to characterize the intensities of the applied ground motion records.

The results of the IDAs for each of the frames were compiled into an IDA curve set displaying the IDA curves for the maximum base shears and the maximum roof drifts for each earthquake record. An average IDA curve was developed from the 28 single IDA curves for the two DMs recorded and then plotted on a single chart illustrated in Fig. 1. The figure shows that the IDA runs for the PR frames with the lowest connection strength terminated at significantly lower levels of spectral acceleration.

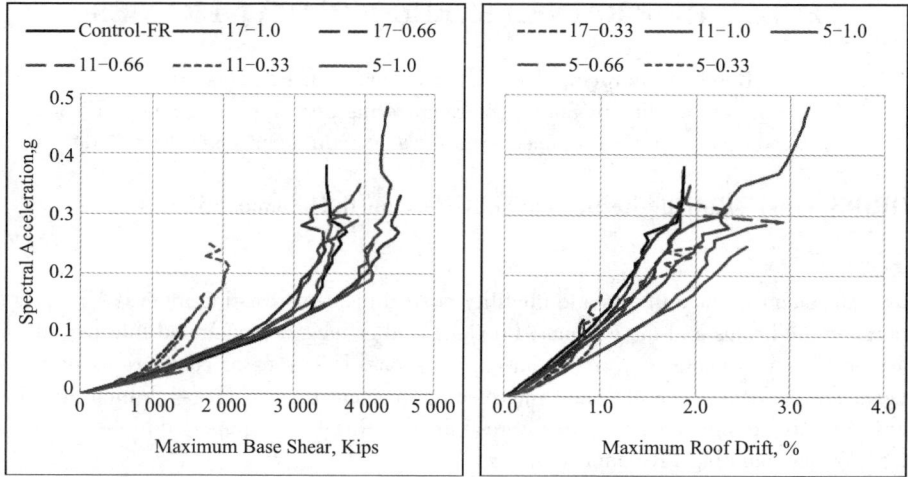

Fig. 1. Average IDA curve comparisons

CONCLUSIONS

The results show that the PR frames with connections strengths less than two thirds of the beam bending strength were unable to achieve the levels of ground motion intensity of the other frames. Therefore, it is recommended that PR connections with strengths greater than or equal to two thirds of the beam bending strength be used. The results also show that the PR frames with the lowest connection stiffness had higher rates of roof drift accumulation resulting in larger ultimate roof drifts at higher levels of ground motion intensity. It is therefore recommended that PR connections with stiffnesses greater than $11EI/L$ be used to limit the amount of excessive drifts to provide sufficient seismic performance. Overall, the average IDA curves show that the PR frames exhibited comparable performance to the control frame suggesting that PR connections shouldn't be categorically excluded from being used in SMFs.

REFERENCES

[1] ANSI/AISC 358-10, "Prequalified Connections for Special and Intermediate Steel Moment Frames for Seismic Applications", American Institute of Steel Construction, Chicago, IL, 2011.

[2] FEMA, "State of the Art Report on Systems Performance of Steel Moment Frames Subject to Earthquake Ground Shaking", *Federal Emergency Management Agency*, FEMA-355C, Washington D. C., 2000.

[3] Maison B. F., Kasai K., Mayangarum A., "Effects of Partially Restrained Connection Stiffness and Strength on Frame Seismic Performance", *SAC Joint Venture*, SAC/BD-99/17, 2000.

[4] FEMA, "Quantification of Building Seismic Performance Factors", Federal Emergency Management Agency, FEMA P695, Washington D. C., 2009.

SEISMIC PERFORMANCE OF CONTROLLED SPINE FRAME WITH ENERGY-DISSIPATING MEMBERS

Xingchen Chen[a], Toru Takeuchi[a], Ryota Matsui[a]

[a]Tokyo Institute of Technology, Architecture and Building Engineering, Japan

chen. x. ad@ m. titech. ac. jp, takeuchi. t. ab@ m. titech. ac. jp,

matsui. r. aa@ m. titech. ac. jp

KEYWORDS: Spine frame, seismic performance, damage distribution, vertical structural irregularity, incremental dynamic analysis(IDA).

ABSTRACT

Catastrophic earthquakes striking beneath Tokyo and along the Nankai trough most likely to occur in the next 30 years, which warns that, the existing buildings in those areas are running a high risk of suffering earthquake shaking beyond the conventional design level. [1] In order to prevent damage concentration and to mitigate residual deformation after major earthquakes, various controlled rocking systems have been proposed[2] (e. g. *Fig. 1* (b)). However, to apply these systems to actual buildings, several obstacles must be overcome, such as the requirement of large, self-centering post-tensioned (PT) strands and special treatment at uplift column bases. As one solution, an innovative non uplifting spine frame(NL) system was proposed in this study(*Fig. 1*(c)).

The new system consists of: 1) a stiff braced steel frame(i. e. spine frame); 2) replaceable energy dissipating members; 3) envelope moment frames. The spine frame plays a key role in distributing damages uniformly to whole stories. The envelope moment frames remain mostly elastic and provide self-centering force.

This system is being employed in an actual building under construction in Japan, which was used as the prototype building herein. Three systems were applied to this building: 1) a conventional shear damper(SD) system[3] (*Fig. 1*(a)); 2) a controlled uplifting rocking(LU) system; 3) the proposed NL system. Their seismic performance and limit-state capacities, particularly damage concentration in specific story, residual deformation, and robustness against both severe earthquakes and vertical structural irregularities, were then compared and discussed by nonlinear dynamic analysis using OpenSEES software[4]. The proposed NL system was proved to have the excellent performance preventing damage concentration in weak stories as well as sufficient self-centering capacity and robustness under major earthquakes.

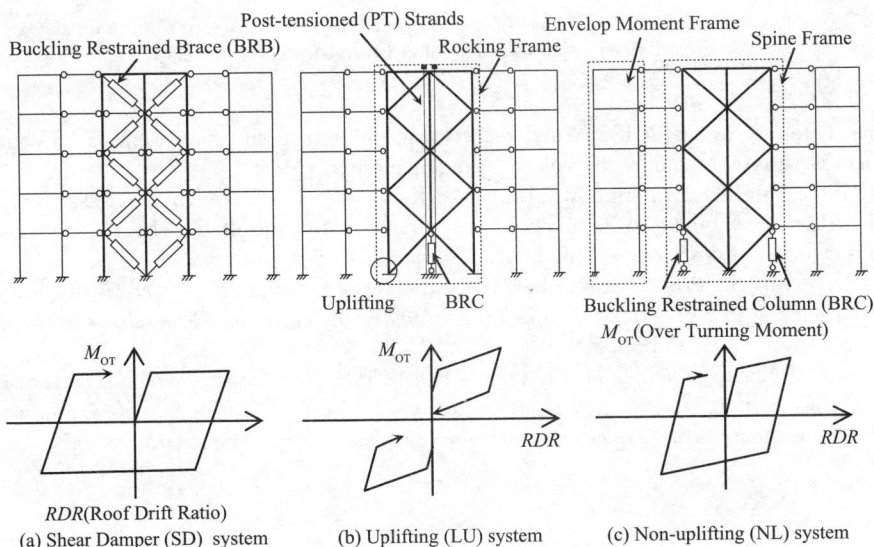

Fig. 1. concept and hysteretic curves of the three structural systems

CONCLUSIONS

A novel non-uplifting spine system without PT strands and with elastic envelope frames was proposed in this study. Its seismic performance was compared with a BRB frame system and an uplifting spine frame system through application to a five story building. The following conclusions were drawn from this study.

（1）In the proposed NL model, the inter-story deformation and damage distributed uniformly over the height of the structure due to the rigid body effect of the spine frame. Even when irregular story existed, the deformation concentration was prevented and the maximum deformation was identical with that of the regular model. The LU model showed similar behavior with the NL model, while severe damage concentration in the irregular story was observed in the SD model. (*Fig.* 2(a))

（2）Sufficient self-centering capacity of the proposed NL model was shown from the residual deformation that was as small as that of the LU model. (*Fig.* 2(b)) It is proved that the elastic restoring force of the moment-resisting frame was large enough to resist the residual force of the plastic members, such as energy-dissipating members.

（3）Moreover, the proposed NL model showed stable seismic performance as earthquake intensity increasing even including the irregular story. By contrast, the first-story irregular LU model exhibited degradation after the bottom diagonal members in the rocking frame yielded. (*Fig.* 2. (c))

In summary, the proposed non-uplifting spine frame system was verified as showing excellent performance in sufficient self-centering capacity and robustness under large earthquakes even without PT strands. This finding is promising and should be explored with more general structural configurations and critical parameters. The proposed system is currently being employed in a school building under construction in Japan. The building is expected to be completed in 2015. The construction process and studies of the simple design method of this structural system will be reported in the near future.

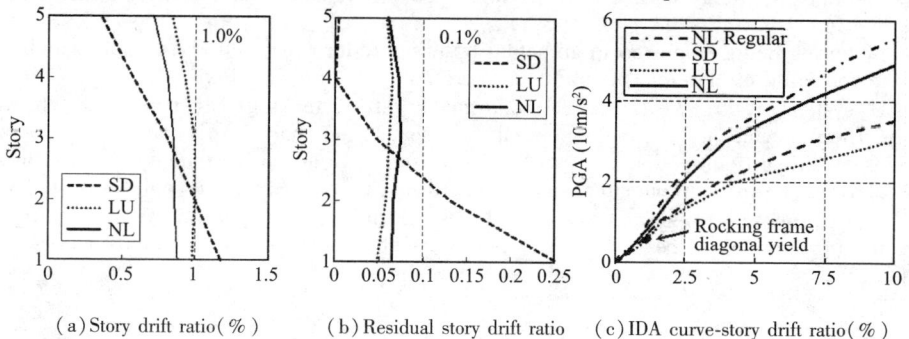

(a) Story drift ratio(%)　　(b) Residual story drift ratio　(c) IDA curve-story drift ratio(%)

Fig. 2. Seismic performance of structural-irregular models

REFERENCES

[1] Cabinet Office of Japan, 2013. "Final report of the investigation group working on countermeasures against Tokyo Inland Earthquakes". (in Japanese)

[2] Deierlein G. , Ma X. , Eatherton M. , Hajjar J. , Krawinkler H. , Takeuchi T. , Kasai K. , Midorikawa M. 2011. "Earthquake resilient steel braced frames with controlled rocking and energy dissipating fuses". *Steel Construction*, Vol. 4, Issue 3, pp. 171-175.

[3] Uriz P. , Mahin S. , 2008. "Toward earthquake-resistant design of concentrically braced steel-frame structures. PEER-2008/08", *Pacific Earthquake Engineering Research Center (PEER)*, University of California, Berkeley.

[4] Mazzoni S. , McKenna F. , Scott M. H. , Fenves G. L. , 2009. "OpenSEES version 2.0 user manual". *Pacific Earthquake Engineering Research Center (PEER)*, University of California, Berkeley. Available from: http://opensees. berkeley. edu(last accessed Feb 2015).

INFLUENCE OF DETAILING OF SHORT LINK ON SEISMIC RESPONSE OF ECCENTRICALLY BRACED FRAMES

Adina Vătăman[a], Daniel Grecea[a,b], Adrian Ciutina[a]

[a] Department of Steel Structures and Structural Mechanics,
Politehnica University of Timisoara, Romania

[b] Romanian Academy, Timisoara Branch, Romania

adina. vataman@ upt. ro, daniel. grecea@ upt. ro, adrian. ciutina@ upt. ro

KEYWORDS: Steel, EBF, links, dissipative, finite element.

ABSTRACT

The Eccentrically Braced Frames are recognised as highly dissipative seismic systems due to the formation of plastic hinges in link elements. In function of their length, they can work in shear (short links), bending (long links) and respectively combined bending and shear (intermediate links). Besides the definition of the link lengths, based on the ratio between the plastic bending and plastic shear resistances of the link element, and the minimum number of stiffeners present on the link, the normative requirements do not differentiate between minimum/maximum required length of the link nor the increased number of stiffeners present on the link. The paper presents the performances of short steel link elements subjected to shear under the form of a parametrical study performed using Abaqus FEM tool. An initial FE model is calibrated based on an existing experimental test performed within the CEMSIG laboratory at the Politehnica University of Timisoara for monotonic loading. In a second step the numerical analysis investigates three parameters that can affect the behaviour of such systems, whilst remaining in the range of short shear links: (i) the use of compact / slender webs; (ii) the influence of the number of web stiffeners and (iii) the influence of link length, by considering five different lengths. The results are presented under the forms of V-γ response curves and judged in function of characteristic parameters, resulting from these.

CONCLUSIONS

The present paper investigates the performances of steel short link elements subjected to shear through numerical analyses by finite element models. This situation corresponds to lateral seismic loads induced in EBF systems. Based on an existing experimental test, an initial calibration was done. The study was followed by a parametric investigation considering three variables: use of a compact/slender web, an increased number of intermediate stiffeners and the variation of the link length. The study revealed the following conclusions:

-the steel experimental tests can be modelled with good accuracy by FE models only by including in the finite modelling the actual behaviour of experimental conditions: in this case the model should consider the slip in braces and lateral supporting conditions;

-high levels of ductility can be achieved by steel shear short links, in the order of 250 mrad, at least under monotonic loading conditions;

-the change in the web slenderness has influence only on the plastic domain as long as the shear area remains similar. Even with smaller shear area, the compact web (HEA 200 profile) proves a better performance both in ultimate resistance and corresponding distortion, as seen in *Fig. 1.* a) ;

-the increase in the number of stiffeners changes the hardening stiffness and the ultimate resistance, without affecting the elastic response. The failure mode begins in the new-formed panel but eventually only one leads to web crippling (see *Fig. 1.* b).) ;

-the increase in the length of the link, while remaining in the range of short link elements, decreases significantly the initial stiffness of the characteristic V-γ curves. Post-elastic behaviour with earlier failure was observed for the specimens with the longest length. The failure is formed in these cases only in a side of the link as seen in *Fig. 2*;

-the design resistance of the link, computed as a web panel according to EC3-1-8 but considering

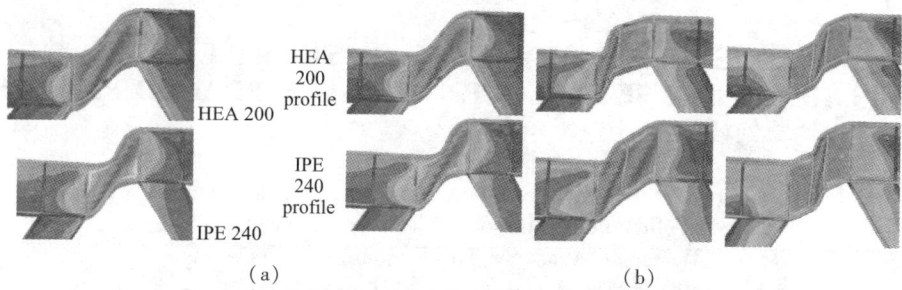

Fig. 1. (a) Deformed link with regard to web slenderness;
(b) Deformed link with regard to web stiffening

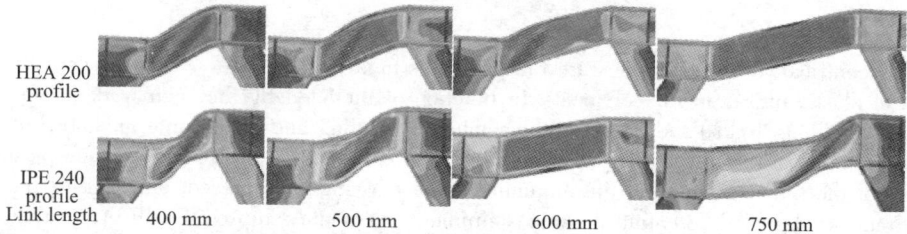

Fig. 2. Deformed shape of the link with regard to link length

the shear area as presented in Eurocode 8-1-1 § 6. 8. remains safe, independent of the link length or number of stiffeners. Despite this, the value of the theoretical initial stiffness, computed on the basis of the column web panels (according to EC3-1-8 prescriptions) is similar to the numerical one only for the long links (750mm).

It is certain that the results of the study should be judged in the context of the initial experimental test. Other loading conditions such as cyclic loading, presence of the concrete slab with/without connection over the link or higher link lengths might affect the proved performances. These issues are considered for further investigation by the research team.

ACKNOWLEDGMENTS

This work was partially supported by the strategic grant POSDRU/159/1. 5/S/137516 of the Ministry of National Education of Romania, co-financed by the European Social Fund-Investing in People, within the Sectorial Operational Programme Human Resources Development 2007-2013.

REFERENCES

[1] EN 1998-1 EUROCODE 8-2003. Design of structures for earthquake resistance, Part 1. General rules, seismic actions and rules for buildings, Brussels, CEN, European Committee for Standardisation.

[2] Danku G. , Ciutina A. , Dubina D. , 2013. , , Influence of steel-concrete interaction in dissipative zones of frames: II-Numerical study" *Steel and Composite Structures*, Vol. 15, No. 3, ISSN 1229-9367, pp: 323-348.

[3] Degée H. , Lebrun N. , Plumier A. , 2010. A. "Considerations on the design, analysis and performances of eccentrically braced composite frames under seismic action", *Proceedings of SDSS 2010 Conference -Stability and Ductility of Steel Structures*, 337-344.

[4] Yurisman Y. , Budiono B. , Moestopo M. , Suarjana M. , 2010. "Behavior of shear link of WF section with diagonal web stiffener of Eccentrically Braced Frame (EBF) of steel structure", *ITB J. Eng. Sci.* ,42(2) ,103-128.

DUAL FRAMES OF HIGH STRENGTH STEEL RHSCF COLUMNS FOR SEISMIC ZONES

Dan Dubina[a,b], Cristian Vulcu[a], Aurel Stratan[a], Adrian Ciutina[a]

[a]Politehnica University of Timisoara, Dept. of Steel Structures and
Structural Mechanics, Romania

[b]CCTFA, Romanian Academy, Timisoara Branch

dan. dubina@ upt. ro; cristian. vulcu@ upt. ro; aurel.
stratan@ upt. ro; adrian. ciutina@ upt. ro

KEYWORDS: dual-steel frames, high strength steel, concrete filled tubes, welded beam-to-column joints, seismic performance

ABSTRACT

Seismic resistant building frames designed as dissipative structures, must allow for plastic deformations to develop in specific members, whose behaviour is expected to be predicted and controlled by proper calculation and detailing. Dual-steel structural systems, optimized according to a Performance Based Design(PBD)philosophy, in which High Strength Steel(HSS)is used in predominantly "elastic" members, while Mild Carbon Steel(MCS)is used in dissipative members, can be reliable and cost efficient. Because current European seismic design codes[1] do not cover this specific configuration, an extensive research project, HSS-SERF-High Strength Steel in Seismic Resistant Building Frames[2][3], was carried out with the aim to investigate and evaluate the seismic performance of dual-steel building frames. On this purpose, and based on a large numerical and experimental program, the following objectives have been addressed by the project:

1. To find reliable structural typologies and joint/connection detailing fordual-steel building frames, and to validate them by tests and advanced numerical simulations;

2. To develop design criteria and PBD methodology for dual-steel structures using HSS;

3. To recommend relevant design parameters(i. e. behaviour factor q, overstrength factor Ω) to be implemented in further versions of the seismic design code, such as EN 1998-1 [1], in order to apply capacity design approach for dual-steel framing typologies;

4. To evaluate technical and economic benefit of dual-steel approach involving HSS.

Univ. of Stuttgart(Germany), Univ. of Liege(Belgium), Univ. of Ljubljana(Slovenia), Univ. "Federico II" of Naples(Italy), VTT Technical Research Centre(Finland), GIPAC Ltd. Design Office(Portugal), RIVAAcciaio S. p. A with Univ. of Pisa as subcontractor(Italy), and Ruukki Construction Oy(Finland), under the coordination of the Politehnica University of Timisoara(Romania), have been involved in the project, which resulted in the following main outcomes and contributions of the project:

-Principles and design recommendations for dual-steel frames(guidelines);

-The investigated frame typologies based on the dual-steel approach with composite columns, are solutions with a high innovative character in the European context;

-Characterisation in terms of global ductility and over-strength demands of dual-steel frames;

-Proposal of a series of innovative beam-to-column joint typologies with composite columns(partially encased-PE, fully encased-FE, and concrete filled tubes-CFT), for which the structural performance was confirmed by experimental and numerical investigations;

-Recommendations for weld details and appropriate component method design approaches.

Additional to the HSS-SERF project summary, the paper presents the research activities conducted by the Politehnica University of Timisoara. Particularly, an experimental program was developed and carried out with the aim to characterize the behaviour of two types of moment resisting joints in dual-steel frames of concrete filled high strength steel rectangular hollow section(CF-RHS)columns and mild carbon steel beams. The parameters considered in the configuration of the joints are given by two joint typologies(reduced beam section RBS, cover plates CP), two steel grades for the RHS tubes (S460, S700), and two failure modes(beam, joint components). Besides, two loading conditions(monotonic, cyclic) were considered, leading to a total number of 16 joint assemblies. The specific detailing and the design approach for RBS and CP joint typologies are presented, as well as the experimental program, test set-up and instrumentation. Test results are shown for each beam-to-column joint configuration, followed by the seismic performance evaluation (see *Fig. 1*).

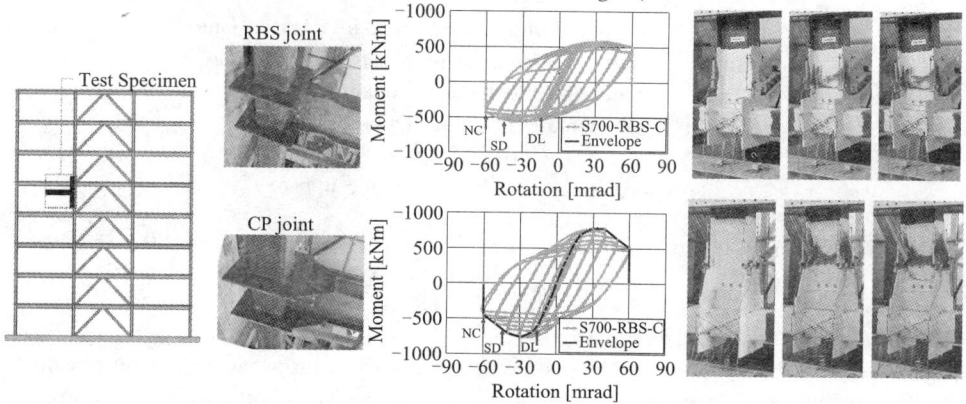

Fig. 1. Dual-steel frame (D-CBF) → choice of members; welded RBS and CP beam-to-CFT column joint specimens; evaluation of seismic performance → envelope curves, performance levels (DL, SD, NC), state of joints

CONCLUSIONS

The current paper summarised HSS-SERF research project with the emphasis on: aim, objectives, partnership, research activities, outcomes and contributions. The results presented herein focused on the research activities conducted by the Politehnica University of Timisoara within HSS-SERF project. Particularly, an experimental program was developed and carried out with the aim to characterize the behaviour of two joint typologies for dual-steel frames of high strength steel CF-RHS columns. As general conclusion, the experimental investigations evidenced a good conception and design of the joints (RBS and CP), justified by: elastic response of the connetion zone, formation of the plastic hinge in the beam, satisfactory response of connection components. In addition, the current study proved the feasibility of using higher steel grades in non-dissipative members (columns) and joint components. Furthermore, the seismic performance of the welded beam-to-CFT column joints was evaluated. Corresponding to the Significant Damage performance level, the RBS and CP joint configurations evidenced rotation capacities larger than the 40 mrad, and therefore the seismic performance of the joints was considered as acceptable.

ACKNOWLEDGMENTS

The present work was supported by the funds of European Project HSS-SERF: "High Strength Steel in Seismic Resistant Building Frames", Grant No RFSR-CT-2009-00024.

REFERENCES

[1] EN 1998-1 (2004). European Committee for Standardization-CEN. Eurocode 8: Design of structures for earthquake resistance-Part 1, General rules, seismic actions and rules for buildings.
[2] High Strength Steel in Seismic Resistant Building Frames (HSS-SERF). D. Dubina et al., Final Report, Directorate-General for Research and Innovation, Unit D. 4-Coal and Steel, RFCS Publications, European Commission, ISBN 978-92-79-44081-6, doi: 10. 2777/725123, 2015.
[3] Proceedings of the International Workshop: "Application of High Strength Steels in Seismic Resistant Structures", 28-29 June 2013, Naples, Italy. Editors: Dubina D., Landolfo R., Stratan A., Vulcu C., "Orizonturi Universitare" Publishing House, ISBN: 978-973-638-552-0, 2014.

NUMERICAL SIMULATION OF PALLET RACK SYSTEMS FAILURE UNDER SEISMIC ACTIONS

Andrei Crisan[a], Dan Dubina[b], Ioan Marginean[a]

[a] Politehnica University Timişoara, dep. of Steel Structures and Structural Mechanics, andrei. crisan@ upt. ro, ioan. marginean@ upt. ro

[b] Romanian Academy, Timisoara Branch, M. Viteazul 24, Timisoara, Romania dan. dubina@ upt. ro

KEYWORDS: Seismic, Pallet Rack Structures, progressive collapse, time-history

ABSTRACT

Pallet rack storage structures are mainly non-dissipative, in the sense that they do not possess sections and member ductility, and the connections are partial strength and semi-rigid in most cases. They are at most low-dissipative when they possess enough design over strength, but then their design has to be predominantly elastic. For ULS criteria, the reference failure models to be considered are either local plastic mechanism in upright members (e. g. failure mode for local buckling induced by bending-compression in members, uprights or braces), or failure of connections. So, robustness of uprights, braces and connections is very important. Transversally (i. e. cross-aisle direction), there are usually two frames, and if are braced and properly connected to each other, a stabilizing effect can be obtained. Longitudinally (i. e. down-aisle direction), the sensitivity to failure is, in principle, higher. When a local failure occurs, a domino effect will generate a progressive collapse, with a quick development.

It can be imagined that a robust rack structure must alternate regular modules with progressive collapse prevention barriers with the mission to prevent inclination and to stop the progressive failure. If proven effective, this can be an innovative and easy to apply solution.

Questioning these assumptions, present paper presents the findings of the numerical analysis performed for a pallet rack storage structure in a high seismic area (Bucharest, Romania). The numerical analysis is an extension of the experimental program and numerical simulations[1]-[3] carried out within the CEMSIG research centre (www. cemsig. ct. upt. ro) within the Politehnica University of Timisoara.

The experimental program was presented in detail in[1]-[3]. The storage rack frames under study consists of members having two types of sections, RS125 and RS95, of perforated-to-brut cross-section ratios, A_N/A_B, of 0. 806 and 0. 760, respectively.

In present study, non-linear time-history analyses were performed using a commercial FEA software in order to observe the failure mechanism development of pallet rack uprights under seismic loading. The initial structure was design in accordance to EN15512:2009 [4] specifications considering the seismic loading in accordance with the specifications of EN 1998-1 [5] design code. The dead load was limited to the self-weight of the structure and an imposed live load of 300 daN/m². The loads were considered fixed on beams (no slippage allowed).

Fig 1 presents the progressive development of overall failure mechanism for the considered structure at the base level, considering a local earthquake recorded motion.

Due to the fact that all FE analysis were aborted due to convergence errors, Applied Element Method (AEM) approach software package was further employed. In the first step, the structure was statically loaded with the imposed live load and element self-weight. The elements which resulted as damaged in the previous FEM analysis, were considered to lose their entire capacity and were removed instantly in a time-history analysis.

The progressive collapse sequence of the structure, obtained using the AEM based software is presented in *Fig. 2*.

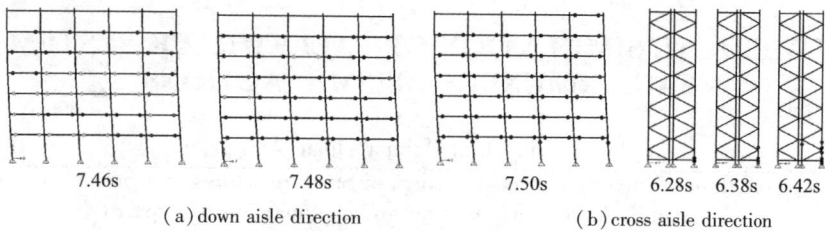

(a) down aisle direction   (b) cross aisle direction

Fig. 1. Time-history results for Vrancea'77 motion

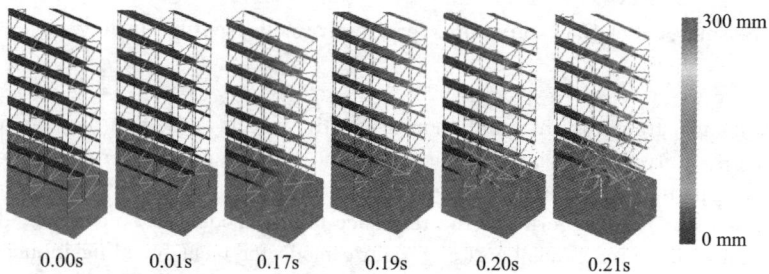

Fig. 2. Progressive collapse sequence

CONCLUSIONS

The behaviour and development of failure mechanisms into the pallet rack storage systems under seismic action was analysed. Time-history nonlinear analysis have proved that these structures, under the effect of earthquake actions develop a progressive failure mechanism, similar with the dynamic instability propagation in case of reticulated structures. Introduction of stiffer modules in to the initial structure, to act as failure barriers, combined with proper horizontal and vertical bracings might stop the development of progressive collapse.

ACKNOWLEDGMENTS

This work was partially supported by the strategic grant POSDRU/159/1. 5/S/137070 (2014) of the Ministry of National Education, Romania, co-financed by the European Social Fund-Investing in People, within theSectoral Operational Programme Human Resources Development 2007-2013.

Partial funding for this research was provided by the Executive Agency for Higher Education, Research, Development and Innovation Funding, Romania, under grant PCCA 55/2012 "Structural conception and COllapse control performance based DEsign of multistory structures under aCcidental actions" (2012-2016).

REFERENCES

[1] Andrei Crisan, Viorel Ungureanu, Dan Dubina: "Behaviour of cold-formed perforated sections in compression. Part 1-experimental investigations, Thin-Walled Structures, Vol. 61, Pages 86-96, 2012

[2] Andrei Crisan, Viorel Ungureanu, Dan Dubina: "Behaviour of cold-formed perforated sections in compression. Part 2-interactive overall-sectional buckling", Thin-Walled Structures, Vol. 61, Pages 97-105, 2012

[3] Andrei Crisan, Viorel Ungureanu, Dan Dubina "Influence of the Shape And Connection Details of Web Members on Out-of-Plane Instability of Storage Rack Upright Frames", Thin-Walled Structures, Vol. 81, Pages 175-184, 2014

[4] EN 15512. Steel static storage systems-Adjustable pallet racking systems -Principles for structural design. Brussels(Belgium): European Committee for Standardization(CEN); 2009.

[5] EN 1998-1: Eurocode 8: Design of structures for earthquake resistance-Part 1: General rules, seismic actions and rules for buildings

SEISMIC PERFORMANCE OF DUAL FRAMES WITH STEEL PANELS

Calin Neagu[a], Florea Dinu[a,b], Dan Dubina[a,b]

[a] Politehnica University of Timisoara, Dept. of Steel Structures
and Structural Mechanics, Romania

calin. neagu@ upt. ro, florea. dinu@ upt. to, dan. dubina@ upt. to

[b] Romanian Academy, Timisoara Branch, Romania

KEYWORDS: Steel panel, seismic performance, non – linear analysis, q factor.

ABSTRACT

Steel plate shear walls (SPSW) are efficient structural systems for resisting lateral loads. They exhibit high initial stiffness and stable cyclic behavior in the plastic range. The seismic response may be improved by using dual systems that are obtained by combining braced and unbraced frames. The application of SPSW system in Europe did not follow a similar trend as in North American or Japan. This can be motivated by a lack of consistent design provisions, e. g. no design rules can be found in seismic code EN 1998-1[1].

In order to investigate the issues presented above, a research program that included experimental and numerical studies was developed at Politehnica University Timisoara, Laboratory of Steel Structures. In the present study, we investigated the performance of two SPSW dual structures with 12 and 18 stories by means of non-linear dynamic analysis. Two sets of ground motions, that are representative for soft and stiff soil conditions, were employed. The dual structural system used in the research program is composed of a moment resisting frame, two additional stanchions that act as vertical boundary elements and two infill panels (Fig. 1) connected to the boundary elements by means of bolts. The structures were designed according to Eurocodes and AISC[2]. According to EN1998[1], the building site is characterized by a peak ground acceleration of 0. 4 g, with a corner period, T_C, of 0. 8 s (Type 1 response spectra and ground type D). For the preliminary design, a q factor of 5 was considered. The geometry and sections of members are shown in Fig. 1.

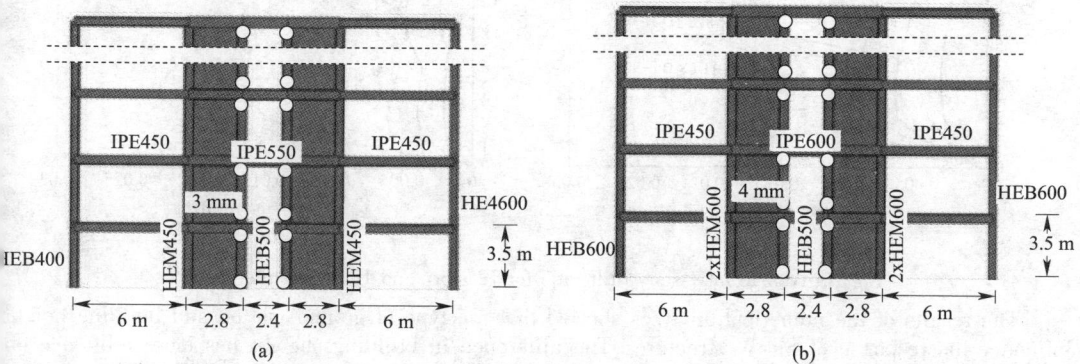

Fig. 1. Geometry and cross sections of structures:
(a) 12 story structure; (b) 18 story structure

The beams, stanchions and columns were designed using S355 while the plates were designed using S235 steel. Cross section of beams, columns and stanchions were made from European IPE and H sections, respectively. In order to estimate structural performance under seismic loads, non-linear incremental dynamic analyses (IDA) were employed. For each type of soil, i. e. soft soil (type D) and stiff soil (type A) (see EN1998[1]), seven ground motions were used in the simulations. Each record was incrementally scaled to multiple levels and then nonlinear dynamic analyses were run.

RESULTS

Fig. 2 and Fig. 3 show the IDA curves for the two structures and type D soil and inter-story drift ratios per building height. Table 1 shows the average value of ductility component q_u for each type of soil and structure.

Table 1. Values of ductility factor q_μ

Str.	Type of soil	q_u
12	A	4.0
18	A	4.1
12	D	3.7
18	D	4.0

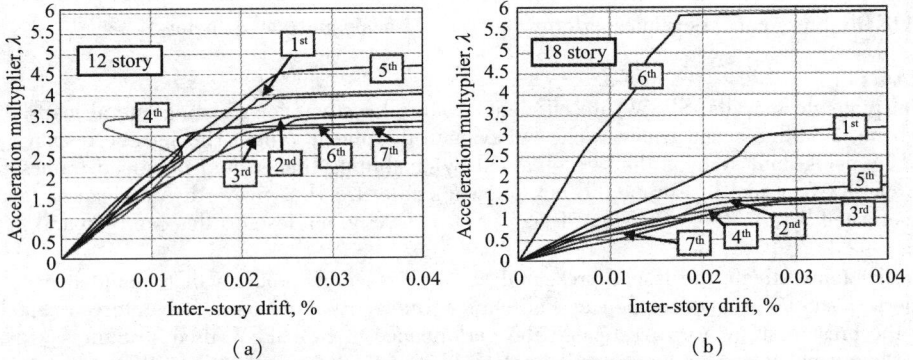

(a) (b)

Fig. 2. IDA curves for ground motions scaled to type D spectrum:
(a) 12 story structure; (b) 18 story structure

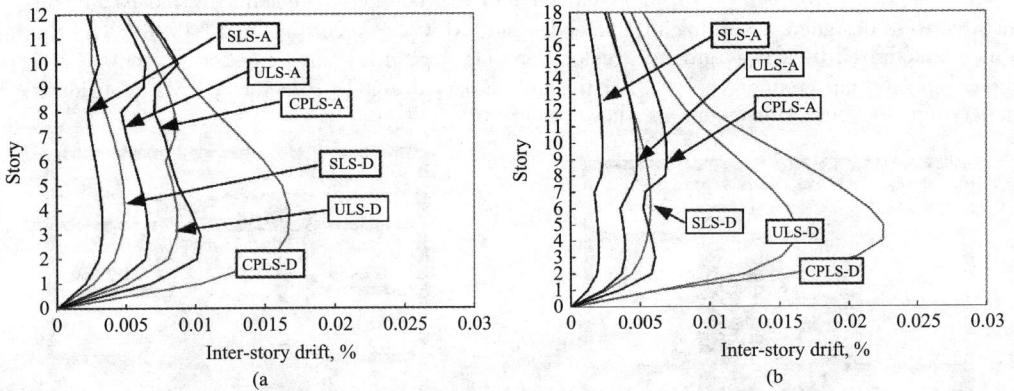

(a (b

Fig. 3. Average inter-story drift ratio for 12 story and 18 story structure

The results of the numerical analyses showed that the type of ground motion and building height influence the response of SPSW structures. The difference in buildings height has more influence on ductility factor q_u factor in case of soft soil conditions than in case of stiff soil conditions. The values of behavior factor obtained in the numerical study are closed to those obtained experimentally and confirm that SPSW structures can be classified as high-ductility structures.

ACKNOWLEDGMENT

This work was partially supported by the strategic grant POSDRU/159/1.5/S/137070(2014) of the Ministry of National Education, Romania, co-financed by the European Social Fund-Investing in People, within theSectoral Operational Programme Human Resources Development 2007-2013.

REFERENCES

[1] EN 1998-1, *Eurocode* 8: *Design of structures for earthquake resistance-Part* 1: *General rules, seismic actions and rules for buildings*, CEN, 2004.
[2] ANSI/AISC 341-10, *Seismic provisions for structural steel buildings*, American Institute for Steel Construction, 2010.

INFLUENCE OF SEMI-RIGID CONNECTIONS ON THE SEISMIC BEHAVIOUR OF BRACED FRAMES WITH BUCKLING RESTRAINED BRACES

Melina Bosco[a], Edoardo M. Marino[a], Pier Paolo Rossi[a] and Paola R. Stramondo[a]

[a]Dept. of Civil Engineering and Architecture, University of Catania,
v. le A. Doria 6, 95125 Catania
mbosco@dica.unict.it, emarino@dica.unict.it,
prossi@dica.unict.it, pstramon@dica.unict.it

KEYWORDS: semi-rigid connections, concentrically braced frames, modelling, seismic response.

ABSTRACT

Buckling Restrained Braces (BRBs) are very effective to provide the structure with seismic resistance. In particular, in occurrence of strong ground motions, BRBs are more effective than conventional steel braces because they dissipates energy through stable tension-compression hysteretic cycles. Research on BRBs was first carried out in Japan. Later, various types of BRBs have been developed mainly in Asia and, after the 1994 Northridge earthquake, in the United States. A review of the state of the art of BRBs can be found in[1-2]. BRBs have been successfully experimented in Japan in the seismic protection of buildings, where they are used as hysteretic dampers combined with moment-resisting frames. In some other countries, BRBs are inserted in low redundant frames in which beams are connected to columns by means of joints with low rotational stiffness. Design methods and behaviour factors have been recently proposed for these structures [3-5]. According to these design methods, the entire seismic force is sustained by the braced frames and beam-to-column connections are assumed perfectly pinned. Further, idealised numerical models with pinned connections are also used for the numerical investigations performed to validate these design methods. The use of a numerical model with beam-to-column pinned connections makes simple both the elastic analysis used for design of steel frames with BRBs and the numerical investigation for the evaluation of their nonlinear response. However, such a simplification ignores aspects of the response of steel structures that can have an important impact on their seismic performance. In fact, as actual beam-to-column connections are semi-rigid, the braced frames retain some lateral stiffness even after yielding of BRBs. This residual stiffness could have a beneficial effect on the seismic response of the frames in that it could promote a more uniform distribution of the BRB ductility demand along the height of the building. On the other hand, because of the rotational stiffness of the beam-to-column connections, an increase in bending moments of columns could occur and cause their yielding or instability before BRBs achieve their ductility capacity. Finally, beam-to-column connections could fail prematurely because their bending moment demand overcomes their capacity.

Based on these considerations, this paper investigates the effect of a more realistic modelling of beam-to-column connections on the seismic response of steel structures with BRBs. The structural scheme is defined by the intersection of two sets of four plane frames that are arranged along two orthogonal directions and have three spans each. The frames located along the perimeter are endowed with BRBs in the chevron configuration and are designed to sustain the entire seismic force of the building. The frames equipped with BRBs are designed by the method proposed by Bosco and Marino[3], later modified in reference [4], while the other frames are endowed with semi-rigid connections and designed to sustain gravity loads only (*Fig. 1*). The seismic action is represented by the elastic response spectrum proposed in EuroCode 8 (EC8) for soft soil (type C) and is characterised by a design peak ground acceleration a_g equal to 0.35g. The design spectrum is derived according to EC8 by means of the behaviour factor q. The value of q is evaluated as a function of the design storey drift angle Δ_d[6]

$$q = 425\Delta_d - 0.25 \tag{1}$$

The design storey drift angle Δ_d is related to the ductility capacity μ_{max} of the BRBs. Two values of

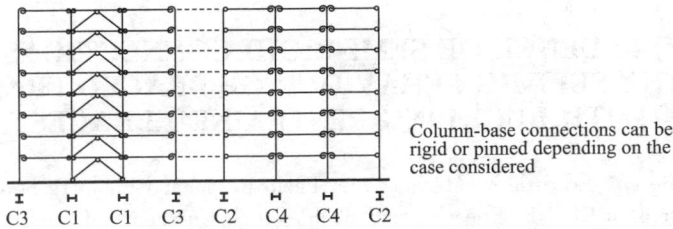

Column-base connections can be rigid or pinned depending on the case considered

$$\underset{C3}{\text{I}} \quad \underset{C1}{\text{H}} \quad \underset{C1}{\text{H}} \quad \underset{C3}{\text{I}} \quad \underset{C2}{\text{I}} \quad \underset{C4}{\text{H}} \quad \underset{C4}{\text{H}} \quad \underset{C2}{\text{I}}$$

Fig. 1. Numerical model of the frames

Δ_d, equal to 1.0% and 2.0%, are considered here to investigate different levels of the ductility capacity of BRBs. These values are representative of BRBs with low and high ductility capacity. Details on the evaluation of μ_{max} and on the relationship between μ_{max} and Δ_d may be found in reference [3].

The seismic performance of these structures is investigated by nonlinear dynamic analysis. The rotational stiffness of beam-to-column and column-base joints is assumed as the main variable of investigation, and the seismic performances of the analysed structures are compared to evaluate the benefit and/or disadvantage resulting from the use of semi-rigid connections.

CONCLUSIONS

The paper analyses the influence of the rotational stiffness of beam-to-column connections on the seismic performance of building structures with BRBs. Beams and columns that do not belong to the braced frames are designed to sustain gravity loads only. Furthermore, as usual in common practice, nominally pinned beam-to-column connections are assumed in design. The analyses show that the use of semi-rigid connections does not improves the seismic performance of the frame in terms of the normalized ductility of BRBs. In fact, even the presence of beam-to-column connections with a not negligible rotational stiffness does not make the BRB ductility demand more uniform along the height of the frame and does not reduce the maximum value. On the other hand, the use of semi-rigid connections aggravates the response of the gravity columns that are generally designed only to support vertical loads and that may experience yielding and/or buckling before BRBs achieve their ductility capacity. The performance of the semi-rigid connections is never acceptable. In fact, even if their moment resistance increases with the rotational stiffness, their bending moment demand increases too. In conclusion, the stiffness of the beam-to-column connections does not improve the seismic performance of BRBs but drastically aggravates the response of the members that do not belong to the braced frames. Hence, the usual assumption of beam-to-column connections made for design of structures with concentric braces is not on the safe side.

REFERENCES

[1] Della Corte G, D'Aniello M, Landolfo R, Mazzolani FM, "Review of steel buckling-restrained braces", *Steel Construction*, 4(2): 85-93, 2011.

[2] Xie Q, "State of the art of buckling-restrained braces", *Journal of Constructional Steel Research*, 61: 727-748, 2005.

[3] Bosco M, Marino EM, "Design method and behavior factor for steel frames with buckling restrained braces", *Earthquake Engineering and Structural Dynamics*, 42(8): 1243-1263, 2013.

[4] Bosco M, Ghersi A, Marino EM, Rossi PP, Stramondo PR "Steel frames with buckling restrained braces: an extension of the Eurocode 8 provision for concentric braces", *Proc. of EUROSTEEL* 2014, September 10-12, 2014, Naples, Italy.

[5] Mahmoudi M, Zaree M, "Evaluating response modification factors of concentrically braced steel frames", *Journal of Constructional Steel Research*, 66: 1196-1204, 2010.

[6] Milazzo S, "Fattore di struttura per telai in acciaio con BRB progettati nel contesto dell' Eurocodice 8", Graduation Thesis, University of Catania, Italy, 2014. (in Italian).

SEISMIC PERFORMANCE AND RE-CENTRING CAPABILITY OF DUAL ECCENTRICALLY BRACED FRAMES WITH REPLACEABLE LINKS

Adriana Ioan[a], Aurel Stratan[a], Dan Dubina[a,b], Mario D'Aniello[c], Raffaele Landolfo[c]

[a] Politehnica University of Timisoara, Dept. of Steel Structures and Structural Mechanics, Timisoara, Romania adriana. ioan-chesoan@ student. upt. ro, aurel. stratan@ upt. ro, dan. dubina@ upt. ro

[b] Romanian Academy, Fundamental and Advanced Technical Research Centre(CCTFA), Timisoara, Romania

[c] University of Naples Federico II, Department of Structures for Engineering and Architecture, Naples, Italy mdaniel@ unina. it, landolfo@ unina. it

KEYWORDS: re-centring capacity, bolted links removal

ABSTRACT

Conventional seismic design philosophy is based on dissipative structural response, which implicitly accepts damage of the structure under the design earthquake and leads to significant economic losses. Repair of the structure is often impeded by the permanent(residual) drifts of the structure. The repair costs and downtime of a structure hit by an earthquake can be significantly reduced by adopting removable dissipative members and providing the structure with re-centring capability. These two concepts were implemented in a dual structure, obtained by combining steel eccentrically braced frames (with removable bolted links) and moment resisting frames. The bolted links provide the energy dissipation capacity and are easily replaceable, while the more flexible moment resisting frames provide the necessary re-centring capability. The solution is validated by full-scale pseudo-dynamic test of a three-storey model of a steel structure with re-centring capability(see accompanying paper [1]) at the European Laboratory for Structural Assessment(ELSA) at the Joint Research Centre in Ispra within the framework of Transnational Access of the SERIES Project.

A numerical model is calibrated based on the experimental results. It is used in order to extend experimental results on current practice structures. Two configurations of a higher-rise dual structure with eccentrically braced frames with removable links(see *Fig. 1*) are designed according to European codes.

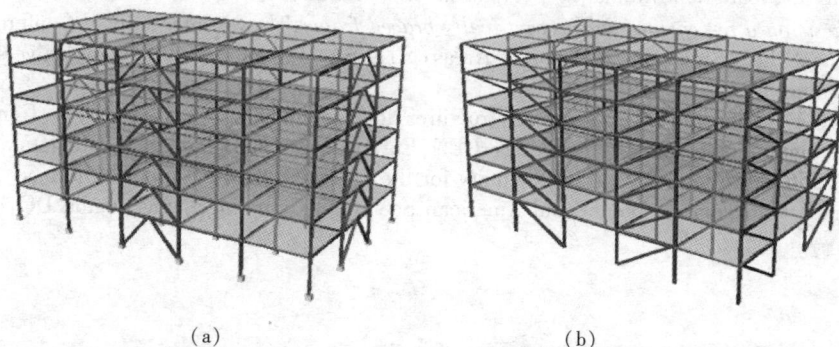

Fig. 1. (a) Configuration A and (b) Configuration B structures

The seismic performance of the two configurations is assessed by means of nonlinear static(push-over) and dynamic(time-history) analyses. A series of numerical simulations are performed in order to validate the re-centring capability of the two configurations and to study different solutions for removing links within a storey and per height of the structure.

CONCLUSIONS

The removable dissipative elements and re-centring capacity concepts, validated through experimental tests[1], were extended to current practice dual structures with eccentrically braced frames that were designed according to European norms(Eurocodes).

For a structure with re-centring capability, the design objective consists in preventing yielding in members other than removable dissipative ones, up to a desired deformation. Ideally the latter should be the ultimate deformation capacity of the removable dissipative member. Following code-based capacity design rules was not enough to accomplish this objective for the investigated structures. Practically, this can be done by adjusting member dimensions using a static(pushover) or dynamic(time-history) nonlinear analysis. Using higher-strength steel in moment resisting frames was shown to be efficient in avoiding yielding in their members.

Numerical simulations were performed in order to investigate solutions for removing links. Three possibilities of links removing order within a storey were studied:(a) firstly removing the links on the longitudinal direction(the least loaded ones) and secondly the ones on the transversal direction(the most loaded ones), (b) vice versa, and(c) in a circular pattern. It was observed that for the first version the link's residual shear force drop is smaller than for the other two studied solutions and the redistribution of forces between the links of the same storey is also smaller.

On the first version(a) of link removal order within a storey, was analysed also the removal of links starting from the most loaded to the least loaded storey(from the lower storey toward the upper one). In this case, was observed a larger interaction between stories, but values of the shear force drop with about 30% ~40% smaller than in the case of eliminating links from the upper storey toward the lower one and smaller redistribution of forces between the links of the same storey.

ACKNOWLEDGMENT

The research leading to these results has received funding from the European Community's Research Fund for Coal and Steel (RFCS) under grant agreement n° RFSR-CT-2009-00024 "High strength steel in seismic resistant building frames" and from the European Community's Seventh Framework Programme [FP7/2007-2013] for access to the European Laboratory for Structural Assessment of the European Commission-Joint Research Centre under grant agreement n° 227887 and was partially supported by the strategic grant POSDRU/159/1.5/S/137070(2014) of the Ministry of National Education, Romania, co-financed by the European Social Fund-Investing in People, within the Sectorial Operational Programme Human Resources Development 2007-2013.

REFERENCES

[1] Stratan A., Ioan A., Dubina D., Poljanšek M., Molina J., Pegon P., Taucer F., 2015. *Large-scale tests on a re-centring dual eccentrically braced frame*, The 8th International Conference on Behaviour of Steel Structures in Seismic Areas(STESSA' 15), 1-3 July, Tongji University, Shanghai, China;

[2] EN1998-1-1, Eurocode 8, Design of structures for earthquake resistance-Part 1, General rules, seismic actions and rules for buildings, CEN, European Committee for Standardization, 2004;

[3] FEMA 356, "Prestandard and commentary for the seismic rehabilitation of buildings", Federal Emergency Management Agency and American Society of Civil Eng., Washington DC, USA, 2000.

SEISMIC PERFORMANCE ASSESSMENT OF A TALL BUILDING HAVING PRE-NORTHRIDGE MOMENT-RESISTING CONNECTIONS

Jiun-Wei Lai[a], Matthew Schoettler[a], Shanshan Wang[b] and Stephen A. Mahin[a, b]

[a]Pacific Earthquake Engineering Research Center, University of California, Berkeley,
U. S. A adrian. jwlai@ berkeley. edu, mschoettler@ berkeley. edu

[b] Department of Civil and Environmental Engineering, University of California,
Berkeley, U. S. A shanonwang@ berkeley. edu, mahin@ berkeley. edu

KEYWORDS: Existing Tall Buildings, Moment-resisting Frames, Seismic Performance, Steel Structures.

ABSTRACT

The Pacific Earthquake Engineering Research(PEER) Center has expanded its Tall Building Initiative project to include assessment of the seismic performance of existing tall steel buildings. Thus far, buildings investigated were 20 stories or more in height, and constructed between the 1960 and 1990. During this period, many tall buildings were constructed in California, but earthquake-resistant design procedures were not fully developed. The number of buildings fitting these criteria in San Francisco, California alone is on the order of 80. From these structures, a 35-story steel building, designed in 1968, and having representative details from that period, was selected for evaluation. In this paper, results from one of four three-dimensional nonlinear analysis models developed to assess the potential seismic performance of this structure are presented. Two earthquake hazard levels based on ASCE 41 recommendations are used for the evaluation. Structural responses are computed using state-of-the-art analysis tools. The structural analysis results are interpreted to assess the impact of suspected deficiencies on seismic response, and the ability of different evaluation guidelines, numerical models and analysis methods to identify seismic vulnerabilities. With this understanding, retrofit strategies can be assessed.

CONCLUSIONS

The computed response of the prototype building does not satisfy the target performance objectives, damage control state under BSE-1E events and limited safety state under BSE-2E events. Several seismic deficiencies are identified during the evaluation procedure. It is found that the prototype building has tendency of forming soft story. Pre-Northridge beam-to-column connection detail results in high percentage of connection failure under BES-2E events, as shown in the dynamic analyses. High column stress demands, fracture vulnerable column splice details, beams and columns not conforming to the strong-column weak-beam criterion and no anchorage between column bases to foundation were also found during the Tier 1 screening. It should be noted that the results presented herein do not consider column splice failures.

From the seismic evaluation results, it is suggested to retrofit the building. Given the large number of difficult to access, beam-to-column connections needing retrofit, upgrading connection details is likely economically prohibitive. Analysis models, not shown, with ideally ductile behaviour still show substantial soft story behaviour and large residual displacements. Thus, methods to reduce beam ends rotational demand and prevent the formation of a weak story mechanism need to be explored. The lack of column attachment to the foundation and the fragility of the columns, pose significant challenges to stiffening the building by adding strong, centralized bracing or wall elements. Retrofit schemes such as adding distributed velocity-or displacement-dependent bracing devices to the current lateral force resisting frames or implementing mid-level isolation systems might be considered

ACKNOWLEDGMENT

This paper is based upon research supported in part by California Emergency Management Agency (CalEMA) under funding No. DR-1884 --Seismic Performance of Existing Tall Buildings and Development of Pilot Internet Database for Post-Earthquake Applications. Any findings, opinions and conclusions or recommendations expressed in the paper are those of the authors and do not necessarily reflect the views of CalEMA. The assistance of Frank McKenna and Andreas Schellenberg in developing OpenSees numerical models is gratefully acknowledged.

REFERENCES

[1] PEER, (2010) "Guidelines for Performance-based Seismic Design of Tall Buildings", PEER 2010/05.
[2] ASCE, (2014) "Seismic Evaluation and Retrofit of Existing Buildings", American Society of Civil Engineers, ASCE/SEI 41-13, Reston, VA.
[3] CBSC, (2013) "California Building Code", California Code of Regulations, Title 24, Part 2, Volumes 1 and 2, California Building Standards Commission, Sacramento, CA.
[4] Department of Building Inspection(DBI) (2013), San Francisco, CA.
[5] ASCE, (2010) "Minimum Design Loads for Buildings and Other Structures", American Society of Civil Engineers, ASCE/SEI 7-10, Reston, VA.
[6] Baker, J. W., (2014) "Ground Motions for PEER Tall Buildings Project", unpublished in-house report.
[7] McKenna, F., Scott, M., and Fenves, G., (2010). "Nonlinear Finite-Element Analysis Software Architecture using Object Composition." Journal of Computing in Civil Engineering, Vol. 24, No. 1, p95-107.
[8] Kanvinde, A., (2012) personal communication through phone message.
[9] Ribeiro, F., Barbosa, A., Scott, M., and Neves, L. (2015). "Deterioration Modeling of Steel Moment Resisting Frames Using Finite-Length Plastic Hinge Force-Based Beam-Column Elements", Journal of Structural Engineering, Vol. 141, Issue 2.
[10] FEMA, (2006) "NEHRP Recommended Provisions: Design Examples, FEMA 451", Building Seismic Safety Council, Federal Emergency Management Agency.
[11] ATC, (2010) "Modeling and Acceptance Criteria for Seismic Design and Analysis of Tall Buildings", PEER/ATC-72-1 (PEER Report 2010/111).
[12] NIST, (2012) "Soil-Structure Interaction for Building Structures", NEHRP Consultants Joint Venture, NIST GCR 12-917-21 (ATC-83 report).
[13] Lai, J. W., Schoettler, M., Wang, S. and Mahin, S. (2015) "Seismic Evaluation and Retrofit of Existing Tall Steel Buildings: A 35-story Steel Building Having Pre-Northridge Connections", (report release pending).
[14] ICBO, (1967) "Uniform Building Code", International Conference of Building Officials, Pasadena, CA.
[15] LATBSDC, (2014) "An Alternative Procedure for Seismic Analysis and Design of Tall Buildings Located in the Los Angeles Region", Los Angeles Tall Buildings Structural Design Council, Los Angeles, CA.
[16] Wang, S., Lai, J. W., Schoettler, M. and Mahin, S. A. (2015) "Seismic Retrofit of a High-rise Steel Moment Resisting Frame using Fluid Viscous Dampers", Proceedings, 8th International Conference on Behavior of Steel Structures in Seismic Areas, Shanghai, China, July 1-3, 2015.

SEISMIC PERFORMANCE EVALUATION OF EXISTING HIGH-RISE STEEL BUILDING SUBJECTED TO LONG-PERIOD GROUND MOTION AND ASSESSMENT OF RETROFIT BY STEEL DAMPERS

D. Sato[a], T. Nagae[b], H. Kitamura[c], M. Nakagawa[d], K. Sukemura[e] and K. Kajiwara[f]

[a] Structural Engineering Research Center, Tokyo Institute of Technology
e-mails: daiki-s@ serc. titech. ac. jp

[b] Disaster Mitigation Research Center, Nagoya University
e-mail: nagae@ nagoya-u. jp

[c] Department of Architecture, Tokyo University of Science
e-mail: kita-h@ rs. noda. tus. ac. jp

[d] NIPPON STEEL & SUMIKIN ENGINEERING CO. , LTD
(Former Graduate Student, Tokyo University of Science)

[e] Maeda Corporation
(Former Graduate Student, Tokyo University of Science)

[f] Hyogo Earthquake Engineering Research Center, NIED
e-mail: kaji@ bosai. go. jp

KEYWORDS: High-rise steel building, Long-period ground motion, Cumulative damage, Retrofit.

ABSTRACT

The design of high-rise buildings and the construction methods used have changed in relation to technology, society, and the economy. The safety of high-rise buildings designed 40 years ago, or longer, was previously examined using observed earthquake data such as El Centro; such studies were based on the maximum response. However, the main structural damage occurring in high-rise steel buildings subjected to long-period ground motions is cumulative inelastic deformation, which is concentrated on the beam-column connections. In this respect, an evaluation of cumulative fatigue damage in the beam-column connections will not determine whether a building could withstand repeated ground motions throughout a large subduction-zone earthquake, such as the predictedNankai Trough mega-earthquake. As many existing high-rise buildings were constructed in cities on the Pacific Ocean side of Japan, where considerable damage is predicted to occur with a massive Nankai Trough earthquake, there are misgivings about the safety of existing high-rise buildings in relation to cumulative damage occurring throughout long-period ground motion.

In this paper, we design 3D models that reflect the characteristics of the investigations made in relation to the existing 1970's high-rise steel buildings in Tokyo and 1980's buildings in Osaka. In addition, a time history response analysis is performed using predicted long-period ground motion based on recent research results. The seismic performance of the existing high-rise steel buildings subjected to long-period ground motions are thus evaluated based not only on the maximum response but also on the cumulative damage. Furthermore, the effect of retrofitting using steel dampers is assessed.

Fig. 1. (a) and *1* (b) illustrate two types of arrangements of steel dampers used for retrofitting in the existing model. *Fig. 1* (a) uses a method of installing steel dampers in all the layers of the building, and is known as a "consecutive layer placement" type of retrofitting. Using this method of retrofitting, as the damper axial tension is transmitted to the lower layer, the number of dampers in consecutive layers is either increased or decreased. In the "tree placement" type of method used for installing dampers, a large number of dampers are installed in the lower layer, as shown in *Fig. 1.* (b). *Fig. 2.* show the time history response analysis results of the maximum story drift angle R, and the cumulative inelastic deformation ratio of the beam$_G$ η. In the lower layer, the maximum story drift angle of the building using the tree placement type of fitting dampers decreased greatly compared to the building without dampers, and the effect of the response reduction was larger than that of the consecutive layer placement type. In addition, in the lower layer, the cumulative inelastic deformation ratio of the beam of the building with the consecutive layer placement type decreased compared to that of the building without dampers; however, the maximum cumulative inelastic deformation ratio of the beam reached close to 13. 5. In all layers of the tree placement type, the cumulative inelastic deformation ratio of the beam was greatly decreased, and no damage occurred to the main frame. *Fig. 3.* shows the absorbed energy by plasticity of the dampers, $_dW_p$; the absorbed energy by plasticity of the main frame, $_fW_p$; the absorbed energy by damping of the main frame, $_fW_h$; and the input energy, E. It can be verified that the energy absorbed by plasticity of the main frame of the building without dampers was larger than that of the building with dampers. In addition, the input energy of the tree placement type and that of the couple arrangement was almost the same. The absorbed energy by plasticity of the dampers in the tree placement type was the same as that in the consecutive layer placement type; however, the absorbed energy by plasticity of the main frame of the tree placement type was smaller than that of the consecutive layer placement type.

These results show that the vibration control performance changes in relation to the damper placement, and that the vibration control performance of the tree placement type is high compared with that of the consecutive layer placement type.

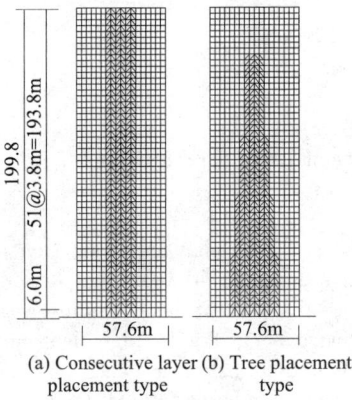

(a) Consecutive layer placement type (b) Tree placement type

Fig. 1. Existing steel high-rise

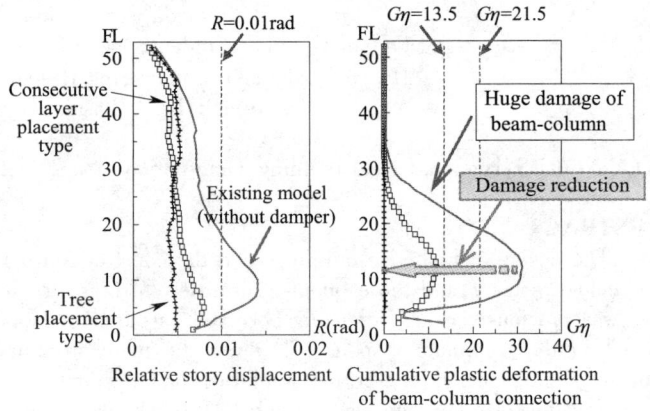

Relative story displacement Cumulative plastic deformation of beam-column connection

Fig. 2. Time history analysis

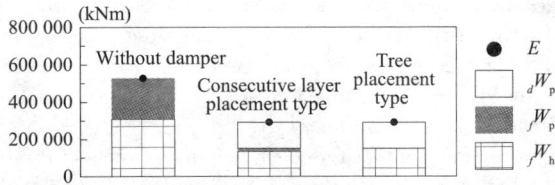

Fig. 3. Energy disrtibution of 3. 2m model

REFERENCES

[1] Sato, D. , Shimada Y. , Ouchi, H. , Nagae, T. , Kitamura, H. , Fukuyama, K. , Kajiwara, K. , Inoue, T. , Nakashima, M. , " Energy Dissipation and Distribution Ratio in a Steel High-rise Building subjected to Long-Period Ground Motions-E-Defense Shaking Table Tests of Partially Extracted Frame -", *Journal of Structural and Construction Engineering*, AIJ, Vol. 75, No. 653, pp. 1217-1226, 2010. 7

[2] Kitamura, H. , Miyauchi, Y. , Uramoto, H. , " Study on Standards for Judging Structural Performances in Seismic Performance Based Design : Evaluation of the Safety Limit Value and Margin I and II Levels in JSCA Seismic Performance Menu", *Journal of Structural and Construction Engineering*, AIJ, No. 604, pp. 183-191, 2010. 7

THE OPTIMIZATION OF STEEL BRACED FRAME STRUCTURE BASED ON HIGH STRENGTH STEEL

Guochang Li, Yuwei An and Zhijian Yang

School of Civil Engineering, Shenyang Jianzhu University. Shenyang, in China. 110168

e-mails: liguochang0604@ sina. com, anyuwei. snqk@ qq. com, faemail@ 163. com

KEYWORDS: High Strength Steel; Optimization Design; High-Rise Steel Frame Supporting Structure; Inter-Story Drift; The Plastic Hinge; Economic Performance

ABSTRACT

In this paper, the MIDAS/GEN finite element software was used to carry out strength calculation on the building structures based on the structural seismic design theory. The steel frame structure was optimized according to the stress ratio. The seismic and economical performance of the structures was analyzed.

The inter-story drift and floor shear under the action of earthquake is shown in *Fig. 1.* and *Fig. 2.* It can be seen that inter-story drift angle of the structure after optimization increased, and floor shear decreased. The largest inter-story drift angle was 1/287 in 12 story before optimization and 1/260 in 12 story after optimization, which can satisfy the requirements of specification.

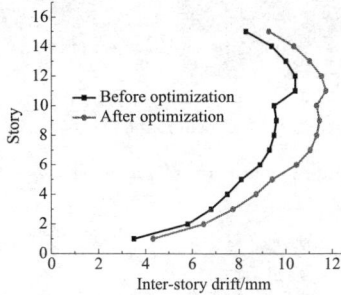

Fig. 1. Inter-story drift

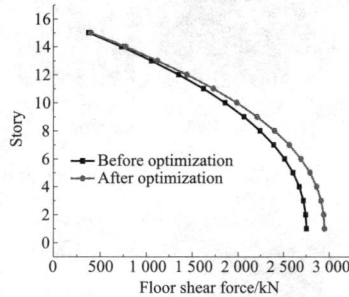

Fig. 2. Floor shear force

The floor shear and inter-story drift is shown in *Fig. 3* and *4.* The largest inter-story drift angle is 1/53 in 4 story before optimization and 1/58 in 7 story after optimization, which can satisfy the requirements of specification. Meanwhile, the sudden change of inter-story drift angle has been effectively adjusted, and it has a good seismic performance.

Fig. 5 Plastic hinges distribution at performance point of Pushover. Based on high-strength steel, structure component cross section sizes were adjusted reasonably. Local stress peak had been adjusted effectively, the overall performance of the structure was improved, and ultimate bearing capacity was higher.

Before optimization, the distribution of plastic hingeswas concentrated on the beam components 2 ~ 6 story, and on 1 ~ 8 story beam components after optimization. But the quantity was greatly reduced after optimization. Plastic hinges were at B-IO safe living stage basically.

Fig. 6 the plastic hinges distribution before and after optimization of dynamic elastic-plastic time history analysis. The plastic hinges on brace were in the damage state, which means that the brace components effectively absorbed a lot of energy of seismic load. The plastic hinge was not formed at the column foot, meeting the requirement of seismic resistance. Because of the high yield strength and uniform stress of the structure after optimization, the number of plastic hinges decreased and the seismic performance was better.

According to the whole distribution and condition of the plastic hinge, plastic hinge of the second state is on 2 ~ 10stories, plastic hinge of the third state is concentrated on 4 ~ 6 stories, and structural weak story occured before optimization. After optimization, plastic hinge distribution was uniform, and the seismic performance was better.

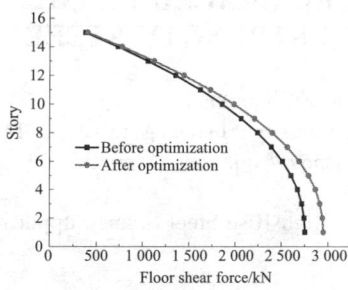

Fig. 3. Layer shear force

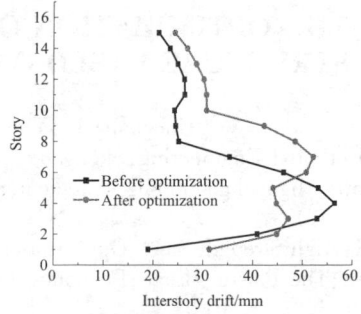

Fig. 4. Inter-story drift

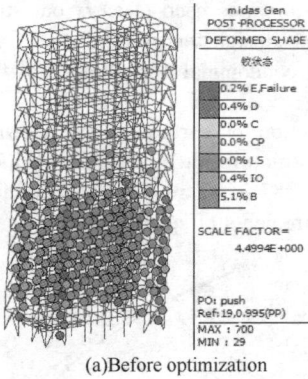

(a)Before optimization

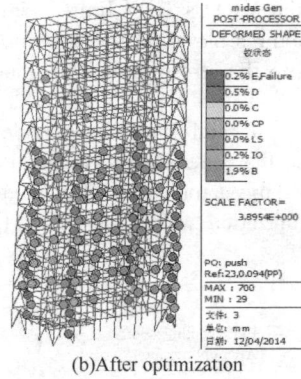

(b)After optimization

Fig. 5. Plastic hinges distribution at performance point of Pushover

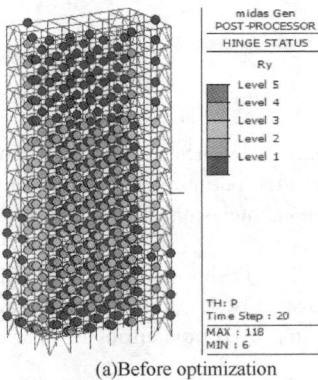

(a)Before optimization

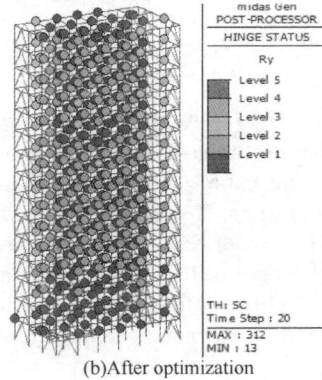

(b)After optimization

Fig. 6. Plastic hinges distribution before optimization of of dynamic elastic-plastic time history analysis.

ACKNOWLEDGEMENTS

This project was supported by the National Science-technology Support program(2012BAJ13B05).

REFERENCES

[1] Xi H. , *The research of high-rise structure housing design*. Hebei university of technology ,2006
[2] Qiu LB , Liu Y , Hou ZX , Chen SR. , "The application research of high strength structural steel in construction" ,*Industrial Building* ,03 ,1-5 + 47 ,2014.
[3] The ministry of construction of the People's Republic of China. *GB50017-2003 Code for design steel structure*. Beijing: China Plans to press ,2003.

CYCLIC LOADING TEST OF SUBSTRUCTURE FRAME WITH NEW COLUMN SUPPORT SYSTEM FOR STEEL MOMENT RESISTING STRUCTURES TO PERFORM BEAM YIELDING MECHANISM

Sachi Furukawa[a], Yoshihiro Kimura[b], Katsunori Kaneda[c], and Akira Wada[d]

[a]Tohoku University, School of Engineering, Japan furukawa. s@ m. tohoku. co. jp

[b] Tohoku University, New Industry Creation Hatchery Center, Japan
kimura@ m. tohoku. ac. jp

[c]Structural Design PLUS ONE Co. Ltd. , Japan
k-kaneda@ sp-plusone. co. jp

[d]Tokyo Institute of Technology, School of Engineeing, Japan
wada@ akira-wada. com

KEYWORDS: Moment-resisting structure, beam yielding mechanism, column base, and cyclic loading test.

ABSTRACT

Steel moment-resisting frames with a conventional column base, such as an exposed and embedded-type column bases, are likely to develop plastic hinges at the first story columns because of different support conditions at the top and bottom ends of the column. The formation of plastic hinges at the first story column may cause the soft story mechanism during large earthquakes[1]. To achieve a beam yielding mechanism for steel moment-resisting frames, the authors have previously proposed the application of a pin connection system to midpoints of the first story columns to control their moment distribution, i. e. , moment demands at the top and bottom ends[2,3]. With the proposed pin connection system, the upper steel column and bottom reinforced concrete column extending from a base beam are connected by an anchor bolt inserted along the centre axis of the column. A round plate (i. e. , a pin plate) is welded to the base plate of the upper steel column to facilitate the rotation of the connection. Analysis showed that the proposed pin connection system reliably prevents yielding of the first story column, and most of the seismic energy is dissipated by the plastic deformation of the beam ends[2,3]. The practical application of this system has been promoted; four steel moment frames were built using the proposed pin connection.

This paper discusses the performance of a steel moment-resisting frame adopting the proposed column supportsystem based on static cyclic loading tests of a substructure frame consisting of a first story column (comprising an upper steel column and a lower reinforced concrete column) and the second floor beams. Three types of pin connections were tested: (A) a sole anchor bolt embedded along the center axis of the bottom reinforced concrete column (Fig. 1. (a-2), (b-2)), (SC) an anchor bolt accompanied by a cap plate covering the top of the reinforced concrete column (Fig. 1. (a-3), (b-3)), and (SP) an anchor bolt welded with shear plates and embedded in the reinforced concrete column (Fig. 1. (a-4), (b-4)). In addition, a steel column with an embedded column base (N type) (Fig. 1. (a-1), (b-1)) was tested for comparison. In total, five specimens were tested.

CONCLUSIONS

The seismic behaviour of the substructure with the proposed connections was examined in cyclic loading tests. The major findings are as follows:

(1) The column base of the substructure with the conventional embedded column base yielded at the drift ratio δ_T/h of 0. 013 and local buckling occurred. In contrast, the proposed connections successfully prevented the column from yielding.

(2) When the type A connection was adopted, the anchor bolt at the connection deformed under flexure after the grout was crushed under the bearing pressure of the upper steel column. A larger diameter improved the performance of the connection and stable hysteresis behaviour was maintained up to the drift ratio of δ_T/h = 0. 03.

(3) When the SC(anchor bolt with a cap plate covering the top of the reinforced concrete column) and SP(anchor bolt welded by shear plates embedded in the base reinforced concrete column) connections were adopted, the anchor bolt remained elastic up to the drift ratio of $\delta_T/h = 0.05$, and the connection sufficiently performed as a pin. The yielding of the columns was successfully prevented, and the hysteresis performance of the substructure was stable. 95% of the hysteretic energy was dissipated by the beams.

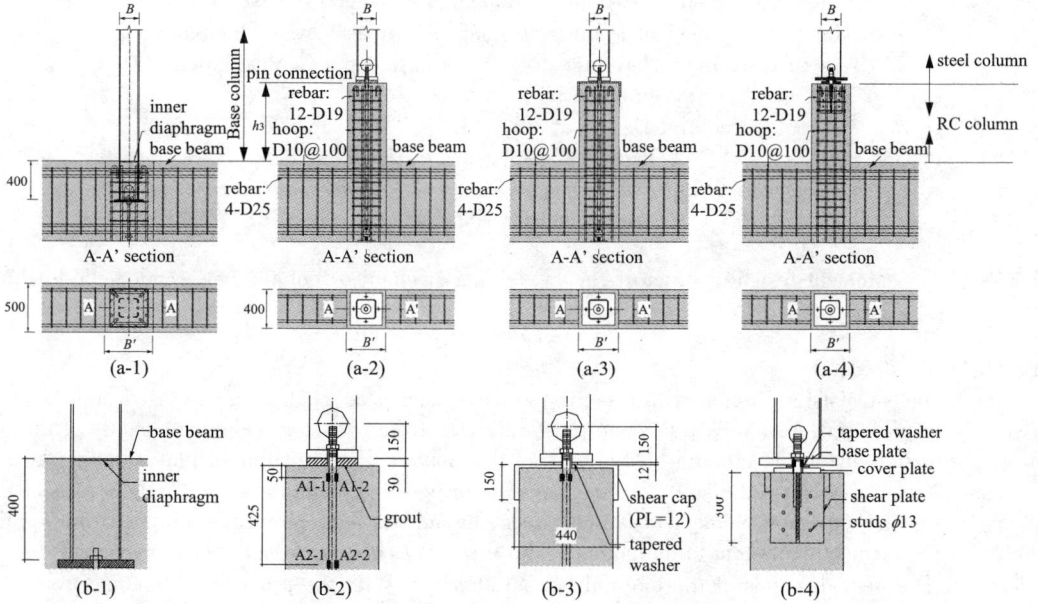

Fig. 1. (a) Specimen, (b) Details of connection

REFERENCES

[1] Keiichiro Suita, Yuichi Matsuoka, Satoshi Yamada, Yuko Shimada, Motoki Akazawa, Motohide Tada, Makoto Ohsaki and Kasuhiko Kasai, "Outline of full-scale 4-story building collapse test (E-Defense Experimental Projects for Steel Buildings -Part 21-24)", *International Journal of Engineering Science*, ID: 22417-22320, 2008.

[2] Katsunori Kaneda, Yoshihiro Kimura, Shinichi Hamasaki and Akira Wada, "Proposal of new column support system for multi-story steel moment resisting structures to perform beam yielding mechanism," *Journal of Structure and Construction Engineering*, Architectural Institute of Japan, Vol. 75, No, 654, pp. 1537-1546, Aug., 2010.

[3] Katsunori Kaneda, Yoshihiro Kimura, Naoki Miyahara and Akira Wada, "Ultimate Seismic capacity of multi-story steel moment resisting frames having new column support system," *Journal of Structure and Construction Engineering*, Architectural Institute of Japan, Vol. 76, No, 661, pp. 649-658, Mar., 2011.

EVALUATION OF LOW-AND MEDIUM-RISE BUILDINGS ENHANCED SEISMIC PERFOMANCE BY HIGH-STRENGTH STEEL AND HYSTERETIC DAMPERS

Yasunari Watanabe[a], Toshiaki Sato[a], Haruyuki Kitamura[a],
Kazuaki Miyagawa[b], Takuya Ueki[c]

[a] Department of Architecture, Tokyo University of Science, Japan
7114667@ ed. tus. ac. jp, sato-t@ rs. noda. tus. ac. jp, kita-h@ rs. noda. tus. ac. jp
[b] JFE Civil Engineering & Construction Corporation, Japan
miyagawa@ jfe-civil. com
[c] JFE Steel Corporation, Japan
t-ueki@ jfe-steel. co. jp

KEYWORDS: High-strength steels, Hysteretic dampers, Low-and medium-rise buildings, Time history response analysis, Static incremental analysis

ABSTRACT

Owing to our experience suffering from previous megathrust earthquakes, we require buildings to withstand these events. Based on this background, it is important to develop low-and medium-rise buildings with high seismic performance. This paper proposes a design method in which beams and columns are determined by long-term stress and seismic loads are sustained by hysteretic dampers. Two proposed methods involve the use of high-strength steels and of larger members. The purpose is to evaluate response behavior and damage of buildings applying the two proposed methods using static incremental analysis and time history response analysis.

This study uses buildings which are parts of distribution warehouses having heavy live loads. A four-story, steel construction is used as the building for analysis. Member of the basic model is decided by long-term stress. This model is called "SD-1.0". The high-seismic-performance model is made by using two methods. One is the method of applying high-strength steels to columns. Another is the method of applying larger members to columns for the same yield stress intensity as high-strength steels without changing the type of steels used. The beams are the same as those in SD-1.0. The high-seismic-performance model made by using high-strength steels is called "HD-1.0". The high-strength steel is H-SA700[1]. The yield stress intensity of H-SA700 is 700 N/mm[2]. Because the yield stress intensity of BCP325 is 325 N/mm[2], the yield stress intensity of H-SA700 is approximately 2.2 times larger than that of BCP325. Therefore, columns of the high-seismic-performance model are made by using larger members applied to columns where the section modulus is approximately 2.5 times larger than in SD-1.0. This model is called "SD-2.5".

The input seismic motion for time history response analysis uses the earthquake motion, which normalizes on the pseudo-velocity response spectrum at 1.6 m/s over the corner period 0.64 s. The phase characteristics use the JMA KOBE NS direction in 1995. *Fig. 1* shows the response obtained.

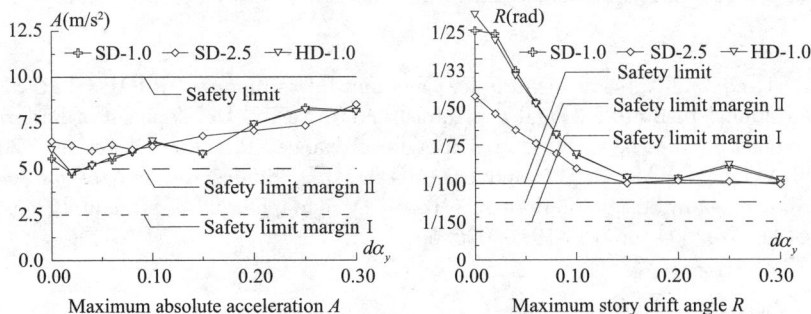

Fig. 1. Evaluation of response

The assessment of seismic performance uses the JSCA seismic performance menu (JSCA menu)[2]. Members of SD-1.0 and HD-1.0 are the same. Therefore, the response of both models does not appear to make a significant difference. Although the maximum absolute acceleration A tends to increase with the yield shear force coefficient of dampers in the first layer $_d\alpha_y$, 10.0 m/s^2 as a safety limit is not exceeded. While, the maximum story drift angle R is difficult to decrease for $_d\alpha_y$ is greater than 0.15. R is more than 1/75 rad over the safety limit for every $_d\alpha_y$. As $_d\alpha_y$ increases, the difference of R for SD-2.5 and HD-1.0 is small.

Fig. 2 shows damage of beams and columns in the case that $_d\alpha_y$ is 0.20.

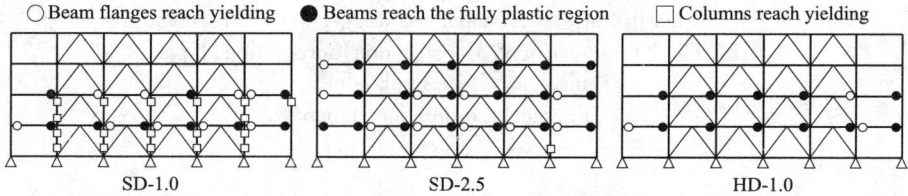

Fig. 2. The damage of beams and columns

The damage of beams of SD-2.5 is the most severe. In addition, one of columns sustained by the axial force from dampers in SD-2.5 reaches yielding. While columns in SD-1.0 reach the yielding condition, columns in HD-1.0 can remain in the elastic region. Because columns in HD-1.0 remain in the elastic region, beams in HD-1.0 are sustained by the larger shear force than those in SD-1.0. Therefore, the number of the plastic hinges in HD-1.0 is larger than that of SD-1.0.

CONCLUSIONS

This paper proposes a design method in which beams and columns are designed according to long-term stress and seismic loads are supported by hysteretic dampers. The building types investigated are low-and medium-rise buildings. The input seismic motion was the earthquake motion, which normalizes on the pseudo-velocity response spectrum at 1.6 m/s. Two methods were proposed: using high-strength steels or larger members for columns. The purpose was to evaluate response behavior and damages conditions of buildings applying these two proposed methods by time history response analysis. The following conclusions were found. In the case that the yield shear force coefficient of dampers in the first layer greatly increases, although the maximum story drift angle is difficult to decrease, the maximum absolute acceleration increases. In the case of inputting the earthquake motion, which normalizes on the pseudo-velocity response spectrum at 1.6 m/s, columns of the model of using high-strength steels remain in the elastic region. In addition, damage of beams is smaller than in the method of using larger members.

ACKNOWLEDGMENT

This study was part of collaborative research with JFE Steel Corp. ,JFE Civil Engineering & Construction Corp and the Kitamura laboratory in the Tokyo University of Science. The study was assisted by Chief Engineer Masato Ishii of Nikken Sekkei Ltd. through providing an analysis program and valuable opinions. The writers gratefully acknowledge their assists.

REFERENCES

[1] Ministry of Land, and National Institute for Land and Infrastructure, "PROJECT RESEARCH REPORT of National Institute for Land and Infrastructure No. 32, Development of Performance Evaluation Methods for Innovative Structures applying Advanced Structural Materials", 2010. 2

[2] KITAMURA, H. , MIYAUCHI, Y. , and URAMOTO, H. , "Study on Standards for Judging Structural Performances in Seismic Performance Based Design", Journal of Structural Construction Engineering, AIJ, No. 604, pp. 183-191, 2006. 6

PRELIMINARY ANALYSIS INTO THE SEISMIC BEHAVIOR OF HIGH STRENGTH STEEL FRAMES

Fang-Xin Hu[a], Gang Shi[a], Yong-Jiu Shi[a]

[a]Key Laboratory of Civil Engineering Safety and Durability of China Education Ministry,
Department of Civil Engineering, Tsinghua University, Beijing 100084, P. R. China
hfx11@ mails. tsinghua. edu. cn, shigang@ tsinghua. edu. cn, shiyj@ tsinghua. edu. cn

KEYWORDS: High strength steel, Seismic behavior, Steel frame.

ABSTRACT

Due to the obvious architectural, structural and economical benefits, high strength steel with nominal yield strength $f_y \geqslant 460$ MPa has been widely used in engineering structures. Static behavior of high strength steel structures has been extensively investigated at present, but there is still relatively little research work on their seismic performance. This paper presents design of several prototype frames, and numerical simulations of high strength steel frames extracted from those frames subject to monotonic and cyclic load by using validated finite-element models. The monotonic and cyclic base shear versus overall drift angle relationship is interpreted in detail. Reduction factors for seismic force are evaluated using both monotonic pushover curve and cyclic pushover envelope, and they are compared with the design values to assess current seismic design procedure and evidence the effectiveness of using high strength steels.

CONCLUSIONS

This paper presents monotonic and cyclic pushover analysis on the seismic behavior of high strength steel frames using validated finite-element models. The analyzed frames have one bay (span length is 6m) and two stories (story height is 2. 7m). A list of member information is shown in *Table 1*.

Table 1. Analyzed frames

Frame	Beam			Column			Total weight(t)
	Steel	Section		Steel	Section		
B345-C345	Q345	H-340 × 170 × 8 × 10		Q345	H-300 × 250 × 10 × 10		1. 215
B345-C460	Q345	H-340 × 170 × 8 × 10		Q460	H-240 × 220 × 10 × 10		1. 116
B345-C550	Q345	H-340 × 110 × 8 × 16		Q550	H-170 × 170 × 12 × 12		1. 058
B460-C460	Q460	H-340 × 110 × 10 × 10		Q460	H-240 × 220 × 10 × 10		1. 065

The monotonic pushover and cyclic pushover analyses have been conducted, and then reduction factor for seismic force can be calibrated using those analysis results and compared with the design value in steel structures design code, as shown in *Table 2*.

Table 2. Summary of analysis results

Frame	Reduction factor q		q_{design}
	monotonic	cyclic	
B345-C345	5. 3	3. 2	5. 7
B345-C460	4. 6	3. 4	5. 7
B345-C550	4. 9	4. 6	5. 7
B460-C460	4. 3	3. 3	4. 4

The primary findings of this preliminary analysis are the following:
1. cyclic loading tended to generate severer strength degradation than monotonic loading at large drift levels, especially for B345-C345, B345-C460 and B460-C460 whose column sections were made of relatively slender plates. The local buckling in column bases in those three frames made them more vulnerable to accumulated plastic deformation resulted from cyclic loading;

2. for all the frames analysed the design reduction factor was larger than the calculated value, and generally the reduction factor evaluated by cyclic pushover envelope was smaller than that using monotonic pushover curve. This result illustrates that cyclic loading has a noticeable negative effect on the ductility exhibited by steel frames, and those reduction factors adopted in current steel structures design code may not be safe enough;

3. B345-C550 with compact column cross section may have the best seismic performance in view of reduction factor calibrated from cyclic pushover analysis. Although steels with higher strength are considered with lower ductility in material properties, high strength steel frames should not be necessarily poor in ductility since the compactness of member cross section can impact the inelastic structural-behavior significantly. It implies that those provisions in Chinese seismic design code to prohibit the application of high strength steels with yield-to-tensile strength ratio larger than 0. 85 seem to be not so rational.

4. This preliminary study shows an opportunity to realize the economical profit by reducing structural weight using high strength steels while the strength andductility can still be balanced to achieve a good seismic performance.

ACKNOWLEDGMENT

The authors acknowledge the support for this work, which was sponsored by the National Natural Science Foundation of China(No. 51478244).

REFERENCES

[1] Shi G, Hu FX, Shi YJ, "Recent research advances of high strength steel structures and codification of design specification in China", International Journal of Steel Structures, 14(4):873-887, 2014.

[2] Shi G, Hu FX, Shi YJ, "Recent research advances of high strength steel structures and codification of design specification", Proceedings of 10th Pacific Structural Steel Conference, Singapore (8-11/10), 606-611, 2013.

[3] Matsui C, Mitani I, "Inelastic behavior of high strength steel frames subjected to constant vertical and alternating horizontal loads", Proceedings of 6th World Conference on Earthquake Engineering, New Delhi, India, 3:3169-3174, 1977.

[4] Dubina D, Stratan A, Dinu F, "Dual high-strength steel eccentrically braced frames with removable links", Earthquake Engineering & Structural Dynamics, 37(15):1703-1720, 2008.

[5] GB 50017-201X, Code for design of steel structures, Beijing, China, 2015. (draft for approval, in Chinese)

[6] ABAQUS, Analysis user's manual(Vol. I-V), Version 6. 10, Providence, RI: Dassault Systèmes Simulia Corp. , 2010.

[7] Ban HY, Shi G, Shi YJ, Wang YQ, "Research progress on the mechanical property of high strength structural steels", Advanced Materials Research, 250-253:640-648, 2011.

[8] ANSI/AISC 341-10, Seismic Provisions for Structural Steel Buildings, Chicago, IL: AISC, 2005.

[9] Nassar AA, Krawinkler H, Seismic demands for SDOF and MDOF systems, Report No. 95, Stanford, California: John A. Blume Earthquake Engineering Center, Department of Civil Engineering, Stanford University, 1991.

[10] GB 50011-2010, Code for seismic design of buildings, Beijing, China: Architecture & Building Press, 2010. (in Chinese)

EFFECTS OF SLAB-BEAM INTERACTION ON THE SEISMIC BEHAVIOUR OF DUAL ECCENTRICALLY BRACED STEEL FRAMES

Horatiu-Alin Mociran[a], Stefan Marius Buru[a]

[a]Technical University of Cluj-Napoca, Faculty of Civil Engineering, Romania
horatiu. mociran@ mecon. utcluj. ro, marius. buru@ mecon. utcluj. ro

KEYWORDS: Dual eccentrically braced structures, Steel structures, Seismic design.

ABSTRACT

The effects of slab-beam interaction on the seismic response of dual frame structures constituted by moment resisting frames(MRFs) and eccentrically braced frames(EBFs) with short links are examined.

A six storey residential building was considered in this research. The building has a square plan of 22m x 22m. The story height equals 3. 5m at all levels. The structure is formed by four frames along X direction and four frames along Y direction. The perimeter frames are eccentrically braced in the middle bays, being dual, while the internal frames are MRFs. The building is located in Bucharest, a high seismic zone for Romania, and was designed in accordance with Romanian seismic design code (P100-1/2013) and EN 1993-1-1, considering the dissipative behaviour[1]. Steel grade S355 was used for beams, links and braces, while S460 was used for columns. The building was designed for all code requirements regarding the Ultimate Limit State (ULS) and Serviceability Limit State (SLS). The beams were considered in two configurations: bare steel and composite. The slab is 150mm thick and made from C20/25 concrete. Shear stud connectors were designed according to EN 1994-1 to ensure a full interaction between the steel beam and the concrete slab.

The 3D building was subjected to a set of seven semi-artificial accelerograms of Vrancea 1977 type, at two intensity level(SLS and ULS), through nonlinear time history analyses.

The average peak response of the 3D structure, with or without the concrete slab is analysed in terms of interstorey drifts and plastic rotations of links.

In *Fig. 1* are shown the mean peak interstoreydrifts/height at SLS for the perimeter dual eccentrically steel braced frames with and without concrete slabs, disposed along X and Y directions.

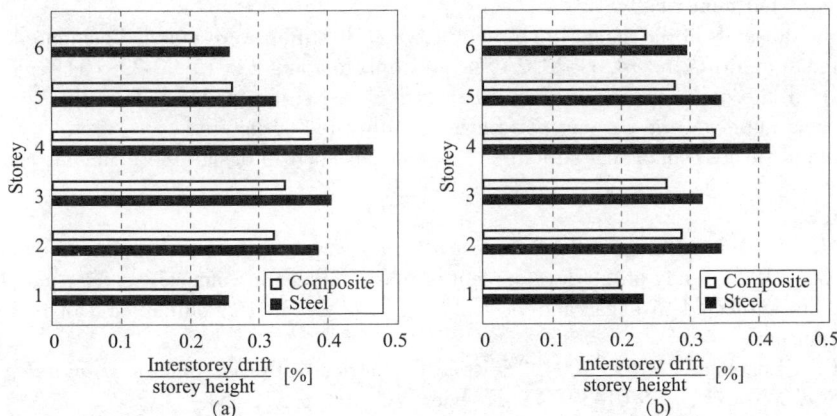

Fig. 1. Interstorey drifts/height for dual eccentrically braced frame with steel or composite beams at SLS: (a) X direction; (b) Y direction

All interstorey drifts/storey height are lower than 0. 5%, the limit imposed at SLS by P100-1/2013 for an earthquake with a reference return period of 40 years.

By considering the composite beam behaviour, the maximum interstorey drift along X direction was reduced by 19. 22% and along Y direction by 18. 89%.

In the considered structure, as intended, yielding occurred only in links of EBFs, because the beams sections of MRFs were governed by the non-seismic design situation and high strength steel grade was used for the columns.

Fig. 2 illustrates the distribution of average peak link rotations over the building height at ULS.

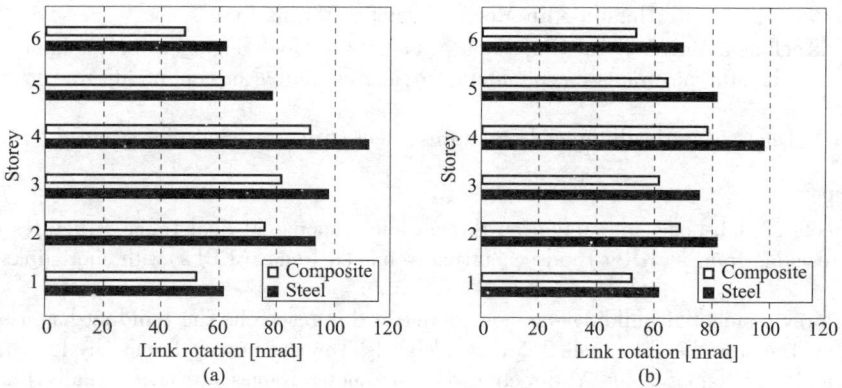

Fig. 2. Link rotations for dual eccentrically braced frame with steel or composite beams at ULS:
(a) X direction; (b) Y direction

Greater rotations took place at mid-stories and smaller were produced at first and top floor. With one exception(112mrad at 4th floor of the steel structure along X direction), links rotations were within FEMA 356 limitation(110mrad). However, experimental tests demonstrated that shear links can reach 110mrad rotations without stiffness and strength deterioration[2,3]. Due to the composite behaviour of links, the maximum link rotation along X direction was reduced by 18. 18% and along Y direction by 20%.

CONCLUSIONS
The main findings from this research are summarized as follows:
- The global seismic performance(interstorey drifts, yielding mechanism) of the steel structure is adequate. The new seismic design provisions of P100-1/2013 regarding the dual eccentrically braced frame were validated numerically.
- Seismic demands of the composite structure, which is stiffer, were reduced compared to the steel structure: interstorey drifts by(14. 41-20. 23)% and link rotations by(15. 00-23. 10)%. The yielding of links occurred later in composite structure than in the steel structure, but follows the same path. The composite beams improved the seismic performance of the steel structure.
- To obtain a more economical structure, it is recommended to design the beams as composite.

REFERENCES
[1] P100-1/2013 ,2013. *Seismic design code. General rules and rules for buildings(In Romanian)*.
[2] Danku G. ,2011. "Study of the development of plastic hinges in composite steel-concrete structural members subjected to shear and/or bending". *Ph, D. Thesis*, Politehnica University of Timisoara, Timisoara.
[3] Popov EP. ,Engelhardt M. D. ,2004. "Seismic Eccentrically Braced Frames". *Journal of Constructional Steel Research*, ISSN 0143-974X, Elsevier, Vol. 10, pp. 321-354.

PRELIMINARY ANALYSIS AND DESIGN OF AN EXPERIMENTAL FACILITY FOR THE PSEUDODYNAMIC EARTHQUAKE TEST OF A REAL SCALE STEEL MOMENT RESISTING FRAME WITH PARTIAL STRENGTH JOINT

Antonella B. Francavilla[a], Massimo Latour[a], Vincenzo Piluso[a] and Gianvittorio Rizzano[a]

[a]University of Salerno, Department of civil Engineering, Italy

e-mails: afrancavilla@ unisa. it, mlatour@ unisa. it, v. piluso@ unisa. it, g. rizzano@ unisa. it

KEYWORDS: Steel joints, Moment Resisting Frames, Pseudodynamic test, Ductility.

ABSTRACT

The reliable prediction of nonlinear structural behaviour during severe seismic events has proven to be an extremely difficult task. Although many nonlinear analysis programs exist, the accuracy of their results depends on the assumptions made in the characterization of member stiffness. Many of these analytical models are calibrated using experimental observations[1]. The PsD method is a hybrid testing method. It combines on-line computer simulation of the dynamic aspects of the problem with experimental information about the structure, acquired quasi-statically, to provide realistic dynamic response histories even for the non-linear behaviour of severely damaged structures. In this work, the preliminary analysis and the design of a testing facility that is inserted into a research program previously initiated at Laboratory of Materials and Structures of the Department of Civil Engineering of Salerno concerning the testing of steel connections of both traditional and innovative types, are illustrated. Therefore, both nonlinear static and dynamic analyses with a computer software SeismoStruct have been performed.

In order to test a real scale two storeys-two bays steel MRFs, the design of the geometrical and mechanical properties of the steel frames results strongly conditioned by the real size of the Laboratory and by the equipment available. The beam-to-column connections have been designed[2-4], as required by EC 3, according to the component method by appropriately identifying the element of the joint where the dissipation happens and conferring on that element a lower plastic threshold. In particular, four types of connections are designed and tested: EEP-CYC02 (a partial strength extended endplate joint), EEP-DB-CYC03 (a full strength extended endplate joint), TS-CYC04 (a partial strength joint with a couple of T-stubs) and TSJ-XS-CYC07 (a partial strength joint with a pair of dissipative T-stubs, with an hourglass shape). It can be observed that the joint EEP-DB-CYC03 is characterized by the most wide envelope up to joint rotation equal to about 0. 04 rad but the other typologies of joint, such as in particular TSJ-XS-CYC07 with dissipative T-stubs, are able to dissipate same or greater amount of energy due to their greater ductility.

In order to evaluate the seismic behaviour of steel frames equipped with the typologies of joints previously described, static and dynamic inelastic analyses have been developed. For both analyses a software SeismoStruct[5] has been adopted. It can be observed that even though the steel frame with dog-bone provides a dissipation of the energy greater than that of the others joint typologies, the scatters between the performances provided by the different joint typologies are moderate evidencing a good behaviour also of the steel frames with partial strength joints.

Incremental dynamic nonlinear analyses have been performed on the four steel frames with the different typologies of joints. In *Fig. 1*, the comparison between the curves base shear-displacement provided by the IDA for the four frames are given while in *Fig. 2* the comparison is shown in terms of energy dissipated by the dissipative zones of the four structural solution. Also in this case, it can be observed that, even though the best behaviour is provided by the dissipation in the beam by means of a dog-bone solution, the partial strength joints are able to provide good structural solutions provided that they are adequately designed. Specially, the solution with dissipative double split tee joints appears interesting. In fact, this solution provides a good dissipation of the seismic input energy due to its higher ductility.

Finally, in order to analyse the re-centring capability of the structures, the average values of the residual displacements of the frames provided by the IDA for the 10 accelerograms have been evalua-

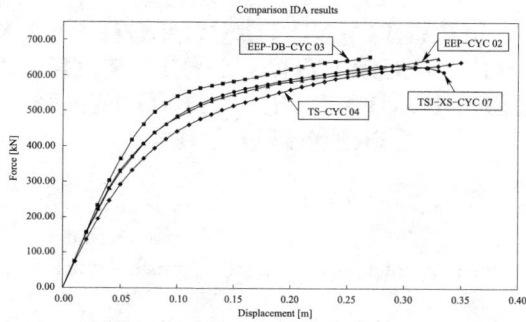

Fig. 1. Comparison between the IDA curves in terms of Force-displacement

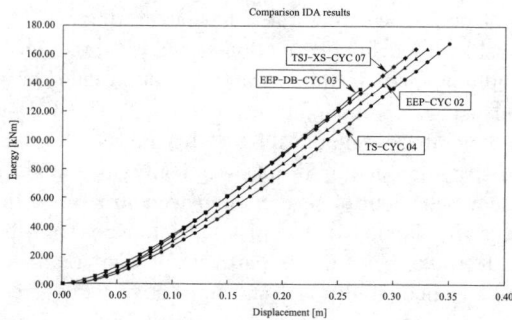

Fig. 2. Comparison between the IDA curves in terms of dissipated energy up to failure

ted. The frame with dissipative double split tee joints is characterized by the best performance with the weakest residual deformations.

CONCLUSIONS

The present work belongs to a more wide theoretical and experimental program aimed to verify, by means of pseudo-dynamic tests on 3D two bays-two storeys steel frames equipped with different joint typologies, the actual possibility to design seismic resistant steel frames with partial strength joints. On the base of the performed analyses, it has been recognised that steel frames with partial strength joint can provide a good performance under severe seismic action provided that the joints are adequately designed. The IDA evidenced that the frame with X-shaped Double Split Tee joints is able to dissipate a great amount of energy and it is characterized by a significant capacity to re-center the structure after a severe earthquake.

REFERENCES

[1] Shing, P. B. and Mahin, S. A. (1984). "Pseudodynamic test method for seismic performance evaluation: Theory and implementation." Earthquake Engineering Research Center UCB/EERC-8'1/01, University of California, Berkeley.

[2] Iannone, F. , Latour, M. , Piluso, V. and Rizzano, G. (2011). "Experimental Analysis of Bolted Steel Beam-to-Column Connections: Component Identification", Journal of Earthquake Engineering, 15:2, 214-244.

[3] Latour, M. , Piluso, V. and Rizzano, G. (2011). "Cyclic Modeling of Bolted Beam-to-Column Connections: Component Approach", Journal of Earthquake Engineering, 15:4, 534-563.

[4] Latour, M. , Piluso, V. and Rizzano, G. (2011). "Experimental Analysis of innovative dissipative bolted double split tee beam-to-column connections", Steel Constructions, 4, No. 2, 53-64.

[5] Seismosoft (2014). "SeismoStruct v7. 0-A computer program for static and dynamic analysis of framed structures"

PROGRESSIVE COLLAPSE OF SEISMIC DESIGNED STEEL MOMENT FRAMES: NONLINEAR STATIC AND DYNAMIC ANALYSIS

Massimiliano Ferraioli[a], Alberto Mandara[a]

[a] Department of Civil Engineering, Design, Building and Environment,
Second University of Naples, via Roma 9, 81031, Aversa (CE), Italy
massimiliano. ferraioli@ unina2. it, alberto. mandara@ unina2. it

KEYWORDS: Nonlinear pushdown analysis, Progressive collapse, Steel moment resisting frames.

ABSTRACT

Changes in design and construction practices over the past several decades have lessened inherent robustness in certain modern structural systems, reducing their reserve capacity to accommodate abnormal loading conditions that have very low frequency of occurrence, but extraordinary consequences resulting from sudden changes to the building's geometry and load-path. The response to this abnormal loading condition is most likely to be dynamic and nonlinear, both geometrically and in the material behaviour. The nonlinear dynamic analysis requires step-by-step integration which is very time consuming. Consequently, simple and reliable analytical tools have been developed in literature. However, the load cases for these static procedures generally require the use of factors to account for nonlinear and inertial effects, and inconsistencies have been found in the way the existing guidelines applied the dynamic amplification factors to their static approaches[1-3].

The study investigates the progressive collapse resisting capacity of earthquake-resistant steel moment-resisting frames subjected to column failure. The aim is to investigate whether these structures are able to resist progressive collapse after column removal, that may represent a situation where an extreme event may cause a critical column to suddenly lose its load bearing capacity (*Fig. 1*). Since the response to this abnormal loading condition is most likely to be dynamic and nonlinear, both nonlinear static and nonlinear dynamic analyses are carried out. The vertical pushover analysis (also called pushdown) is applied with two different procedures. The first one is the traditional procedure generally accepted in current guidelines that increases the load incrementally to a specified level after column has been removed. The second procedure tries to reproduce the timing of progressive collapse and, for this reason, gravity loads are first applied to the undamaged structure, after which the column is removed. The load-displacement relationships obtained from pushdown analyses are compared with the results of incremental nonlinear dynamic analyses. The effect of design variables such as number of stories, number of bays, level of seismic design load, is investigated. The results are eventually used to evaluate the dynamic amplification factor to be applied in pushdown analysis for a more accurate estimation of the collapse resistance.

The application of incremental dynamic analysis confirms that, assuming a DAF equal to 2, the nonlinear static analysis approach leads to anoverconservative estimation of progressive collapse resistance for a column-removed building. In this context, the appropriate value of the dynamic amplification factor (DAF) to be applied in pushdown analysis for a more accurate estimation of the collapse resistance has been evaluated. To this aim, the root-mean-square error (RMSE) has been used as a measure of the differences between the values predicted by pushdown static analysis and the values obtained by incremental dynamic analysis. In particular, the RMSE has been calculated using the values of the load factors from IDA and pushdown analysis corresponding to the same vertical displacement. In *Fig. 2*a the RMSE is plotted as a function of the dynamic amplification factor (DAF) for the frames designed with: a) PGASLV = 0. 15g; b) PGASLV = 0. 25 g; c) PGASLV = 0. 35g. The results show that the minimum error occurs when DAF = 1. 3 in 3-bay frames, while in 5-bay frames the minimum occurs for values of DAF lower than 1. 3. In *Fig.* 2b the averaged RMSE for all the structures examined is plotted, showing that a value in the range of 1. 25 ÷ 1. 3 seems to better represent the dynamic effect for all the examined steel moment resisting frames.

CONCLUSIONS

This paper has investigated the response to progressive collapse of steel moment resisting frames designed according to the Italian Seismic Code. The main purpose of the study has been to verify whether the current provisions adopted for seismic design enable a satisfying performance in case one of the supporting column is suddenly removed from the structure, that is the case in which the progressive collapse potential is involved. Both static nonlinear and incremental dynamic approaches have been followed in order to estimate the load bearing capacity of the structure. The study has led to conclude that, based on acceptance criteria given in current guidelines, the structures dealt with show a great potential of progressive collapse. At the same time, a critical analysis of observed collapse mechanisms suggests that different acceptance criteria should be considered, other than the simple rotational capacity of structural members.

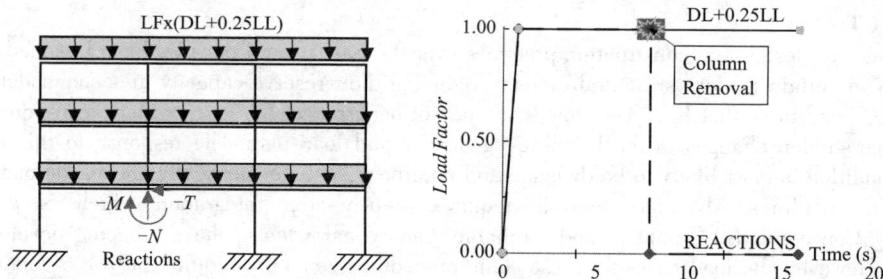

Fig. 1. Applied load for dynamic analysis and time-history function of applied load for dynamic analysis.

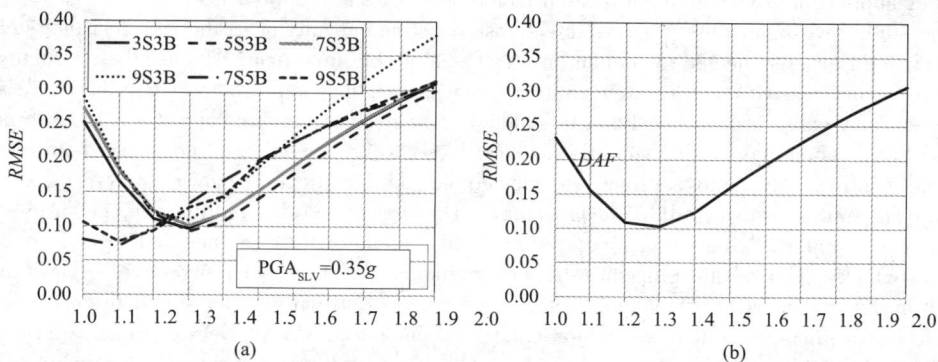

Fig. 2. Root-mean-square error(RMSE) of pushdown analysis as a function of dynamic increase factor(DAF): a) Design PGA = 0.35g; b) Average value of RMSE for all structures

REFERENCES

[1] Ferraioli M., Avossa A. M., Mandara A., "Assessment of Progressive Collapse Capacity of Earthquake-Resistant Steel Moment Frames Using Pushdown Analysis", *The Open Construction and Building Technology Journal*, vol. 8, pp. 324-336, 2014.

[2] Ferraioli M., Avossa A. M., "Progressive collapse of seismic resistant multistory frame buildings", Life-Cycle and Sustainability of Civil Infrastructure Systems, Proceedings of the 3rd International Symposium on Life-Cycle Civil Engineering, IALCCE 2012, pp. 2048-2055, 2012.

[3] Ferraioli M., Lavino A., Mandara A., "Behaviour factor of code-designed steel moment-resisting frames", *International Journal of Steel Structures*, vol. 14(2), pp. 243-254, 2014.

ASSESSMENT OF ADAPTIVE PUSHOVER PROCEDURES FOR EARTHQUAKE-RESISTANT STEEL MOMENT FRAMES

M. Ferraioli[a], A. M. Avossa[a], A. Lavino[a] and A. Mandara[a]

[a] Department of Civil Engineering, Design, Building and Environment,
Second University of Naples, via Roma 9, 81031, Aversa (CE), Italy
massimiliano. ferraioli@ unina2. it, albertomaria. avossa@ unina2. it, a. lavino@ hotmail. it,
alberto. mandara@ unina2. it

KEYWORDS: Adaptive capacity spectrum method, Modal pushover, Nonlinear analysis, Steel moment resisting frames.

ABSTRACT

The more rational approach for seismic evaluation of buildings is based on inelastic displacement rather than elastic forces because structural damage is directly related to local deformations. As a consequence, there has been a growing interest in displacement-based seismic assessment and several articles on the subject can be found in literature[1-3]. In this approach the displacement or interstorey drift is considered as the basic demand parameter in the design, evaluation and rehabilitation of structures. While nonlinear response history analysis (RHA) is the most rigorous procedure to estimate seismic demands, static pushover analysis is extensively employed to determine the deformation demands with acceptable accuracy without the complex modelling and computational effort of RHA. The two major weaknesses of the conventional pushover methods are: 1) the higher mode effects are neglected; 2) the changes in the dynamic properties of the structures and, consequently, in the loading pattern are ignored. This is mainly due to the fact that inertia force distribution changes continuously under earthquake ground motion due to higher mode contribution and stiffness degradation. This paper investigates the reliability of advanced nonlinear static procedures (NSPs) in predicting the response characteristics of typical earthquake-resistant steel frames under seismic loads. In particular, the traditional pushover analysis methods such as N2 and FEMA-440 are compared to Adaptive Capacity Spectrum Method (ACSM) and Modified Modal Pushover Analysis (MMPA). The first procedure considers the effects of the time-varying earthquake-induced inertial forces by means of a continuous updating of the lateral load distribution during the analysis according to modal properties, softening of the structure, its period elongation, and the modification of the inertial forces due to spectral amplification. The second procedure was developed to include higher modes effects with acceptable accuracy especially in prediction of plastic hinge rotations. The reliability of these NSPs and their ability to overcome the limitations of traditional pushover analysis are evaluated through comparison with benchmark responses obtained from a comprehensive set of nonlinear response history analyses. The effect of several variables, such as number of stories, number of bays, level of earthquake load, and irregularities in elevation is finally investigated.

In order to evaluate the accuracy of the mentioned approaches in predicting seismic demands, the results of considered pushover procedures have been compared with nonlinear dynamic analysis assumed as a reference solution. In *Fig. 1* a typical scatter plot comparing total drift ratio from NSP and RHA is shown. *Fig. 2* shows the total error for regular frames as a function of peak ground acceleration. The results show that higher errors generally occur for PGA = 0. 20g and for PGA = 0. 8g. In the first case (PGA = 0. 20g), as previously stated, all frames respond elastically and the higher error of NSPs compared with nonlinear dynamic analysis derives from the analysis method (Linear Static Analysis for N2-EXT, ACSM1, ACSM2; Linear Modal Analysis for MMPA and Linear Time-history analysis for RHA). In the second case (PGA = 0. 80g), the use of NSPs based on invariant load pattern (N2-EXT, MMPA) generally implies more significant errors than in the case of adaptive pushover procedures, especially for the higher frames. As said before, this result comes from the overestimation of interstorey drift in the lower storeys obtained applying the NSPs based on invariant load patterns. The

N2-EXT method produces higher error-rate when compared with the other NSPs, and this error tends to increase with peak ground acceleration and irregularity in the structure.

CONCLUSIONS

Some advanced pushover procedures taking into account the frequency content of response spectra, higher mode effects, progressive changes in the modal properties due to structural yielding and interaction between modes in the inelastic range have been compared. Pushover methods accounting for higher mode effects along the elevation provide more accurate estimation of seismic demands when compared with traditional pushover methods based on load pattern using first mode. Although higher modes have been found to be significantly participating in the global dynamic behaviour of the structure, the estimation of their effect could be inadequate, especially at the lower storeys, if an invariant lateral load pattern is assumed in the analysis. The resulting interstorey drift profiles show that, in general, the accuracy of nonlinear static procedure based on invariant load patterns at the lower storey levels is worse, especially for higher or irregular frames. On the contrary, at the upper storey levels the accuracy is comparable with that obtained from adaptive procedures. This means that the effects of higher modes are well interpreted by correction factors pro-posed in the extended N2 method and by the elastic response spectrum superposition suggested in the modified version of the Modal Pushover Analysis.

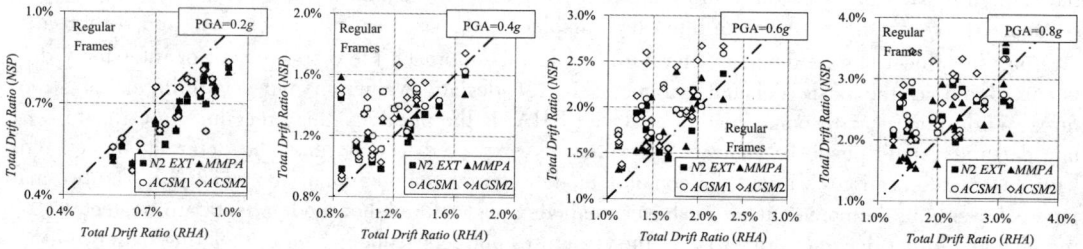

Fig. 1. Typical scatter plot comparing total drift ratio from NSP and RHA

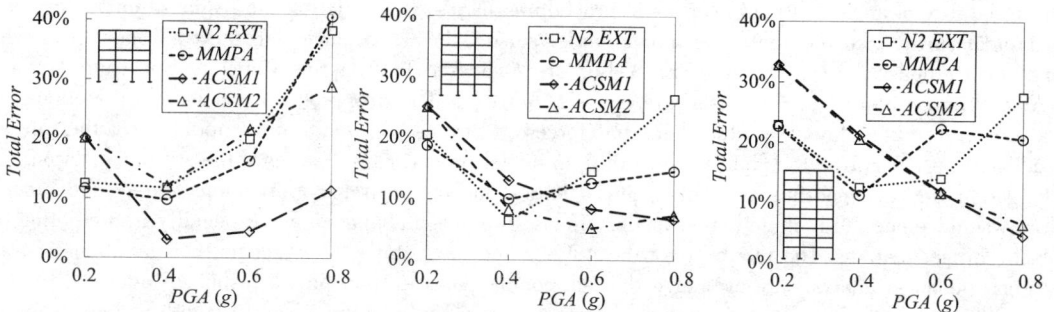

Fig. 2. Total error of the mean interstorey drift of the NSPs with respect to the mean interstorey drift of the RHA.

REFERENCES

[1] Ferraioli M. , Avossa A. M. , Lavino A, Mandara A. , "Accuracy of Advanced Methods for Nonlinear Static Analysis of Steel Moment-Resisting Frames", *The Open Construction and Building Technology Journal*, vol. 8, pp. 310-323, 2014.

[2] Ferraioli M. , Lavino A. , Avossa A. M. , Mandara A. , "Displacement-based seismic assessment of steel moment resisting frame structures" In Proceedings of the 14th World Conference on Earthquake Engineering, Beijing, China, October 12-17, 2008.

[3] Ferraioli M. , Lavino A. , Mandara A. , "Behaviour factor of code-designed steel moment-resisting frames", *International Journal of Steel Structures*, vol. 14(2), pp. 1-12, 2014.

INFLUENCE OF SEISMIC DETAILING ON THE PROGRESSIVE COLLAPSE OF STEEL MOMENT FRAMES

David Cassiano[a], Mario D'Aniello[b], Carlos Rebelo[a], Raffaele Landolfo[b],
Luís Simões da Silva[a]

[a] ISISE-University of Coimbra, Faculty of Sciences and Technology, Portugal
dcassiano@ student. dec. uc. pt, crebelo@ dec. uc. pt, luisss@ dec. uc. pt
[b] University of Naples "Federico II", Department of Structures for Engineering and
Architecture, Italy
mdaniel@ unina. it, landolfo@ unina. it

KEYWORDS: Robustness, Pushdown, RRSR, Ductility, MRF.

ABSTRACT

In previous decades, many steel moment frame buildings were built in seismic zones, when seismic codes were at its early stages of development, and as such, these structures were often designed solely to resist lateral wind loads without providing an overall ductile mechanism. On the contrary, the current seismic design criteria based on hierarchy of resistance allows enhancing the structural ductility and controlling the structural plastic behaviour. Therefore, seismic design criteria might be also beneficial to improve the structural robustness. In order to investigate this issue for steel Moment Resisting Frame (MRF) in the present paper a parametric study based on pushdown analysis and the Energy Balance Method is described and discussed. With this regard the following cases are examined: (i) MRF structures not designed for seismic actions, (ii) MRF structures designed for moderate seismic actions. The investigated parameters are (i) the number of storeys, (ii) the interstorey height, (iii) the span length, (iv) the building plan layout and (v) the column loss scenario. This study has enabled to quantify the effectiveness of seismic detailing in limiting the progressive collapse under column loss scenarios.

CONCLUSIONS

A parametric study based on pushdown analyses was conducted investigating the behaviour of 48 MRF structures for three column loss scenarios (namely a total number of 144 analysis cases). The robustness was measured through the Residual Reserve Strength Ratios (RRSR), defined as the ratio between the system's ultimate capacity in the damaged configuration and the dynamically amplified force for which the system reached equilibrium. Three types of global collapse mechanisms were identified, providing different ductility levels.

It was verified that 8 storey buildings and wind-designed structures are more prone to semi-ductile or brittle global failure modes. The numerical results also showed that 4 storey-10m span structures are prone to progressive collapse, whereas for the 8 storey -10m span structures, experienced no collapses. In general, the medium rise 8 storey structures provided significantly higher values of robustness than low rise structures, indicating that the number of elements above the removed column that can be mobilized through Vierendeel action is a key parameter in arresting a progressive collapse.

The analysis of the DIF values indicated that 4 storey structures are sensitive to post yield structural ductility while 8 storey strong beam-weak column structures tend to remain elastic.

The importance of seismic detailing for robustness was investigated. Results show that structures designed according to the Capacity Design methodology do not present the highest values of RRSR, but the failure modes are more predictable and present lower dispersion of RRSR values. The strong beam-weak column structures tend to remain in the elastic domain after column loss. Given the comparatively higher RRSR values and considering that these structures present larger beam cross section dimensions, robustness levels appear to be very dependent on the capacity of the beam elements above the directly affected zone.

The resistance to progressive collapse appears to depend significantly on local behaviour, namely on bay/MRF layout, local beam and column capacities and number of elements above the removed column which can be mobilized through Vierendeel action.

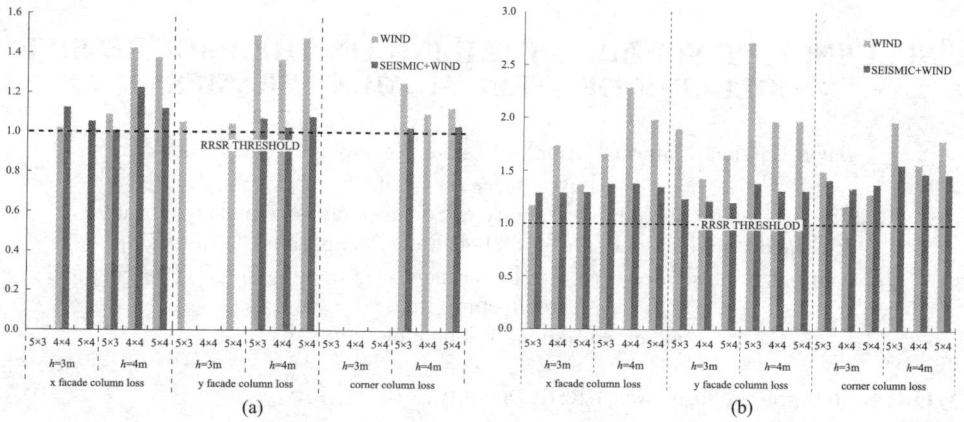

Fig. 1. (a) RRSR for 4 storey 10m span structures; (b) RRSR for 8 storey
10m span structures

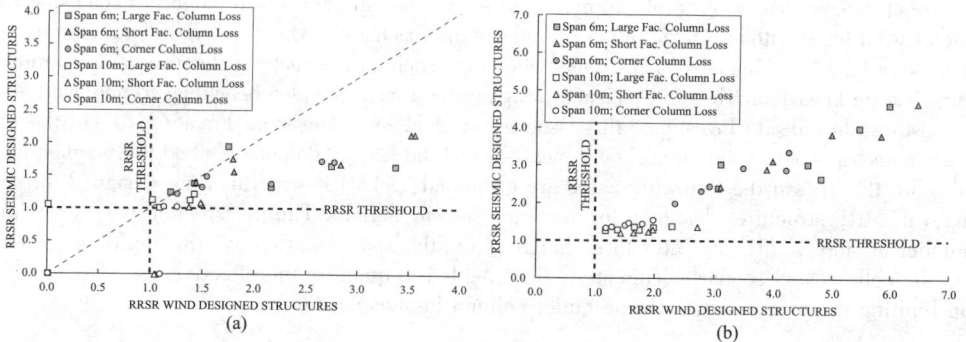

Fig. 2. (a) RRSR comparison for 4 storey buildings; (b) RRSR comparison
for 8 storey buildings

REFERENCES

[1] CEN,2004. "EN 1998-1-Eurocode 8-Design of buildings for earthquake resistance-Part 1: General rules, seismic actions and rules for buildings".

[2] Elghazouli, A. Y. (Ed.),2009. "Seismic Design of Buildings to Eurocode 8", Oxon: Spon Press.

[3] El-Tawil, S. , Li, H. , Kunnath, S. ,2013. "Computational Simulation of Gravity-Induced Progressive Collapse of Steel-Frame Buildings: Current Trends and Future Research Needs", *Journal of Structural Engineering*, 140, Special Issue: Computational Simulation in Structural Engineering, A2513001.

[4] United States of America Department of Defense,2009. "United Facilities Criteria (UFC)-Design of buildings to resist progressive collapse".

[5] Izzuddin, B. A. , Vlassis, A. G. , Elghazouli, A. Y, Nethercot, D. A. ,2008. "Progressive Collapse of Multi-Story Buildings due to Sudden Column Loss-Part I: Simplified Assessment Framework", Engineering Structures,30,1308-1318.

[6] American Society of Civil Engineers,2000. "Prestandard and commentary for the seismic rehabilitation of buildings-FEMA 356", Washington D. C. : Federal Emergency Management Agency.

[7] Khandelwal, K. , El-Tawil, S. , Kunnath, S. , Lew, H. ,2008. "Macromodel-Based Simulation of Progressive Collapse: Steel Frame Structures", Journal of Structural Engineering, 134 (7), 1070-1078.

[8] Starossek, U. , Haberland, M. ,2008. "Approahes to measures of structural robustness", IABMAS'08,4th International Conference on Bridge Maintenance, Safety and Management, Seoul, Korea.

[9] Xu, G. , Ellingwood, B. ,2011. "Probabilistic Robustness Assessment of Pre-Northridge Steel Moment Resisting Frames", Journal of Structural Engineering,137,925-934.

RANDOM SEISMIC RESPONSE EVALUATION OF MID-RISE BUILDINGS WITH STIFFNESS IRREGULARITY CONSIDERING SOIL-STRUCTURE INTERACTION EFFECTS

H. Shakib[a]   F. Homaei[b]

[a]Professor, School of Civil and Environmental Engineering Tarbiat Modares University, Tehran, Iran shakib@ modares. ac. ir
[b]Ph. D student, School of Civil and Environmental Engineering Tarbiat Modares University, Tehran, Iran f. homaei@ modares. ac. ir

KEYWORDS: Non-geometric vertical irregularities, Non-stationary random ground motion, Soil-structure interaction, Seismic response, Shear buildings

ABSTRACT

Seismic response of vertically irregular buildings with non-uniform distribution of stiffness along the height is evaluated. As foundation flexibility may affect the response of structure in several ways (Shakib and Fuladgar 2004; Moghaddasi, Cubrinovski et al. 2011; Rajeev and Tesfamariam 2012), soil-structure interaction(SSI) effects are also considered. Since the applied loads are random in many engineering cases, comparison and evaluation of structural responses are performed in the framework of random excitation.

The model of superstructure was considered as an elastic shear building; a ten story steel frame structure with concentrated mass at each story level which has no eccentricity in plane(eccentricity distance of CM(center of mass) to CR(center of stiffness) was assumed zero). The amount of irregularity was limited to 40% compare to the regular structure and its location varies through the structure height.

Equations of motion of the superstructure resting on an elastic homogeneous half-space were established. Frequency-independent spring and dashpot set in parallel(Richart, Hall et al. 1970) was used for the interaction forces at foundation level. Through a series of non-stationary random processes, structural response and its components obtained.

Curves of mean square value of structural responses have been developed(*Fig. 1*). It is shown that regardless to the position of irregularity, displacement demands have been concentrated on upper and lower stories compare to middle ones. For regular structure, the portion of demands at these stories is fairly equal; meanwhile, for irregular structures, a large amount of displacement demands are concentrated at bottom stories.

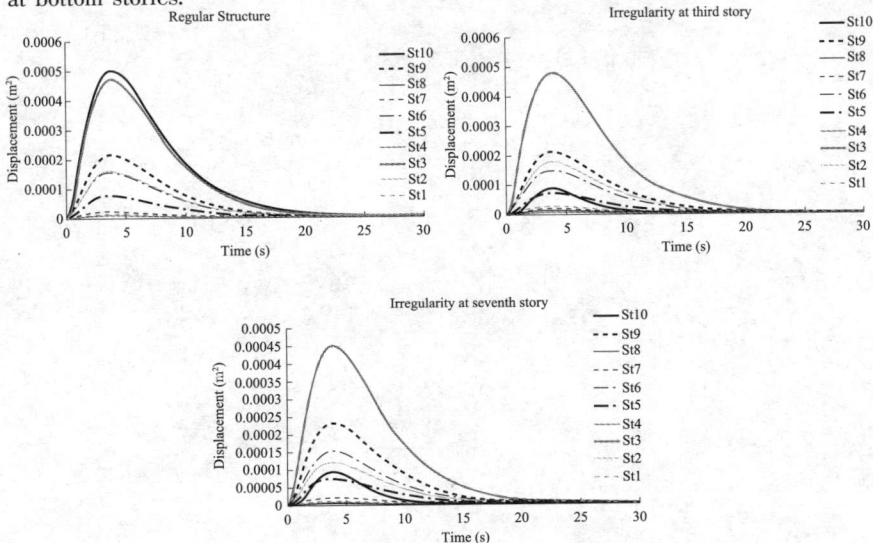

Fig. 1. Displacement response of structures

REFERENCES

[1] Moghaddasi, M. , M. Cubrinovski, et al. (2011). "Effects of soil-foundation-structure interaction on seismic structural response via robust Monte Carlo simulation. " Engineering Structures **33** (4): 1338-1347.

[2] Rajeev, P. and S. Tesfamariam (2012). "Seismic fragilities of non-ductile reinforced concrete frames with consideration of soil structure interaction. "Soil Dynamics and Earthquake Engineering **40**(0): 78-86.

[3] Richart, F. E. , J. R. Hall, et al. (1970). "Vibrations of soils and foundations. "

[4] Shakib, H. and A. Fuladgar(2004). "Dynamic soil-structure interaction effects on the seismic response of asymmetric buildings. "Soil Dynamics and Earthquake Engineering **24**(5): 379-388.

SEISMIC RESPONSE OF SPECIAL CONCENTRIC BRACED FRAMES WITH STAGGERED ARRANGEMENT OF BRACES

P. C Ashwin Kumar, Abhay Kumar and Dipti Ranjan Sahoo[a]

[a]Department of Civil Engineering, Indian Institute of Technology Delhi, India

abhay9411@ gmail. com, ashwin85pc@ yahoo. co. in, drsahoo@ civil. iitd. ac. in

KEYWORDS: Concentric braced frame, Design basis earthquake, Nonlinear dynamic analysis.

ABSTRACT

Concentric braced frames (CBFs) are used in regions of high seismicity to provide the structural system with adequate strength and ductility[1]. This helps the structure to survive the seismic forces but, on the other hand, the bracing bent columns are subjected to high uplifting forces that may damage the structural integrity. One of the best ways to mitigate this concentrated uplifting force is the use of staggered arrangement of braces in the braced frames. The objective of this paper is to understand the behaviour of the staggered arrangement of braces as compared to the single stack arrangement, when subjected to design-basis earthquake ground motions. A regular 5 story building was considered with braces provided at the periphery of the building. The design base shear at each floor of the braced frame was found out using the equivalent lateral force method. The design of the braces, beams and columns followed the guidelines as provided in the code[2,3]. Five separate staggered brace combinations were used and were compared with the response of the regular brace arrangement (single stack) used in the construction industry (Fig. 1). Two configurations had the total number of braces less than the regular arrangement and three cases had more braces in number. Various combinations of support conditions (i. e., pinned/fixed) was considered in the design cases to understand their influence on the overall seismic behaviour. The beam-column joint locations where the brace intersect were modelled as fixed joints to cater for the fixity induced by the gusset plate. It was also made sure that the cross sectional sizes of the sections used as braces, beams and columns complied with the slenderness and compactness criteria as set by the code[2]. Non-linear static and Non-Linear Dynamic analysis was performed on these configurations to understand the performance of these systems. Non-linear static analysis showed a common trait of excessive deformation in the top story due to lower number of braces and hence lower stiffness. The results also showed that by using more number of braces in the staggered arrangement the system develops more strength and ductility. The non-linear time history analysis in which a series of 20 ground motions representing the design basis earthquake was applied to the configurations. The staggered brace configurations with more number of braces showed performance level matching the single stack arrangement whereas the configurations with lesser number of braces showed varied response (Fig. 2) and the average of the maximum inter story drift ratio for all the configurations did not reach the target drift of 2. 5% [4]. Further, staggered configurations used less amount of steel as compared to the single stack arrangement. The minimum variation of the total member weight with respect to the single stack conventional arrangement is about 6. 45% and the maximum is about 12. 8% . The column forces at the base of the structure were also compared and it was found that the staggered arrangement substantially reduced the axial force in the column members. Even though the results favour the use of staggered brace configurations with higher brace numbers, it needs to be understood that without a set guidelines to select a staggered configuration and more studies on braced frame with higher number of staggered braces it cannot be convincingly be implemented in the design of building. Although, this type of staggered arrangement convincingly reduces the axial forces in the columns and is economical, the implementation of these depends on rigorous analysis and inclination of architects to accept this arrangement.

CONCLUSIONS

The following conclusions can be drawn from the present study:

– Staggered arrangement can match the force and deformation response of a single stack of brace

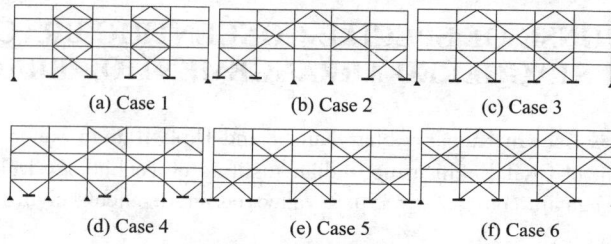

(a) Case 1 (b) Case 2 (c) Case 3

(d) Case 4 (e) Case 5 (f) Case 6

Fig. 1. Brace configurations considered for the analysis

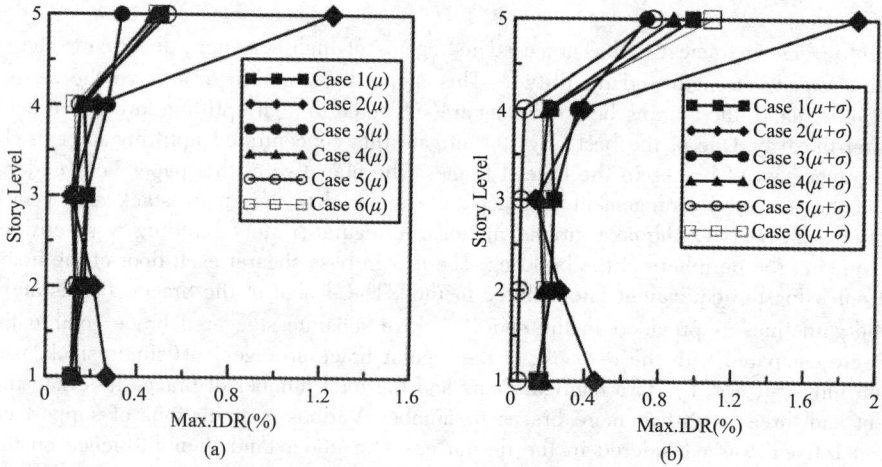

(a) (b)

Fig. 2. Comparison of inter-story drift response

arrangement that has been followed in the current practice. The amount of material consumed in beams, columns and brace is found to be lower by using staggered arrangement of braces in comparison to single stack arrangement of braces in order to have a similar seismic response.

– The column forces for staggered arrangement are much smaller than those for the single stack of brace arrangement. This may eliminate premature failure of members and base plates of braced frame structures. The inability of the present code to cater for the design requirements of the staggered arrangement, especially when two consecutive floors use same size of brace, warrants the need for a revised design procedure. The orientation of this new design method may well be directed towards performance-based design concept.

– Although more number of braces increases the redundancy of the structure, the improved performance cannot be attributed singly to the increase in number of braces. It needs to be explored beyond this point by using different configuration so that the dependency of this improvement on other factors, possibly story stiffness, can be highlighted in a better way.

REFERENCES

[1] Bruneau, M. , Uang, C. M. , and Sabelli. Ductile Design of Steel Structures, Second Edition, McGraw Hill, New York, 2011.

[2] ANSI/AISC 341-10. Seismic Provisions for Structural Steel Buildings. American Institute of Steel Construction, Chicago, IL, U. S. A, 2010.

[3] ANSI/AISC 360-10. Specification for Structural Steel Buildings. American Institute of Steel Construction, Chicago, U. S. A. , 2010.

[4] Sabelli R. , Roeder C. W. , and Hajjar J. F. Seismic Design of steel special concentrically braced frame systems: A guide for practicing engineers, NEHRP Seismic Design Technical Brief No. 8, Gaithersburg, MD, NIST GCR 13-917-24, 2013.

AN ACCURATE MODELING APPROACH FOR CALCULATING THE VIBRATION CHARACTERISTICS OF STEEL FRAMED STRUCTURES WITH SEMI-RIGID CONNECTIONS

Halil F. Ozel[a] and Afsin Saritas[a]
[a]Middle East Technical University, Ankara, Turkey
firat. ozel@ metu. edu. tr, asaritas@ metu. edu. tr

KEYWORDS: Steel framed structures, Finite element modeling, Semi-rigid connection, Vibration characteristics.

ABSTRACT
Vibration characteristics of steel framed structures are affected by the presence of semi-rigid connections. In this study, a mixed formulation frame finite element is developed from force method, where the variational form of the element bases on the use of three-fields Hu-Washizu-Barr principle. Consistent mass matrix of the element is obtained such that determination of vibration frequencies of members with varying geometry and material distribution as well as the presence of semi-rigid connections at any section on the element is accurately captured without any need for specification of different displacement shape functions for each individual case. The element response does not necessitate further discretization due to the presence of semi-rigid connections. Benchmark numerical examples for a steel I-beam verifies the accuracy of proposed element with and without semi-rigid connections.

CONCLUSIONS

A cantilever I-section steel beam with rigid/semi-rigid connections at base are solved for varying length to depth aspect ratios in order to assess the accuracy of proposed element's response.

Table 1. ANSYS Results for Cantilever I-Beam with Rigid Connection

L/d	1st Bending(Hz)	2nd Bending(Hz)	1st Axial(Hz)
10	42. 154	227. 13	468. 35
5	155. 66	649. 25	937. 13
2	671. 28	—	2358. 7

Table 2. SAP2000 and Proposed Model Results for Cantilever I-Beam with Rigid Connection

L/d	Mode Type	SAP 2000		Proposed Model		
		$N_{el} = 4$	$N_{el} = 32$	$N_{el} = 1$	$N_{el} = 4$	$N_{el} = 32$
10	1st Bending(Hz)	41. 3104	42. 4160	42. 4325	42. 1005	42. 0848
	2nd Bending(Hz)	213. 8580	232. 2880	388. 6018	229. 2061	226. 9499
	1st Axial(Hz)	464. 9001	467. 9457	515. 9358	470. 9140	467. 9490
5	1st Bending(Hz)	154. 2258	157. 8283	157. 9986	155. 5062	155. 3225
	2nd Bending(Hz)	643. 5006	692. 5208	1031. 8716	671. 8748	659. 1511
	1st Axial(Hz)	930. 2326	935. 4537	1202. 4073	941. 8281	935. 8980
2	1st Bending(Hz)	692. 5208	702. 2472	718. 4257	686. 5626	684. 1043
	2nd Bending(Hz)	2136. 7521	2277. 9043	2579. 6789	2080. 1213	2030. 6245
	1st Axial(Hz)	2325. 5814	2341. 9204	3118. 2153	2354. 5701	2339. 7449

Rigid case results obtained from 3d solid element analyses with ANSYS are given in *Table 1*, and proposed element and SAP2000 results are shown in *Table 2*. Proposed element is able to determine the first bending and axial modes as well as the second bending mode for a cantilever I-beam rigidly connected at base with great accuracy when compared with ANSYS results, while the same cannot be said for SAP2000.

Modal analyses results obtained for the cantilever beam with semi-rigid connection with stiffness ratio of λ as a multiple of EI/L of beam are given for SAP2000 and proposed element solutions in Tables 3 and 4, respectively. Comparison of 32 element results of SAP2000 and proposed model solutions for both the long and short beam cases demonstrate that the results from proposed approach provide the accurate and reliable vibration frequencies since the same level of difference was also present in *Table 2* for the rigid case results.

Table 3. SAP2000 Results for Cantilever I-Beam with Semi-Rigid Connection

Nel	Mode	L/d	10			5			2	
		$\lambda=2$	$\lambda=11$	$\lambda=20$	$\lambda=2$	$\lambda=11$	$\lambda=20$	$\lambda=2$	$\lambda=11$	$\lambda=20$
4	B1(Hz)	24.29	35.68	37.92	94.66	135.34	142.94	508.13	644.33	664.89
	B2(Hz)	178.86	197.90	203.54	596.66	624.61	631.71	2136.75	2136.75	2136.75
	A1(Hz)	444.44	459.14	464.25	930.23	930.23	930.23	2325.58	2325.58	2325.58
32	B1(Hz)	24.75	36.52	38.86	96.39	138.22	146.11	515.20	653.17	673.86
	B2(Hz)	192.75	214.04	220.46	640.21	671.14	678.89	2277.90	2277.90	2277.90
	A1(Hz)	467.95	467.95	467.95	935.45	935.45	935.45	2341.92	2341.92	2341.92

Table 4. Proposed Model Results for Cantilever I-Beam with Semi-Rigid Connection

N_{el}	Mode	L/d	10			5			2	
		$\lambda=2$	$\lambda=11$	$\lambda=20$	$\lambda=2$	$\lambda=11$	$\lambda=20$	$\lambda=2$	$\lambda=11$	$\lambda=20$
1	B1(Hz)	41.43	42.05	42.19	153.81	156.51	157.09	699.33	712.83	715.15
	B2(Hz)	352.54	373.32	378.90	1031.87	1031.87	1031.87	2579.68	2579.68	2579.68
	A1(Hz)	515.94	515.94	515.94	1149.86	1182.72	1190.30	3137.40	3123.86	3121.53
4	B1(Hz)	25.10	36.54	38.76	97.03	137.22	144.62	505.54	639.38	659.39
	B2(Hz)	199.00	215.61	220.42	637.96	658.35	663.43	2062.07	2074.72	2076.94
	A1(Hz)	470.91	470.91	470.91	941.83	941.83	941.83	2354.57	2354.57	2354.57
32	B1(Hz)	24.99	36.49	38.72	96.51	136.91	144.36	501.29	636.50	656.70
	B2(Hz)	189.09	209.88	215.95	610.68	640.00	647.23	2020.76	2027.91	2029.05
	A1(Hz)	467.95	467.95	467.95	935.90	935.90	935.90	2339.75	2339.75	2339.75

Overall, proposed element is able to determine the vibration characteristics for an I-section steel beam with or without semi-rigid connections with great accuracy. Proposed element accurately captures stiffness and mass distributions without the need for the description of the displacement field by the use of the formulation approaches by[1] and[2]. Accuracy obtained from proposed approach is enhanced by the use of a shear correction factor suggested by[3] for I-section steel beams.

ACKNOWLEDGMENT

The authors thank for the support provided by Scientific and Technological Research Council of Turkey(TUBITAK)under Project No: 113M223.

REFERENCES

[1] Saritas, A. , Filippou, F. C. , 2009, "Inelastic axial-flexure-shear coupling in a mixed formulation beam finite element", *International Journal of Non-Linear Mechanics*, 44(8) :913-922.

[2] Molins, C. , Roca, P. , Barbat, A. H. , 1998, "Flexibility-based linear dynamic analysis of complex structures with curved-3D members", *Earthquake Engineering & Structural Dynamics*, 27 (7) : 731-747.

[3] Charney, F. A. , Iyer, H. , and Spears, P. W. , 2005, "Computation of major axis shear deformations in wide flange steel girders and columns", *Journal of Constructional Steel Research*, 61: 1525-1558.

ON THE WEAK STOREY BEHAVIOUR OF CONCENTRICALLY BRACED FRAMES

Daniel B. Merczel[a,b], Jean-Marie Aribert[a], Hugues Somja[a], Mohammed Hjiaj[a], János Lógó[b]

[a]University INSA de Rennes, Dept. of Civil Engineering, France
dmerczel@ insa-rennes. fr, aribert@ insa-rennes. fr, hsomja@ insa-rennes. fr,
mhjiaj@ insa-rennes. fr

[b]Budapest University of Technology and Economics, Dept. os Structural Mechanics, Hungary
logo@ ep-mech. me. bme. hu

KEYWORDS: concentrically braced frame, seismic design, weak storey behaviour, wavelet transformation.

ABSTRACT

Concentrically braced frames (hereinafter referred to as CBF-s) are among the most common structural systems for resisting horizontal forces. For wind actions, low return period seismic actions or for seismic actions in buildings of a high importance class, such as power plants, the design behaviour is elastic. Due to their geometry CBF-s counteract the horizontal forces by truss action that entails primarily axial forces in the members. The large stiffness limits the lateral drifts and vibrations of low frequencies in the braced buildings, therefore provides occupancy comfort and impedes damages in the non-structural parts.

In classical design CBF-s are expected to develop a global plastic mechanism when subjected to stronger seismic actions. In the global mechanism all the braces undergo plastic deformations while the columns and beams are non-dissipative elements. Conversely, recent seismic events and numerical simulations pointed out, that some CBF-s develop an unfavourable plastic collapse mechanism when subjected to earthquake excitation. This effect is often referred to as the weak storey mechanism.

Though Eurocode 8 design standard imposes structure-specific requirements to evade the occurrence of weak storeys, these are mostly not efficient enough, as it is shown in several examples in the article. The adverse behaviour basically results the localization of lateral displacements and dissipation, but eventually it may also lead to early failure.

A series of nonlinear time history analyses (NTHA) were conducted on buildings of various storey numbers designed according to Eurocode 8 provisions. These examples were mostly found to be inadequate in developing a well-distributed dissipation and also in resisting the design strength earthquake. The analysis of the response of structures with developed weak storeys as well as the analysis of imperfect braced frame models, see *Fig 1*, gives explanations to the gradual development of the weak storeys. The gradual cyclic deterioration of the braces does not only affect the resistance of the members but also fundamentally modifies the dynamic behaviour of the building and its response to excitation.

In the article the change of the modal behaviour due to the inelastic brace deformations is introduced and identified in the actual response by Morlet wavelet transformation.

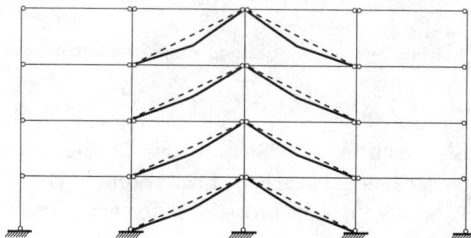

Fig. 1. Residual deformation of the braces after earthquake

CONCLUSIONS

The deterioration of the braces causes a significant decrease of the lateral stiffness on the weak storey. This result in a qualitative change of the modal response of the structure, see *Fig. 2*.

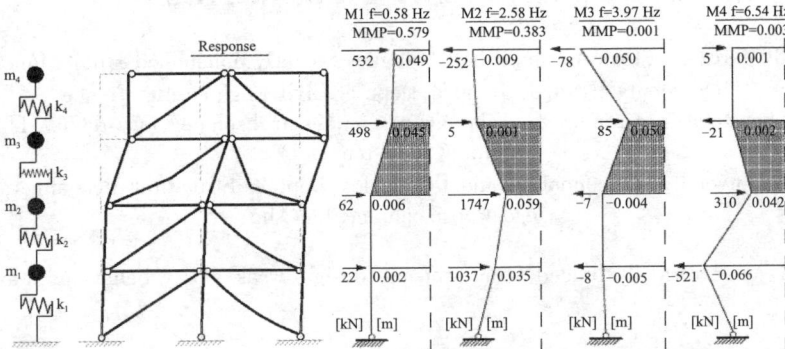

Fig 2. Response of CBF with developed weak storey

The gradual change of the frequency and thus the development of the weak storey can be identified in the actual response by means of wavelet transformation. By evaluating *Eq.* (1), we can map the instantaneous frequency components of the lateral displacement time series.

$$W(a,b) = \frac{1}{\sqrt{a}}\int g^* \left(\frac{t-b}{a}\right) \cdot x(t)dt \qquad (1)$$

where $x(t)$ is the lateral displacement signal
     a parameter of dilation
     b parameter of translation

The result of the wavelet transformation is the scalogram, which shows the gradual change of the instantaneous frequency components of the motion with a colormap (*Fig* 3.) and therefore proves that the inelastic response of a CBF is as it has been described by *Fig* 2.

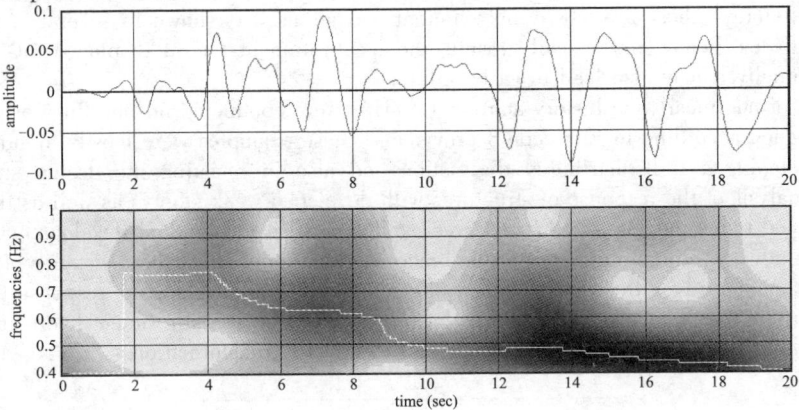

Fig 3. Wavelet transform of lateral displacement time series

REFERENCES

[1] Nip, K. H., Gardner, L., Elghazouli, A. Y. (2010). Cyclic testing and numerical modeling of carbon steel and stainless steel tubular bracing members. Engineering Structures 32: 424-441.

[2] Elghazouli, A. Y. (2003). Seismic design procedures for concentrically braced frames. Structures & Buildings 156: 381-394.

[3] Büssow, R. (2007). An algorithm for the continuous Morlet wavelet transform. Mechanical Systems and Signal Processing 21: 2970-2979.

INFLUENCE OF RESIDUAL STRESSES ON THE PERFORMANCE OF SPECIAL CONCENTRICALLY BRACED FRAMES

Taylor C. Steele and LydellD. A. Wiebe

Department of Civil Engineering, McMaster University, Hamilton, ON, Canada

e-mails: steeletc@ mcmaster. ca, wiebel@ mcmaster. ca

KEYWORDS: special concentrically brace frames, residual stresses, capacity design, nonlinear time history analysis, inelastic buckling.

ABSTRACT

Special concentrically braced frames (SCBFs) are stiff ductile steel lateral force resisting systems. The bracing members in SCBFs may be hot-rolled wide flange sections, which have residual stresses that reduce their compressive resistance[1]. Past studies have not modelled residual stresses in the braces during nonlinear analysis of SCBFs [2,3,4]. In this study, a three-story structure is designed and analysed for design-level events. When residual stresses are included in the model, the braces buckle earlier, leading to slightly larger roof and interstory drifts. The peak forces in the other frame members are not significantly different from those in the model without considering residual stress. However, fracture of a bottom-level brace is predicted for five of the records when residual stresses are neglected, compared to only one when they are modelled. The behavior of both models is concerning and suggests that even more refined collapse modelling of low-rise SCBFs is unlikely to demonstrate reliable collapse prevention for earthquakes beyond the design level.

CONCLUSIONS

A special concentrically braced frame(SCBF) was designed for a site in the Western United States for a design basis earthquake(DBE) using a force reduction factor of $R = 6$. The structures were modelled with and without residual stresses in the members, and were subjected to seven ground motions scaled to match the DBE elastic design spectrum. Based on the analyses, including residual stresses in the numerical model generally had the most significant influence on the first one or two cycles of the response; the difference in subsequent cycles was considered to be negligible. *Fig. 1* displays the median peak member forces during the non-linear time history analyses, and *Fig. 2* shows the median peak floor drifts, interstory drifts, residual drifts, and floor accelerations. For most response quantities, the peak global response occurred after the cycles that were influenced by residual stresses, so there was very little difference between the peak responses of the two models. Although the peak brace forces were approximately 15% less when residual stresses were considered, the beam bending moments and column axial forces were within 1% ~ 2% for both models, while the beam axial forces were within 2% ~ 8%.

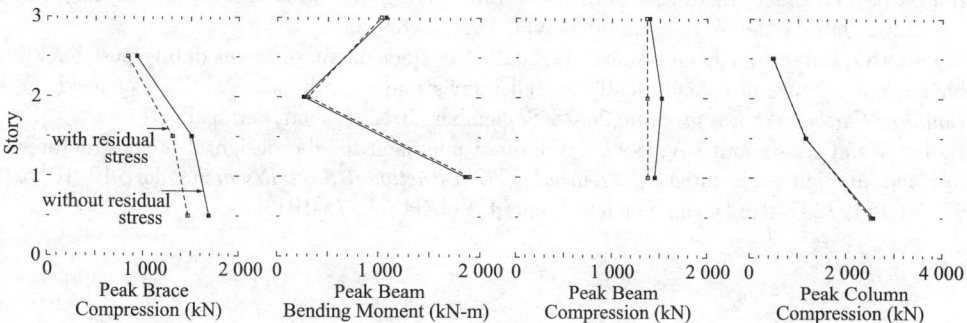

Fig. 1. Median of the peak member forces in the chevron braced frame

The most significant influence of residual stresses was on the trigger of the low-cycle fatigue frac-

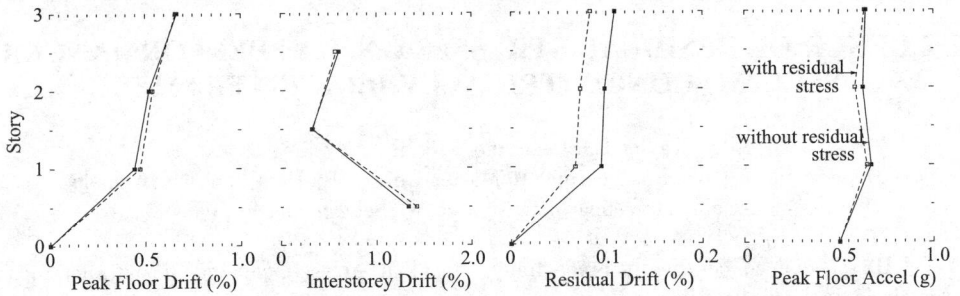

Fig. 2. Median of the peak member forces in the chevron braced frame

ture model. For four records, modelling residual stresses prevented the fracture of the braces, where one brace in the model without residual stresses fractured for each of these records after approximately six large cycles. For one other record, both models had one brace fracture, but the brace in the model with residual stresses fractured later within the same cycle. This difference is attributed to the use of the same low-cycle fatigue model for both cases, even though the model had been calibrated using a model without residual stresses[2,5]. When residual stresses are modelled, the initial compressive strains reduce the amount of plastic tensile strain experienced at the extreme fibers, thereby delaying low-cycle fatigue. Although the brace fracture did not lead to collapse during the DBE-level records that were considered, it suggests that MCE-level analyses would likely show that the design method was unconservative.

This study was limited to one braced frame configuration that was three stories in height. The study could be extended to consider taller structures, different ground motion intensities, and configurations where the force and deformation demands are distributed differently along the height of the structure. Further study would also be needed to calibrate the parameters used to model low-cycle fatigue when residual stresses are also modelled. However, the overall similarity of the results for the two models suggests that even more refined modelling is unlikely to demonstrate reliable collapse prevention at a level greater than the DBE.

REFERENCES

[1] Beedle LS and Tall L, 1959. "Residual stress and the compressive properties of steel: Basic column strength," Fritz Laboratory Report No. 220A. 34, Department of Civil Engineering, Lehigh University, Bethlehem, PE.

[2] Uriz P and Mahin SA, 2008. "Towards earthquake-resistant design of concentrically braced steel-frame structures," PEER Report 2008/08, Pacific Earthquake Engineering Research Center, University of California, Berkeley United States.

[3] Hsiao PC, Lehman DE and Roeder CW, 2010. "A model to simulate special concentrically braced frames beyond brace fracture," *Earthquake Engineering and Structural Dynamics*, doi: 10. 1002/ eqe. 2202, John Wiley & Sons Limited, Vol. 42, pp. 183-200.

[4] Lignos DG, Karamanci E and Martin G, 2012. "A steel database for modeling post-buckling behavior and fracture of concentrically braced frames under earthquakes," *Proc. of the* 15*th World Conf. on Earthquake Engineering*, 24-28 September, 2012, Lisbon, Portugal.

[5] Ballio G and Castiglioni CA, 1995. "A unified approach for the design of steel structures under low and/or high cycle fatigue," *Journal of Constructional Steel Research*, doi: 10. 1016/0143-974X(95)97297-B, Elsevier Science Limited, Vol. 34, pp. 75-101.

SEISMIC PERFORMANCE OF RC STRUCTURE RETROFITTED WITH STEEL BUCKLING-RESTRAINED BRACED FRAME

An-Chien Wu[a], Kuan-Yu Pan[b], Keh-Chyuan Tsai[c], Chao-Hsien Li[a], Pao-Chun Lin[a],
Kung-Juin Wang[d], Chi-Hsuan Yang[b]

[a] Associate Researcher, National Center for Research on Earthquake
Engineering, Taipei, Taiwan
acwu@ ncree. narl. org. tw, chli@ ncree. narl. org. tw, pclin@ ncree. narl. org. tw

[b] Graduate student, Department of Civil Engineering, National Taiwan
University, Taipei, Taiwan
r01521219@ ntu. edu. tw, r01521231@ ntu. edu. tw

[c] Professor, Department of Civil Engineering, National Taiwan University, Taipei, Taiwan
kctsai@ ntu. edu. tw

[d] Technologist, National Center for Research on Earthquake Engineering, Taipei, Taiwan
kjwang@ ncree. narl. org. tw

KEYWORDS: seismic retrofit, concrete frame, bearing block, buckling-restrained braced, post-installed anchor.

ABSTRACT

This research evaluates the seismic performance of two strong-beam-weak-column RC frames retrofitted using the WT steel frames and buckling-restrained braces (BRBs). High-strength mortar bearing blocks constructed at the four inner corners of the RC frame are adopted to transfer the force to the BRB. Test results show that the lateral strength of the retrofitted frame continued to increase until reaching 3% rad. drift ratio, and developed the story shear 2.6 times that of the bare RC frame before the BRB core fractured at the 5% rad. drift ratio. This study concludes that the retrofitting approach is efficient in increasing the strength while achieving good ductility and energy-dissipating capacities. The shear failure in the RC column due to the BRB compressive force can be effectively predicted by the softened strut-and-tie model.

CONCLUSIONS

The conclusions of this research can be summarized as follows:

1. From the test results of the BRB-D-WTF specimen, the lateral strength continued to increase until reaching 3% drift ratio, and developed 891kN story shear (2.6 times of the bare RC frame) before the BRB core fractured at the 5% drift ratio.

2. From the test results of the BRB-S-WTF specimen, when the brace is in compression, it imposes a compressive force near the tip of the frame corner. This compressive force may cause the shear failure in the RC column. This failure mode can be evaluated using the softened strut-and-tie model.

3. The pinchingbehavior in the force-deformation response of the BRB-S-WTF specimen results from the gaps produced at the corners after the bearing blocks crushed. Reinforcing the bearing blocks with wire meshes could reduce the pinching behavior.

4. The axial forces in the struts are primarily subjected to compressive forces when the brace is in tension. This matches the expectedbehavior.

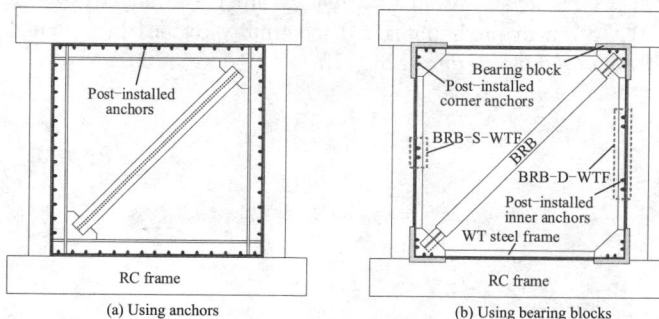

(a) Using anchors   (b) Using bearing blocks

Fig. 1. Schematics of two different load transfer methods

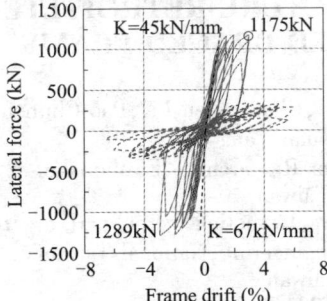

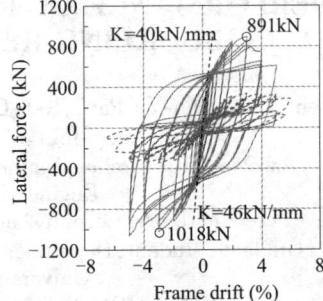

(a) BRB-S-WTF   (b) BRB-D-WTF

Fig. 2. Axial forces in the struts

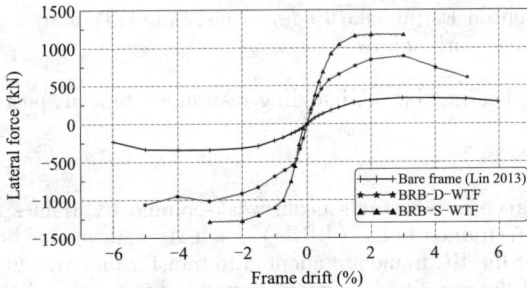

Fig. 3. Response envelopes of the specimens

Fig. 4. End of the test of Specimen BRB-S-WTF

REFERENCES

[1] Maheri MR, Kousari R, Razazan M, "Pushover tests on steel X-braced and knee-braced RC frames", *Engineering Structures*, 25 (13): 1697-1705, 2003.

[2] Lin RY, "Connection between retrofitting steel braces with framework and existing RC frame", *Master thesis*, National Taiwan University, Taipei, Taiwan, 2013.

[3] Mahrenholtz C, Lin PC, Wu AC, Tsai KC, Hwang SJ, Lin RY, Bhayusukma MY, "Retrofit of reinforced concrete frames with buckling-restrained braces", *Earthquake Engineering and Structural Dynamics*, 44 (1): 59-78, 2014.

[4] Paret TF, "Brace for the Big One", *Modern Steel Construction*, August, 2007.

[5] Tsai KC, Wu AC, Wei CY, Lin PC, Chuang MC, Yu YJ, "Weld end-slot connection and debonding layers for buckling-restrained braces", *Earthquake Engineering and Structural Dynamics*, 43 (12): 1785-1807, 2014.

[6] Pan KY, "Seismic retrofit of reinforced concrete frames using buckling-restrained braced frames", *Master thesis*, National Taiwan University, Taipei, Taiwan, 2014.

[7] Hwang SJ, Lee HJ, "Strength prediction for discontinuity regions by softened strut-and-tie model", *Journal of Structural Engineering*, 128 (12): 1519-1526, 2002.

BACKWARD SEISMIC ANALYSIS OF STEEL TANKS

Patricio A. Pineda[a] and G. Rodolfo Saragoni[a]

[a] University of Chile, Santiago, Chile

patricio. pineda@ ing. uchile. cl, rsaragon@ ing. uchile. cl

KEYWORDS: Backward, Sliding, Tanks, Asperities, Subduction, Anchored.

ABSTRACT

Steel tanks have presented repeated failures during last large earthquakes, mainly buckling shell, horizontal sliding and some collapse, despite being designed with codes widely used in the world such as the standards API 650-E, AWWA D100, NZSEE and Chilean Standard NCh2369. Of2003. In this study, the backward analysis of the performance of tanks for the following three Chilean earthquakes is considered: Central Chile, 1985, Tocopilla, 2007 and Maule, 2010.

In this study has been noticed that one of more common seismic failures of tanks is horizontal sliding. These sliding are assumed due to inertial pressure forces on the tanks, however according the backward analysis of tanks located in coastal areas of subduction large earthquakes; the sliding can be due to the coseismic no vibratory displacement of meters of the coast measured by GPS. *Table 1* summarizes the observed slide of tanks.

Table 1. Principal observed horizontal tanks sliding.

Earthquake	Magnitude	Plate Fault	Horizontal Sliding(mm)	$D(m)$	$H(m)$
Alaska	9. 2	Subduction	1524	3. 2	9. 144
Tocopilla	7. 7	Subduction	80	35	14. 5
Landers	7. 3	Cortical	70-80	16. 5	7. 3

From *Table 1* can be appreciated that sliding of tanks can reach values of meters in coast areas of subduction earthquakes.

CONCLUSIONS

The following preliminary formula to estimate the sliding S of tanks in coastal areas of subduction zones in terms of magnitude M is proposed:

$$S[m] = -5.47 + 0.76M \tag{1}$$

The sliding in subduction earthquake is in the direction perpendicular to the coast or in the convergence of the subducted plate. Considering this formula, anchors of tanks will be required in some cases despite API 650-E recommendation in order to avoid damage in bottom plates as well as piping. The sliding also produce seismic forces not consider in API650-E and other design codes for unanchored tanks.

ACKNOWLEDGMENT

Is highly recognized the cooperation of Bio Bio ENAP Refinery and "Compañía de Acero del Pacífico(CAP)".

REFERENCES

[1] API Standard 650, "Welded Tanks for Oil Storage", Twelfth Edition March 2013, Addendum 1 September 2014, Errata 1 July 2013, Errata 2 December 2014, *American Petroleum Institute*, 2013.

[2] Official Chilean Standard NCh2369. Of2003, "Earthquake-resistant design of industrial structures and facilities", *National Institute of Normalization*(in Spanish), 2003.

[3] Pineda P, Arze E, "Seismic Design of Large Tanks", Thesis for the degree in Civil Engineering,

Faculty of Physical Sciences and Mathematics, Central University of Chile, Santiago, Chile, December(in Spanish) ,2000.

[4] Pineda P, Saragoni R, Arze E, " Performance of Steel Tanks in Chile 2010 and 1985 Earthquakes", *Proceedings of the 7th International Conference on Behaviour of Steel Structures in Seismic Areas*, Santiago, Chile, pp. 337-342, 9-11, 2012.

[5] Housner G, "The dynamic behaviour of water tank", *Bulletin of the Seismological Society of America*, vol. 53(2), 381-87, 1963.

[6] Clough D, "Experimental Evaluation of Seismic Design Methods for Broad Cylindrical Tanks", *Earthquake Engineering Research Center*, Report to the Sponsor Administrator: Chevron Oil Field Research Company, Report NO. UCB/EERC-77/10, College of Engineering, University of California-Berkeley, California, 1977.

[7] Rinne J E, "Oil Storage Tanks", in volume II-A: The Prince William Sound, Alaska earthquake of 1964 and Aftershocks, *Environmental Science Services Administration*, *U. S. Coast and Geodetic Survey*, Washington: Government Printing Office, pp. 245-252, 1967.

[8] Cooper T W, "A Study of the Performance of Petroleum Storage Tanks During Earthquakes, 1933-1995", NIST GCR 97-720, *United States Department of Commerce Technology Administration*, *National Institute of Standards and Technology*, 1997.

[9] Peyrat S, Madariaga R, Buforn R, Campos J, Asch G, Vilotte J P, "Kinematic Rupture process of the 2007 Tocopilla earthquake and its main aftershocks from teleseismic and strong-motion data", *Geophysical Journal International*, 2009.

[10] Barrientos S, "Slip distribution of the 1985 Central Chile earthquake", *Tectonophysics*, 145, 225-241, 1988.

[11] Boroschek R, Soto P, Leon R, "Maule region earthquake", February 27, Mw = 8. 8, *RENADIC Report 10/08*, *Published electronically at http://www. cec. uchile. cl/ ~ renadic/red. html.* (last access 18/08/2011) ,2010.

[12] Vera N, "Buckling Design of Storage Tanks", Thesis for the degree in Civil Engineering, *Faculty of Engineering*, *Technical University Federico Santa Maria*, Valparaiso, Chile, May(in Spanish) , 1992.

Analytical and Experimental Methods

AN ADVANCED HYBRID SIMULATION MODEL BASED ON PHENOMENOLOGICAL AND ARTIFICIAL INTELLIGENCE APPROACHES TO PREDICT THE RESPONSE OF STRUCTURES UNDER SEISMIC LOADS

Syed Murtuza Abbas[a], Gian Andrea Rassati[a]

[a]Department of Civil and Architectural Engineering and Construction Management,
University of Cincinnati, USA
abbassm@ mail. uc. edu, rassatga@ ucmail. uc. edu

KEYWORDS: Hybrid Simulations, Beam-to-column Connections, Phenomenology, Seismic Loading, Artificial Intelligence.

ABSTRACT

Hybrid simulation technology is being widely used in the field of structural engineering for testing of structural systems to study their dynamic behaviour under seismic loads. It involves coupling of experimental laboratory testing of complex parts of a system with computational models of the remaining parts of the system whose behavior can be simulated with confidence in a finite element program. A hybrid engine helps the experimental and computational modules to interact with each other in real-time under seismic loading, and gives the overall response of the entire system as a whole. An extremely important aspect that has been overlooked since the introduction of hybrid testing is the fact that even in an ordinary structure, the number of sub-structures that would require physical testing because of their difficult, or even not-yet-accomplished simulation, greatly exceeds the hybrid simulation performing capabilities of any single laboratory or even a consortium of laboratories. With an increased complexity of the structure, the problem becomes even worse. This issue has not yet been investigated by researchers even long after the hybrid simulation technology attained maturity, and has mostly been addressed in research by using simplified models[1] for running simulations at the expense of the level of accuracy of results obtained.

Phenomenological models to predict the hysteresis response of structural components to seismic loads that do not have any mechanistic basis but are purely rule-based[2] have been developed in the past. The advanced hybrid simulation (AHS) model removes the current limitations of hybrid simulation technology by employing such phenomenological models. It engages a single experimental module per type of sub-structure that is complex enough to require experimental testing, and predicts the hysteretic response of all similar sub-structures present in the entire structural system using phenomenology and artificial intelligence. This, coupled with the response of computational models of rest of the system at every increment, provides highly realistic and economical results by drastically cutting down the number of experimental tests required for hybrid testing.

The AHS model is independent of the material and geometry of the sub-structure, as it just requires inputs from the experimental response of a sub-structure at every load increment to predict the response of all similar sub-structures to any type of loading. The AHS model feeds on the results coming directly from the simultaneously running experimental testing of a complex structural detail, and extracts information from it using an artificial intelligence algorithm to automatically generate characteristic inputs required by the phenomenological model to make response predictions. The AHS model eliminates the necessity of one complete set of experimental test results as was required by some phenomenological models developed in the past. Unlike the previously developed phenomenological models that required huge amounts of user inputs every time they were engaged to make response predictions for a new structural detail, the AHS model is completely automated. There is no requirement of any user input to run the AHS model, and the model itself extracts all the information it needs for response prediction from the results of the experimental test from the start. Even though the AHS model was utilized to make predictions of moment-rotation behaviour of beam-to-column connections in this work, the model is equally capable of predicting the moment-rotation or force-displacement response of any structural detail that exhibits a hysteretic behaviour under cyclic loading. A comparison of moment-rotation response prediction made by the AHS model using automatically generated characteristic inputs with experimental data for a connection exhibiting hardening followed by softening is shown in *Fig. 1*.

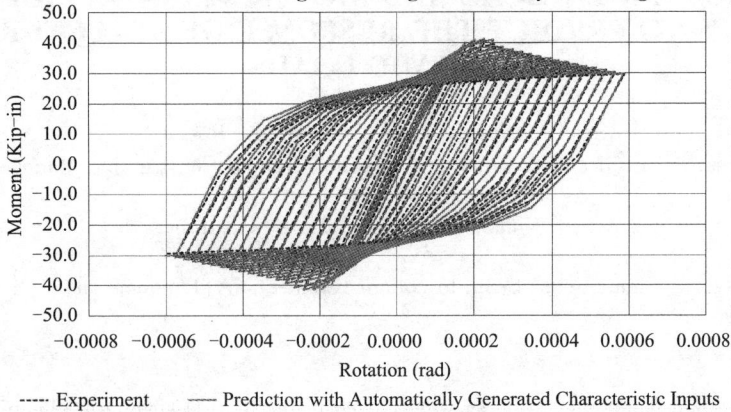

----- Experiment ——— Prediction with Automatically Generated Characteristic Inputs

Fig. 1. Comparison of moment-rotation response prediction made by the AHS code usingautomatically generated characteristic inputs with experimental data for a connection exhibiting hardening followed by softening

CONCLUSIONS

The advanced hybrid simulation model overcomes the current limitation of hybrid simulation technology by employing phenomenology and artificial intelligence. This new testing paradigm allows relatively inexpensive, and yet extremely detailed investigations on structural systems and components. The AHS model reduces the number of physical experimental tests required to conduct hybrid simulations of real-life structures from as many complex structural details that are present in the structure to one experimental testing per family of complex structural details present in the structure. This virtually removes the complexity of a structure as a deterring factor for running hybrid simulations allowing the widespread application of this innovative hybrid simulation technology to test structures whose hybrid testing was not possible before.

The developed model is computationally inexpensive, so as not to create a potential bottleneck for real-time or quasi-real-time hybrid testing protocols, andrequires much less run-time compared to a finite element model of the structural detail. This feature of the AHS model allows the hybrid testing of highly sophisticated structures with large numbers of complex structural details, e. g. , a high-rise building with multiple energy dissipating braces on several stories. Structural details have predominantly been tested under standardized loading histories in the past, which is not necessarily representative of seismic demands. The AHS model offers a viable approach to study the response of complex structural systems subjected to seismic loading in a hybrid testing environment.

REFERENCES

[1] Kwon, O. S. , Nakata, N. , Elnashai, A. , Spencer, B. 2005. "A framework for multisite distributed simulation and application to complex structural systems. " *Journal of Earthquake Engineering*, Vol. 9(5) , pp. 741-753.
[2] Chalasani, R. , Rassati, G. , Kukreti, A. , Kumar, P. , 2009. "A Phenomenological Model for the Prediction of the Hysteretic Response of Structural Components to Seismic Loads. " *Behavior of Steel Structures in Seismic Areas*, STESSA, CRC Press.

EVALUATION OF TWO SCALING METHODS IN ASSOCIATION WITH A NEW ANDPRACTICAL RECORD SELECTION PROCEDURE

Leila Haj Najafi[a], Mohsen Tehranizadeh[b]

[a]Amirkabir University of Technology, Department of Civil and Environmental
Engineering, Iran
lila_najafi@ aut. ac. ir, lilanajafi@ yahoo. com

[b] Amirkabir University of Technology, Department of Civil and Environmental Engineering, Iran
dtehz@ yahoo. com

KEYWORDS: Record Selection, Record Scaling, Uniform Hazard Spectrum, Efficiency, Soil Condition

ABSTRACT

Development of a systematic method to select and modify from current ground motion databases is a critical step in nonlinear dynamic analysis which provides a collection of earthquake motions that can realistically represent important aspects of the motion controlling nonlinear response of structural and-nonstructural components. This paper provides and evaluates a very simple and practical procedure for selecting ground motions in addition to compare two common scaling methods based on the uniform hazard spectrum(UHS) method and presents scale factors of the selected ground motions associated with these methods.

The procedure proposed in this paper is the employment of random selection by consideration of minimizing deviations around the geometric mean of natural logarithmic spectral acceleration values to reduce the effect of record-by-record variations in structural responses. In addition, two common methods have been applied for scaling recorded earthquakes according to different classes of soil. The outcomes could be directly used as the scaling ratio in related researches. The outcomes could be directly used as the scaling ratio in related researches.

The first that has been recommended by the ATC-58-1 and ASCE05-7 is also recommended by many provisions like IBC2006 and CBC2007. For two-dimensional analysis of symmetric-plan buildings, ASCE05-7 requires intensity-based scaling of ground motion records using appropriate scale factors so that the average value of the 5 percent-damped response spectra for the set of scaled records is not less than the design response spectrum over the period range from $0.2T_1$ to $1.5T_1$. The design value of an engineering demand parameter(EDP) is taken as the average value of the EDP over seven(or more) ground motions, or its maximum value over all ground motions, if the system is analysed for fewer than seven ground motions[1]. The ASCE05-7 scaling procedure does not insure a unique scaling factor for each record; obviously, various combinations of scaling factors can be defined to insure that the average spectrum of scaled records remains above the design spectrum(or amplified spectrum in case of 3-D analyses) over the specified period range[2] & [3].

The second method that is very frequently used by designers and also has been applied in ATC-58-1 example section is scaling the ground motion only in the fundamental period of the structure. Early quantitative investigations into ground motion scaling indicated that a suite of ground motions may be safely scaled to the suite's median spectral acceleration value, at a period T, without biasing the median response of a structure having the same first-mode period T[4] & [5].

In this paper, the scale factors were provided for a short-rise building with a fundamental period equal to one. After providing scale factors in two methods, the evaluation and comparison of these two methods will be done and considering all the conditions these scaling factors could be employed directly in other studies.

Scaling procedure comprises four soil conditions as well as the records were selected from a very frequently applied list of ground motions; consequently, the obtained scale factors could be utilized directly from this paper in the other studies in this field without any excessivecalculational attempts.

The obtained scale factors for different types of soil according to the two methods have been present in *Tables 1* and *2*.

Table 1. Scale factors for different types of soil according to Design and Provision methods for a building by $T = 1.0(s)$

Record ID		Design Method 2% in 50				Provision Method 2% in 50		
	Site	Site B	Site C	Site D	Site A	Site B	Site C	Site D
BM68elc	4.68	4.66	6.05	6.98	4.68	5.85	7.15	7.87
CO83c05	3.80	6.83	8.88	10.24	3.80	4.75	5.81	6.40
IV79e13	5.28	6.43	8.36	9.64	5.28	6.60	8.06	8.87
LP89slc	1.78	1.40	1.83	2.11	1.78	2.22	2.72	3.00
MH84g02	5.76	8.00	10.40	12.00	5.76	7.19	8.80	9.68
PS86psa	4.37	4.36	5.68	6.55	4.37	5.45	6.67	7.34
NR94del	4.58	5.91	7.68	8.86	4.58	5.72	7.00	7.70
PM73phn	4.92	4.61	5.97	6.89	4.92	6.16	7.52	8.28
SF71pel	3.21	4.66	3.11	5.38	3.21	4.01	4.90	5.40
SH87pls	2.96	6.53	4.35	7.53	2.96	3.71	4.53	5.00
WN87wat	4.40	6.54	4.36	7.55	4.40	5.50	6.71	7.40

Table 2. Scale factors for different types of soil according to Design and Provision methods for a building by $T = 1.0(s)$

Record ID		Design Method 10% in 50				Provision Method 10% in 50		
	Site	Site B	Site C	Site D	Site A	Site B	Site C	Site D
BM68elc	2.48	3.10	4.03	4.66	3.12	3.90	4.77	5.25
CO83c05	3.64	4.55	5.92	6.83	2.54	3.17	3.87	4.26
IV79e13	3.43	4.28	5.57	6.43	3.52	4.40	5.37	5.91
LP89slc	0.75	0.94	1.22	1.41	1.19	1.48	1.81	2.00
MH84g02	4.27	5.33	6.93	8.00	3.84	4.80	5.86	6.45
PS86psa	2.33	2.91	3.78	4.37	2.91	3.63	4.45	4.89
NR94del	3.15	3.94	5.12	5.91	3.05	3.81	4.66	5.13
PM73phn	2.45	3.06	3.98	4.60	3.28	4.10	5.02	5.52
SF71pel	1.91	2.39	3.11	3.59	2.14	2.67	3.27	3.60
SH87pls	2.68	3.35	4.35	5.02	1.98	2.47	3.02	3.32
WN87wat	2.68	3.35	4.36	5.03	2.93	3.66	4.48	4.93

CONCLUSIONS

This research proposed a simple and practical method for selecting required records for nonlinear time history analysis of a model based on the least standard deviation in natural logarithmic acceleration spectral values. In addition, this paper employs two common methods for scaling recorded earthquake data based on provisions requirements and designers experiences according to diverse class of soils. The results could be directly used as the scaling factors in related researches. Evaluation and comparison of the results deduce that by reduction in shear wave velocity in the soil classes (going from class A (hard rock) to D (stiff soil)) the differences between two methods increased and provision method becomes more conservative.

REFERENCES

[1] ASCE. Minimum Design Loads for Buildings and Other Structures, ASCE/SEI 7-10. American Society of Civil Engineers, Reston, Virginia, 2010.

[2] International Building Code, International Conference of Building Officials (ICBO), Whittier, CA, 2006.

[3] California Building Code, International Conference of Building Officials (ICBO), Whittier, CA, 2007.

[4] Iervolino, I. and Cornell, C. A. Record Selection for nonlinear seismic analysis of structures. Earthquake Spectra, 21 (3), 685-713, 2005.

[5] NIST, Soil-Structure Interaction for Building Structures, NIST/GCR 11-917-14. NEHRP Consultants Joint Venture for the National Institute of Standards and Technology, Gaithersburg, Maryland, 2011.

EXPERIMENTAL ANALYSIS OF DUAL-STEEL BOLTED T-STUBS UNDER MONOTONIC AND CYCLIC LOADING

Andreas Kleiner[a], Ulrike Kuhlmann[a]

[a]University of Stuttgart, Institute of Structual Design, Germany

Andreas. Kleiner@ ke. uni-stuttgart. de, Sekretariat@ ke. uni-stuttgart. de

KEYWORDS: High Strength Steel, Mild Carbon Steel, T – stub, cyclic behaviour, ductility.

ABSTRACT

According to modern codes the seismic design of steel or composite buildings are based on the concept of dissipative structures, where specific zones of the structures should be able to develop plastic deformation, mainly in ductile members, in order to dissipate the seismic energy.

On the contrary, under seismic action the non-dissipative zones should behave elastically in order to avoid the sudden collapse of the building. In order to achieve this design purpose the use of different steel grades for either dissipative or non-dissipative elements is an effective and actually viable solution. This consideration is the basis of the "Dual-Steel Concept", where High Strength Steel (HSS) is used for non-dissipative members and Mild Carbon Steel (MCS) for dissipative ones. The use of different steel grades may allow rationalizing the design, reducing the size of structural members and implies an overall structural weight with relevant cost savings.

In the framework of the research project "High Strength Steel in Seismic Resistant Building Frames" (HSS-SERF), an experimental programme devoted to the monotonic and cyclic response of 56 T-stubs have been carried out in order to evaluate and investigate the performance, the load carrying capacity as well as the ductility under monotonic and cyclic loading conditions. For the experimental investigations two T-stub series (unstiffened & stiffened T-stub) were subjected to monotonic quasi-static (mq-s) and cyclic quasi-static (cq-s) loading conditions.

The specimens have been designed aiming at a mode 2 failure according to the analytical predictions for plastic resistance derived from EN 1993-1-8[1]. One group of T-stubs was designed being close to mode 1 (ductile) and one group respectively close to mode 3 (brittle) by varying the thickness and the steel grade (S460 / S690) of the endplate and the distance of the bolts. Failure of the T-stubs in mode 2 combines both adequate design resistance and sufficient rotational capacity of the joint, which tends to be an optimal solution in seismic design approach. Specimens were tested under a modified cyclic loading sequence based on the ECCS recommendations[2].

CONCLUSIONS

In case of bolted T-stub connections and respectively bolted beam-to-column joints, the major contributions of the overall T-stub deformation are the endplate deformation and the tension bolt elongation, as far as its plastic failure mechanism was governed by mode 2. The present paper demonstrates that the endplate can deliberately be designed, taken into account the thickness and the steel grade, in order to achieve sufficient ductility as requested by code provisions.

The obtained results can be classified under two different aspects related to differences between the monotonic and cyclic behaviour of the specimens: load carrying capacity and ductility.

For what concerns the first aspect *Fig. 1* and *Fig. 2* show the comparison between monotonic quasi-static (mq-s) and cyclic quasi-static (cq-s) loading for unstiffened and stiffened T-stubs, designed to fail in mode 2-1 (ductile) and mode 2-3 (brittle). As it can clearly be recognised ductile failure behaviour differed completely from brittle failure where the ultimate behaviour is remarkably varying. In general a clear influence can be observed for T-stubs subjected to ductile behaviour related to the combined bending and tension forces subjected to the bolts after bending deformation of the endplate. The strength and the ductility of the T-stub specimens directly correlate to the stress distribution of the bolts. The ultimate strength for stiffened T-stubs (see *Fig. 2*) are nearly coincident with the ultimate strength of the bolts due to bolt governed collapse.

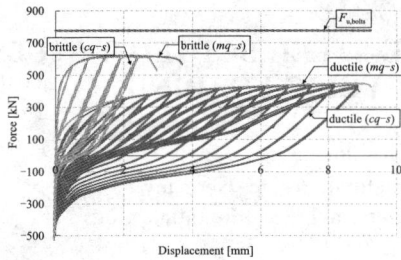

Fig. 1. F-D-curve: Unstiffened T-stub(series 100)   Fig. 2. F-D-curve: Stiffened T-stub(series 200)

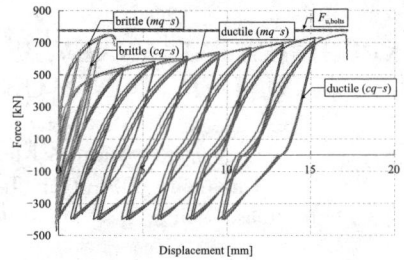

Regarding the development of strength it is evident to underline that for stiffened T-stubs the strength increases continuously up to failure due to membrane effects in the endplate whereas unstiffened T-stubs show a nearly constant behaviour up to large plastic deformations ending up in sensible strength degradation only in ultimate conditions.

This yield region is more pronounced for specimens failing in mode 2-1 where large bending deformations develop in the endplate, when compared to tests failing in mode 2-3(see *Fig. 2*).

A sudden drop of load was observed for stiffened T-stubs failing in mode 2-3, which characterises a brittle failure type. The curves for unstiffened T-stubs subjected to ductile behaviour are characterised by different amounts of pinching, caused by localized plastic deformations of the T-stub endplates[3]. In conclusion, the degree in which cyclic loading affected the ductility of T-stubs is much dependent on the failure mode. Specimens failing by mode 2-3 were characterized by an important decrease of ductility coupled with a significant load carrying capacity. On the other hand, ductility of specimens failing in mode 2-1 was much more pronounced.

ACKNOWLEDGMENT

The work presented here has been carried out within a research project by different European partners and with a financial grant from the Research Fund for Coal and Steel(RFCS)of the European Community under grant agreement n° RFSR-CT-2009-00024. The authors like to thank their partners for cooperation and gratefully acknowledge the financial support from the Research Fund for Coal and Steel(RFCS). Acknowledgement is also given to FUCHS Schraubenwerk GmbH for supplying the bolts.

REFERENCES

[1] EN 1993-1-8:2010-12-*Eurocode* 3: *Design of steel structures-Part* 1-8: *Design of joints*; German version EN 1993-1-8:2005 + AC: 2009.

[2] ECCS-European Convention for Constructional Steelwork, Technical Committee 1, Structural Safety and Loadings; Working Group 1. 3,*Seismic Design: Recommended Testing Procedure for Assessing the Behaviour of Structural Steel Elements under Cyclic Loads*,1st ed. ,1986.

[3] Coelho A. ; Bijlaard F. ; Gresnigt N. ,da Silva L. ,"*Experimental assessment of the behaviour of bolted T-stub connections made up of welded plates*",Journal of Constructional Steel Research 60 (2004)269-311.

EFFECT OF STRENGTH AND STIFFNESS OF SINGLE-STOREY STEEL BUILDINGS ON CONTENT SLIDING RESPONSE IN EARTHQUAKES

Trevor Z. Yeow[a], Gregory A. MacRae[a], and Rajesh R. Dhakal[a]

[a]Department of Civil and Natural Resources Engineering, University of Canterbury, Christchurch, New Zealand

Trevor. Yeow@ pg. canterbury. ac. nz, Gregory. MacRae@ canterbury. ac. nz, Rajesh. Dhakal@ canterbury. ac. nz

KEYWORDS: content sliding, performance-based design, strength, stiffness, non-structural

ABSTRACT

One of the common arguments against building stronger and/or stiffer buildings is that the sliding response of building contents during strong shaking events is more severe in these buildings due to its higher peak total acceleration response. This may not be true however, as recent studies have shown that peak total floor acceleration alone not a good indicator of the extent of content sliding. This study aims to assess the validity of this assumption for contents located within single-storey steel buildings of different strength and stiffness using content sliding response history analysis. A procedure is also proposed in this study to account for content sliding in layout design.

CONCLUSIONS

The 10% probability of exceedance in 50 years maximum sliding displacement values obtained from response history analysis, $\delta_{S,RHA}$, is shown in *Fig. 1*. It was found that contents located within stronger buildings (with lower force reduction factors, R) generally experience a larger sliding response compared to that within weaker buildings. This is true regardless of the period of the building, T. However, the $\delta_{S,RHA}$-T trends are affected by R. For the case where the friction coefficient, μ, is 0.1 as shown in *Fig. 1(a)*, it can be seen that $\delta_{S,RHA}$ actually increases with T for the period range considered when $R = 1$. While $\delta_{S,RHA}$ continues to increase slightly up until approximately 0.75 s for the $R = 5$ case, it can be seen that $\delta_{S,RHA}$ decreases rapidly for T greater than 0.75 s. Similar trends were observed for higher values of μ.

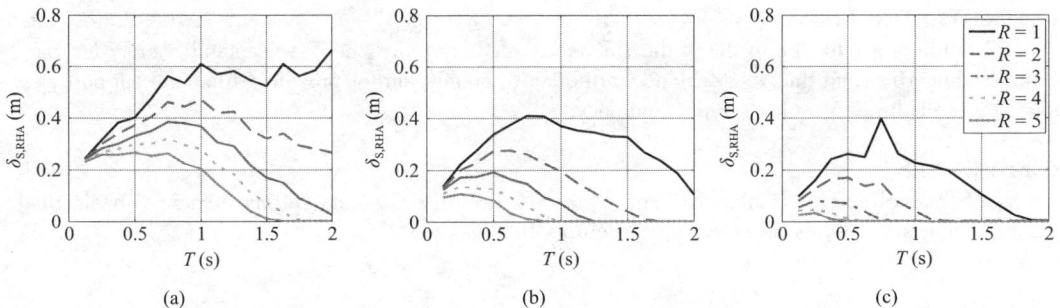

Fig. 1. 10% in 50 year content sliding response; (a)$\mu = 0.1$, (b)$\mu = 0.2$, (c)$\mu = 0.3$

In addition, parametric prediction equations based on a concept by Lin et al. [1], in that the maximum sliding displacement of contents in elastically responding buildings can be approximated by the first sliding excursion of contents subjected to sinusoidal floor motion, was tested against the yielding steel buildings from this study. These equations are shown in *Eq. 1*, where A_{FT} and V_{FT} are the peak total floor acceleration and velocity respectively.

$$\delta_{S,PARA} = V_{FT}^{2} (\pi - 10.27\mu/A_{FT} + 7.13 \ (\mu/A_{FT})^{1.446})/A_{FT}g \quad \text{if } \mu < A_{FT} \quad (1a)$$

$$\delta_{S,PARA} = 0 \quad \text{if } \mu > A_{FT} \quad (1b)$$

The median$\delta_{S,RHA}$ and $\delta_{S,PARA}$ relationships are shown in *Fig. 2* for a range of different cases. It can be seen that the ratio between these parameters is approximately $1:1$, which indicates that *Eq. 1* is sufficient to estimate the median maximum sliding displacement response.

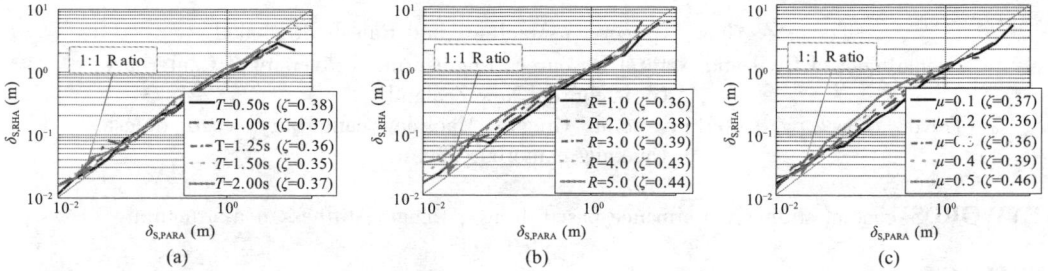

Fig. 2. Median $\delta_{S,RHA}$-$\delta_{S,PARA}$ relationships($R = 1, T = 1.25s$, and $\mu = 0.1$ unless otherwise stated); (a) varying T, (b) varying R, and(c) varyting μ

The lognormal dispersion of the$\delta_{S,RHA}$-$\delta_{S,PARA}$ relationship, ζ, is generally around 0.4 based on *Fig. 2*. ζ can therefore be used together with *Eq. 1* for design purposes, where it may be important to specify a minimum safe distance, $\delta_{S,DES}$, between large contents and escape paths so that occupants have a clear path to exit the building following a strong shaking event. This can be done using *Eq. 2*, where Z indicates the number of standard deviations $\delta_{S,DES}$ is from $\delta_{S,PARA}$ on a natural log scale. This can be read from probability tables to obtain the target percentile(e. g. $Z = 1.65$ represents the 95th percentile). An example using the 2011 February Christchurch earthquake is provided within the full paper, and demonstrates that this procedure is conservative and simple to perform.

$$\delta_{S,DES} = \exp(\ln(\delta_{S,PARA}) + 0.4) \tag{2}$$

Insummary, it was found from response history analyses of content sliding movement that:
- Content sliding response is generally more severe in stronger buildings.
- Content sliding response is not necessarily more severe in stiffer buildings. In fact there are cases that show the opposite trend.
- Existing parametric equations can be used to predict the maximum content sliding response in yielding steel buildings, and can therefore be used for design purposes.

ACKNOWLEDGEMENTS
The authors would like to thank the University of Canterbury, the New Zealand Society for Earthquake Engineering, and the New Zealand Earthquake Commission for providing financial support at various stages of the first author's PhD candidacy.

REFERENCES
[1] Lin SL, MacRae GA, Dhakal RP, and Yeow TZ. Building contents sliding demands in elastically responding structures. Engineering Structures In-Press.

EXPERIMENTAL STUDIES OF ECCENTRICALLY BRACED FRAME WITH ROTATIONAL BOLTED ACTIVE LINKS

Hoi Kit Leung[a], G Charles Clifton[b], Hsen Han Khoo[c], Gregory A MacRae[d]

[a]University of Auckland, BGT Strucutres New Zealand

hleu585@ auckland. ac. nz

[b] Univeristy of Auckland

c. clifton@ auckland. ac. nz

[c] KTA (Sarawak) Sdn Bhd, Malaysia, Unitec Institution of Technology, New Zealand

hkhoo@ unitec. ac. nz

[d] University of Canterbury

gregory. macrae@ canterbury. ac. nz

KEYWORDS: Rotational Bolted Active Link, Symmetric Friction Connection, Rotational Friction

ABSTRACT

Friction connections are an efficient way to dissipate energy in seismic resisting systems. This paper proposes the Rotational bolted active link (RBAL) for application in eccentrically braced frames (EBF) and reinforced concrete(RC) shear wall as coupling beams. The inelastic deformation and energy dissipation of the RBAL is achieved through rotational sliding of symmetry friction connections. Experimental studies on eight full scale RBAL specimens have been conducted using mild steel and abrasion resistant steel capping plates. The shear friction resistance increased during cyclic loading when mild steel capping plates were used. This is due to material wear, which increased the bolt clamping length and bolt tension. The use of abrasion resistant steel capping plates reduced material wear, and produced stable hysteresis behaviour while maintaining strength and stiffness over large cycle up to 0. 09 rad link rotation. The experimental results shows the RBAL can achieve high ductility and is considered as a low damage alternative to conventionally designed EBFs and RC shear walls.

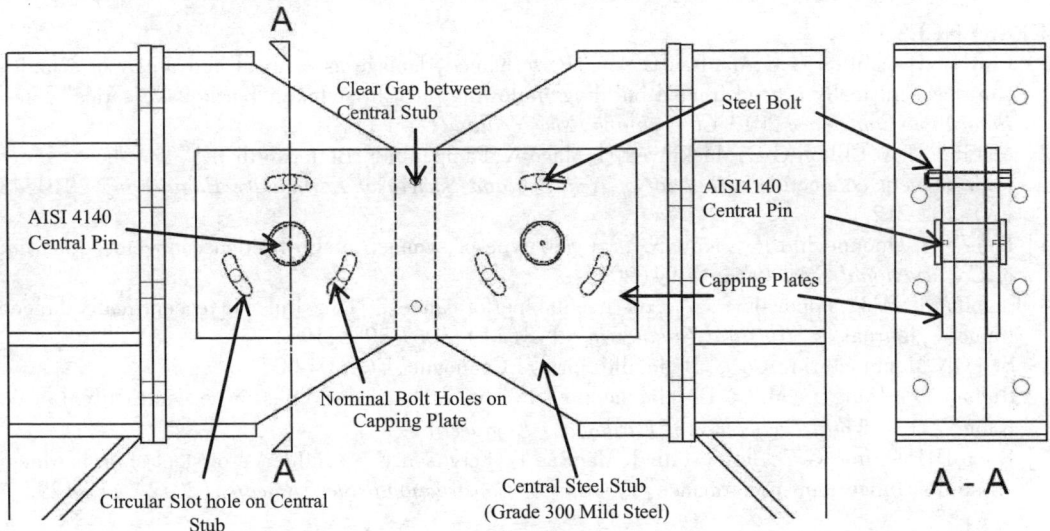

Fig. 1. The Rotational Bolted Active Link Layout

CONCLUSIONS

1. The stiffness and strength of the RBAL with both abrasion resisting steel and mild steel capping plates were maintained over a large number of cycles. The experimental studies show that the perform-

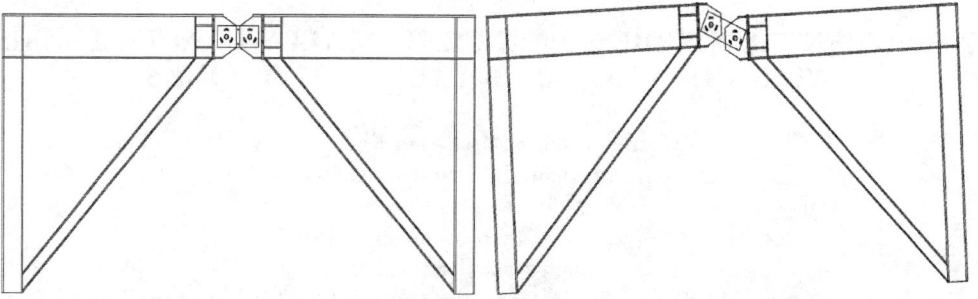

Fig. 2. (Left) EBF with RBAL, (Right) EBF with RBAL under Lateral Imposed Load

ance of the RBAL is highly dependent on the hardness of the capping plates. The use of abrasion resisting steel capping plates produced stable and predictable hysteresis while the hysteresis of mild steel capping plates gradually increased as the RBAL approached larger rotations. It is recommended that abrasion resisting steel is used for the capping plates of the RBAL.

2. For the RBAL with abrasion resisting steel capping plates, a 10% bolt tension loss was observed throughout the experiment. However, the loss in bolt tension has essentially no effect on the capacity of the RBAL. For the RBAL with mild steel capping plates,30% bolt tension loss was observed up to 0. 05 rad rotation followed by an increase in bolt tension. The wearing of particles due to excessive galling on the sliding surface contributes to the constant increase in friction sliding capacity. This is undesirable as the friction sliding capacity is difficult to predict.

ACKNOWLEDGMENT

The authors gratefully acknowledge the financial support of the New Zealand Earthquake Commission(EQC). The first author would like to acknowledge BGT structures, New Zealand for their support. All opinions expressed in this paper are those of the authors and not necessarily the views off the sponsors.

REFERENCES

[1] Gardiner S, Clifton G C, MacRac G A. , "Performance, damage assessment and repair of a multi-storey eccentrically braced framed buildings following the Christchurch Earthquake series", *Steel Innovation Conference* 2013 *Christchurch, New Zealand.*

[2] MacRac G A, Clifton G C, Mackinven H, Mago N, Pampanin S, Butterworth J. , "The sliding hinge joint moment connection", *Bulletin of New Zealand Society of Earthquake Engineering* 2010,43 (3) :202-212.

[3] Loo W Y, Quenneville P, Chouw N. , "A new type of symmetric slip friction connector", *Journal of Constructional Steel Research* 94(2014).

[4] Engelhardt M D, Popov E P. , "Experimental performance of long links in eccentrically braced frames", Journal *of Structural Engineering*, 118(11) ,3067-3088 ,1992.

[5] SOLON Manufacturing Co. , "Belleville Springs Catalogue", *Cat* 04-007.

[6] Richard P, Uang C M. , " Development of testing protocol for links in eccentrically braced frames" ,13th *World Conference of Earthquake Engineering.*

[7] Khoo HH, Clifton G C, Butterworth J, MacRae G, Ferguson G. , "Influence of steel shim hardness on sliding hinge joint performance", *Journal of Constructional Steel Research*(2012) 119-129.

EARTHQUAKE SEQUENCE EFFECTS ONSTEEL BUILDINGS

Ali A. Rad[a], Gregory A. MacRae[a], Trevor Z. Yeow[a], Desmond Bull[a]

[a]Department of Civil and Natural Resources Engineering, University of Canterbury,
Christchurch ali. abdolahirad@ pg. canterbury. ac. nz, gregory. macrae@ canterbury. ac. nz,
trevor. yeow@ pg. canterbury. ac. nz, des. bull@ canterbury. ac. nz

KEYWORDS: Christchurch Earthquake Sequence, Inelastic Time History Analysis, Peak and Residual Drift.

ABSTRACT

Inelastic dynamic response history analysis was conducted on a series of simple 2-D shear-type building structures subjected to a series of earthquake events to evaluate their total response in terms of peak and residual inter-story drift. The buildings were designed according to the Equivalent Static Method in NZS1170. 5 (2004). Ground motion recordings from the Botanical Gardens in Christchurch, New Zealand, from the September and February Christchurch earthquake sequence records were considered. It was assumed that there was no repair or structural intervention between events. Structures considered have design ductility (response reduction factor) of 3, 4, and 5, target maximum allowable inter-story drifts of 1% and 2%, and the number of stories (N) of 6, and 12. It is shown that that the average peak and residual inter-story drifts for the two event sequence were more than that under the February event alone. The earthquake sequence effect is shown to increase with higher design ductility.

CONCLUSIONS

Inelastic response history analyses were conducted of shear-type structures of 6 and 12 stories with design ductility (response reduction factor) of 3, 4, and 5, and design drifts of 1% and 2%, in order to evaluate the change in residual and peak inter-story drift ratio under September-February Canterbury earthquake sequence compared to that from the February event alone. The main findings are:

1) Considering the proceeding September-February events together led to average maximum increases of residual and peak inter-story drift response up to 15% compared to considering the February event alone.

2) Thetotal residual and peak inter-story drift ratio was found to increase most strongly with increasing design ductility (response reduction factor). They were less sensitive to design drift and story heights. For low design ductility, there is no effect of the September event on the February response.

ACKNOWLEDGEMENTS

The authors would like to thanks the University of Canterbury for a Canterbury Scholarship to fund this research.

REFERENCES

[1] Ruiz-García, J. , " Mainshock-Aftershock Ground Motion Features and Their Influence in Building's Seismic Response", Journal of Earthquake Engineering, 16, 719-737, 2012.

[2] Goda, K. and Taylor C. A. , "Effects of aftershocks on peak ductility demand due to strong ground motion records from shallow crustal earthquakes", Earthquake Engineering and Structural Dynamics. 41, 2311-30, 2012.

[3] Fragiacomo, M. , Amadio, C. , and Macorini, L. , "Seismic response of steel frames under repeated earthquake ground motions", Engineering Structures 26, 2021-2035, 2004.

[4] Hatzigeorgiou, G. D. and Liolios, A. A. , "Nonlinear behaviour of RC frames under repeated strong ground motions, " Soil Dynamics and Earthquake Engineering 30, 1010-25, 2010.

［5］ Ruiz-García，J.，Mainshock-Aftershock Ground Motion Features and Their Influence in Building's Seismic Response，Journal of Earthquake Engineering，16，719-737，2012.

［6］ Goda，K. and Taylor C. A.，"Effects of aftershocks on peak ductility demand due to strong ground motion records from shallow crustal earthquakes"，Earthquake Engineering and Structural Dynamics. 41，2311-30，2012.

AN APPROACH FOR EVALUATING THE DAMAGE-CONTROL BEHAVIOR OF STEEL FRAMES WITH BUCKLING RESTRAINED BRACES BASED ON ENERGY BALANCE CONCEPT

Ke Ke[a], Xiu-Zhang He[a], Yi-Yi Chen[a]

[a] State Key Laboratory of Disaster Reduction in Civil Engineering, Tongji University, Shanghai 200092, China

1987keke@ tongji. edu. cn, xiuzhanghe@ gmail. com, yiyichen@ tongji. edu. cn

KEYWORDS: buckling restrained braces, damage-control structures, energy-balance concept, damage index

ABSTRACT

Recent disasters imply that the conventional plastic design methodology may lead to costly repair work, manifesting the significance of structural resilience. In this respect, much effort has been made focusing on the innovation of structural systems and design approaches. Recent research works indicate that the implementation of buckling restrained braces into steel frames can significantly improve the seismic performance of the system, and the damage-control behavior that plastic deformation is restricted within the braces while frame members stay damage-free in a certain range can effectively enhance the seismic resilience. In this regard, the evaluation of the damage-control behavior is a critical issue for the structural resilience design. Compared with the nonlinear response history analysis, an applicable approach that retains the computational efficiency and conceptual simplicity is of great significance in engineering practice.

In this study, an evaluation approach for damage-control behavior assessment based on the energy-balance concept is presented and applied in systems of steel frame with buckling restrained braces (BRBs). First, the energy balance of the system is established, and the energy factor considering the exact inelastic behavior of the systemis derived. Then, a procedure for evaluating the damage-control behavior of steel frames with BRBs based on the energy factor associating with the pushover analysis is presented in detail. A damage index that quantifies the damage-control behavior is also proposed. Afterwards, the procedure is applied in the example structure and nonlinear response history analysis is also performed for validation, in which the feature of plastic energy is used to identify the structural damage. The results show that the proposed approach shows attractive accuracy, and can be used to evaluate the damage-control behavior practically.

CONCLUSIONS

Thisresearch focuses on damage-control assessment approach of steel frames with buckling restrained braces(BRBs). Based on the modified energy-balance concept, an energy-based approach is proposed. The developed procedure based on the energy-balance considering the interaction of systematic nonlinear behavior and ground motions can account for the damage-control behavior quantitatively with provided input excitations. The comparison of the responses of the example structure and the prediction of the proposed approach shows attractive accuracy, demonstrating the applicability of the proposed approach.

Essentially, the system of steel frames with BRBs can be evaluated with a more rational approach focusing on the damage-control behavior. Though the proposed approach in this research is established based on the fundamental mode with assumptions, it is thought to be applicable in evaluating low to medium rise buildings, retaining attractive efficiency before nonlinear dynamic analysis. In addition, as the approach regarding the damage-control behavior check is established considering the dynamic properties of system, it is also believed to be applicable in various structures for damage-control assessment and resilience enhancement. research work regarding this part are in progress and they will provide full validation of the approach.

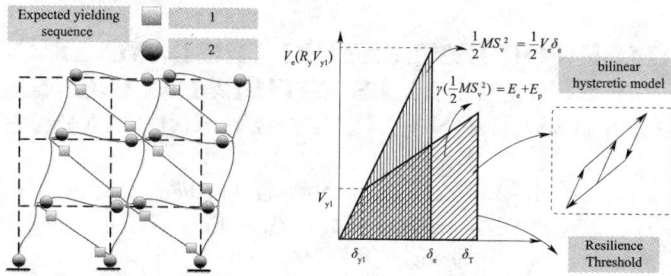

Fig. 1. The yielding sequence and the energy balance of steel frames with BRBs

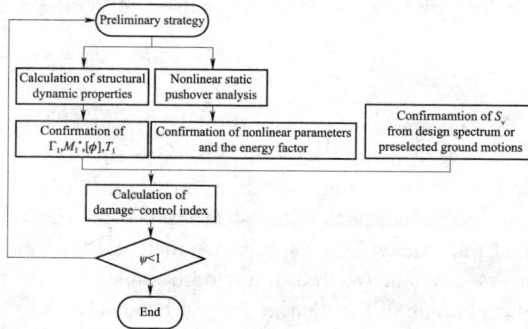

Fig. 2. The evaluation approach for damage-control assessment of steel frames with BRBs

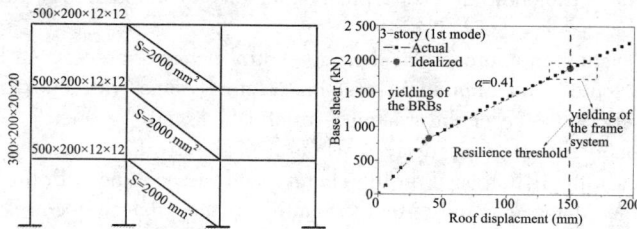

Fig. 3. The section arrangements and pushover curve of the example structure

ACKNOWLEDGMENTS

This research is supported by the National Science Foundation of China(Grant No. 51038008). Any o-pinions,findings,and conclusions presented in this paper are those of the authors and do not necessarily reflect the views of the sponsors.

REFERENCES

[1] Pekcan G,Itani AM,Linke C,"Enhancing seismic resilience using truss girder frame systems with supplemental devices", *Journal of Constructional Steel Research*, Vol. 94,pp. 23-32.

[2] Wada A,Connor JJ,Kawai H,et al,"Damage Tolerant Structure",*5th US-Japan Workshop on the Improvement of Building Structural Design and Construction Practice*,1992.

[3] Vargas R,Bruneau M,"Analytical response and design of buildings with metallic structural fuses. I",*Journal of Structural Engineering*,Vol. 135,pp. 386-393.

[4] Leelataviwat S,Saewon W,Goel SC,"Application of energy balance concept in seismic evaluation of structures",*Journal of Structural Engineering*,Vol. 135,pp. 113-121.

[5] Chintanapakdee C,Chopra AK,"Evaluation of modal pushover analysis using generic frames", *Earthquake Engineering and Structural Dynamics*,Vol. 32,pp. 417-442.

[6] Hernandez-Montes E,Kwon OS,Aschheim MA,"An energy-based formulation for first-and multi-ple-mode nonlinear static (pushover) analyses",*Journal of Earthquake Engineering*,Vol. 8, pp. 69-88 .

MODELLING ON POST-LOCAL BUCKLING DEGRADATION BEHAVIOR OF SQUARE HOLLOW STEEL SECTION BEAM-COLUMNS

Yong-Tao Bai[a,b], Masahiro Kurata[b], and Masayoshi Nakashima[b]

[a] JSPS Postdoctoral Fellow, Japan

baiyongtao@ gmail. com

[b] Disaster Prevention Research Institute, Kyoto University, Japan

kurata. masahiro. 5c@ kyoto-u. ac. jp, nakashima@ archi. kyoto-u. ac. jp

KEYWORDS: Fiber element model, Stress-strain relationship, Strength degradation, Local buckling, Steel columns.

ABSTRACT

This paper presents a fiber element model that is capable of tracing post-local buckling behavior of square hollow steel section with consideration to strength and stiffness degradation, and axial contraction. The degradation behavior was incorporated explicitly in stress-strain curves and was assumed to concentrate on a local region where plasticity and local buckling occur. The developed model was first applied to the results of cyclic tests on square hollow steel section (HSS) beam-columns that were subjected to cyclic loading with constant axial force. In particular, post-peak degradation ratio and axial contraction were compared. An influence of a plastic hinge length in the model on the post-local buckling behavior was also examined. The model accurately tracked the post-local buckling behavior of the test results, and would be useful to a collapse simulation of high-rise steel frames.

INTRODUCTION

The seismic behavior of steel and steel-concrete composite structures is dominated from strength, deformation capacity and local failure as well. This paper evaluates a post-local buckling degradation behavior of square HSS beam-columns and presents a modelling of such behaviour using fiber elements[1]. A simple-meshed fiber element model is developed by defining degrading uniaxial stress strain relationship based on the test results of square HSS beam-columns conducted in the past.

STRESS STRAIN MODEL FOR LOCAL BUCKLING

A uniaxial stress-strain material model incorporating the strength and stiffness degradation after local buckling is developed. This model utilizes a Giuffré-Menegotto-Pinto (M-P) nonlinear relationship with isotropic strain hardening. In the M-P relationship, Young's modulus E, yield strength of steel σ_y, and ultimate tensile strength of steel σ_u are determined based on coupon tests.

An onset of local buckling triggers a negative slope in the skeleton curve at a critical stress σ_{lb}. The critical stress σ_{lb} is determined by its associated strain ε_{lb}. The negative slope after local buckling is bilinear with two defining factors, τ_{lb} and τ_{re}. The value of the second negative slope τ_{re} should be a small negative value, for instance -0.005 times Young's modulus to keep some amount of residual strength considering contact between severely buckled plate elements.

CYCLIC TESTS FOR MODEL VERIFICATION

A series of cyclic tests on square HSS beam-columnswas conducted with width-to-thickness ratios and axial load ratios as test parameters[2]. The deformation at the top of the beam-column specimen was controlled by a quasi-static loading system.

Fig. 1 (a) shows the comparisons for the hysteresis curves of test results with those of simulation results with deterioration effects for the specimens with a width-thickness ration of 17 and axial load ratio of 0. 3. The strength started deteriorating at the rotation of 0. 04 rad for the two specimens with different axial load ratios, 0. 1 and 0. 3. However, the negative slope after the onset of local buckling was steeper with larger axial load. The specimen S-1703 lost its axial-load carrying capacity at a rotation of 0. 06. The model with deterioration successfully predicted the rotation at the initiation of local bucking and the strength degradation along with the increase in loading amplitudes up to extremely large deformation at approximately 8 ~ 10 times the yielding rotation.

Comparisons on the axial contractions for S1703 is shown in *Fig. 1*(b). The strength degradation had close relation with the increase of axial contraction. By appropriately accounting for the degradation in stiffness and strength in the strain-stress relationship, the fiber element model successfully tracked the increase in axial contraction. *Fig. 1*(c) illustrates a typical buckling mode of the S1701 specimen at column end. The yield lines deformed extremely large and that both sides of the steel plates along the yield line nearly touched with each other.

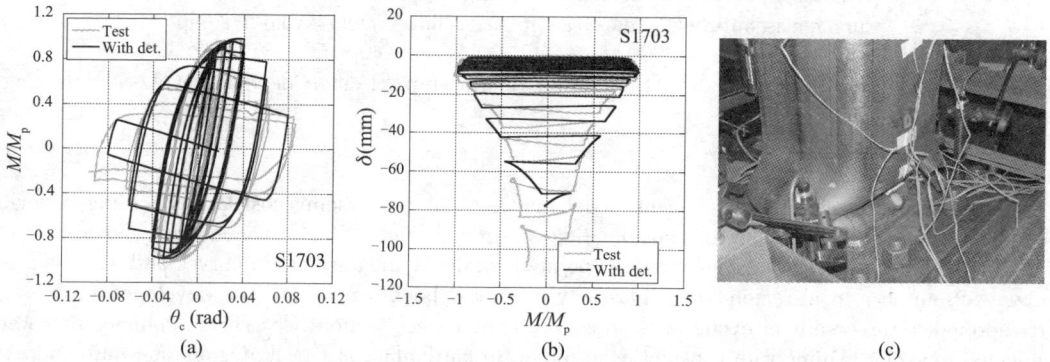

Fig. 1. (a) Normalized moment vs. end-rotation relationships; (b) Normalized moment vs. axial contraction relationships; and (c) Buckling mode at column end

CONCLUSIONS

This study presents a fiber element model for simulating post-local buckling behavior of square HSS beam-columns. The deterioration performance was considered at a stress-strain relationship by defining degradation parameters referencing test results.

The model was validated using the quasi-static cyclic loading test results of beam-column with various width-to-thickness ratios and axial load ratios. The simulation reasonably predicted the test behaviors up to extremely large deformation. The model also tracked well the axial contractions of the specimens while the parameters in the strain-stress relationship was tuned only to degradation in strength and stiffness. This was found as a great advantage of the developed model, which is expected to be effective in simulating seismic collapse of steel buildings and powerful in the simulation for large-scale structures like as high-rises.

ACKNOWLEDGMENT

The author Yongtao Bai was supported by the research fellowship from the Japan Society for the Promotion of Science(JSPS) (No. P14059). The works presented in this paper were supported by Grant-in-Aid for JSPS Fellows(26 · 04059).

REFERENCES

[1] Bai, Y, Kawano, A, Odawara K, Matsuo S. , "Constitutive Models for Hollow Steel Tubes and Concrete Filled Steel Tubes Considering the Strength Deterioration. " Journal of Structural and Construction Engineering(AIJ), 77(677): 1141-1150, 2012.

[2] Kurata M, Nakashima M, Suita K. , "Effect of Column Base Behaviour on the Seismic Response of Steel Moment Frames", Journal of Earthquake Engineering, 9(2): 415-438, 2005.

LARGE SCALE COLLAPSE EXPERIMENTS OF WIDE FLAGE STEEL BEAM-COLUMNS

Yusuke Suzuki[a] , Dimitrios G. Lignos[b]

[a] PhD Candidate, Dept. of Civil Engineering and Applied Mechanics,
McGill University, Canda
yusuke. suzuki@ mail. mcgill. ca

[b] Assistant Professor, Dept. of Civil Engineering and Applied Mechanics,
McGill University, Canada
dimitrios. lignos@ mcgill. ca

KEYWORDS: Wide flange column, collapse experiment, loading protocol, local buckling.

ABSTRACT

This paper discusses the findings of an experimental program that focuses on the cyclic behaviour of wide flange steel columns as part of moment resisting frames (MRFs) near collapse. The goal is to assess the seismic performance of wide flange steel columns at large deformations associated with dynamic collapse of steel MRFs subjected to earthquakes. Ten cantilever wide flange steel column specimens that cover a wide range of local flange and web slenderness ratios and represent typical first story interior and end columns of steel MRFs, including W14x53, W14x61 and W14x82 are tested. To replicate realistic loading conditions for interior and end first story steel columns in terms of axial load coupled with lateral deformation demands, a "near-collapse" lateral loading protocol is employed as part of the experimental program. The lateral loading protocol is coupled with axially constant and varying loading protocols that represent interior and end column behaviour, respectively, and represents the ratcheting behaviour that a steel column experiences as part of a steel MRF prior to collapse. The test results demonstrate that a routinely used symmetric lateral loading protocol is not realistic to assess the collapse performance of a steel column as part of a steel MRF. Notable differences of achieved lateral drifts, strength deterioration and column axial shortening at large deformations are also observed.

CONCLUSIONS

The post-buckling behaviour of steel columns used in the seismic design of steel moment resisting frames (MRFs) is investigated through an extensive experimental program. The major findings are summarized as follows:

1) All the specimens fully deteriorated in flexural and axial strength due to local buckling. *Fig. 1* shows a typical failure mode of a W14x53 steel column. Local buckling developed at about $0.5d$ (d: is the corresponding column depth) from the fixed end of the steel column. The formation of a second buckling wave at about $1.0d$ triggered lateral torsional buckling.

Fig. 1. Failure mode of W14x53 steel column
(W-6-34-C1-C specimen)

2）The steel column flexural strength and stiffness under a symmetric cyclic lateral loading proto-col[1] deteriorates much more than the case that a near-collapse or a monotonic lateral loading proto-col[2] is employed. *Fig. 2* shows a comparison of the moment and axial shortening versus chord rotation diagrams of identical W14x51 steel columns under various loading protocols. The achieved lateral drift of a steel column subjected to a near-collapse lateral loading protocol is in average twice larger than the corresponding drift of the same steel column, which is subjected to a symmetric reversed cyclic lat-eral loading protocol. This indicates that the post-buckling deterioration parameters of a numerical model should be identified from a combination of the proposed near-collapse and a monotonic lateral loading protocol in order to reliably trace the steel column strength and stiffness deterioration both in flexural and axial strength of a steel column.

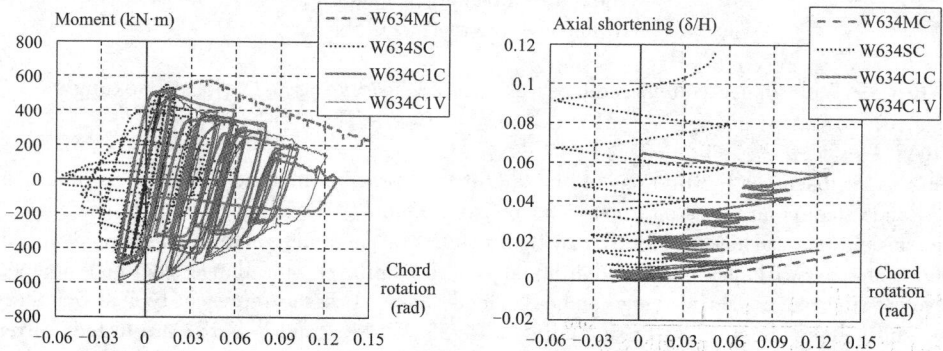

Fig. 2. Comparison of the hysteretic performance of identical W14x51 steel columns under various loading protocols

3）At fairly small lateral deformation amplitudes, the flexural strength of a steel column under va-rying axial load deteriorates more rapidly than that of the same column under constant compressive axi-al load due to the high axial compressive load. However, at large lateral deformations, the effect of cu-mulative damage on the buckled flange of steel columns under varying axial load is typically smaller than that of identical steel columns subjected to a constant compressive axial load ratio. This difference implies that interior steel columns would typically lose faster their flexural strength compared to end columns within the same story of a steel MRF.

4）The axial shortening of an end column（i. e., varying axial load）varies linearly with respect to the column chord rotation. The axial shortening of an interior column is typically 6 to 7 times larger than that of an end column under the same lateral loading protocol. Therefore, differential axial short-ening is likely to develop between adjacent steel columns within the same story of a MRF. This implies that a fully restrained moment connection near an end column will be subjected to both flexural and tensile axial load. This loading condition is currently not considered for collapse prevention of steel MRFs designed in seismic regions. However, the differential axial shortening diminishes in case that a highly compact cross section isconsidered.

ACKNOWLEDGMENT

This study is based on work supported by Nippon Steel & Sumitomo Metal Corporation in Japan. The financial support is gratefully acknowledged.

REFERENCES

［1］ Clark P, Frank K, Krawinkler H, Shaw R. , "Protocol for fabrication, inspection, testing, and docu-mentation of beam-column connection tests and other experimental specimens", SAC Steel Project Background Document, SAC/BD-97/02, 1997.
［2］ Suzuki Y, Lignos DG. , "Development of loading protocols for experimental testing of steel col-umns subjected to combined high axial load and lateral drift demands near collapse", Proceedings of Tenth U. S. National Conference on Earthquake Engineering, Anchorage, AK, 2014.

ANALYTICAL STUDY ON THE YIELD STRENGTH OF ROOF BRACE AND THE IN-PLANE DEFROMATION OF STEEL-GYMNASIUM ROOF

Yuka Matsumoto[a] and Marie Suzuki[b]

[a]Yokohama National University, Japan

yk-mtsmt@ ynu. ac. jp

[b] Kumagai Gumi Co. , Ltd, Japan

marie. suzuki@ ku. kumagaigumi. co. jp

KEYWORDS: Steel gymnasium, In-plane deformation, Roof brace, Seismic response analysis

ABSTRACT

The roof structure of the steel gymnasium should be designed to have the sufficient strength to transmit the inertial force generated by the earthquake to the lateral force resisting system. When the strength of the roof structure is not sufficient and the elongation or fracture of the roof brace occurs, the excessive in-plane deformation of the roof possibly causes the damage of the non-structural components. A school gymnasium is commonly designated as a shelter in case of a severe earthquake and the serviceability after the event is highly demanded. However, the damage of roof brace or non-structural components often spoils the serviceability, even if the damage of major structural components is prevented.

The aim of this study is to investigate the relation between the yield strength and the in-plane deformation of the roof structure and examine the accuracy of the current criterion in the seismic evaluation in Japan. For this purpose, a series of seismic response analyses was conducted on the 3-dimentional frame models.

Analysis model is shown in *Fig. 1*. It is a steel structure composed of braced frames in the X-direction and moment resistant frames in the Y-direction. The moment resistant frames are built in parallel and connected using beams and roof braces. The layout of roof brace is shown in *Fig. 2*. The layout and the section of roof brace are employed as analysis parameter.

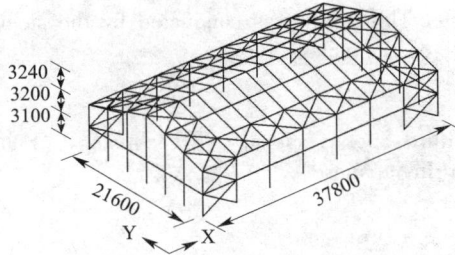

Fig. 1. Analysis model of steel gymnasium

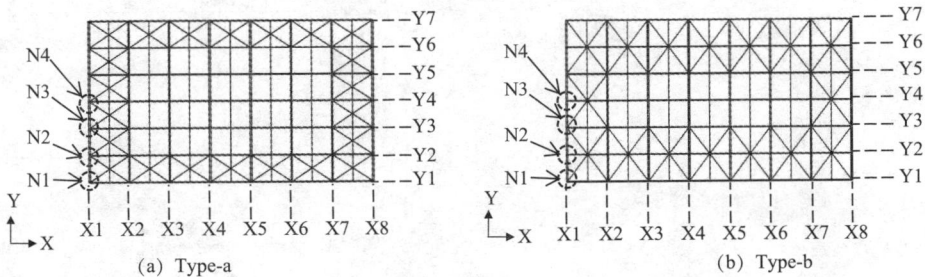

(a) Type-a   (b) Type-b

Fig. 2. Layout of roof braces

Ministry of Education, Culture, Sports, Science and Technology in Japan has issued a standard for the seismic evaluationof the school gymnasium and the computation method of the shear force in the roof section are proposed[1]. Based on the method, the shear force in the roof section is computed for each analysis model and the ratio of the computed shear force to the yield strength is obtained. The analytical result is compared with the computation.

Whenthe shear force ratio is about 1, the entire roof braces in the critical section get yielded in the seismic response analysis. The maximum deformation angle of the roof section is about 1/60 and the elongation of the roof brace is from 0.4 to 0.8%.

When the shear force ratio is from 0.5 to 0.7, the roof braces in the critical section partly get yielded. The maximum deformation angle of the roof section is about 1/200 and the elongation of the roof brace is less than 0.3%. When the shear force ratio is less than 0.3, the roof braces in the critical section completely remains elastic.

The analytical shear force in the roof section exceeds the computed value in most cases and a certain margin is desirable at the design stage.

CONCLUSIONS

A series of seismic response analyses was conducted on the 3-dimentional model of steel gymnasium. Based on the method proposed in the current seismic evaluation standard, the shear force in the roof section was computed for each analysis model and the ratio of the computed shear force to the yield strength was obtained. From the comparison of analytical result and the computation, the following conclusions can be made.

(1) The yielding condition of the roof brace depended on the shear force ratio. When the shear force ratio is about 1, the entire roof braces in the critical section get yielded in the seismic response analysis. The maximum deformation angle of the roof section is about 1/60 and the elongation of the roof brace is from 0.4 to 0.8. When the shear force ratio is from 0.5 to 0.7, the roof braces partly get yielded. When the shear force ratio is less than 0.3, the roof braces completely remains elastic.

(2) The analytical shear force in the roof section exceeds the computed value in most cases and a certain margin is desirable at the design stage.

ACKNOWLEDGMENT

The Japan Building Disaster Prevention Association organized the committee on the sophistication of the seismic evaluation method. This study was conducted by the steel working group (supervisor: Prof. S. Yamada, Tokyo Institute of Technology).

REFERENCES

[1] Ministry of Education, Culture, Sports, Science and Technology, "Standard for the seismic evaluation of gymnasium", 2006 (in Japanese).

EXPERIMENTAL DETERMINATION OF BASE SHEAR FROM FULL-SCALE SHAKE TABLE TESTING OF TWO COLD-FORMED STEEL FRAMED BUILDINGS

Kara D. Peterman[a] and Benjamin W. Schafer[b]

[a] Northeastern University, Dept. of Civil and Environmental Engineering, USA

k. peterman@ neu. edu

[b] Johns Hopkins University, Dept. of Civil Engineering, USA

schafer@ jhu. edu

KEYWORDS: cold-formed steel, seismic, shake table testing, base shear.

ABSTRACT

The work presented herein is based on the full-scale experimental work conducted as part of the CFS-NEES effort, in which two cold-formed steel(CFS) framed buildings were tested on the twin shake tables at the University of Buffalo Structural Engineering and Earthquake Simulation Laboratory (SEESL). The buildings were designed in accordance with the North American specification and modern design and construction preferences. The first building examined the performance of the structural system only; gravity framing, sheathed CFS shear walls, and sheathed floor and roof diaphragms were tested absent of any non-structural elements. The second building was outfitted with gravity wall and floor sheathing(oriented strand board and gypsum), interior partition walls, staircases, and exterior weatherproofing sheathing. An extensive series of non-destructive and destructive dynamic tests were performed, from which building system behavior was determined. The performance of key subsystems, including the lateral force resisting system, floor and roof diaphragms, ledger framing system, the effect of openings and windows, and the contribution of the gravity and non-structural systems were also determined from the experimental data. This work examines base shear and presents several approaches for determining base shear experimentally. Comparisons to design and computational analyses are presented along with recommendations for future experimental efforts and engineering design.

INTRODUCTION

The CFS-framed building specimens were built to the state-of-the practice, in accordance with the North American specifications as prescribed in AISI-S213[1] and modern construction preferences. A series of 141 tests, destructive and non-destructive, were conducted on two separate buildings, one framed with structural components only(phase 1), the other progressively outfitted with non-structural components(phase 2).

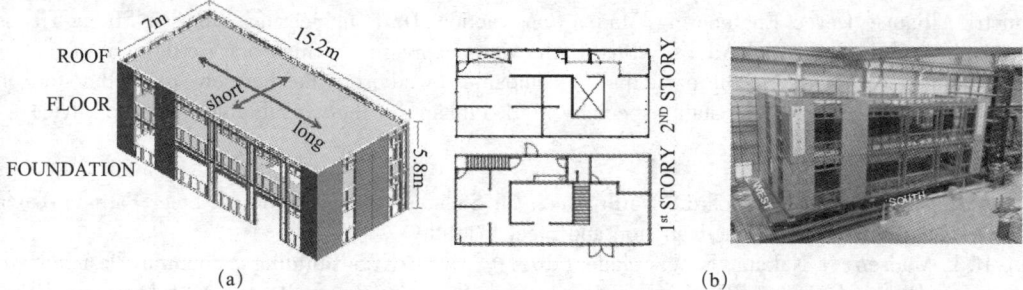

Fig. 1. (a) Specimen terminology and dimensions
(b) First and second story floor plans showing partition wall layout for Phase 2
specimen. Photograph at right shows Phase 1 building specimen as constructed on shake tables

These building specimens were outfitted with 168 different sensors including 20 load cells at each shear wallholdown, 20 strain gauges at each shear wall tie, 74 1DOF accelerometers placed throughout the building,and 54 strain gauges on the building exterior and on the shear walls and openings. Utilizing the load cells and strain gauges installed on these critical shear wall details(the anchor component and the tie component,connecting one story to another),base shear is estimated via four experimental methods.

CONCLUSIONS

Base shear estimated via four experimental methods. Method 1 is based upon the equivalent lateral force assumption in ASCE 7-05[4] and uses data from strain gauges and load cells. Method 2 reliessoley on strain gauge data from shear wall ties. Method 3 is determined based on the load cell data alone. Finally,method 4 is a theoretical determination based on the behavior of a two story shear building model without damping. The four methods are compared in *Table 1* below for the long and short directions of the test building.

Table 1. Comparison across base shear calculation methods

	16% CNP Phase 1		16% CNP Phase 2a		16% CNP Phase 2b		16% CNP Phase 2c		16% CNP Phase 2d		16% CNP Phase 2e		16% RRS Phase 2e	
Method —	Long (kN)	Short (kN)	Long (kN)	Short (kN)	Long (kN)	Short (kN)	Long (kN)	Short (kN)	Long (kN)	Short (kN)	Long (kN)	Short (kN)	Long (kN)	Short (kN)
1	-0.91	-7.12	-11.47	2.76	-14.89	-7.70	6.56	8.57	9.42	15.64	10.76	14.18	10.21	18.60
2	-4.30	-3.24	5.09	-3.58	-0.33	0.21	-0.69	-0.19	0.38	0.00	-0.30	-0.44	1.08	-0.44
3	-3.46	-4.84	-3.55	3.51	-9.02	-5.04	4.13	6.97	6.35	9.67	6.46	9.17	7.47	11.44
4	26.35	-18.24	32.16	-18.08	34.45	20.93	6.03	20.39	30.65	21.04	-1.73	19.10	67.70	31.47

	100% CNP Phase 1										100% CNP Phase 2e		100% RRS Phase 2e	
Method —	Long (kN)	Short (kN)									Long (kN)	Short (kN)	Long (kN)	Short (kN)
1	-84.32	-79.05									-5.40	-13.47	-30.43	-13.47
2	-41.19	-16.40									13.17	5.72	13.17	5.72
3	-34.20	-46.26									4.18	-7.04	-9.62	-7.04
4	158.46	-88.29									227.74	102.19	406.99	153.32

Direct comparisons to methods 1-3 at the elastic level(ground motions at 16% of full-scale)demonstrate that the superposition of shear wall base shears are within the values predicted by method 4, providing an upper bound of the perfectly elastic system.

ACKNOWLEDGMENTS

The authors would like to thank the National Science Foundation(NSF-CMMI #1041578),American Iron and Steel Institute(AISI),ClarkDietrich,Steel Stud Manufacturers Association,Steel Framing Industry Alliance,Devco Engineering,Mader Construction,DSi Engineering,Simpson Strong-Tie and the members of the Industrial Advisory Board. The views expressed in this work are those of the authors and not NSF,AISI,or any of the participating companies or advisors. Furthermore,the authors are immensely grateful to the SEESL staff,especially Mark Pitman,for their invaluable help and advice.

REFERENCES

[1] AISI S213-07: AISI Standard "North American Standard for Cold-Formed Steel Farming-Lateral Design",2007 edition. American Iron and Steel Institute.

[2] R. L. Madsen, N. Nakata, B. W. Schafer (2011) "CFS-NEES Building Structural Design Narrative",Research Report,RR01,access at www. ce. jhu. edu/cfsness,October 2011,revised RR01b April 2012,revised RR01c May 2012.

[3] Peterman,K. D. ,(2014). Behavior of full-scale cold-formed steel buildings under seismic excitations. Johns Hopkins University,Baltimore,Maryland May 2014.

[4] ASCE 7-05:ASCE Standard [ASCE/SEI 7-05] "Minimum Design Loads for Buildings and Other Structures. " 2005 edition. American Society of Civil Engineers.

SUBSTRUCTURE ONLINE HYBRID TEST ON A STEEL FRAME INSTALLED WITH METALLIC DAMPERS

Tao Wang[a], Yufeng Du[a], Jinzhen Xie[a] and Haoran Jiang[a]
[a]Key Laboratory of Earthquake Engineering and Engineering Vibration, Institute of
Engineering Mechanics, CES, Harbin 150080, China
e-mails: wangtao@ iem. ac. cn, dyf@ 126. com, xjz@ iem. ac. cn, hrjiang@ gmail. com

KEYWORDS: Steel moment frame; seismic-reduction performance curve; steel plate damper with slits; substructure online hybrid test; story drift concentration

ABSTRACT

A four-story four-bay steel moment frame was studied which is characterized by a weak-story failure mode. To enhance its seismic performance, the typical design procedure proposed by Kasai was employed to determine the additional stiffness and strength. The seismic-reduction performance curves were developed by use of the standard artificial ground motion BCJ-L2. Steel plate dampers with slits were then designed using the demanded additional stiffness and strength. To evaluate the seismic performance of the passive controlled structure, substructure online hybrid tests were conducted.

An existing steel moment frame was used as the example. It is a four-story structure with four bays. The columns have box sections, while beams are wide-fange I-shaped steel members. Details can be found in *Table. 1*. The steel of all members are SN490 (JIS) with the nominal yielding strength of 325 MPa. Each bay of the planar frame has a width of 6 m. The height of the first story is 6 m, and the rest stories are 4 m high.

The method proposed by Kasai et al was used to find the design parameters of additional structures. The corresponding seismic reduction performance curves in terms of the displacement reduction rate and shear force reduction rate are plotted *in Fig. 1*, togther with the equivalent damping and period.

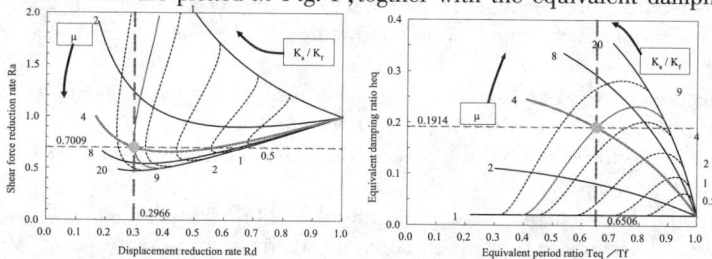

Fig. 1. Seismic reduction performance curves (kasai et al)

To use the above curves, the displacementof the existing frame is first predicted by uses of an equivalent single-degree-of-freedom system considering a linear vibration mode. The drift ratio is calculated as 0. 0225 rad. The target drift ratio is set to be 1/150 (0. 00667 rad) so that the frame almost remain elastic. The target displacement reduction rate R_d is 0. 2966. Next is to find the minimum additional stiffness ratio K_a/K_f at the same R_d value from Fig. 3. It is found K_a/K_f reaches minimum of 5. 45 when the ductility ratio is close to 4. At this moment the shear force reduction rate R_a is about 0. 7009.

Assume the controlled structure deform uniformly along the height when being subjected to the lateral seismic force protocol using Ai distribution. That is, all of the four stories have identical story drift angle and the ductility ratio.

A substructureonline hybrid test was conducted to verify the seismic performance of the enhanced steel frame installed with steel dampers. *Fig. 2* shows the substructuring scheme. A 1/3 scaled model was developed using Q235B steel.

The ground motion recorded at JR Takatori station during 1995 Nanbu earthquake was used to conduct the substructure online hybrid test. Its peak ground acceleration is 741 gal,. Four tests were conducted in sequence with the input PGA as 0. 15g, 0. 30g, 0. 45g, and 0. 75g, respectively, corresponding to 20%, 40%, 60%, and 100% of original Takatori ground motion.

Fig. 3 shows the biggest inter-story drift angles. At the level of 1 ~ 3, the biggest inter story drift

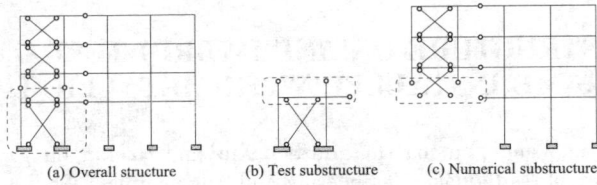

(a) Overall structure   (b) Test substructure   (c) Numerical substructure

Fig. 2. Substructure online hybrid test

angle don't exceed 0. 02 rad. At Level-4, the story drift concentration is observed.

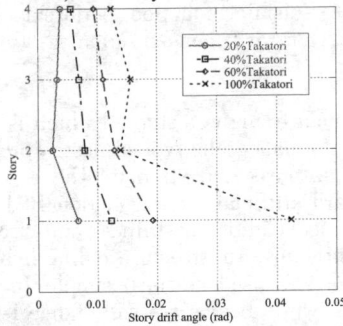

Fig. 3. The biggest inter-story drift angle of each level

CONCLUSION

It can be concluded as follows:

1) The steel moment frame enhanced bythe steel plate damper with slits has good seismic perform-ance. The biggest inter-story drift angle of the structure is less than 1/50 undergoing the Level-3 (0. 45g) earthquake.

2) The enhanced steel moment frame collapses because of the too large inter-story drift angle of the first floor when undergoing the Level-4(0. 75g) earthquake.

3) After experiencing theTakatori earthquakes, the damper doesn't crack and fail, it still has good capacity of energy dissipation.

4) The discrepancy between the target drift ratio of 1/150 and the actual Level-2 drift of 1/50 comes from the selected ground motion which contains significantly velocity pulse. Moreover, the story drift concentration cannot be avoided by the selected design method. These two questions need further study in the future.

REFERENCES

[1] Yamada,S. ,Suita,K. ,Tada,M. ,Kasai,K. ,Matsuoka,Y. ,and Shimada,Y. Collapse experiment on 4-story steel moment frame: part 1 Outline of test results. Proceedings of 14th World Conference on Earthquake Engineering,October,2008,Beijing,China.

[2] Ricky W. K. Chan, Faris Albermani. Experimental study of steel slit damper for passive energy dissipation[J]. Engineering Structures. 2008 ,30(4) :1058-1066.

[3] Kasai. K. Fu,Y. M. ,Watanabe,A. ,Passive Control Systems for Seismic Damage Mitigation,J. of Structural Engineering,ASCE. Vol. 124,No. 5 ,1998. 501-512.

[4] Toko Hitaka and Chiaki Matsui. Experimental of Study on Shear Wall with Slits. Journal of Structural Engineering [J]. 2003(5) : 586-594.

[5] JSSI. Handbook of Structural Design and Construction of Passive Control 2nd.

[6] Mazzoni,S. ,McKenna,F. ,Scott,H. M. and Fenves,L. G. Open System for Earthquake Engineering Simulation User Command-Language Manual. Pacific Earthquake Engineering Research Center,University of California,Berkeley,2006,http://opensees. berkeley. edu.

[7] BCJ. Structural Provisions for Building Structures,Building Center of Japan,1997,Tokyo,Japan.

INFLUENCE OF MODELLING OF STEEL LINK BEAMS ON THE SEISMIC RESPONSE OF SINGLE-STOREY EBFs

Melina Bosco[d], Aurelio Ghersi[b], Pier Paolo Rossi[b] and Paola Stramondo[b]

[d]Department of Building, Civil and Environmental Engineering, Concordia University, Canada

e-mail: mbosco@ encs. concordia. ca

[b]Dept. of Civil Engineering and Architecture, University of Catania, Italy

e-mail: aghersi@ dica. unict. it, prossi@ dica. unict. it, pstramon@ dica. unict. it

KEYWORDS: link, steel, modelling, isotropic hardening, kinematic hardening.

ABSTRACT

The seismic response of links in eccentrically braced structures has often been modelled as elastic plastic with kinematic hardening or, even more simplistically, as elastic perfectly plastic. These models were based on the Euler-Bernoulli beam and considered springs at the beam ends to simulate the inelastic shear and flexural response of the link. The use of these models has not raised particular criticism because of the lack of models able to represent the isotropic hardening with acceptable accuracy. Recently, the writers have proposed a simple but refined link model[1] in which the response of the hinges is dictated by the uniaxial material model proposed by Zona and Dall'Asta for buckling restrained braces[2]. This model takes into account both kinematic and isotropic hardening. Several numerical tests[1] have proved that the proposed model is able to match the cyclic response of short and long links if the main parameters of the model are properly calibrated. The mean values of the parameters obtained from the results of laboratory tests have also proved to be suitable in case the cyclic response of the links is not known a priori.

The proposed link model (named M1) is used here to assess the effectiveness of two common and simpler link models (M2 and M3) in which the effect of the isotropic hardening is not taken into account explicitly. All the link models consist of elements connected in series (Fig. 1). The central element (EL0) has the same length and moment of inertia of the link and simulates the flexural elastic behaviour of the link. The two elements on each end of the link (EL1 and EL2) are zero length and connect the beam segments outside the link to the central element EL0. The first of these two elements (EL1) simulates the elastic and inelastic shear behaviour of half a link, while the second (EL2) simulates the inelastic flexural behaviour of the ending part of the link. The nodes of EL1 are allowed to have only relative vertical displacements; those of EL2 may have only relative rotations.

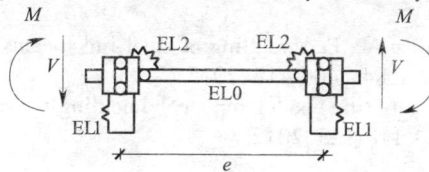

Fig. 1. Elements of the link model

The models differ because of the inelastic response of the elements EL1 and EL2. In the second model (M2) (Fig. 2) the behaviour of elements EL1 is elastic-plastic with kinematic hardening. An equivalent kinematic strain hardening is used to include the effects of both isotropic and kinematic hardening. The elastic stiffness and the plastic shear force of element EL1 are equal to those assigned to model M1. The post elastic stiffness is such that the shear force corresponding to a plastic rotation angle equal to 0.08 rad is equal to that provided by the model M1 at the same plastic rotation angle. Elements EL2 are not included in the model because element EL0 is a beam with hinge element. The plastic hinges of element EL0 are characterised by a length e_{pl} equal to $1/100\ e$ and by an elastic-plastic with kinematic hardening moment-curvature relationship. This moment-curvature relationship is characterised by a plastic bending moment equal to M_p, an elastic flexural stiffness equal to EI

and a post elastic stiffness such that the bending moment corresponding to a plastic rotation angle e-qual to 0.02 rad is equal to that provided by the model M1 at the same plastic rotation angle. Like the second model, the third model(M3) considers an element EL0 modelled by a beam with hinge element and elements EL1 characterised by an elastic-plastic with kinematic hardening behaviour. However, in this case the yield shear force and the yield bending moments are equal to the values corresponding to the fully saturated isotropic hardening condition(i. e. $V_y = V_{y,max}$ and $M_y = M_{y,max}$ in *Fig. 2*). In addi-tion, the post yield stiffness $k_{L1,1}$ of element EL1 is equal to the corresponding stiffness of model M1 while the post yield stiffness of the plastic hinge of element EL0 is such that the bending moment cor-responding to a plastic rotation angle equal to 0.02 rad is equal to that provided by the model M1 at the same plastic rotation angle.

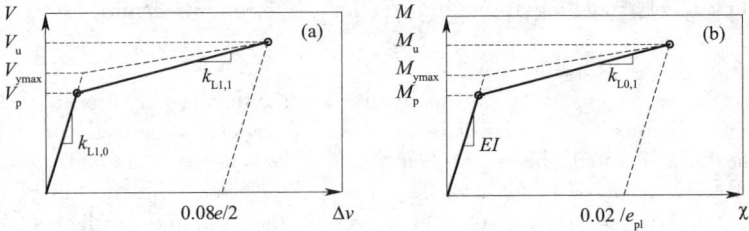

Fig. 2. Model 2: response of(a)element EL1 and(b)plastic hinge of element EL2

The comparison is made on single-storey systems characterized by different mechanical length of the links and subjected to artificial accelerograms. Focus is paid to both global and local response pa-rameters, namely maximum transient and permanent displacements and plastic rotations of links.

CONCLUSIONS

The numerical tests show that when the effect of isotropic hardening is represented by increasing the kinematic hardening(model M2)or by increasing the value of the yielding shear force and bending moments(model M3)the peak ground accelerations corresponding to prefixed values of the plastic rota-tion angle of the links are close to those corresponding to the benchmark model. The prediction of re-sidual drifts is appreciably influenced by the model adopted. Specifically, model M3 overestimates the residual drifts while minor differences are recorded if model M2 is adopted.

The writers note thatthese results are valid for single-storey systems and that numerical analyses on multistorey frames are necessary to obtain more general and reliable results.

REFERENCES

[1] Bosco M. ,Marino E. M. ,Rossi P. P. Modelling of steel link beams of short, intermediate or long length. *Engineering Structures*,84: 406-418,2015.
[2] Zona A. , Dall'Asta A. Elastoplastic model for steel buckling-restrained braces. *Journal of Con-structional Research* **68**(1): 118-125,2012.

INFLUENCE OF DAMPING ON THE PREDICTION OF DYNAMIC RESPONSE OF MOMENT FRAMES BY NONLINEAR STATIC METHODS

Francesca Barbagallo[a], Melina Bosco[a], Aurelio Ghersi[a], Edoardo M. Marino[a]

[a]Dept. of Civil Engineering and Architecture, University of Catania,
v. le A, Doria 6, 95125 Catania
fbarbaga@ dica. unict. it, mbosco@ dica. unict. it, aghersi@ dica. unict. it,
emarino@ dica. unict. it

KEYWORDS: Seismic response, Nonlinear staticmethod, Damping.

ABSTRACT

The nonlinear static method of analysis is allowed by modern seismic codes and it has become the most popular method for the prediction of seismic response of existing buildings. In literature several proposals of nonlinear static methods of analysis can be found. Among these, N2 method proposed by Fajfar et al. [1] was adopted by the European seismic codes (EC8) [2], while Capacity Spectrum (CS) method, developed by Freeman[3,4], was recommended by the ATC 40[5]. Both of them are organized in two steps. Firstly, the performance curve of the structure is determined by means of pushover analysis. Secondly, the displacement demand is obtained through the analysis of an equivalent SDOF system subjected to a seismic input defined by the response spectrum. According to CS method, the equivalent SDOF system is an elastic system with equivalent damping ratio higher than the nominal one. Furthermore, its stiffness is assumed equal to the slope of the line connecting the origin of the pushover to the inelastic peak limit of the performance curve. Instead, according to N2 Method the equivalent SDOF system is an elastoplastic system. Its stiffness is equal to the slope of the elastic branch of the bilinearised performance curve and its damping ratio is equal to the nominal value. Moreover, N2 method adopts the equal displacement rule. In this paper, CS method has been applied by employing different formulations for the evaluation of the equivalent viscous damping ratio ξ_{eq}, which is determined as a function of the ductility demand μ. In particular, the formulation proposed by Rosenblueth[6] provides the highest increase of ξ_{eq} with respect to the nominal value. The equation developed by Priestley[7] for the steel members provides a larger increase of damping compared to the formulation for concrete elements. Finally, the equations proposed by Takeda[6] and the values proposed by Freeman[4] provide the lowest increase of ξ_{eq}.

The influence of damping on the prediction of the dynamic seismic response of steel moment frames provided by CS method is investigated. The authors carried out a numerical investigation on a set of 24 steel moment frames, partly drown from[8]. The 24 analysed frames are six-storey high and have three equal spans. They differ in type of collapse mechanism (global mechanism and soft storey mechanism), fundamental period, size of cross-section used for beams (from IPE220 to IPE330), gravity loads (high or low) and span length (4.50 and 5.50 m). The numerical analyses involve two stages. First, the displacement demands of the considered frames are determined by incremental nonlinear dynamic analysis (IDA). These displacements are assumed here as the actual ones. Second, the performance curve of each frame is determined by a pushover analysis and each of its points is related to the corresponding a_g by N2 method and by CS method considering different formulations of ξ_{eq}. The displacement demands obtained by N2 and CS method are compared to those determined by IDA.

CONCLUSIONS

CS method generally leads to results closer to IDA than N2 method. This may be due to the fact that N2 method assumes a fixed equivalent damping ratio equal to the nominal value. On the contrary, CS method assumes a large equivalent damping ratio to account for the effect of the energy dissipation in reducing the displacement demand. This compensates for the smaller stiffness considered with respect to N2 method.

CS method has been applied by employing different damping laws available in literature. Almost all these formulations lead to lower top displacement demands than CS method applied with the values of ξ_{eq} proposed by Freeman. In particular, when Priestley's equation for steel members and Rosenblueth's equation are used, CS method provides a top displacement demand very close to that obtained by IDA, both for the frames with global mechanism and with soft storey mechanism(*Fig. 1*). In order to examine the response of structures that undergo different levels of plastic deformationthe top displacement demands are evaluated for three values of a_g (0. 28 g, 0. 35 g and 0. 6 g). According to EC8, in high seismicity regions, these values represent seismic events having probability of exceedance of 50% , 10% and 2% in 50 years, respectively. For a_g = 0. 28 g, N2 method and CS method applied with different formulations lead to similar results. This may be explained because the frames are not well excited in the inelastic range and the increase of equivalent damping is not significant yet. For larger values of a_g, the structures show a strong inelastic behaviour and the value of ξ_{eq} increases. As a consequence, N2 Method overestimates the top displacement demands and CS Method applied with different damping laws provide different results. For both the frames with global and soft storey mechanism, the equations proposed by Takeda and the values of equivalent damping proposed by Freeman lead to conservative results compared to IDA. Instead, the formulations of ξ_{eq} proposed by Rosenblueth and by Priestley provide results which are very close to the actual ones. In particular, for a_g = 0. 35 g and a_g = 0. 60 g, CS method applied with Rosenblueth's formulation slightly underestimates the top displacement demand, especially in case of soft storey mechanism. Instead, CS method applied with Priestley's equation for steel members provides results generally close to the IDA prediction.

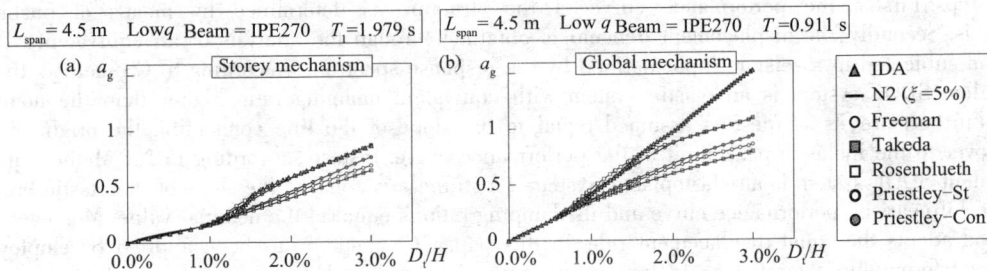

Fig. 1. Peak ground acceleration and top displacement relationship

REFERENCES

[1] Fajfar P, "Capacity spectrum method based on inelastic demand spectra", *Earthquake Engineering and Structural Dynamics*, 28: 979-993, 1999.

[2] CEN. EN 1998-1, "EuroCode 8: Design of structures for earthquake resistance-Part 1: General rules, seismic actions and rules for buildings", European Committee for Standardization, Bruxelles, 2003.

[3] Freeman SA, "The capacity spectrum method as a tool for seismic design", Proc. of the 11th European Conference on Earthquake Engineering, Paris, France, 1998.

[4] Freeman SA, "Review of the development of the capacity spectrum method", *ISET Journal of Earthquake Technology*, 41: 1-13, 2004.

[5] ATC(1996). "Seismic Evaluation and Retrofit of Concrete Buildings(Report SSC 96-01 of California Seismic Safety Commission)", Report ATC-40, Applied Technology Council, Redwood City, California, U. S. A.

[6] Blandon CA, "Equivalent viscous damping equations for direct displacement based design", *Dissertation submitted in partial fulfilment for Master Degree in Earthquake Engineering*, Rose School, Pavia, 2004.

[7] Priestley MJN, "Myths and fallacies in earthquake engineering, Revisited", The Mallet Milne Lecture, 2003. IUSS Press, 2003, Pavia, Italy.

[8] Bosco M, Ghersi A, Marino EM, "On the Evaluation of Seismic Response of Structures by Nonlinear Static Methods", *Earthquake Engineering and Structural Dynamics*, **38**: 1465-1482, 2009.

DERIVATION OF DUCTILITY-EQUIVALENT VISCOUS DAMPING RELATIONSHIPS FOR STEEL MOMENT-RESISTING FRAMES WITH PARTIAL STRENGTH JOINTS

Hugo Augusto[a], José Miguel Castro[b], Carlos Rebelo[a], Luís Simões da Silva[a]

[a] ISISE, University of Coimbra, Department of Civil Engineering, Portugal

hugo. augusto@ dec. uc. pt, crebelo@ dec. uc. pt, luisss@ dec. uc. pt

[b] University of Porto, Department of Civil Engineering, Portugal miguel. castro@ fe. up. pt

KEYWORDS: Equivalent viscous damping, joints, partial-strength, seismic, modification factor.

ABSTRACT

In current seismic codes ductile structures are allowed to develop plastic deformation in order to dissipate energy, aiming to reduce the seismic forces and consequently allowing for a more economical structure design, mainly in regions of high seismicity. Notwithstanding it is recognized that, for the ductile structures, structural and non-structural damages are inflicted by the deformations and displacements for relatively constant forces. This justifies the development of new or improved procedures based on displacements that allies the performance based requirements to the economical and rational seismic design of structures. The existing displacement based procedures use the structure stiffness and the energy dissipated during the event for a predefined performance level, normally related to the building importance class. It is therefore important to have some key features that relate the displacements, ductility and energy dissipation, as inputs for the procedures. This is a key feature of the Direct Displacement-Based Design procedure (DDBD)[1] that uses effective stiffness, ductility-equivalent viscous damping relationships and period-displacement relationships. In the case of the steel moment-resisting frame(MRF) structures with partial-strength beam-to-column end-plate bolted joints, the key features are being derived to take into account the shift of the plastic hinges from the beams to the joints.

This paper seeks to determine ductility-equivalent viscous damping relationships for steel moment-resisting frames with partial-strength end-plate bolted joints. An improvement of an existing expression, recommended by Priestley *et. al.* for steel frame buildings, is proposed. To this end, a finite element model of a beam-to-column sub-assemblage characterized by steel end-plate is developed in ABAQUS. The model, which is validated against monotonic and cyclic experimental data obtained by the research team, is employed to carry out non-linear time-history analyses, using real records scaled to achieve several levels of ductility. The results obtained are used in the derivation of ductility-equivalent viscous damping relationships that are of key importance in several analysis and design procedures such as for example the Direct Displacement-Based Design procedure.

CONCLUSIONS

A large parametric study containing almost 1000 NLTH analyses was conducted, using finite elements models, see *Fig. 1* (a). The models were subjected to successive nonlinear time history(NLTH) analyses for a set of twenty real records, several levels of ductility demand and effective periods. The results allow for the modification of existing expressions proposing new ones, *Eq.* (1) and proposing new coefficients to characterize the ductility-equivalent viscous damping relationships, see *Fig. 1*(b), and also proposing new expressions for the modification factors for the displacement response spectrum, *Eq.* (2). The best correlation with the inelastic displacement reduction factors was obtained when the *Eq.* (1) was used in combination with the *Eq.* (2). Hence, after manipulation of the two expressions it is possible to obtain the final equation described in *Fig. 1* (d). These are the relationships needed to be implemented in the Direct Displacement Based Design procedure to account for partial-strength joints, according to the three dissipative plastic mechanisms found in the Eurocode 3[2], where CWP stands for the component column web panel in shear; EP-Mode1 or EP-Mode2 stands for the components end plate in bending mode of failure 1 or 2, respectively

$$\xi_{eq} = 0.03 + C\left(\frac{\mu - 1}{\mu\pi}\right) \qquad (1)$$

$$R_{\xi} = (6.0/(3 + \xi_{eq}))^{0.5} \qquad (2)$$

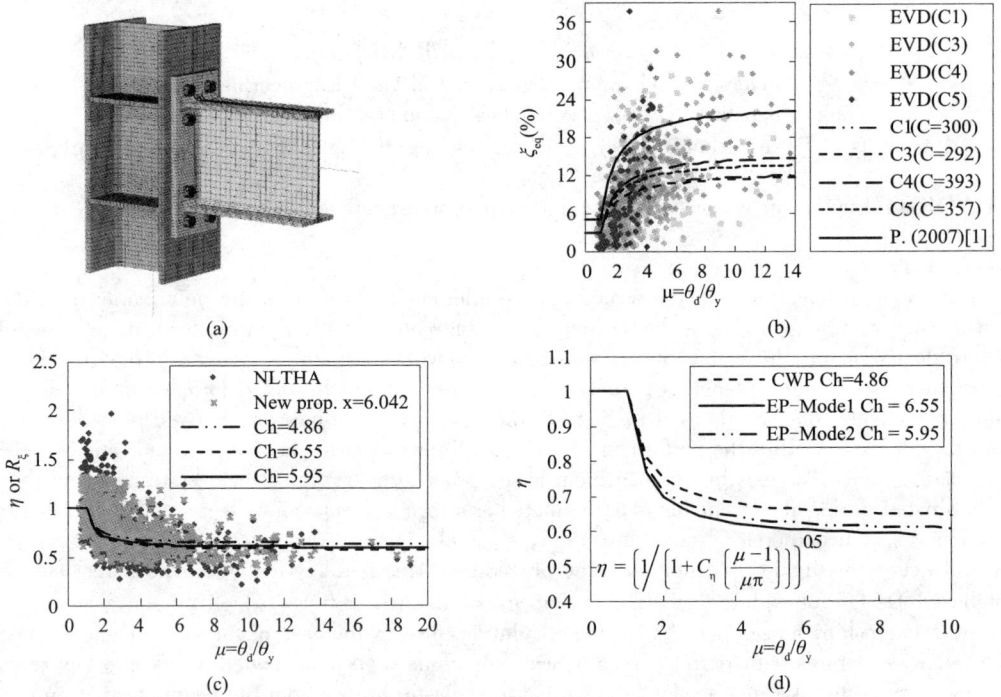

(a)

(b)

(c)

(d)

Fig. 1. Procedure to determine the needed relationships for DDBD procedure in the presence of MRF with partial-strength joints.

ACKNOWLEDGMENT

The authors gratefully acknowledge that the research leading to these results has received funding from the European Community's Research Fund for Coal and Steel(RFCS)under grant agreement n° RFSR-CT-2010-00029. Also the financial support from the Portuguese Ministry for Education and Science(Ministério da Educaçãoe Ciência)(through the agency *Fundação para a Ciência e a Tecnologia* (FCT))within QREN-POPH-Typology 4. 1-Advanced Education, reimbursed by the European Social Fund and Portuguese national funds MEC, under contract grants SFRH/BD/91167/ 2012, for Hugo Augusto, is gratefully acknowledged.

REFERENCES

[1] Priestley, MJN, Calvi, GM, Kowalsky, MJ, 2007. *Displacement-Based Seismic Design of Structures.* IUSS Press, Pavia, Italy.

[2] Eurocode 3, 2005. *Design of steel structures — Part* 1-8: *design of joints,* EN 1993-1-8. Belgium: Brussels.

CALIBRATION OF STRENGTH AND STIFFNESS DETERIORATION HYSTERETIC MODELS USING OPTIMIZATION ALGORITHMS

Miguel Araújo[a], Luís Macedo[a], José Miguel Castro[a]

[a]Department of Civil Engineering, Faculty of Engineering, University of Porto, Portugal

maraujo@ fe. up. pt, luis. macedo@ fe. up. pt, miguel. castro@ fe. up. pt

KEYWORDS: Steel members, Cyclic hysteric response, Deterioration moodes, Harmony search optimization.

ABSTRACT

Under earthquake loading conditions, excessive levels of deformation in one or several members of a structural system may result in local strength and stiffness deterioration and, consequently, loss of structural integrity, which may be characterized by significant damage or even collapse of a building. Reliable collapse or damage-state assessments of existing buildings are greatly dependent on the accuracy of the analytical models adopted and their ability to capture the expected level of degradation within a member. The deformation capacity of steel elements, and its impact in the collapse prediction of existing buildings, are key issues when conducting performance-based analysis and have been the focus of very recent research studies.

Currently, existing hysteretic models that incorporate strength and stiffness deterioration of members, such as the one proposed by Ibarra et al. [1], have taken a step further in providing more realistic collapse assessments of steel structures. However, the parameters required when using such models have been calibrated based on typical American cross-section profiles[2] and hence, no information is available for steel profiles of common use in European construction practice. Furthermore, although recent studies have shown the influence of the axial load on the deformation capacity of deep wide-flange steel members[3], these parameters have not been calibrated yet to account for this effect.

The present study aims to calibrate the parameters of the Ibarra et al. [1] hysteretic model based on the behaviour of European steel profiles(IPE, HEA, HEB and HEM) assessed by Araújo and Castro[14] using detailed FE analyses under combined uniaxial bending and axial compression[4,5]. Robust Harmony Search optimization algorithms[6] are employed to conduct the calibration of those parameters. The database of results obtained in this work can be used by both the scientific community and practitioners when dealing with the seismic performance and collapse assessment of steel structures composed by European sections.

CONCLUSIONS

It was concluded that the calibration of the cyclic responsesof steel members made of different European cross-section profiles using the Harmony Search(HS) optimization algorithm provides very consistent optimum deterioration parameters of the Ibarra et al. [1] hysteretic model, which perfectly capture the strength, post-capping and stiffness modes of deterioration of the HEA280 and IPE400 members, as depicted in Fig. 1. a) and b). A sample of 20000 analysis was carried out 20 times to assess the stability of the calibrated parameters. Highly stable deterioration parameters(*Fig. 1.* c) and d)) were obtained, with coefficients of variation(CoV) typically lower than 0. 10. The parameters obtained were seen not to be in full agreement with those evaluated according to expressions available in the literature, pointing out therefore to the need for further research in this topic.

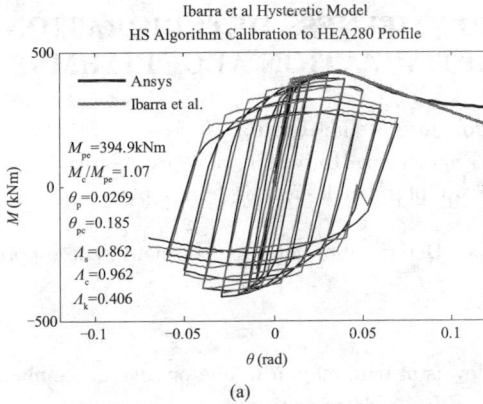

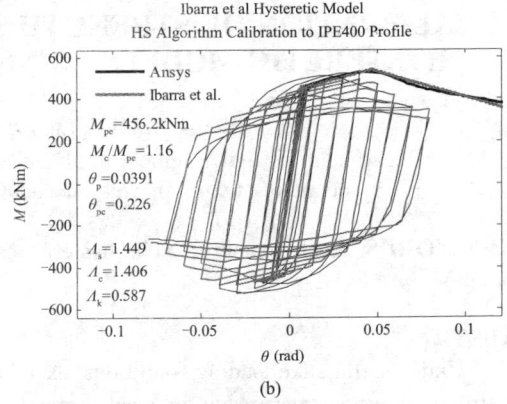

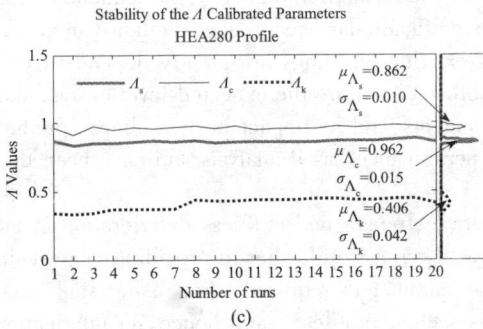

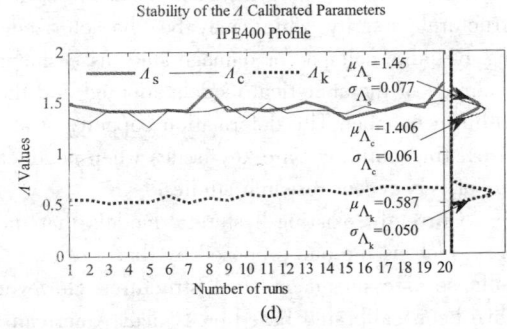

Fig. 1. Calibration of the cyclic responses of steel members with HEA280 and IPE400 profiles((a)and (b))using the HS algorithm and stability of the calibrated deterioration parameters of the Ibarra *et al.* [1] model ((c)and(d)).

ACKNOWLEDGMENT

The authors would like to thank Dr. Raffaele Landolfo and Dr. Mario D'Aniello from the University of Naples "Federico II" and Dr. Vicenzo Piluso from the University of Salerno for providing the experimental results adopted in the calibration of the models of the present work.

REFERENCES

[1] Ibarra LF, Medina RA, Krawinkler H. "Hysteretic models that incorporate strength and stiffness deterioration", *Earthquake Engineering and Structural Dynamics*, 34:1489-1511, 2005.

[2] Lignos DG, Krawinkler H, "Deterioration modelling of steel components in support of collapse prediction of steel moment frames under earthquake loading", *Journal of Structural Engineering*, 137:1291-1302, 2011.

[3] Elkady A, Lignos DG, "Analytical investigation of the cyclic behaviour and plastic hinge formation in deep wide-flange steel beam-columns", Bulletin of Earthquake Engineering, 13: 1097-1118, 2015.

[4] Araújo M, Castro JM, "Numerical assessment of the deformation capacity of steel members subjected to monotonic and cyclic loading", *Proceedings of the International Workshop on High Strength Steel in Seismic Resistant Structures*, Naples, Italy, 2013.

[5] Araújo M, Macedo L, Castro JM, "On the estimation of the deformation capacity of European steel members. How adequate are the EC8-3 deformation limits?", *Journal of Constructional Steel Research*, 2015(under submission).

[6] Macedo L, Araújo M, Castro JM, "Assessment and calibration of the harmony search algorithm for earthquake record selection", *Proceedings of the Vienna Congress on Recent Advances in Earthquake Engineering and Structural Dynamics*, Vienna, 2013.

QUASISTATIC EXPERIMENTAL TESTING OF VULNERABLE CONCENTRIC BRACED FRAMES

Barbara G. Simpson[a], Stephen A. Mahin[a]

[a]Department of Civil & Environmental Engineering, University of California, Berkeley, USA
simp7@berkeley.edu, mahin@berkeley.edu

KEYWORDS: Braced Frames, Retrofit, Full-Scale Testing, Concrete-Filled Braces, Strongback.

ABSTRACT

Prior to the 1988 Uniform Building Code, very few regulations existed for the seismic detailing of concentrically braced structures, generating concern over the potential vulnerability of older braced frames during strong earthquake ground motions. Despite this concern, relatively few studies on the vulnerability and behavior of these older systems exist. This study presents the experimental results of three quasistatic tests of a two-story, one bay concentric braced frame. The first two test specimens utilized square HSS braces placed in a "chevron" configuration with one column oriented in strong axis bending and the other in weak axis bending. The first specimen (NCBF-B-1) was designed according to the 1982 Uniform Building Code and did not satisfy current United States seismic design requirements. These inadequacies were typical of vintage construction and included high brace width-to-thickness ratios, non-ductile gusset connections lacking adequate fold lines, a weak beam designed without consideration of an unbalanced load that could arise from brace buckling, and no capacity design considerations in proportioning members or connections. This specimen formed a soft story in the second floor, while the rest of the frame experienced only minor yielding and little permanent damage. Both second story braces buckled-demonstrating considerable local buckling at the brace midpoint-and then fractured within a few additional cycles. Since the imposed story drifts were modest, the frame was subsequently repaired, replacing the fractured second story braces and gussets with the same sections. The braces were filled with low strength concrete to help postpone local buckling and fracture, and net section reinforcement was added at all the brace-to-gusset connections. Testing of the second upgraded concrete-filled specimen (NCBF-B-2) resulted in a soft story mechanism in the lower story. Only one brace fractured, causing the frame to behave like an eccentrically braced frame with a long link beam providing a relatively weak and flexible energy dissipating mechanism. Many local failure mechanisms were observed during subsequent loading cycles, including nearly-complete fracture at one column-to-baseplate interface, significant local buckling in the beam, and multiple connection failures. The third phase of testing examined a "strongback" retrofit (NCBF-B-3) intended to alleviate the soft story mechanisms observed in both the first and second test specimens. The test utilized a "lambda" configuration comprised of a buckling restrained brace (BRB) acting as an energy dissipating member and two HSS braces acting as a strong elastic mast intended to spread interstory drifts over the entire story height, preventing damage from concentrating in one particular story.

CONCLUSIONS

This sequence of tests has shown that NCBFs are highly susceptible to early fracture and soft story behaviour. The high width-to-thickness ratio of the bracing members contributes to early and severe local buckling that leads to early brace fracture compared to previously tested, similar SCBF specimens. The first baseline NCBF-B-1 specimen formed a weak story mechanism that would likely concentrate future story drifts during an actual earthquake leading to severe damage and possibly collapse. Filling the braces with low strength concrete (NCBF-B-2) in an attempt to upgrade behaviour did not alleviate the problem of early brace fracture as much as expected. While the concrete helped to delay local buckling, brace fracture occurred at the same roof drift as the hollow braces. The second specimen also exhibited weak story behaviour in the first story. The "strongback" retrofit (NCBF-B-3SB) was successful in completing the loading protocol required for testing a BRB frame intended for new construction. The SB system had sufficient displacement capacity to satisfy basic code requirements and was successful in mitigating soft or weak story behaviour. Further studies are necessary to adequately evaluate the performance of each test.

ACKNOWLEDGMENT

This project would not have been possible without the substantial contributions of Schuff Steel and StarSeismic who donated the steel fabrication and the BRB for the NCBF-B-3SB test specimen. This research is supported by National Science Foundation (NSF) under grant number CMMI-1208002: Collaborative Developments for Seismic Rehabilitation of Vulnerable Braced Frames. The overall Principal Investigator for this project is Professor Charles Roeder of the University of Washington, Seattle. The findings, opinions, and recommendations or conclusions in this paper are those of the author alone and do not necessarily reflect those of the National Science Foundation.

REFERENCES

[1] Lee S, Goel S, "Seismic behavior of hollow and concrete filled square tubular bracing members," Research Report UMCE 87-11, Department of Civil Engineering, University of Michigan, Ann Arbor, MI, 1987.

[2] Rai D, Goel S, "Seismic evaluation and upgrading of chevron braced frames," *Journal of Constructional Steel Research*, 59:971-994, 2003.

[3] Farzad N, The Seismic Design Handbook, New York City, NY: Structural Engineering Series, 1989.

[4] Martini K, Amin N, Lee P, Bonowitz D, "The Potential Role of Non-linear Analysis in the Seismic Design of Building Structures," in *Proceedings, 4th US National Conference on Earthquake Engineering*, Palm Springs, CA, 1990.

[5] Tremblay R, Merzouq S, "Dual Buckling Restrained Braced Steel Frame for Enhanced Seismic Response," in *Proceedings, Passive Control Systems*, Yokohama, Japan, 2004.

[6] Mar D, "Design Examples Using Mode Shaping Spines for Frame and Wall Buildings," in *9th US National and 10th Canadian Conference on Earthquake Engineering*, Toronto, Canada, 2010.

[7] Macrae GA, "The Continuous Column Concept-Development and Use," in *Proceeding of the 9th Pacific Conference on Earthquake Engineering*, Aukland, New Zealand, 2011.

[8] Lai J, Mahin S, "Strong-back System: A Way to Reduce Damage Concentrataion in Steel Braced Frames," *American Society of Civil Engineers, AISC J. Struct. Eng.*, 2014.

[9] Lai J, Mahin S, "Experimental and Analytical Studies on the Seismic Behavior of Conventional and Hybrid Braced Frames," Berkeley, CA, 2012.

A METHOD TO AVOID WEAK STOREY MECHANISMS IN CONCENTRICALLY BRACED FRAMES

Daniel B. Merczel[ab], Jean-Marie Aribert[a], Hugues Somja[a], Mohammed Hjiaj[a], János Lógó[b]

[a]University INSA de Rennes, Dept. of Civil Engineering, France

dmerczel@ insa-rennes. fr, aribert@ insa-rennes. fr, hsomja@ insa-rennes. fr, mhjiaj@ insa-rennes. fr

[b]Budapest University of Technology and Economics, Dept. of Structural Mechanics, Hungary

logo@ ep-mech. me. bme. hu

KEYWORDS: concentrically barced frame, seismic design, weak storey, plastic analysis

ABSTRACT

The use of a vertical truss as the load bearing structure for horizontal loads is a common and wide-spread practice in high-rise steel buildings. The bracing of one or two bays of a frame provides a large lateral stiffness and load bearing capacity. One possible brace arrangement is the concentric bracing of the frame.

In classical design concentrically braced frames (hereinafter referred to as CBFs) are expected to develop a global plastic mechanism when subjected to earthquakes, see *Fig. 1/*a. In the global mechanism all the braces undergo plastic deformations while the columns and beams are non-dissipative elements. Conversely, recent seismic events and numerical simulations pointed out, that some CBF-s develop an unfavourable plastic collapse mechanism when subjected to earthquake excitation even if the design has been made for such effects via the application of the corresponding standards such as Eurocode 8. This effect is often referred to as weak storey mechanism, *Fig. 1/*b.

The aim of the article is to propose a new redesign method of CBF-s that is expected to prevent the development of weak storeys. With this purpose a plasticity based calculation method has been elaborated that aims to provide adequate reserve on every storey of the CBF to develop the desired global mechanism, regardless of the excitation and the initial occurrence of plastic deformations. The method is therefore called the Robust Seismic Brace Design method (RSBD). The RSBD method fundamentally requires every partial weak storey collapse mechanism to have a larger lateral ultimate resistance than the desired global plastic mechanism. In addition another criterion is imposed to the participation of the braces in the weak storey mechanism by the so called Brace Participation Ratio (BPR). The two criteria together provide an adequate number of conditions to design both the braces and the columns so that the occurrence of weak storeys can be prevented.

A series of nonlinear time history analyses (NTHA) were conducted on buildings of various storey numbers designed according to Eurocode provisions and redesigned with the developed method. The comparative results fully confirm the validity of the RSBD method as illustrated hereafter in *Fig 2*.

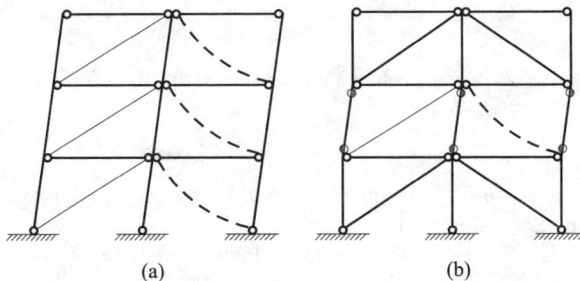

(a)   (b)

Fig 1. (a) Global plastic mechanism; (b) Weak storey mechanism

CONCLUSIONS

To prevent the occurrence of unfavourable weak storey mechanism and weak storey collapse, the method imposes imposes requirements to the structure that can be evaluated by geometry and material characteristics and without the definition of the actual seismic action. The load independence results a structure which is robust in terms of exhibiting a distributed dissipation and resistance to any particular earthquake not exceeding the design level. Therefore the method is called Robust Seismic Brace Design (henceforth RSBD).

The RSBD method is a plasticity-based redesign method of CBF structures obtained by the application of Eurocode 8 provisions. For load arrangements defined by inertia forces of possible weak storey mechanisms it has to be ensured that the load parameter corresponding to the formation of a local storey mechanism, λ_{loc}, is not smaller than the parameter of the global mechanism, λ_{glob}.

$$\frac{\lambda_{loc,i}}{\lambda_{glob,i}} \geq 1.0 \tag{1}$$

A second criterion of the RSBD method aims to promote auniform drift distribution in the inelastic response of every storey. The Brace Participation Ratio(BPR) is proportional to the anticipated inelastic drift on every storey, therefore it is intended to be quasi constant on every floor. As a second design criterion, additional to Eq. (1), it is required that the maximum and minimum BPR do not differ by more than 0.1.

$$BPR_i = \frac{\lambda_{br,i}}{\lambda_{glob,i}} \tag{2}$$

$$BPR_{min} + 0.1 \geq BPR_{max} \tag{3}$$

where $\lambda_{br,I}$ is the load multiplier corresponding to the yield of the brace on the ith storey,

To verify the performance of the proposed method an incremental dynamic analysis program has been carried out on various CBF buildings. In the diagram below the identification number corresponds to the number of storeys and the following letter indicates different structural system. The diagram shows the difference of the lateral drifts between the storeys in the buildings. The Eurocode 8 designs generally exhibit larger differences than the reinforced buildings. The RSBD method results designs where the difference is moderate in every case, therefore the proposed method is efficient in preventing the occurrence of the weak storey behaviour.

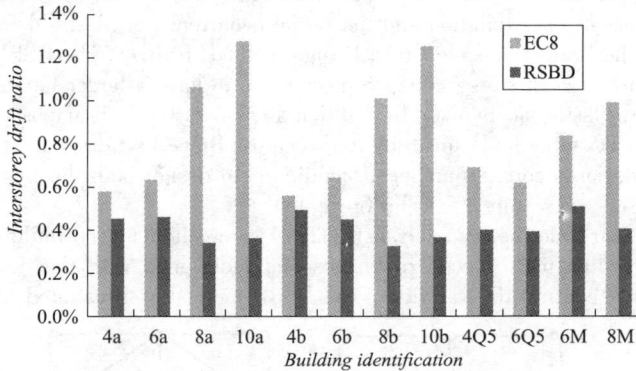

Fig. 2. Performance of Eurocode 8 and RSBD redesigns

REFERENCES

[1] EN 1998-1:2004 Eurocode 8 "Design of structures for earthquake resistance-Part 1: General rules, seismic actions and rules for buildings"

[2] Merczel D. B, Somja H., Aribert J-M, Lógó, J, "On the behavior of concentrically braced frames subjected to seismic loading", Periodica Politechnica Civil Engineering, 57/2, 113-122, 2013.

DEFORMATION AND STRAIN HISTORIES IN SHELL-TO-BASE JOINTS OF UNANCHORED STEEL STORAGE-TANKS DURING SEISMIC LOADING

Clemens Tappauf[a], Andreas Taras[a]

[a]Graz University of Technology (TUG), Institute for Steel Structures, Austria

c. tappauf@ tugraz. at, taras@ tugraz. at

KEYWORDS: Steel storage-tanks, earthquake, dynamic calculation, ultra-low cycle fatigue.

ABSTRACT

n this paper, the alternating uplift behaviour of unanchored steel storage tanks undergoing seismic base excitation will be investigated on the basis of realistic numerical simulations. The uplift behaviour of tanks is of interest for design due to the risk of ultra-low cycle fatigue (ULCF) damage in the tank base (wall-to-bottom) connection, which experiences very large multiaxial plastic strains during the up-lift motion. Thus, a correct prediction of uplift heights, frequencies and cycles is of relevance for developing correct design rules for tanks of this type. The study presented in this paper is part of an international research project, which sees the collaboration of the authors' university, the Swiss Federal Institute of Technology Lausanne-Steel Structures Laboratory (EPFL-ICOM) as well as Karlsruhe Institute of Technology-Research Center for Steel, Timber and Masonry (KIT), set out to study the phenomena of (local) plastic cyclic strains at the welded connection.

At the current stage of knowledge, the uplift behaviour of unanchored tanks is commonly studied on the basis of quasi-static models (e. g. in Eurocode 8-EN 1998[1]) in order to estimate the uplift height and length, as well as simplified dynamic single mass & spring models (e. g. [3]), in order to be able to estimate the uplift height, frequency and total number of uplift cycles.

In order to obtain realistic deformation histories and to verify (respectively qualify) these simplified methods, additional global 3D FEM models were created as shown in Fig. 1a, *using finite shell beam and mass elements. In these models, the mass of the liquid was appropriately taken into account as distributed masses acting on the tank walls, while the hydrostatic smoothening action and the liquid weight on the tank bottom were taken into account as quasi-static pressures. The distribution of the equivalent masses (representing the liquid filling) on the tank wall followed the distribution of hydrodynamic pressures for the rigid impulsive mode according to EC8 in vertical direction.*

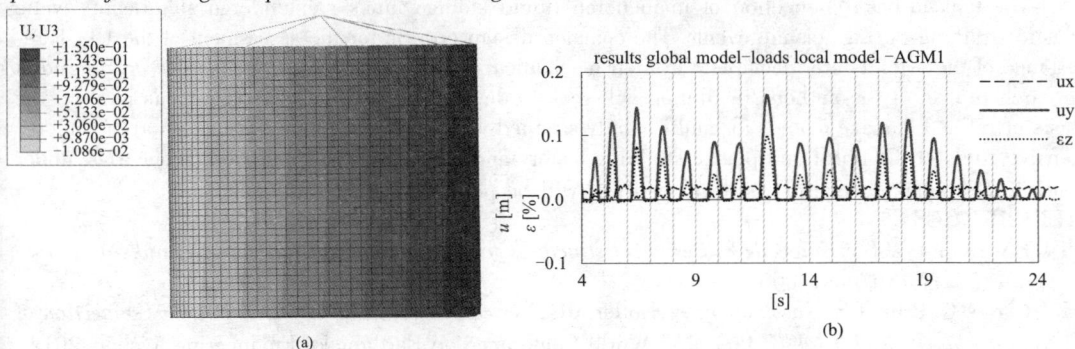

Fig. 1. (a) global shell FEM model; (b) global shell FEM model

Thanks to the use of the shell FEM model, not only the uplift histories, but also information regarding the relatively large strains in transversal (i. e. circumferential) direction could be retrieved from the global model, see Fig 1b. In order to obtain detailed information on the local strain histories caused by these deformations, the deformation/strain data was consequently applied to a sub-model of the welded joint as imposed deformations, see *Fig. 2.*

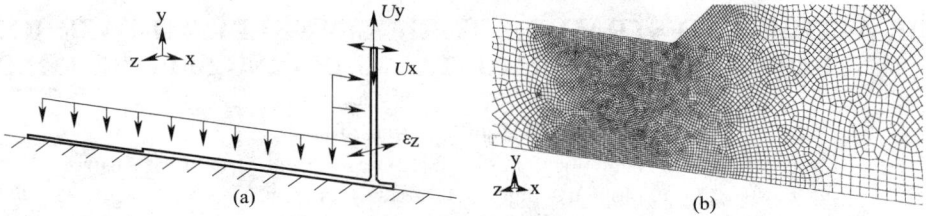

Fig. 2. (a) sketch uplift high-uplift length accto EC8 [1]; (b) model of the wall-to-base detail and plastichinge development acc. to EC8

Finally, plastic strain histories and straining accumulationare discussed for the welded tank wall-to-base connection. The main type results of the numerical simulations for the local FEM model are exemplified in *Fig. 3. Fig. 3*a shows the (accumulated, sign-independent) equivalent plastic strain (PEEQ), while *Fig. 3*b shows the magnitude of the plastic strains (PEMAG) over time.

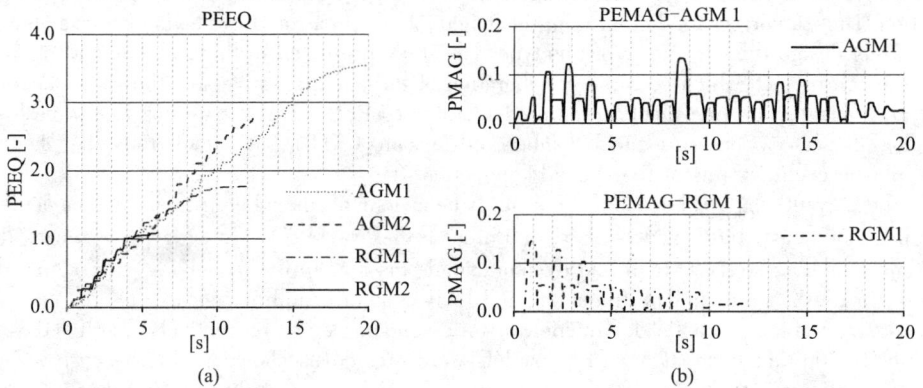

Fig. 3. Results-strain (a) PEEQ ground motions AGM1, AGM2, RGM1, RGM2; (b) PEMAG ground motions AGM1, RGM1

CONCLUSIONS

The tank-to-base connection of unanchored liquid storage tanks can undergo significant, cyclic plastic straining during seismic events. The common design criteria for the assessment of the ULCF resistance of these joints are incomplete in their mechanical derivation and consistency. The ongoing joint research project at the authors' institution will seek to develop new design criteria for this joint on the basis of refined damage models for multi-axially strained welded joints. As part of this effort, a comprehensive study of the uplift and plastic straining history and behaviour of various tank geometries, under realistic earthquake scenarios, is developed and will be published in[3].

REFERENCES

[1] EN 1998-4:2006, *Eurocode 8. Design of structures for earthquake resistance. Silos, tanks and pipelines*, CEN, Brussels, 2006.

[2] Cortes G, Prinz GS, Nussbaumer A, Koller MG, "*Cyclic Demand at the Shell-Bottom Connection of Unanchored Steel Tanks*", Proc. 15th World Conference on Earthquake Engineering, Lisbon, 2012.

[3] Tappauf, C., *On the global and local behaviour of unanchored steel liquid storage tanks in seismic loading (working title)*, PhD thesis, Graz University of Technology.

BEHAVIOUR OF ECCENTRICALLY BRACED STRUCTURES WITH VERTICAL TRUSS ELEMENTS

Helmuth Köber[a]

[a] Technical University of Civil Engineering Bucharest, Romania, Steel Structures Department
koberhelmuth@ yahoo. de

KEYWORDS: Eccentrically braced frames, vertical truss elements, dynamic nonlinear analyses, uniform distribution of plastic deformations, steel consumption

ABSTRACT

The present paper is intended to relieve the behaviour under severe seismic actions of some eccentrically bracing systems equipped with additional vertical connection elements (vertical truss elements between the ends of all dissipative members that are not connected directly to columns) compared to the seismic behaviour of traditional eccentrically bracings as shown in *Fig. 1*. Three eccentrically bracing systems were considered with and without additional truss elements. The analyzed frames had ten storeys of 3. 5m and two spans of 6. 6m. All the frames (with and without truss elements) had short dissipative elements with a length of 1. 2m.

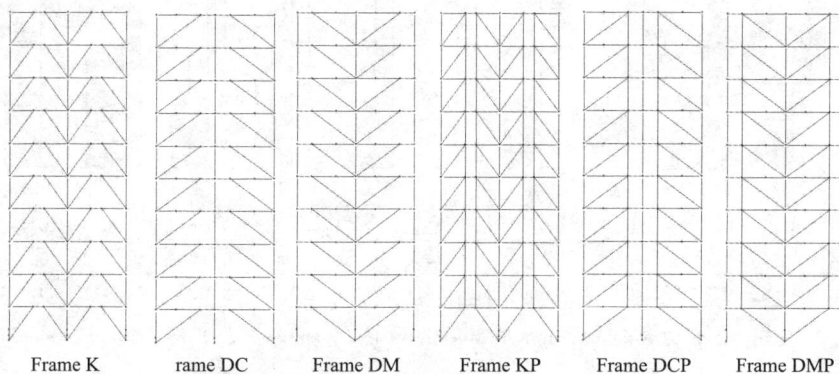

Frame K   rame DC   Frame DM   Frame KP   Frame DCP   Frame DMP

Fig. 1. Analyzed frames

The six analyzed frames were sized for the forces produced by the same horizontal seismic load, evaluated according to the Romanian code P100-1/2013[1] and the provisions of the European standard Ec 8, EN 1998-1 : 2004[2]. All kind of structural elements (columns, diagonals, vertical trusses, dissipative members and adjacent beam segments) had built-up I-shaped cross-sections, designed according to Ec 3, EN1993-1-1 : 2005[3]. Each frame was subjected to dynamic nonlinear analyses using Vrancea earthquakes acceleration records, calibrated for a peak ground acceleration value of about 0. 3 times the acceleration of gravity. Rayleigh damping was considered[4,5].

For the frames without vertical trusses, the distribution of plastic deformations in the dissipative members along the height of the building is quite different from a dynamic analysis to another (see the graphics in *Fig. 2*). For the frames equipped with additional truss elements, more uniform distributions of inelastic deformations in the links along the height of the frames could be noticed in case of all considered acceleration records.

CONCLUSIONS

The values of the estimated steel consumption are quite the same for the eccentrically braced frames equipped or not with vertical truss elements. For the frames with vertical trusses, generally greater axial forces were recorded in all kind of structural elements and smaller bending moments could be noticed in the columns and beam segments placed outside the dissipative members.

Maximum link axis rotations – frame DC

(story)	1	2	3	4	5	6	7	8	9	10
☐ VN77	0.01044	0.02473	0.03293	0.03712	0.03629	0.0328	0.02914	0.02295	0.01699	0.00644
■ VN86	0.00524	0.01079	0.01157	0.00945	0.00545	0.00518	0.01255	0.01419	0.01331	0.00413
☐ VN90	0.00334	0.00764	0.00971	0.00982	0.01142	0.01507	0.01984	0.01729	0.01814	0.00672

Maximum link axis rotations – frame DCP

(story)	1	2	3	4	5	6	7	8	9	10
☐ VN77	0.01923	0.02222	0.02321	0.02394	0.02446	0.02474	0.0258	0.02691	0.03075	0.03111
■ VN86	0.00896	0.00982	0.01012	0.00993	0.00945	0.00884	0.00917	0.00917	0.00992	0.00989
☐ VN90	0.00682	0.00782	0.00878	0.00985	0.01084	0.01156	0.01264	0.01334	0.01411	0.01414

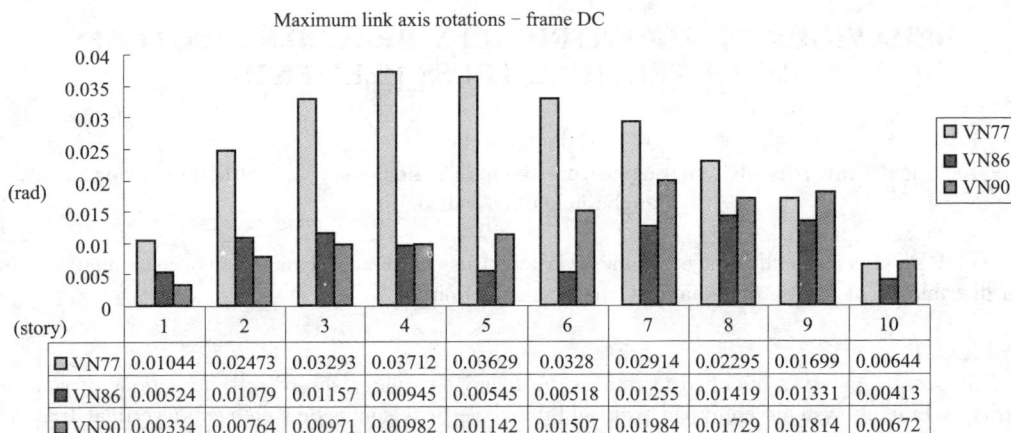

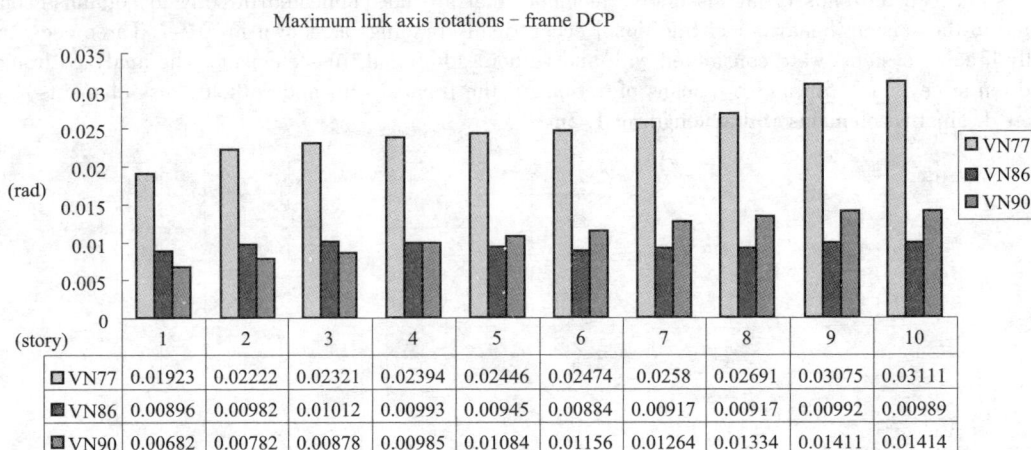

Fig. 2. Maximum plastic deformations in the dissipative members of frame DC and frame DCP

The main advantage of the eccentrically braced frames with vertical truss elements consists in the more uniform distribution of inelastic deformations along the height of the structure. The plastic deformation demand of all dissipative members is in the same range and so the dissipative members placed at different storeys, participate in a comparable manner at the dissipation of energy through plastic deformations.

The main disadvantage of providing additional vertical trusses in eccentrically braced frames consists in the difficulty of the emplacement of door and window openings in the braced bays.

REFERENCES

[1] Ministry of Transportations, Public Works and Territory Planning, 2013, Code for aseismic design-Part I-Design prescriptions for buildings, P100-1/2013, Bucharest, Romania.

[2] Eurocode 8, EN 1998-1 : 2004, Design of structures for earthquake resistance, Part 1 : General rules, seismic actions and rules for buildings.

[3] Eurocode 3, EN 1993-1-1 : 2005, Design of steel structures-Part 1-1 : General rules and rules for buildings.

[4] Tsai K. C. ; Li J. W. , 1994, Drain2D + A General Purpose Computer program for Static and Dynamic Analyses of Inelastic 2D Structures.

[5] Chopra A. K. 2004, Dynamics of Structures. Theory and Applications to Earthquakes Engineering-3rd Edition.

INFLUENCE OF GRAVITY LOAD RESISTING SYSTEM ON THE APPLICATION OF THEORY OF PLASTIC MECHANISM CONTROL FOR MOMENT RESISTING FRAMES

A. Longo[a], R. Montuori[a] and V. Piluso[a]

[a]Department of Civil Engineering -University of Salerno

alongo@ unisa. it, r. montuori@ unisa. it, v. piluso@ unisa. it

KEYWORDS: Mechanism Control, Moment Resisting Frames, Gravity Load Resisting System.

ABSTRACT

In this paper the Theory of Plastic Mechanism Control(TPMC) is briefly summarized and applied with reference to Moment Resisting Frames(MRFs). In particular, an innovative application of TPMC is herein presented aiming to account also for the contribution of the columns of the gravity load resisting system. A worked example is reported and the results of non linear static and dynamic analyses are discussed. In addition, with reference to a 5-storey building, a comparison between the design results obtained by means of TPMC and those obtained by Eurocode 8 is provided.

CONCLUSIONS

Structures designed according to TPMC[1,2] are able to develop a collapse mechanism of global type. A further improvement of the method is made up by a closed form solution gained by introducing new considerations regarding the collapse mechanism typologies to be analysed. In this paper, an additional step forward of the method has been introduced. In particular, it is well known that, typically, MRFs are preferably located along the perimeter of the building, whereas the internal part of the structural scheme is designed to withstand gravity loads only. Despite the leaning part of the building has been designed to withstand vertical loads only, during the seismic action its contribution to the seismic resistance could be not negligible. In fact, the plastic moments of the leaning columns can constitute an additional source of energy dissipation capacity and provide an additional internal work in most of the possible collapse mechanisms. Such contribution can be properly taken into account by TPMC, allowing the design of a lighter seismic resistant perimeter system. Starting from background gained on plastic mechanism control, the application of TPMC has been properly extended to include in the design process the contribution due to Gravity Load Resisting System(GLRS).

In this paper, TPMC including the contribution of GLRS is briefly summarized and applied with reference to a 5-storey building. By means of non linear static analyses, a comparison in terms of seismic performances is presented with reference a prototype building designed with and without taking into account the contribution of GLRP. In both cases, TPMC is applied. Finally, the same structures have been also designed according to the seismic provisions given by Italian Seismic Code[11] and the results obtained are herein presented and discussed. Also dynamic non-linear analyses have been carried out, but, for sake of shortness, are not herein reported.

The results have shown that the contribution of leaning columns, leading to an additional internal work corresponding to the undesired kinematic mechanisms, allows to design a lighter seismic resistant part. In addition, in order to check the accuracy of the design methodology, push over analyses have also been carried out. The obtained pattern of yielding has shown that structures designed according TPMC, with and without including the influence of GLRS, fails in a global mode. Therefore, analyses's results have shown the accuracy of the design procedure which is able to prevent the development of undesired partial mechanisms assuring the development of a global type collapse mechanism. Finally, the same structures have been designed according to the seismic provisions given by Italian Seismic Code, the results show that in the collapse conditions, the structure designed according with Italian Seismic Code exhibits a partial mechanism therefore the suggested design rules, i. e. the beam-column hierarchy criterion also suggested in Eurocode 8, is not able to prevent undesired partial mechanisms.

REFERENCES

[1] Mazzolani, F. M. , Piluso, V. , "Theory and Design of Seismic Resistant Steel Frames", E&FN Spon, an imprint of Chapman & Hall, 1996.

[2] Mazzolani, F. M. , Piluso, V. , "Plastic Design of Seismic Resistant Steel Frames", Earthquake Engineering and Structural Dynamics, Vol. 26, pp. 167-191, 1997.

[3] Longo A. , Montuori R. , Piluso V. , "Theory of Plastic Mechanism Control of Dissipative Truss Moment Frames", Engineering Strucutures, Vol. 37, pp. 63-75, 2012.

[4] Longo A. , Montuori R. , Piluso V. , "Failure Mode Control and Seismic Response of Dissipative Truss Moment Frames", Journal of Structural Engineering, Vol. 138, pp. 1388-1397, 2012.

[5] Longo, A. , Montuori R. , and Piluso V. "Theory of plastic mechanism control for MRF-CBF dual systems and its validation. " Bulletin of Earthquake Engineering (2014): Vol. 12, pp. 2405-2418.

[6] Mastrandrea, L. , Piluso, V. (2009): "Plastic Design of Eccentrically Braced Frames, II: Failure Mode Control", Journal of Constructional Steel Research, Volume 65, Issue 5, pp. 1015-1028, 2009.

[7] Conti, M. A. , Mastrandrea, L. , Piluso, V. (2009): "Plastic Design and Seismic Response of Knee Braced Frames", Advanced Steel Construction, Volume 5, Issue 3, pp. 343-366, 2009.

[8] Montuori, R. , Nastri, E. , Piluso, V. (2014): "Theory of Plastic Mechanism Control for the Seismic Design of Braced Frames Equipped with Friction Dampers", Mechanics Research Communications , Vol. 58, pp. 112-123, 2014.

[9] Montuori, R. , Nastri, E. , Piluso, V. (2014): "Theory of Plastic Mechanism Control for Eccentrically Braced Frames with inverted Y-scheme", Journal of Constructional Steel Research, Vol. 92, January 2014, pp. 122-135.

[10] Montuori, R. , Nastri, E. , Piluso, V. (2014): "Advances in theory of plastic mechanism control: closed form solution for MR-Frames", Article first published online: 20 Octobre 2014, DOI: 10. 1002/eqe. 2498, Earthquake Engineering and Structural Dynamics.

[11] NTC 2008-D. M. 14 Gennaio 2008: 《Nuove Norme Tecniche per le Costruzioni》(Italian Seismic Code).

[12] CSI 2007. SAP 2000: Integrated Finite Element Analysis and Design of Structures. Analysis Reference. Computer and Structure Inc. University of California, Berkeley.

[13] Giugliano M. T. , Longo A. , Montuori R. , Piluso V. , "Failure Mode and Drift Control of MRF-CBF Dual Systems", The Open Journal of Construction and Building Technology, Vol. 4, pp. 121-133, 2010.

[14] Longo, A. , Montuori R. , and Piluso V. , "Moment Frames-Concentrically Braced Frames Dual System: Analysis of Different Design Criteria", accepted for publication on Structure and Infrastructure Engineering, 2015.

[15] Longo, A. , Montuori R. , Nastri E. and Piluso V. , "On the Use of HSS in Seismic Resistant Structures", Journal of Constructional Steel Research, Vol. 103, pp. 1-12, 2014.

[16] Longo A, Montuori R, Piluso V. 2008. Plastic design of seismic resistant V-Braced frames. Journal of Earthquake Engineering, vol. 12: 1246-1266.

[17] Longo A, Montuori R, Piluso V. 2008. "Failure Mode Control of X Braced Frames under Seismic Actions" Journal of Earthquake Engineering, vol. 12: 728-759, 2008.

[18] Giugliano M. T. , Longo A. , Montuori R. , Piluso V. "Plastic design of CB-frames with reduced section solution for bracing members"-Journal of Constructional Steel Research-Vol. 66, 2010 p. 611-621 ISSN: 0143-974X, doi: 10. 1016/J. JCSR. 2010. 01. 001.

[19] Longo A, Montuori R, Piluso V. (2008c). Influence of design criteria on the seismic reliability of X-braced frames. Journal of Earthquake Engineering; 12, 406-431, DOI: 10. 1080/13632460 701457231.

[20] Longo A, Montuori R, Piluso V. (2009). Seismic reliability of V-braced frames: influence of design methodologies. Earthquake Engineering and Structural Dynamics, vol. 38: 1587-1608, DOI: 10. 1002/eqe. 919.

COMPARISON OF MODELLING STRATEGIES FOR STEEL STRUCTURES UNDER CYCLIC LOADS

Luís Macedo[a], Miguel Araújo[a] and José Miguel Castro[a]
[a] Department of Civil Engineering, University of Porto, Portugal
luis. macedo@ fe. up. pt, maraujo@ fe. up. pt, miguel. castro@ fe. up. pt

KEYWORDS: Steel members, Cyclic hysteretic response, optimization, Modelling.

ABSTRACT

Nonlinear dynamic analysis is fast becoming a common procedure when designing or assessing earthquake-resistant buildings. There are two important requirements for an accurate evaluation of the behaviour of steel buildings under seismic loads: an accurate numerical modelling and an accurate definition of the seismic input. The latter requirement was the subject of study by the authors[3], whereas the modelling issues are known to assume an important role in the assessment of steel structures[4]. Phenomena such as local buckling, strength and stiffness deterioration of members[5] can have a significant influence on both the local and global response of the structures. Material nonlinearity of beam-column elements can be modelled with two different approaches: concentrated plasticity approach and distributed plasticity approach. Recently, [5] proposed moment-rotation relationships that can be used in concentrated plasticity models, aiming at a realistic representation of the behaviour of steel beam-column elements. This represents an important step in comparison with other existing lumped plasticity and distributed plasticity modelling strategies. This paper discusses different modelling approaches for steel moment-resisting frames. 3D detailed finite element analysis as well as distributed and concentrated plasticity approaches are adopted to simulate the experimental test conducted by Hong-Sik et al. [1]. The calibration of the moment-rotation backbone curve of the concentrated plasticity model is carried out on the basis of the procedure proposed by[2] which involves detailed 3D finite element analysis and an optimisation process(Fig. 1 a)b)). The numerical analyses allow concluding that the concentrated plasticity model with calibrated parameters is able to accurately capturethe test response (Fig. 2 a)b)). The performance of the tested frame is then evaluated through Incremental Dynamical Analysis(IDA) using different modelling strategies. The results reveal the importance of realistically incorporating the deterioration phenomena in the numerical models and the influence of the modelling technique on the collapse evaluation of steel structures.

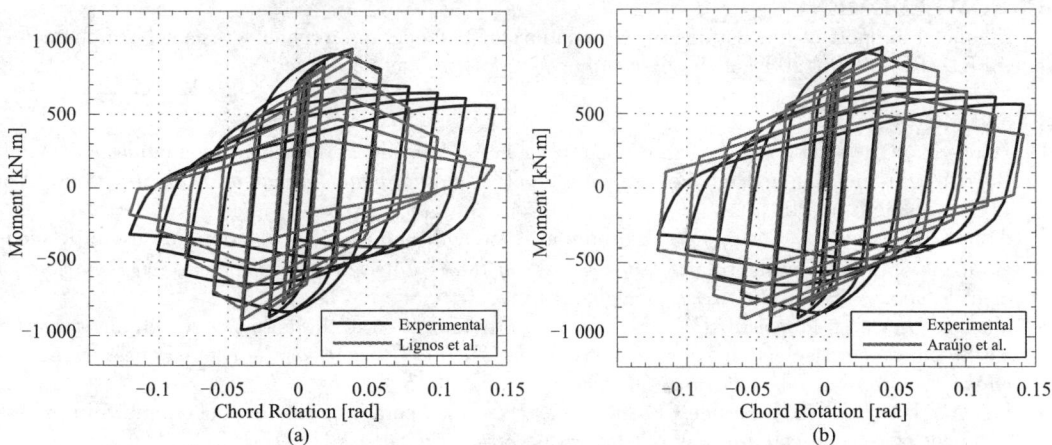

Fig. 1. Column calibration: (a) Lignos and Krawinkler[5] (b) Araujo et al. [2]

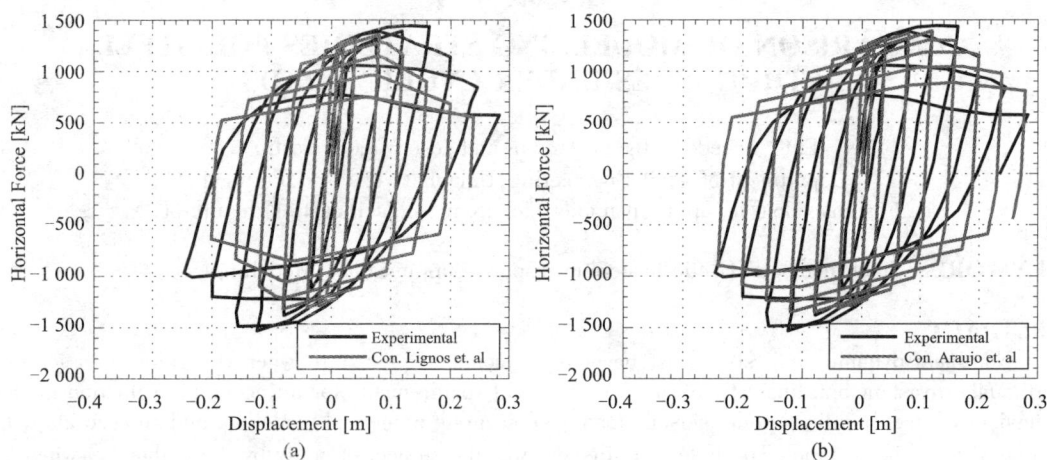

Fig. 2. Frame cyclic response: (a) Conc. Plast. Lignos *et al.* [5] (b) Conc. Plast. Araújo *et al.* [2]

CONCLUSIONS

In this paper the simulation of a full scale test using different modelling strategies was conducted with the aim of investigating the accuracy of the different modelling approaches. The calibration procedure proposed by[2] was used and its ability to accurately estimate the deterioration parameters of the steel members was demonstrated. The comparison between the experimental and the numerical results confirmed the critical importance of accounting for both strength and stiffness deterioration in the numerical models. Whilst the distributed plasticity model exhibited large differences in comparison with the experimental results, the concentrated plasticity models, which account for strength and stiffness deterioration, were able to accurately reproduce the full scale test. Moreover, the numerical model defined using the deterioration parameters calibrated according to the procedure proposed by[2] was seen to lead to accurate and reliable estimates of the frame response. The performance of the frame was then evaluated through IDA analysis and the collapse fragility curves were derived. These curves allowed identifying the limitations associated with the use of distributed plasticity models, which exhibited higher levels of seismic performance due to the non-consideration of cyclic deterioration effects. On the other hand, both concentrated plasticity models showed similar collapse capacities, although slight deviations were observed for higher levels of seismic intensity.

ACKNOWLEDGMENT

Financial support of the Portuguese Foundation for Science and Technology, namely through PhD grant SFRH/BD/60796/2009 of the first author, is gratefully acknowledged.

REFERENCES

[1] Hong-Sik Ryu, Sang-Hoon Oh, and Moon-Sung Lee. "Seismic Performance Evaluation of Moment Resisting Frames Using High Performance Steel", *International Journal of Steel Structures*; Vol 11, No 2, 235-246, 2011.

[2] Araújo M, Macedo L, Castro JM, "Calibration if strength and stiffness deterioration hystertic models using optimization algorithms", *Proceedings of the 8th International Conference on Behaviour of Steel Structures in Seismic Areas*, Shanghai, China, 2015.

[3] Macedo L, Araújo M, Castro JM, "Assessment and calibration of the harmony search algorithm for earthquake record selection", *Proceedings of the Vienna Congress on Recent Advances in Earthquake Engineering and Structural Dynamics*, Vienna, 2013.

[4] Liel AB, Haselton CB, Deierlien GG and Baker JW. "Incorporating modeling uncertainties in the assessment of seismic collapse risk of buildings", *Structural Safety*, 31: 197-211, 2008.

[5] Lignos DG, Krawinkler H, "Deterioration modelling of steel components in support of collapse prediction of steel moment frames under earthquake loading", *Journal of Structural Engineering*, 137: 1291-1302, 2011.

A REFINED THEORETICAL MODEL FORPREDICTING THE ULTIMATE BEHAVIOUR OF BOLTED T-STUBS

Antonella B. Francavilla[a], Massimo Latour[a], Vincenzo Piluso[a], Gianvittorio Rizzano[a]
[a]University of Salerno, Department of civil Engineering, Italy
e-mails: afrancavilla@ unisa. it, mlatour@ unisa. it, v. piluso@ unisa. it,
g. rizzano@ unisa. it

KEYWORDS: Connections, T-stubs, Mechanical model, Ductility.

ABSTRACT

The prediction of the behaviour of beam-to-column connections can be obtained by means of the so-called "component method", largely used in research studies and currently codified in Eurocode 3. Theoretically, this approach is very general and it is able to describe the behaviour of any kind of connection, provided that the basic components of deformability and strength are accurately identified and modelled. To date, even though the component method codified in EC3 is already very advanced and provides to designers significant information regarding the joint behaviour, it still provides some drawbacks especially dealing with the prediction of the ductility supply and the prediction of the cyclic behaviour. In fact, even though some authors have already investigated some aspects related to prediction of the ductility supply[1,2] and cyclic behaviour of connections[3] past experimental and theoretical researches have often focused their attention mainly on predicting stiffness and resistance of joints. In this paper, aiming to provide a contribution towards the codification of a procedure for the prediction of the ductility supply to be included in EC3, the attention is focused on bolted connections. In such connection typologies, usually, the most important components, such as the column flange or the end plate in bending, are modelled by means of equivalent T-stubs, i. e. two equal T-shaped elements connected through the flanges by means of one or more bolt rows. Therefore, in order to propose a theoretical approach for predicting the whole force-displacement response up to failure of bolted T-stubs a new refined model is presented, taking properly into account the existing literature.

In this paper, the mechanical model proposed aims to define the T-stub behaviour up to failure accounting for the following effect: the contact forces are considered applied in a point in between the tip of the plate and the edge of the bolt head, the bolt forces are considered uniformly distributed under the bolt head, the failure of the T-stubs is modelled by checking the ultimate strain on the basic materials composing the plate and the bolt, the compatibility condition between the displacements of the plate and the uplift of the bolt is taken into account and the displacements of the T-stub are evaluated step-by-step as the sum of the elastic and plastic part.

As far as the kinematic mechanism is defined and the mathematical laws to be used in order to evaluate the rotations of the plastic hinges are given, it is possible to calculate the ultimate displacement of the bolted T-stub. For a fixed value of the bending moment Mj acting in correspondence of the T-stub web, the known parameters are five: the force of the T-stub(F), the prying force(Q), the value of the distributed load corresponding to the action provided by the bolt head(q), the ratio between the bending moment acting at the bolt line and that arising at T-stub web(ψ) and the location of the prying forces in the contact zone(n^*). The values of these parameters can be obtained by considering five equations: three equilibrium equations(one translational equilibrium, two rotational equilibrium), and two compatibility equations.

As far as the force-displacement curve of the T-stub is obtained by progressively increasing the bending moment acting on the flange, at the end of each loading step it is possible to check also for the deformation state of plastic hinges and bolt. In this way, it is possible to control if the rotations and the elongations are compatible with the ductility supply provided by the basic materials.

The theoretical model has been validated by means of a comparison with experimental tests carried out at Material and Structure Laboratory of the Department of Civil Engineering of Salerno University[4].

For all the specimens, the multilinear model of F-δ curve has been carried out and compared with the corresponding experimental curve. As an example, in *Fig. 1* the comparison experimental-predicted curve in the cases of T-stub 1 and T-stub 4 is given. The comparison with experimental evidence shows a satisfactory agreement in term of ductility and resistance between the theoretical model and the experimental results.

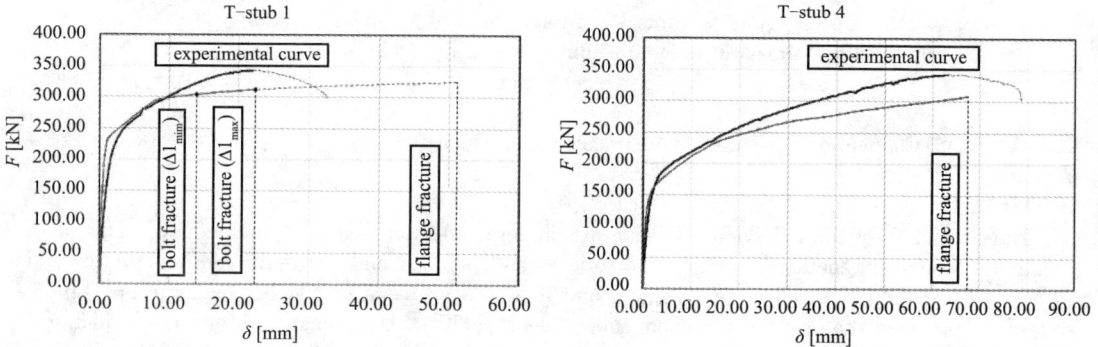

Fig. 1. Comparison between experimental results and theoretical predictions

CONCLUSIONS

In this paper, a theoretical model for predicting the whole force-displacement curve of the T-stubs has been presented. Such a procedure, starting from the definition of proper compatibility and equilibria equations defines the behaviour of the T-stub by means of an incremental procedure.

The comparison with experimental tests carried out by the same authors at laboratory on materials and structures of Salerno University has shown a good accuracy of the model. The obtained results are really encouraging about the possibility of predicting accurately the ductility supply of T-stub by means of a theoretical approach.

REFERENCES

[1] Girão Coelho, A. M., da Silva, L. S. & Bijlaard, F. S. K., 2004. Characterization of the nonlinear behavior of single bolted T-stub connections, Proc., 5th Int. Workshop on Connection in Steel Struct., Amsterdam, 53-64.
[2] Beg, D., Zupancic, E. & Vayas, I., 2004. On the Rotation Capacity of Moment Connections. Journal of Constructional Steel Research, Volume 60, pp. 601-620.
[3] Latour, M., Piluso, V. & Rizzano, G., 2011b. Cyclic Modeling of Bolted Beam-to-Column Connections: Component Approach. Journal of Earthquake Engineering, 15(4), pp. 537-563.
[4] Piluso V., Faella C. & Rizzano G., 2001. Ultimate behaviour of bolted T-Stubs. II: Model validation, Journal of Structural Engineering ASCE, 127(6):694-704.

SEISMIC RESPONSE OF EBFS:
SPLIT K-SCHEME VS INVERTED Y-SCHEME

Rosario Montuori[a], Elide Nastri[a] and Vincenzo Piluso[a]

[a] Department of Civil Engineering, University of Salerno, Italy

e-mails: r. montuori@ unisa. it, enastri@ unisa. it, v. piluso@ unisa. it

KEYWORDS: EBFs, eccentrically braced frames, seismic, kinematic theorem, Eurocode 8

ABSTRACT

The work herein presented is devoted to the evaluation of the seismic performance of Moment Resisting Frame-Eccentrically Braced Frame dual systems (MRF-EBF dual systems) with particular reference to the split K-scheme and the inverted Y-scheme of the EBF part. As it is known EBFs constitute a suitable compromise between seismic resistant MR-frames and concentrically braced frames. In fact, they exhibit both adequate lateral stiffness, due to the high contribution coming from the diagonal braces, and ductile behaviour, due to the ability of the links, in developing wide and stable hysteresis loops. Therefore, the coupling of MRF and EBF constitute an excellent dual system where the primary structural system is constituted by the EBF part, and a secondary fail-safe system is constituted by the MRF part. Because of the inherent lateral stiffness of the EBF part, the main dissipative system is constituted by the link members which can be horizontal (K-scheme) or vertical (inverted Y-scheme). In this paper, MRF-EBF buildings with split K-scheme and inverted Y-scheme, with both 6 and 8 storeys have been designed using either TPMC or EC8 for a total of 8 study cases. TPMC design approach is based on a rigorous approach which exploit thekinemathic theorem of plastic collapse extended to the concept and assures a collapse mechanism of global type[1-2]. Conversely, the design procedure suggested by EC8[3] constitutes only a rough application of the capacity design principles and does not assure a predefined collapse mechanism. The aims of this paper are, on one hand, to provide a further validation of the proposed design procedure based on TPMC, and, on the other hand, to compare the seismic performance of MRF-EBF dual systems designed by means of TPMC with those occurring when Eurocode 8 design criteria are applied. The validation of the proposed design procedure is carried out by means of both push-over and Incremental Dynamic Analyses (IDA). The main purpose of such analyses is to check the fulfilment of the design goal of TPMC, i. e. the development of a pattern of yielding consistent with the collapse mechanism of global type. Such mechanism is universally recognized as the one leading to the highest energy dissipation capacity. In case of MRF-EBF dual systems, it is characterized by the yielding of all the links and all the beams at their ends. Conversely, columns and diagonal braces remain in elastic range. Obviously, exception is made for the base sections of first storey columns. In particular, two study cases are analysed which are characterized by a different number of storeys. A wide discussion is provided in[4].

CONCLUSIONS

In this paper, the same structural systems, a 6-storey MRF-EBF dual system and a 8-storey MRF-EBF dual system with K-scheme and inverted Y-scheme, have been designed according to two different procedures. The first one is the Theory of Plastic Mechanism Control whose robustness is based on the kinematic theorem of plastic collapse and its extension to the concept of mechanism equilibrium curve. The second one corresponds to the combined application of Eurocode 8 provisions devoted to moment-resisting frames and to eccentrically braced frames. Both push-over and dynamic non-linear analyses have pointed out the different seismic performances which can be obtained by means of the investigated design procedures. In particular, the results of both push-over and IDA analyses have pointed out the accuracy of TPMC based approach. In the examined study cases, with reference to the collapse prevention limit state of FEMA 356, on average, K-scheme structures exhibit worse performances compared to the inverted Y-scheme structure, given the approach. Conversely, given the structural scheme, the buildings designed by means of TPMC exhibit better performance compared to the EC8 structures.

REFERENCES

[1] Mazzolani F. M. , Piluso V. (1997) : "Plastic design of seismic resistant steel frames" , *Earthquake Engineering and Structural Dynamics* , Vol. 26 , pp. 167-191.

[2] Piluso , V. , Nastri , E. , Montuori , R. , "Advances in Theory of Plastic Mechanism Control : Closed Form Solution for MR-Frames" , accepted for publication on *Earthquake Engineering and Structural Dynamics* , 2015.

[3] CEN (2005) : "EN 1998-1-1 : Eurocode 8-Design of Structures for Earthquake Resistance. Part 1 : General Rules , Seismic Actions and Rules for Buildings" , *Comite Europeen de Normalisation* , CEN/TC 250.

[4] Nastri E. "Theory of Plastic Mechanism Control for Eccentrically Braced Frames : Closed Form Solution" , *PhD Thesis* , 2015.

SEISMIC RESPONSE ANALYSIS UNDER TRAVELING WAVE EFFECT OF AN ARCH TRUSS ACROSS ABANDONED MINE PIT

Jian Zhou[a], Dong-Ya An[a], Yao-Kang Zhang[a], Jia-Chun Cui[a]

[a] East China Architectural Design & Research Institute Co, Ltd

Jian_zhou@ ecadi. com; dongya_an@ xd-ad. com. cn; yaokang_zhang@ ecadi. com;
jiachun_cui@ xd-ad. com. cn

KEYWORDS: Traveling wave effect, Large-mass method, Arch truss, Abandoned mine pit.

ABSTRACT

Seismic ground motion is a complex time-space process[1]. Traveling wave effect analysis is a seismic response analysis method which takes the influence of seismic wave traveling direction and speed into account. For conventional building structures, because the foundation and basement of the building tie the columns together and coordinate their displacement, traveling wave effect has little impact on the seismic response of superstructure, except the internal forces of circumference members of the superstructure would be increased. But in terms of bridge like structures, which span across a deep pit and lay its foundations along the edge of the pit, the effect of seismic wave would be significant. Taking the structure of Changsha Ice World Project as the background, this paper studies on traveling wave effect on this kind of structures and develops the design instructions.

Changsha Ice World Project is constructed on an abandoned mine pit. The whole building is supported on the side of 80m deep pit cliff with arch truss structure. Span of the main arch trusses varies from 45m to 180m and the rise of the arch varies from 8m to 32. 5m.

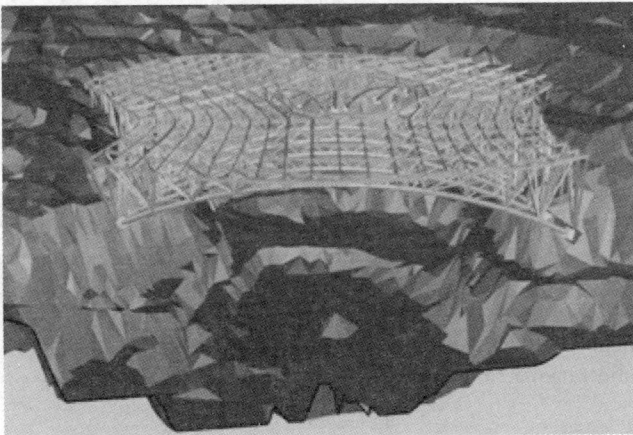

Fig. 1. Relation between arch contour and mine pit cliff

The structure is large in span and complicated in load distribution. Spatial distribution of supports is irregular both horizontally and vertically. Degree of freedom on some supports is released on certain directions to realize rational force on the arch structure. The most important thing is that the structure spans deep pit and bases between supports are inconsecutive. In this case, seismic response of the structure will be more complicated than normal large span spatial structures and principles for normal large span projects cannot be applied. So it is necessary to perform specialized traveling effect analysis and research on this project.

Large mass method is used in this paper to simulate multi-dimensional and multi-support seismic excitation. General finite element program "ANSYS" is used as the analysis program.

Arch structure of the IceWorld Project is arranged at three directions in the space, based on which four directions for seismic wave input have been defined. Calculating the "relative delay" of seismic excitations arriving at each support is a critical issue in analysis. Determination of this parameter depends on two factors, i. e. the travelling speed of the seismic wave and the relative spatial position of the supports. Travelling speed of the seismic wave is taken as 970m/s according to the site seismic safety evaluation report. To prevent irrational "data hopping" of adjacent supports in seismic excitation input, which might cause catastrophe of internal force between these supports, this paper has developed specialized computing input control program which precisely compute "relative delay" according to spatial position of each support, as well as the travelling direction and speed of the seismic wave, and apply different seismic excitations on each support according to its "relative delay". Another complicated feature of the arch structure support is that there's no fully rigid connection at certain directions for some supports, so degree of freedom needs to be released, which brings challenges to adoption of the "large mass method". Because mass blocks typically have no direction, so when the support releases degree of freedom along one direction. Final solution to this problem is to use two functions of the ANSYS program, i. e. "mass with direction" element and "joint local coordinate system".

This paper adopts three directional seismic inputs, with ratio of acceleration peaks at the primary and secondary directions of the horizontal direction and at the vertical direction being 1. 0 ： 0. 85 ： 0. 65.

CONCLUSIONS

This paper has discussed and achieved conclusion on basic regularities of structure responses to traveling wave effect. The conclusion shows that, negative impact of the traveling wave effect shall be reasonably considered during design of the structure. Performing only coincident excitation analysis cannot ensure seismic safety of the structure. Conclusions to the Ice World are detailed as follows:

1. *General response*

Through comprehensive analysis, compared to coincident excitation analysis not considering traveling wave effect, multi-point input analysis taking traveling wave effect into consideration shows reduced general seismic force at the primary input direction; which also shows increased seismic force at the secondary direction. However, seismic response at the secondary direction is smaller than that at the primary direction and could not dominate; and the vertical seismic response has the main trend of increasing.

2. *Internal force of members*

Axial force of theupper chord, lower chord and web member under non-coincident excitations is larger; for different members, non-coincident earthquake causes different increasing extent of seismic force; the upper member has the largest increase followed by the lower chord (arch) and then by the web member; increasing extent of members in different regions under non-coincident excitations are different.

3. *Support reaction force*

Undernon-coincident excitations, reaction force of the support is greater than that under coincident excitations, with the ratio ranging from 0. 85 to 2. 0.

REFERENCES

[1] Yang Qingshan, Liu Wenhua, Tian Yuji, "Response analysis of national stadium under specially variable earthquake ground motions" [J]. *China civil engineering journal*, 2008,41 (2) :35-41.

SHAKING TABLE TEST ON 1000kV UHV TRANSMISSION TOWER-WIRE COUPLING SYSTEM

QiangXie[a], Yun-ZhuCai[b], Song-TaoXue[a]

[a]Tongji University, State Key Laboratory of Disaster Reduction in Civil Engineering, China
qxie@ tongji. edu. cn, xue@ tongji. edu. cn
[b]Tongji University, College of Civil Engineering, China
2014caiyunzhu@ tongji. edu. cn

KEYWORDS: Transmission tower-wire coupling system, Shaking table test, Similarity theory, Seismic responses.

ABSTRACT

Seismic response of transmission tower-wire system is of less concerns for its seldom earthquake damage and evident wind sensitivity. While, with the demands to higher voltage and longer span, transmission lineshave to across high-intensity earthquake region and then deserve more attention to their seismic research. In 1999 Taiwan Chi-Chi Earthquake, the damage to transmission lines caused complete failure in power system in the north of Zhanghua[1]. In 2008 WenchuanEarthquake, numerous transmission towers were seriously damaged resulting in great financial loss[2].

For studying the seismic behavior of tower-wire coupling structure, a shaking table test on a two-span scaled model consisting of asteel tube tower, two equivalent towers, 8-bundled conductors and ground wires is carried out, taking a 1000kV UHV transmission project as background. The towers are designed depending on the elastic similitude law and checked by numerical simulation. Due to strong non-linearity ofwires and limitedtable size, traditional elastic similarity theory does not apply to conductors and ground wires. Thus, a similarityparameter separation method is proposed for solving the technical difficulty(i. e. natural frequency and inertial effect of wires are considered separately). The numerical resultverifies the validity of the modeling method for tower-wire coupling system.

In the purpose of exploring the effect of wires on towers under earthquake, three types of models are considered in the test, the single tower, the tower with lumped masses used to simulate the gravity effect of wires and the tower-wire system. EL-Centro wave, Taft wave, Pasadena wave and Shanghai ArtificialWave (SHW) are selected as the shaking table excitations and both unidirectional and bidirectional tests are implemented. Furthermore, three classes of loading are involved, the frequent one, the moderate one and the rare one with the degree of 8.

Experimental results show that compared with the frequency of single tower, the first natural frequency of the transmission tower with wires was reduced by 16% and 18% in the line direction and cross-line direction respectively. The results also indicate that the dynamic responses of the transmission tower with wires is 6% ~ 30% less than that of the tower with lumped masses, which illustrates that the nonlinear vibration effect of wires increase the seismic resistant capacity of the transmission tower.

CONCLUSIONS

For researching the seismic properties of transmission tower-wire system, a shaking table test for scaled model is implemented. During the wire model design, a similarity parameter separation method is proposed, which is proved to be a good solution for solving the difficulties of wire's strong geometric nonlinearity and size limitation of the shaking table by numerical checking.

Aimed at studying the wire's influence in the tower's dynamic property and seismic responses, three types of test models are involved, the single tower, the tower with lumped masses and tower-wire system. According to relevant results, it is demonstrated:

1. The coupling effect between the wire and tower helps reduce the natural frequency of the transmission towerand such effect is more obvious in the cross-line direction than in the line direction (Table 1).

Table 1. Comparison of the first frequency of three models

direction	single tower	tower with masses		tower-wire system	
	first frequency/Hz	first frequency/Hz	drop ratio	first frequency/Hz	drop ratio
line direction(x)	7.77	7.32	5.8%	6.51	16.2%
cross-line direction(y)	7.58	6.43	15.2%	6.18	18.5%

2. Due to the wire's existence, tower's peak seismic responses decrease and when seismic action comes from cross-line direction, the reduction is more obvious than when action is from the line direction (*Fig. 1*). Compared with the tower with lumped masses model, the tower's dynamic responses of tower-wire system decrease especially under the intensity of 8 rare (*Fig. 2*). Thus, nonlinear vibration effects in conductors increase the seismic capacity of the transmission tower.

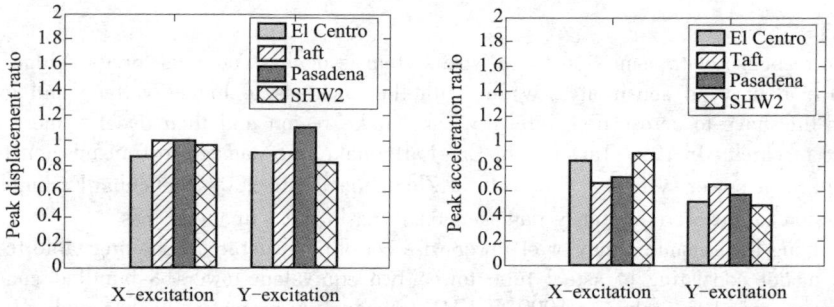

(a) Displacement ratio of toner top   (b) Acceleration ration of tower top
Fig. 1. Seismic response ratios between tower-wire system and single tower

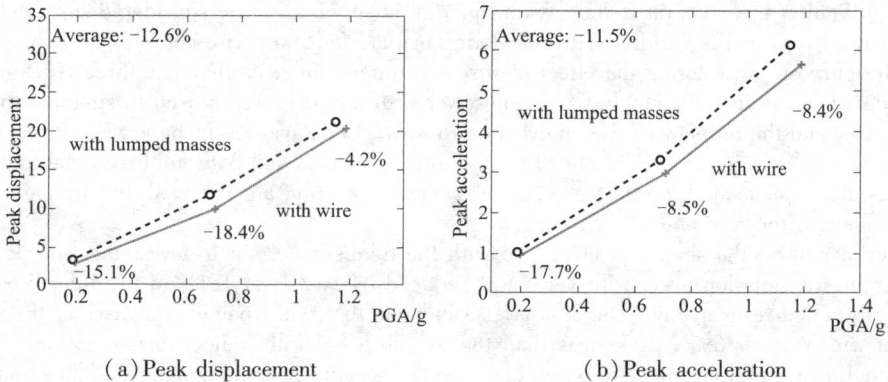

(a) Peak displacement   (b) Peak acceleration
Fig. 2. Comparison of peak responses between tower with lumped masses model and tower-wire system

3. Seismic responses of the tower with lumped masses model is larger than that of tower-wire system, which means it is feasible to just take the mass effect of conductors into consideration and ignore the nonlinear effect of conductors when designing transmission towers.

REFERENCES

[1] Tai Wan Center for Research on Earthquake Engineering, "Chichi earthquake disaster streamline reporting comprehensive survey" [R]. *NCREE-99-033*, 1999. (in Chinese)
[2] Xie Q, Zhu R., "Earth, Wind, and Ice" [J]. *Power and Energy Magazine*, 2011, 9(2): 28-36.

SEISMIC PERFORMANCE OF A NEW TYPE FISH-BONE BRB: AN EXPERIMENTAL STUDY

Liang-Jiu Jia[a,b], Hanbin Ge[c], Rikuya Maruyama[c], Kazuki Shinohara[c]

[a]Tongji University, Research Institute of Structural Engineering and Disaster Reduction, China
[b]Meijo University, Advanced Research Center for Natural Disaster Risk Reduction, Japan
[c]Meijo University, Department of Civil Engineering, Japan
lj_jia@ tongji. edu. cn, gehanbin@ meijo-u. ac. jp
rikuya. maruyama@ gmail. com, 110425157@ ccalumni. meijo-u. ac. jp

KEYWORDS: Bucking-restrained brace, All-steel, Fish-Bone, Cyclic, Seismic.

ABSTRACT

Buckling-restrained brace (BRB), as a passive controlled strategy, has been proposed to absorb energy by plastic yielding of metallic core plate during an earthquake. According to a number of experimental and analysis studies, the maximum ductility demand for a concentrically braced frame (CBF) under a seismic input with a 2% exceedance probability in 50 years in the US, ranged from 20 to 25[1]. Based on available experimental results to date, e. g., [1-4], there are a number of BRBs which cannot meet the maximum ductility demand of a CBF. Moreover, the maximum ductility demands for BRBs employed in bridge engineering, e. g., [5], can be much larger than those for a CBF building, and thus the corresponding maximum ductility demands are higher. Besides, strong aftershocks after a strong earthquake were reported in recent years, and higher ductility demands are required. In this study, a novel steel fish-bone shaped BRB (FB-BRB) illustrated in *Fig. 1* aimed to obtain high ductility capacity is proposed. Experimental studies utilizing three, small-scale FB-BRBs were conducted to investigate its failure modes and corresponding mechanisms, such as ductile fracture, core plate buckling etc. A conventional BRB was also experimentally investigated as a comparison of cyclic performance with the FB-BRBs. The geometrical parameters of the specimens are listed in *Table 1*.

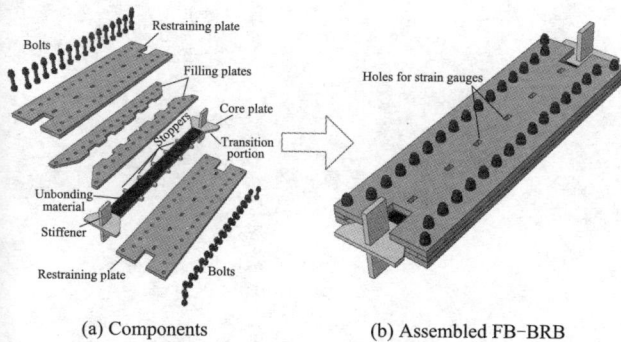

(a) Components   (b) Assembled FB-BRB

Fig. 1. Proposed fish-bone buckling-restrained brace

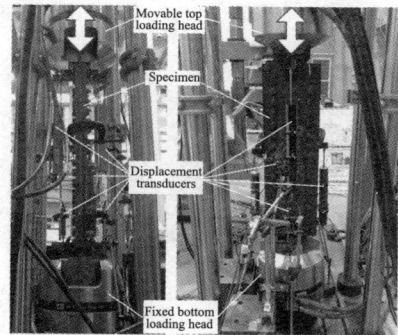

(a) Side view   (b) Front view

Fig. 2. Test setup

Table 1. Dimensions of the specimens

No.	Specimens	L(mm)	d(mm)	d_0(mm)	S	E_s(%)	D(mm)
1	Common-BRB	670	1	2	0	—	0
2	FB-BRB-S5-E8-D0	670	1	2	5	8	0
3	FB-BRB-S2-E8-D0	670	1	1	2	8	0
4	FB-BRB-S2-E6-D3	670	1	1	2	6	3

Note: L = length of yielding portion of the core plate; d = gap between the core plate and restraining plate; d_0 = gap between the core plate and restraining plate; S = number of stoppers; E_s = allowable maximum elongation of the segment between two adjacent stoppers; D = half of the reduced width at the center of each segment.

CONCLUSIONS

In this study, a new type of light-weight fish-bone shaped buckling-restrained braces (FB-BRBs) are proposed, aimed to achieve high seismic performances such as large maximum ductility capacity, μ_{max}, cumulative ductility capacity, μ_c and energy dissipation capacity, ΣE_p. The core plate of the FB-BRB is divided into several segments by multiple pairs of stoppers. The characteristic of the FB-BRBs is that the post-necking straining capacity of the material (steel) can be more effectively employed through generating several necked regions in the multiple segments. Four scaled BRBs with different details were experimentally tested, where a conventional BRB was utilized to compare with the other three FB-BRBs. The average stress-strain curves of the specimens are given in *Fig. 2*, and the values of μ_{max}, μ_c and ΣE_p are also listed in *Table 1*. Experimental results show that FB-BRBs can have a favorable seismic performance than the conventional BRB. Typical failure modes of the specimens are illustrated in *Figs. 3* and *4*, which are induced by several possible factors such as strain concentration, excessive necking deformation, lack of shear strength of the stoppers and premature cracking at the edges of the stoppers. The experimental results indicate that there is a great potential to further improve seismic performance of the newly proposed FB-BRBs.

Table 2. Ductility capacity and energy dissipation of specimens

No.	Specimen	μ_{max}	μ_c	ΣE_p (kJ)
1	Common-BRB	26.1	437	80.8
2	FB-BRB-S5-E08-D0	29.9	484	93.9
3	FB-BRB-S2-E08-D0	29.9	488	94.6
4	FB-BRB-S2-E06-D3	33.6	591	115.0

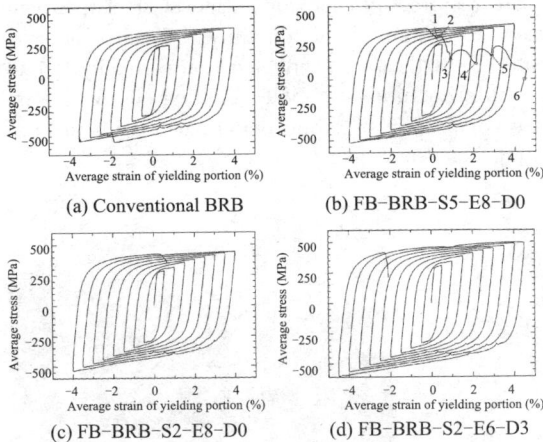

(a) Conventional BRB

(b) FB-BRB-S5-E8-D0

(c) FB-BRB-S2-E8-D0

(d) FB-BRB-S2-E6-D3

Fig. 3. Stress-strain curves of specimens

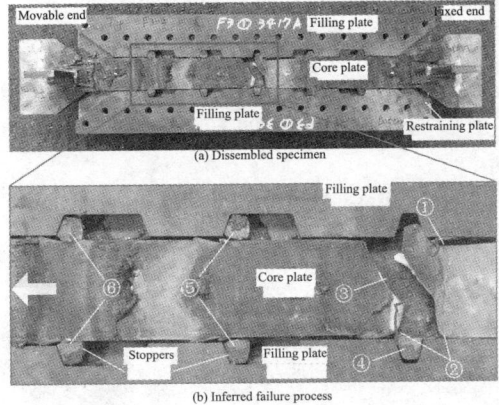

(a) Dissembled specimen

(b) Inferred failure process

Fig. 4. Typical failure mode

REFERENCES

[1] Fahnestock L. A., Sause R., Ricles J. M., Lu L.-W., 2003. "Ductility demands on buckling-restrained braced frames under earthquake loading". *Earthquake Engineering and Engineering Vibration*, Vol. 2, No. 2, pp. 255-268.

[2] Black C., Makris N., Aiken I., 2004. "Component testing, seismic evaluation and characterization of buckling-restrained braces". *J Struct Eng (ASCE)*, Vol. 130, No. 6, pp. 880-894.

[3] Wang C-L, Usami T, Funayama J., 2012. "Evaluating the influence of stoppers on the low-cycle fatigue properties of high-performance buckling-restrained braces". *Eng Struct*, Vol. 41, pp. 167-176.

THE ELISSA PROJECT: PLANNING OF A RESEARCH ON THE SEISMIC PERFORMANCE EVALUATION OF COLD-FORMED STEEL MODULAR SYSTEMS

Luigi Fiorino, Ornella Iuorio, Vincenzo Macillo, Maria Teresa Terracciano, Tatiana Pali, Bianca Bucciero, Raffaele Landolfo

Department of Structure for Engineering and Architecture, University of Naples "Federico II", Naples, Italy lfiorino@ unina. it, ornella. iuorio@ unina. it, vincenzo. macillo@ unina. it, mariateresa. terracciano@ unina. it, tatiana. pali@ unina. it, bianca. bucciero @ hotmail. it, landolfo@ unina. it

KEYWORDS: Cocoon system, Knauf systems, sheathed-braced systems, shaking table tests, lightweight sustainable constructions.

ABSTRACT

The project named "Energy Efficient LIghtweight-Sustainable-SAfe-Steel Construction" (Project acronym: ELISSA) is funded by European Commission under the Seven Framework Programme (www. elissaproject. eu). It is devoted to the development and demonstration of cold-formed steel (CFS) modular systems. In particular, these systems are nano-enhanced prefabricated lightweight CFS skeleton/dry wall construction with improved thermal, vibration/seismic and fire performance, resulting from the inherent thermal, damping and fire spread prevention properties.

The partners of the consortium are: National Technical University of Athens (Greece, Coordinator), STRESS SCARL (Italy), Farbe SPA (Italy), Woelfel Beratende Ingenieure GmbH &Co KG (Germany), Ayerisches Zentrum fur Angewandteenergieforschung ZAE EV (Germany), Knauf Gips GK (Germany), University of ULSTER (United Kingdom), Haring Nepple AG (Switzerland), University of Naples Federico II (Italy), Knauf of Lothar Knauf SAS (Italy), VA-Q-TEC AG (Germany).

The architectural project of "ELISSA house" aims to be expression of a real-life solution, which could potentially incorporate in the full testing phase all the facilities required for a residential housing. Therefore, has been selected as case study a house for single with a great care for details. The dwelling is composed by three rectangular modules of plan dimensions 2.5×4.5 m and area equal to 34 m^2. (Fig. 1).

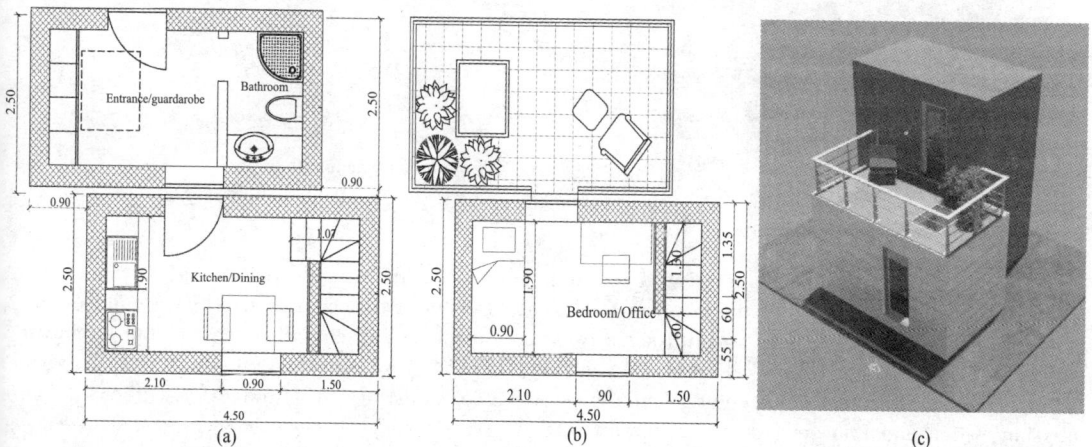

Fig. 1. (a) 1st Floor architectural plan; (b) 2nd Floor architectural plan; (c) Perspective view

The structure of the "ELISSA house" is the "Transformer" system by COCOON (by Haring Nepple AG) [1], upgraded for seismic loads by a sheathed-braced CFS solution [2], in which the seismic resistant elements are made of CFS stud shear walls laterally braced by gypsum-based panels (by Knauf Gips GK). On the base of the structural design, the experimental program including tests on microscale level, meso-scale level and macro-scale level is defined (Table 1). The experimental activity will be mainly carried out at the laboratory of Structures for Engineering and Architecture Department of University of Naples. The tests on micro-scale level consists of tests on main wall components, with particular reference to the connections. In particular, the shear behaviour of clinched steel-to-steel connections and panel-to-steel connections will be evaluated by means monotonic and cyclic tests. In addition, meso-scale tests consisting of monotonic and cyclic tests on full-scale seismic resistant systems (shear walls) will be conducted. Finally, in order to evaluate the global seismic response of the ELISSA house, a shaking table test on two-storey module (macro-scale level) will be performed.

Table 1. Experimental program

Level	Specimens	No. tests
Micro-scale	Steel-to-steel connections	15
	Panel-to-steel connections	25
Meso-scale	Shear walls	6
Macro-scale	"ELISSA house" mock-up	1

The dynamic tests will be carried out on one of the two shaking tables at the laboratory ofStructures for Engineering and Architecture Department of University of Naples, shown in *Fig. 2*. Since the plan dimensions of the mock-up (2.5 × 4.0 m) are greater than maximum allowable specimen dimensions (3.0 × 3.0 m) of the shaking table, a specific structure has been designed with the aim of extending the base of the shaking table. The extension structure consists of a very stiff 3D reticular steel structure made with RHS profiles that allows to test specimens with dimensions up to 3.0 × 4.5 m (*Fig. 3*). The test will be performed in only one horizontal direction by applying the different acceleration input, in order to identify the basic dynamic properties of the mock-up and to evaluate the response under different earthquake intensities.

Fig. 2. Shaking table at University of Naples Federico II

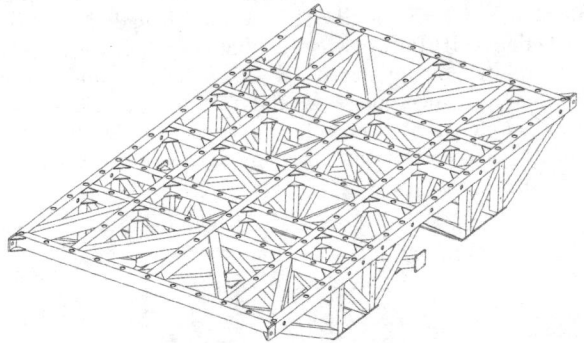

Fig. 3. Extension structure

CONCLUSIONS

The paper presents the planning of a research program on the seismic performance evaluation of cold-formed steel modular systems focusing the attention on the experimental activity on connections, single seismic resistant systems and full-scale two-storey mock-up, which will be carried out at University of Naples Federico II from February 2015 to July 2016. The research program is part of FP7 project named "Energy EfficientLIghtweight-Sustainable-SAfe-Steel Construction" funded by European Commission, which is acknowledged for its support.

REFERENCES
[1] ETA-11/0105, European Technical Approval: System Cocoon "Transformer, DIBt, 11.04.2011.
[2] Fiorino, L. , Iuorio, O. , Macillo, V. , Landolfo, R. "Performance-based design of sheathed CFS buildings in seismic area". Thin-Walled Structures, Elsevier Science. Vol. 61, pp. 248-257. 2012.

EXPERIMENTAL TESTING OF A DOUBLE ACTING RING SPRING SYSTEM FOR USE IN ROCKING STEEL SHEAR WALLS

Gary S. Djojo[a], G. Charles Clifton[a], Richard S. Henry[a], Gregory A. MacRae[b]

[a]University of Auckland, Department of Civil and Environmental Engineering, New Zealand
gdjo001@ aucklanduni. ac. nz, c. clifton@ auckland. ac. nz, rs. henry@ auckland. ac. nz
[b] University of Canterbury, Department of Civil and Natural Resurces Engineering, New Zealand
gregory. macrae@ canterbury. ac. nz

KEYWORDS: steel shear wall, rocking system, srlf-centring, ring spring, experimental testing

ABSTRACT

A rocking steel shear wall is being developed utilisinga centrally located rocking mechanism and energy dissipation devices. This rocking wall is designed to remain essentially elastic during a severe earthquake, undergoing controlled rocking at the base in the centre of the wall, and is expected to return to its original position after an earthquake. The centralised rocking pivot and V brace at the bottom storey of the wall permits the columns at the edges of the wall to move upward and downward during earthquakes, with half the magnitude of vertical movement at the wall edges for a given rotation compared with a conventional wall that rotates about the corners. The energy dissipation devices are designed for the base of the columns and are also intended to provide a restoring-force to centre the wall back to the original position. Therefore, double acting systems are proposed.

Ringfeder ®, compression only friction ring springs can be arranged to work as double acting spring systems to accommodate the expected vertical movement of the columns during the earthquake. In addition, this spring system can be preloaded to have a high initial stiffness up to a defined force level which is necessary for this wall system that is required to be stiff under wind loads and serviceability limit state earthquakes and to rock under more severe earthquakes.

Two double acting ring spring systems are proposed for this rocking wall, as follows:

1. Double acting ring spring type I with a parallelogram hysteresis curve. This system comprises two stacks of ring springs arranged in series at top and bottom of a column base plate and prestressed to 50% of the spring capacity with a high tensile grade AISI 4140 threaded rod connecting top plate to the foundation.

2. Double acting ring spring type II with a flag-shaped hysteresis curve. This system comprises a stack of ring spring with customised endplates at its both ends is put into a cartridge. A high tensile grade AISI 4140 threaded rod is centrally passed through the ring spring and endplates and is fastened to connect between a column base plate and a bottom endplate. Then, the cartridge is sealed by a clamping plate which is bolted to the flanges of the cartridge. The perimeter of the cartridge base is welded to the base plate.

The results of experimental testing of double acting ring spring type I are presented. Only type I system was tested because similar configurations of type II system have been previously studied both analytically and experimentally[1,2]. The type I test specimen as shown in *Fig. 1* was subjected to symmetric and asymmetric cyclic loading with a range of loading rates from 0. 05 mm/s to 1. 34 mm/s to observe the hysteresis curve, the initial force to initiate the elastic spring of the system, the behaviour of the system after it is locked up, and the self-centring capability.

The test results showed that as the system were having vertical movement, each stack operated individually. When the column was in compression, the bottom stack was compressed while the top stack relaxed and when the column was in tension, the top stack was compressed and it compressed the top plate and generated tensile force in the rod while the bottom stack relaxed. When the top spring was loaded, the bottom spring was unloaded, which cancelled one-third of the initial prestressing force on both sides. Hence, this system generated a parallelogram hysteresis curve. A symmetric cyclic loading that was applied to the specimen and a test result of a parallelogram hysteresis curve are shown in *Fig. 2*.

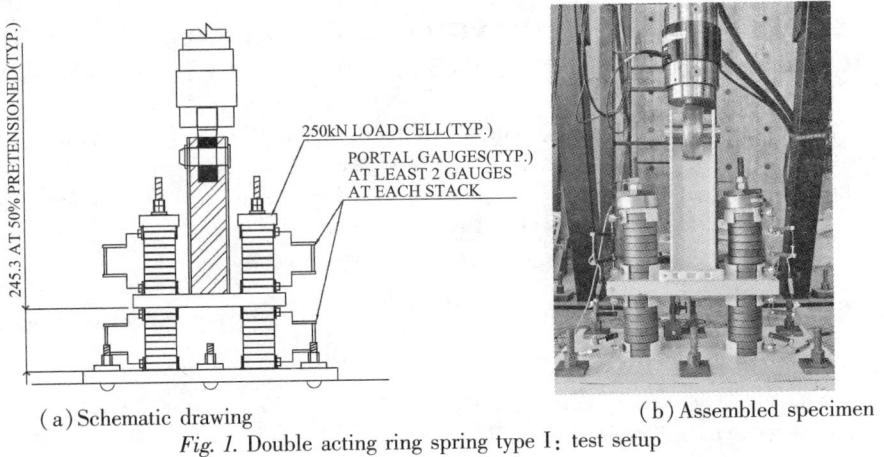

(a) Schematic drawing   (b) Assembled specimen

Fig. 1. Double acting ring spring type I: test setup

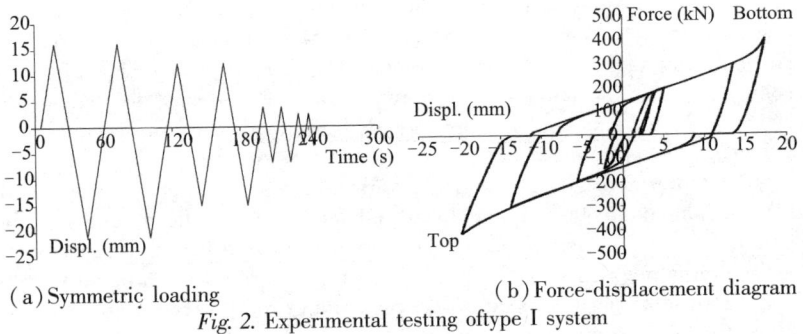

(a) Symmetric loading   (b) Force-displacement diagram

Fig. 2. Experimental testing of type I system

CONCLUSIONS

Experimental testing of a double acting ring spring system using two stacks of ring springs assembled in series has been conducted under symmetric and asymmetric loading. The results showed that parallelogram hysteresis curves with stable and repeatable loops were obtained and considerable energy was dissipated during a cycle of loading and unloading. When the top spring was loaded and the bottom spring was unloaded, partial cancellation about one-third of the initial prestressing force resulted in system movement commencing at a lower level of force compared to that from a single spring prestressed at the same level. The loading stiffness of the spring was the combination of loading stiffness of the spring and unloading stiffness of the opposite spring which shortened the effective spring travel compared to travel of single spring at the same prestressing level. The magnitude of the diminishing oscillating force before it became too low to compress the springs determined the extent of actual self-centring of this system. Under diminishing actions at the end of an earthquake record, it came close to fully self-centring. The experimental testing of a double acting ring spring system using two stacks of ring springs assembled in series showed that the experimental results correspond to its numerical predictions.

REFERENCES

[1] Filiatrault A, Tremblay R, Kar R, "Performance evaluation of friction spring seismic damper", *Journal of Structural Engineering*, 126:491-499, 2000.
[2] Khoo HH, "Development of the low damage self-centring sliding hinge joint", *PhD Thesis*, University of Auckland, New Zealand, 2012.

ANALYSIS OF HYBRID DAMPING DEVICE WITH SELF-CENTRING

R. Kordani[a], G. W. Rodgers[a], J. G. Chase[a]

[a] University of Canterbury, Mechanical Engineering Department, New Zealand

reza. kordani@ pg. canterbury. ac. nz, geoff. rodgers@ canterbury. ac. nz,
geoff. chase@ canterbury. ac. nz

KEYWORDS: Hybrid dampers; Seismic response; Rocking systems; Inherent self-centering.

ABSTRACT

Lead extrusion dampers have been used to dissipate seismic energy in structures and can contribute to damage avoidance design (DAD) rocking connections. In rocking connections that utilises unbound post-tensioned tendons, re-centering of the overall structure is typical. However, the lead extrusion dampers alone are strictly dissipative, have no inherent self-centering and without careful integration into a structural system can lead to residual story drifts. In this study a modified version of High Force-To Volume(HF2V) extrusion damper is introduced to overcome the lack of inherent re-centring, while maintaining the energy absorption capability. The new device is a combination of HF2V and ring spring dampers to provide an overall device with large energy dissipation and inherent self-centering. Response spectral analysis for multiple, probabilistically scaled earthquake suites are used to delineate the displacement reduction factors due to the added damping. Hysteresis analysis of the device under a variety of seismic loadings are also performed and design plots are provided for different sized dampers. Overall, the results indicate an important trade-off between force contributions from the HF2V and ring spring components. Moreover, increasing the ring spring participation force level leads to less residual displacement in exchange for less reduction in peak displacement. This approach of larger ring spring contributions shows less dependence on the structural period, indicating a robustness of the design to a broad spectrum of ground motion inputs.

CONCLUSIONS

The concept of a hybrid energydissipator is introduced. Comprehensive simulation response of structures with a wide range of natural period is undertaken across a range of earthquake suites so the influence of the seismic loading pattern and structural stiffness on the device response are characterized. The analysis of the response spectra shows a significant reduction in the maximum and residual displacement. The reduction factors show the device performs better across low suite ground motions. Dampers with small ring spring force participation(less than 50%)do not perform effectively in reducing the residual displacement especially for higher structural periods(bigger than $T=3$ s). However, even in cases where residual displacements are seen they are not excessive.

While the hybrid device analysis shows significant promise, the static design method used in the research needs to be replaced by a more robust method. Simple design based on a normalised design force does not accurately capture the full velocity dependence and dynamic force contribution. Installing the damper impacts the natural period of the structure and also one device can provide different force capacity based on the loading pattern. These areas are of particular interest for future work.

ACKNOWLEDGMENT

The support of the University of Canterbury Quake Centre(UCQC)to the first author is gratefully acknowledged.

REFERENCES

[1] G. W. Rodgers, "Next Generation Structural Technologies: Implementing High Force-To-Volume

Energy Absorbers," 2009.

[2] K. E. Hill, "THE UTILITY OF RING SPRINGS IN SEISMIC ISOLATION SYSTEMS," University of Canterbury, 1995.

[3] G. Pekcan, J. B. Mander, and S. S. Chen, "NON-LINEAR VISCOUS DAMPERS," vol. 1425, no. June, pp. 1405-1425, 1999.

NON-LINEAR SEISMIC ANALYSIS AND BEHAVIOUR OF CBF-V

Beatrice Faggiano, Antonio Formisano, Carmine Castaldo, Luigi Fiorino, Vincenzo Macillo,
Federico M. Mazzolani

Department of Structures for Engineering and Architecture, University of Naples "Federico II"
Piazzale Tecchio 80, 80125 Naples

e-mails: [1]faggiano@ unina. it, [2]antoform@ unina. it, [3]carmine. castaldo@ unina. it,
[4]lfiorino@ unina. it, [5]vincenzo. macillo@ unina. it, [6]fmm@ unina. it

KEYWORDS: Seismic resistant steel structures, Seismic design criteria, Concentrically V-braced frames, Behaviour factor, Over-strength factor, Non-linear static analyses.

ABSTRACT

The critical review of the design methodologies provided by the European seismic code (Eurocode 8) for steel Concentrically Braced Frames with chevron (or inverted V) diagonals (CBF-V) has been the spark for deepening the seismic behaviour of such typical steel seismic resistant structures, aiming at providing more efficient design criteria able to ensure adequate safety levels under earthquake actions. As reference case studies, common structural configurations of CBF-V are designed according to the Eurocode 8 provisions. Each case study is designed through both the Linear Static (LS) and Dynamic (LD) analysis. For braces either Circular Hollow Sections (CHS) or HE profiles are used. The results presented in this paper are framed into a more wide research activity, aiming at the optimization of seismic design criteria for steel braced structures.

CONCLUSIONS

According to the European seismic rules[1], the resistance against seismic actions of Concentrically Braced Frames (CBF) is provided by the contribution of both tensile and compression braces. Therefore, for dissipative braced systems, the ideal design ultimate condition is the simultaneous buckling of the compressed bracings and yielding of the tensile ones, the braces being the dissipative elements. Beams and columns should be designed according to the capacity design criterion to remain in elastic range and, therefore, they should have an adequate over-strength as respect to braces and, to this purpose, the over-strength factor Ω is applied.

Nevertheless, these rules appear to be in some point poorly effective as respect to the objectives of the design that they should achieve[2]. To this purpose, some structures have been preliminary designed according to EC8.

The study structures belong to a regular building, with 3, 6 and 10 floors, the inter-storey height being $h = 3.5$m (at the ground floor $h_{gf} = 4$m) and the bay span $L = 6$m[3]. The preliminary design according to EC8, by using both Linear Static and Linear Dynamic analyses, has led towards columns with welded box sections, beams with HE profiles and braces with either HE or CHS profiles.

The design criteria, which the structural sizing of the examined frames is based upon, show in general the following critical issues[4]:

– High values of Ω_{min}, ranging from 1.86 to 3.4 and 2.46 to 4.43 for CHS and HE brace sections, respectively, are obtained. These values are conditioned from the limited number of available commercial profiles and imply a significant increment of axial forces in the columns, which requires the adoption of welded box sections. Therefore, the weight of structures is significantly huge, without offering any significant improvement to the seismic capacity.

– The Ω variation ratio is always governed by the top storey braces, whose Ω values are generally larger than those at lower storeys. This is due to the use of brace cross-sections which are subjected to low seismic actions, but should respect contemporary also the standard slenderness limit. As a consequence, a large over-sizing of structural members is necessary, it producing just an unjustified weight increase.

– When HE profiles are used for diagonal braces, Ω_{min} is generally larger than the behaviour factor q used in the design phase(2. 5). This is not acceptable, since a design spectrum greater than the elastic one given by the code should be used.

    – The assumption $\gamma_{pb} = 0.3$ for beam design appears to be not conservative. Indeed, this assumption implies the failure mechanism of beams, which contrary should remain elastic under the design earthquake.

    Theabove results have shown that the weight variation of non-dissipative members increases with the number of storeys. The brace cross-section shape also significantly influences the weight, because it affects the Ω_{min} values: CHS braces give rise to lighter structures than those with HE braces. This occurs thanks to the larger choice of CHS profiles than the HE ones. It has been already noted that, as expected, the design through the LS analysis leads to structural weights larger than those obtained by LD, since the LS design forces are larger than the ones form the LD analysis, but this increase is different whether CHS or HE braces are used(in the case of 10 storeys it is 20% for CHS braces and 10% for HE braces). In addition in the latter case, the scarcely exploited braces at the upper floors give rise to high Ω values and then to oversize both beams and columns[5]. Another important design issue is the variation of the column cross-sections along the height, which causes marked stiffness change among the storeys. This can generally produce damage localized to a given storey, with a consequent reduction of energy dissipation capacity.

    Finally, from the achieved results it has been shown that she simplified computational models of CBF-V should be enhanced in order to be able to catch the main aspects of the structural behaviour of these seismic-resistant systems. In general, the critical analysis of the results briefly presented in this paper can be useful to identify a correct revision trend for the existing provisions; however they require a more deep elaboration for comprehensively depicting the impact of the design criteria on the seismic performance of investigated structures. This allows to plan a further extensive campaign of both experimental and numerical investigations aiming at both optimizing the calculation models and providing "ad hoc" simplification in the design procedures.

REFERENCES

[1] CEN. "EN 1998-1 :2005. Eurocode 8: Design of structures for earthquake resistance-Part 1: General rules, seismic actions and rules for buildings". *European Committee for Standardization, Bruxelles*, 2005.

[2] Faggiano B. , Fiorino L. , Formisano A. , Macillo V. , Castaldo C. , Mazzolani F. M. "Assessment of the Design Provisions for Steel Concentric X Bracing Frames with Reference to Italian and European Codes". *The Open Construction and Building Technology Journal*, 8, (Suppl 1: M3): 208-215, 2014.

[3] De Lucia T. , Formisano A. , Fiorino L. , Faggiano B. , Mazzolani, F. M. "Optimization of design criteria for seismic steel structures with chevron concentric bracing frames" (in Italian). *Proc. of the XV Anidis Congress "L'Ingegneria Sismica in Italia"*, Padova, 30 June-4 July, Braga, F. & Modena, C. (eds.), *Padova University Press Publisher*, ISBN 978-88-97385-59-2, Paper n. F7 on CD-Rom, 2013.

[4] Macillo V. , Castaldo C. , Fiorino L. , Formisano A. , Faggiano B. , Mazzolani F. M. , "Critical review of the NTC2008 design criteria for steel concentric bracing frames". *Proc. of the International Workshop High Strength Steel in Seismic Resistant Structures(HSS_SERF)*, Naples, Italy, 28-29 June 2013.

[5] Decree of the Minister Council Presidency 03/05/2005 n. 3431(OPCM). "*Further modifications and integrations to the Decree of the Minister Council Presidency 20/03/2003 n. 3274*" (in Italian), 2005.

Mixed and Composite Structures

CYCLIC LOADING TEST ON THE SHEARING BEHAVIOR OF WELDED BOX SECTION COLUMNS WITH CONCRETE FILLED

Zhiqiang Li[a], Yiyi Chen[b], Wei Wang[b]

[a]Tongji University, College of Civil Engineering, China

lzq19841983@ 163. com

[b] Tongji University, State Key Laboratory of Disaster Reduction in Civil Engineering, China

yiyichen@ tongji. edu. cn, weiwang@ tongji. edu. cn

KEYWORDS: Shearing behaviour, Cyclic loading test, Shear-span ratio, Failure modes, Seismic performance.

ABSTRACT

Focused on the shearingbehaviour, cyclic test researches were conducted on welded box section columns with concrete filled by using a specially designed loading device. The influence of two important parameters on the shearing performance, shear-span ratio ranged from 0. 5 to 2. 5 and axial compression ratio altered from 0 to 0. 32, was investigated in the test. The test revealed that shearing failure, bending failure and bending-shearing failure patterns occurred according to the difference of shear-span ratio, while the axial compression ratio seemed no evident influence on the failure modes. Based on the test phenomena, criteria for the identification of predominant failure modes were proposed. It is noticed that the specimen with shearing failure has an excellent seismic performance according to the test.

CONCLUSIONS

Focused on the bending and shearing behaviour, hysteretic test was conducted on three welded steel box section columns with concrete filled by using a specially designed loading device. Reversed and equivalent moment was exerted on both ends of specimen. According to the test results, some conclusions can be drawn as follows:

(1) Bending failure, shearing failure and bending-shearing failure may occur as the shear-span ratio changes;

(2) The criterion based on the test phenomena can be employed to identify the predominant failure mode;

(3) Specimen with shearing failure has an excellent seismic performance.

ACKNOWLEDGMENT

The work presented in this paper was supported by the Key Projects of Natural Science Foundation of China(NSFC) through Grant No. 51038008.

REFERENCES

[1] Lu F W, Li S P, Li D W, et al, "Flexural behavior of concrete filled non-uni-thickness walled rectangular steel tube", *Journal of Constructional Steel Research*, 63(8):1051-1057, 2007.

[2] Han L. , "Flexural behaviour of concrete-filled steel tubes", *Journal of Constructional Steel Research*, 60(2):313-337, 2004.

[3] Han L, Lu H, Yao G, et al, "Further study on the flexural behaviour of concrete-filled steel tubes", *Journal of Constructional Steel Research*, 62(6):554-565, 2006.

[4] Lu Y Q, Kennedy D J L, "The flexural behaviour of concrete-filled hollow structural sections", *Canadian journal of Civil Engineering*, 1(21):111-130, 1994.

[5] Shuli Guo, Zhong Tao, "Experimental study of square concrete-filled steel tube columns subjected to shearing", *Journal of Fujian University of Technology*, 2011(6):550-554, 2011. (in Chinese)

[6] Jian Cai, Weisheng Liang, Hui Lin, "Experimental study on shear resistance performance of concrete filled square steel tubular columns", *Journal of Shenzhen University Science and Engineer-*

ing,29(3):189-194,2012. (in Chinese)

[7] Yong Huang, Weigang Chen, Li Duan, "Experimental study on shear resistance performance of concrete filled steel tube shear block stub columns", *Journal of Building Structures*,32(12):178-185,2011. (in Chinese)

[8] Han L,Tao Z,Yao G,"Behavior of concrete-filled steel tubular members subjected to shear and constant axial compression", *Thin-walled Structures*,2008(46):765-780,2008.

[9] CECS 159-2004,"Technical specification for structures with concrete-filled rectangular steel tube members", *China Planning press*,2004. (in Chinese)

DIAPHRAGM BEHAVIOR OF DECONSTRUCTABLE COMPOSITE FLOOR SYSTEMS

Lizhong Wang[a], Mark D. Webster[b], Jerome F. Hajjar[a]

[a] Northeastern University, Department of Civil and Environmental Engineering, USA
wang. l@ husky. neu. edu, jf. hajjar@ neu. edu
[b] Simpson Gumpertz & Heger Inc. , USA
mdwebster@ sgh. com

KEYWORDS: Diaphragm behavior, Design for Deconstruction, Composite floor systems.

ABSTRACT

This study investigates the seismic behavior of deconstructable composite steel/concrete floor systems consisting of precast concrete planks and deconstructable clamping connections attaching the planks to the steel floor beams, see *Fig. 1*. Grouting concrete planks and placing a cast-in place concrete topping, which help to tie all the planks together in conventional precast concrete construction, are eliminated to achieve deconstructability of the proposed system. *Fig. 2* illustrates unbonded threaded rods used to connect the planks in-plane before being clamped to the steel beams. The system is proposed to promote sustainable design of composite steel/concrete floor systems in steel buildings via reuse of the structural steel and concrete components at the end of the useful life of buildings.

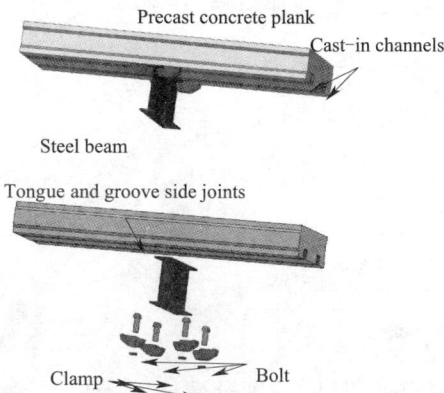

Fig. 1. Deconstructable composite beam prototype

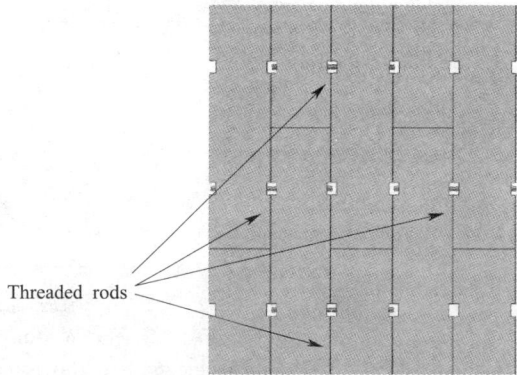

Fig. 2. Precast concrete plank connections

As shown in *Fig. 3*, finite element models representing full-scale diaphragms were developed and analyzed to investigate the failure modes and establish the behavior of the floor diaphragms. The parameters for the analytical models are listed in *Table 1*, including a compressive in-plane force representing the connections of the threaded rods between planks, the number of clamps on the main collector girder.

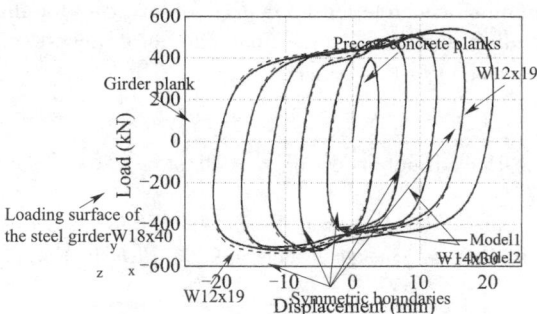

Fig. 3. Finite element model for the diaphragm

Table 1. Parameters for the analytical models

Model number	Compressive stress (MPa)	Number of shear connectors
1	1.5	20
2	1.5	28
3	3.0	20
4	3.0	28
5	6.0	20
6	6.0	28

CONCLUSIONS

The cyclic load-displacement curves are plotted in *Figure 4*. The limit state for the first four models is joint sliding due to diaphragm shear; therefore, the compression between adjacent planks is directly related to the ultimate strength. The hysteresis loops are almost identical for Model 1 and Model 2, even though the number of shear connectors varies. A similar conclusion, however, cannot be reached for Model 3 and Model 4. The stiffness of Model 4 is smaller than that of Model 3, because the failure mode for Model 4 is actually a combination of joint sliding and clamp slipping. Distinct load-displacement curves are plotted for Model 5 and Model 6, as their limit state is slip of the clamps between the steel girder and the girder plank. The number of shear connectors affects the ultimate strength of the diaphragms. All the diaphragms demonstrate ductile behavior with no strength and stiffness degradation.

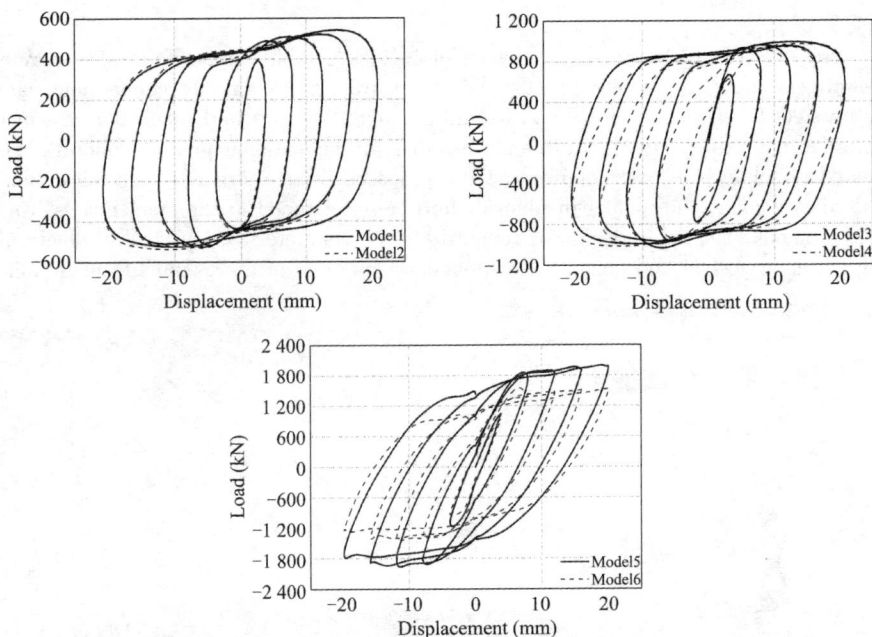

Fig. 4. Cyclic load-displacement curves

In thediaphragm models, the moment at the symmetric boundaries is shared by the steel chords and the concrete slab in accordance with their relative stiffness. It is found that the majority of the external force follows the stiffer load path and flows into the concrete slab, and the rest goes through the steel chords that are bent about their weak axis.

While diagonal cracking may be seen in monolithic concrete diaphragms[1], this failure mode is uncommon in precast concrete diaphragms, since the joint between the diaphragms provides a weak link in the system. This argument is validated by minimal concrete tensile damage observed in the diaphragm models. Joint opening due to diaphragm bending, another potential limit state for a precast concrete diaphragm[2], did not occur for the models developed in this paper.

REFERENCES

[1] Easterling, W. S. , & Porter, M. L, 1994. "Steel-Deck-Reinforced Concrete Diaphragms. I. " *Journal of Structural Engineering*, 120, 560-576,.

[2] Fleischman, R. B. , Naito, C. J. , Restrepo, J. , Sause, R. , & Ghosh, S. K. 2005. "Seismic Design Methodology for Precast Concrete Diaphragms Part 1: Design Framework. " *PCI Journal*, 50 (5),68.

HYSTERETIC BEHAVIOUR OF CONCRETE-FILLED DOUBLE-SKIN STAINLESS STEEL TUBE BEAM-COLUMNS

Ying-FeiLi[a], Feng Zhou[a,b]

[a] Department of Structural Engineering, School of Civil Engineering, Tongji University, China
[b] Tongji University, State Key Laboratory of Disaster Reduction in Civil Engineering, China
2012liyingfei@ tongji. edu. cn, zhoufeng@ tongji. edu. cn

KEYWORDS: Stainless steel; Hysteretic behaviour; Concrete-filled steel tube; Double-skin tube.

ABSTRACT

Recently, an innovative concept called concrete-filled double-skin steel tube (CFDST) was proposed. The advantages of double-skin over conventional single-skin concrete-filled tube include higher earthquake resistance due to reduced self-weight, good local stability due to the interaction of the three components, higher global stability due to the larger section modulus and good fire resistant. CFDST has potential application in offshore structures, high-rise buildings and towers.

Compared to carbon steel, stainless steels have superior characteristics such as corrosion resistance, ductility and formability, improved fire resistance and ease of maintenance, however, the initial cost of stainless steel structural products is about four times that of equivalent carbon steel products. Concrete infilling in stainless steel tube is an effective measure to reduce the cost.

Experimental investigation on the hysteretic behaviour of the concrete-filled double-skin stainless steel tube beam-columns is presented in this paper. A total of five cyclic loading tests were conducted in the study to investigate the effects of the geometric dimension of the stainless steel tubes on the hysteretic behaviour of the concrete-filled double-skin stainless steel tube beam-columns including strength, ductility and energy dissipation capacity. Both the outer and inner skins were austenitic stainless steel type 304 circular hollow sections (CHS). The diameters of the outer and inner tubes were 130mm and 75mm, respectively. The stainless steel tube wall thickness varied from varied from 3mm to 5mm.

Tensile coupon tests were carried out to determine the material properties of the stainless steel tubes. Self-consolidating concrete was designed to pour into the gap between the outer and inner tubes. The material properties of concrete were determined from standard cubic tests. The mean value of the measured compressive concrete cubic strength is 46.5 MPa.

The specimen was connected rigidly with the reaction block on the bottom end and pinned on the upper end to transmit the axial and horizontal force. The lateral supports were considered to prevent the out-of-plane instability of the specimen. Each test specimen was subjected to a constant axial load and a cyclically increasing lateral load. Displacement control was used to drive the hydraulic actuator at a constant speed of 5 mm/min for all test specimens.

The failure mode, envelope curves, ductility coefficient, dissipated energy and rigidity degradation were analysed, it is found from the test results that the concrete-filled double-skin stainless steel tube beam-columns exhibit favourable energy dissipation and ductility.

CONCLUSIONS

This paper provides new test data of thehysteretic behaviour of concrete-filled double-skin stainless steel tube beam-columns. The main conclusions from the investigation are summarized as follows.

1. Concrete-filled double-skin stainless steel tube beam-columns show excellent hysteretic behaviour including strength, ductility and energy dissipation capacity. All of the load and displacement loops were plump, the peak lateral load keep stable with a large lateral displacement increment, and the descent part is smooth. *Fig. 1* shows typical force-displacement hysteresis curves.

2. It is found that, in the range of test parameters, the horizontal bearing capacity and capacity of energy dissipation are sensitive to the thickness of outer tube, while that of inner tube have no obvious influence.

3. It is found that stiffness declines rapidly before lateral displacement reached $3\Delta_y$, then tend to be gentle in the later cycles, and finally maintained at about 30% of the initial section rigidity for all test specimens.

4. The ductility coefficients of each specimen and other test characteristic values were summarized in *Table 1*.

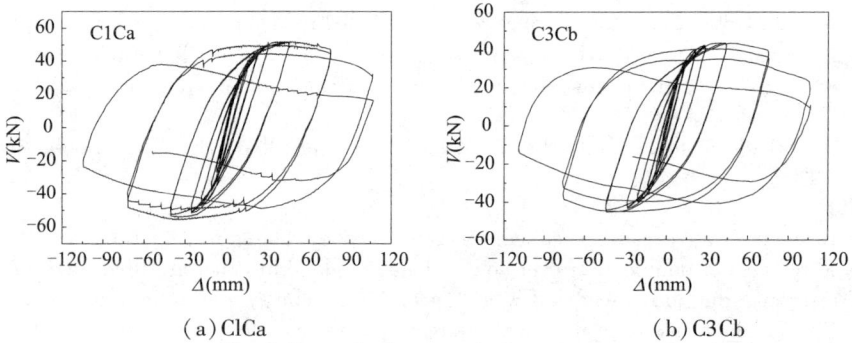

(a) C1Ca   (b) C3Cb

Fig. 1. Force-displacement hysteresis curves

Table 1. The main seismic properties of specimens

Specimens	Loading Direction	Yield		Maximum		Ultimate		μ	E_s (kN · m)
		V_y (kN)	Δ_y (mm)	V_u (kN)	Δ_p (mm)	$0.85V_u$ (kN)	Δ_u (mm)		
C1Ca	Positive	36.6	13.3	51.9	42.4	44.1	86.55	6.51	35.63
	Negative	-42.6	-13.6	-53.6	-31.5	-45.6	-79.98	5.89	
C1Cb	Positive	42.7	14.9	55.4	43.9	47.1	78.28	5.25	43.75
	Negative	-38.9	-14.1	-52.6	-41.0	-44.7	-75.76	5.36	
C1Cc	Positive	37.0	13.4	49.9	38.0	42.4	73.60	5.50	40.85
	Negative	-38.7	-13.9	-51.1	-32.5	-43.5	-78.08	5.62	
C2Cc	Positive	33.3	13.3	43.8	43.7	37.3	85.23	6.43	32.43
	Negative	-34.5	-13.6	-44.9	-43.9	-38.2	-75.38	5.54	
C3Cc	Positive	25.0	11.2	34.7	39.8	29.5	80.75	7.24	24.80
	Negative	-28.7	-11.9	-36.8	-40.6	-31.3	-82.03	6.92	

REFERENCES

[1] X. -L. Zhao and L. -H. Han, "Double skin composite construction," *Progress in Structural Engineering and Materials*, vol. 8, pp. 93-102, 2006.

[2] S. -K. H, "Composite columns of double-skinned shells," *Journal of Constructional Steel Research*, vol. 19, pp. 133-52, 1991.

[3] L. Gardner, "The use of stainless steel in structures," *Progress in Structural Engineering and Materials*, vol. 7, pp. 45-55, 2005.

[4] "ASTM E8/E8M-11 Standard test methods for tension testing of metallic materials, ASTM volume metals-mechanical testing; Elevated and Low-Temperature Tests," in *ASTM volume metals-mechanical testing*, ed, 2010.

[5] K. J. R. Rasmussen, "Full-range stress-strain curves for stainless steel alloys," *Journal of Constructional Steel Research*, vol. 59, pp. 47-61, 1// 2003.

[6] A. T. Council, "ATC-24. Guidelines for Cyclic Seismic Testing of Components of Steel Structures," ed. Redwood City(CA), 1992.

[7] L. -H. Han, H. Huang, Z. Tao, and X. -L. Zhao, "Concrete-filled double skin steel tubular(CFDST) beam-columns subjected to cyclic bending," *Engineering Structures*, vol. 28, pp. 1698-1714, 10// 2006.

[8] A. Elremaily and A. Azizinamini, "Behavior and strength of circular concrete-filled tube columns," *Journal of Constructional Steel Research*, vol. 58, pp. 1567-1591, 12// 2002.

EFFECTS OF OUT OF PLANE STRENGTH AND STIFFNESS OF COMPOSITE FLOOR SLABS ON THE INELASTIC RESPONSE OF ECCENTRICALLY BRACED FRAME STRUCTURES

Amin Momtahan[a], Charles Clifton[a]

[a] University of Auckland, Department of Civil, and Environmental Engineering, New Zealand
amom002@ aucklanduni. ac. nz
c. clifton@ auckland. ac. nz

KEYWORDS: Eccentrically braced frame, Out of plane stiffness, Composite floor slab.

ABSTRACT

The Eccentrically Braced Frame (EBF) structure is a cost-effective and commonly used seismic resisting framing system that can satisfy strength, stiffness and ductility requirements for high seismic zones. The high elastic stiffness is provided by the bracing members and high ductility capacity is a-chieved by transmitting the brace axial forces to another brace, or to an adjacent column, through a short length of beam known as an active link that is typically designed and detailed to undergo inelastic response in shear in a severe earthquake. The inelastic deformation of the active link can result in per-manent deformation and building residual drift. The Christchurch earthquake series from 4 September 2010 to 23 November 2011 comprised six damaging earthquakes. In particular, the Mw 6. 3 Earthquake on 22 February 2011 severely impacted the city, with significant damage to residential and commercial structures. Detailed analyses of the strong motion data recorded indicated that the intensity of this event was 2 to 3 times more than the Ultimate Limit State (ULS) values specified by the New Zealand seismic loading standard (i. e. NZS 1170. 5) over the period range of 0. 5 to 4 seconds.

These large seismic forces were able to load EBF structures into their inelastic range; the first time worldwide that this has happened. Despite this, the EBF building structures have performed signif-icantly better than would have been expected given the intensity of loadings. From the total of four EBF building structures that were investigated in detail after the earthquakes, it was evident that almost all the active links were strained into their inelastic range. The active links exhibited paint flaking and Lüders lines. However, they were free of residual distortions. It was also reported that in most cases less than 50% of the cumulative strain capacity of the yielded active links was utilised and therefore the active links had sufficient post-earthquake capacity not to require replacement. Comparison of the measured inelastic response of the 12 storey Club Tower building for 22 February 2011 event (based on sliding marks on the stairs) with the predicted response of the design models, showed that this building exhibited an apparent stiffness increase of between 2 to 2. 5 times that of the structural model. The re-sults have also indicated that the building has post-earthquake residual drift of only 0. 14%. This indi-cated that this building has self-centred to within construction tolerances, despite the severity of the e-vent. That self centering is particularly interesting as there was specialist self centering systems in-stalled in the building.

All investigated EBF structures had been constructed compositely with the floor slabs. Floor slabs have also shown very minor damage in the form of hairline cracking in the areas directly above active links; no repair of these floors has been required. This led to consideration that the observed increase in the lateral stiffness in the EBFs in Christchurch and enhancement of their active links in the resist-ance to yielding, is in part due to the out-of-plane stiffness of the floor slabs to which the collector beams are compositely connected.

In this study, the slab contribution has been quantified through an advanced numerical model of a single EBF frame with composite floor slab. Numerical time-history analyses of a 10-storey prototype building with and without slab contribution have also been carried out to establish the increase in later-al stiffness and level of reduction in residual drift of the EBF buildings when composite slab contribu-tion is taken into account. This contribution is significant, especially when considered in conjunction with the beneficial impact of Soil Foundation Structure Interaction (SFSI) also reducing the open field intensity of seismic actions into the building superstructure. The slab contribution to seismic loading levels with and without SFSI is presented.

CONCLUSIONS

From the results presented and discussed in this study, when the out of plane stiffness of the floor slabs is taken into account, the EBF building exhibits self-centring by showing a tendency to go back to its original position. The reduction in residual drift is caused by the floor slab pushing and pulling the ends of the collector beam, which force the entire structural system back towards its former alignment after the earthquake. For the majority of EQ records used in in this research, the ratio of the reduced displacement between "No Slabs" and "Slabs" cases was greater at the upper stories. This is because the influence of the slab springs was more prominent towards the upper levels of the structure where the added stiffness from the slab is comparable to the stiffness of the active links.

Observations from Northridge 1994, Kobe 1995, Kocaeli 1999 and Christchurch 2010-11 earthquakes have shown that significant nonlinear action in the soil and soil-foundation interface can be expected due to high levels of seismic excitation. Therefore, it is very important to consider the influence of soil-foundation interface nonlinearity on the response of the structures. Neglecting such effect prohibits the influence of energy dissipation due to soil yielding, large foundation deformation and foundation toppling on the structural response[3]. It was however discussed in[1,2] that soil nonlinear deformation alone is not a significant influence on the response of the structure-foundation system, the main effect is the changes in the foundation stiffness that are induced by the detaching and reattaching from the underlying soil. This means that uplift of the foundation can have a significant influence on the earthquake response of buildings on shallow foundations. Comparison between seismic responses of fixed-base structures with that of the structures with nonlinear SFSI showed that SFSI significantly reduces the peak acceleration of the structure in all cases. A reduction in peak acceleration means that the forces transmitted to the structure are reduced and suggests improved structural performance.

In conclusion, results from this study indicate that the slab out of plane strength and stiffness beneficially contribute to the building overall response and hence building's self-centring. Also, enhanced contribution from the slab towards the self-centring of the building is expected when SFSI effects are considered. However, quantifying the SFSI influence on superstructures founded on shallow foundation is still at an early stage and more numerical and experimental analyses are yet to be implemented.

REFERENCES

[1] Pender, M., Integrated design of structure-foundation systems: the current situation and emerging challenges. 2014.
[2] Storie, L. and M. Pender, Nonlinear spring-bed modelling for earthquake analysis of multi-storey buildings on shallow foundations. 2014.
[3] Moghaddasi Kuchaksarai, M., Probabilistic Quantification of the Effects of Soil-Shallow Foundation-Structure Interaction on Seismic Structural Response, in Civil and Natural Resources. 2012, University of Canterbury: University of Canterbury. Civil and Natural Resources.

BEHAVIOR OF THE COMPOSITE STEEL-TIMBER STRUCTURE WITH SEMI-RIGID JOINT

Masanori FUJITA[a], Tomomichi HAYASHI[b], Yuki OKOSHI[c], Mamoru IWATA[d]

[a]Yamaguchi University, Graduate School of Science and Engineering, Japan
fujitam1@ yamaguchi-u. ac. jp
[b]Yamaguchi University, Graduate School of Science and Engineering, Japan
p040fh@ yamaguchi-u. ac. jp
[c] Nice Corporation Ltd. , Japan
ookoshi. yuuki@ purple. plala. or. jp
[d]Kanagawa University, Dept. of Architecture and Building Engineering, Japan
Iwata01@ kanagawa-u. ac. jp

KEYWORDS: Steel, Timber, Composite structure, Seismic, Semi-rigid

ABSTRACT

In light of the global environment issues, the authors proposed a building system comprising steel and timber structure (Hereafter referred to as CSTS) and its design method, which consist of rolled section steel and timber. CSTS assumed mid-rise story building steel structures. This design method use concept of a damage controlled structure and its mechanical model is established based on the available experimental data. When a pure rigid-frame structure is employed for a composite steel-timber structural member, rigid joints are difficult to use for beam-to-column connections, as the degree of integration differs between the steel and the timber. In this paper, behavior of CSTS using seismic response control members is shown for the frame in order to use a semi-rigid joint for beam-to-column connections by a static incremental analysis and a dynamic response analysis. As a result, effectiveness of design method of CSTS with the semi-rigid joint are verified.

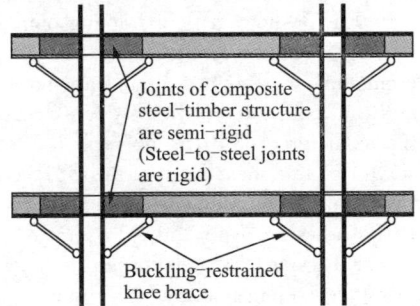

Fig 1. Structural frame

DESIGN CONDITIONS OF CSTS
CSTS structural members

When CSTS structural members are used for a pure rigid-frame structure, a damage-controlled structure with buckling-restrained braces installed as knee braces is used so that semi-rigid joints may be used for beam-to-column connections in*Fig. 1*, as it is difficult to design the details so that tensile stress is transmitted between timber structural members. The CSTS structural members are configured with composite structural members in which the outer surface of the steel members is covered with timber. *Fig. 2* shows a beam and a column of the CSTS. Assuming that the CSTS steel structural members are to be reused, rolled H-section steel and rectangular steel pipe, are employed for the beams and columns, respectively. To ensure the fire-resistance performance of the CSTS, the covering depth of timber is determined taking into consideration the fire-resistant design used for semi-fireproof construction of timber structures, except for the case where a fireproof building is required. In the CSTS, the cross-sectional area of steel is made smaller than that of timber to allow the steel to yield before the timber and therefore semi-rigid joints can be used for column-to-beam connections and beam joints.

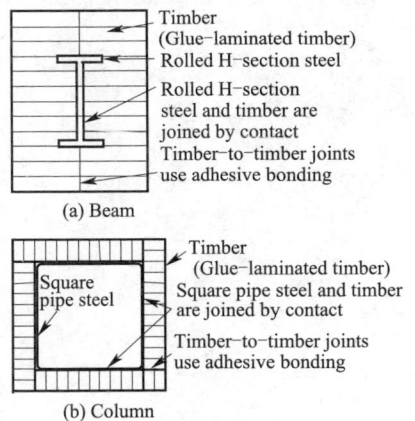

Fig 2. Section of the CSTS

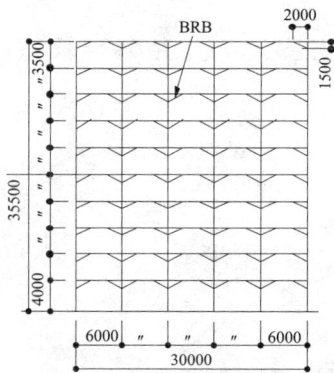

Fig 3. Analytical model

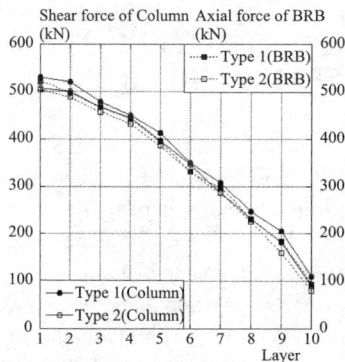

Fig 4. Shear force of column and axial force of BRB

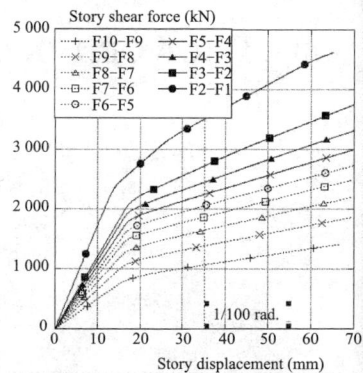

Fig 5. Incremental analysis Results(Type 1)

Types of design condition

Two types of models are set as shown to examine the effects of whether or not tensile stress is transmitted between timber structural members on the behavior of the CSTS using a semi-rigid joint. The Type 1 model represents the case where the timber of the CSTS column does not bear tensile stress, the Type 2 model represents the case where the timber bears tensile stress. These models have the same initial stiffness.

BEHAVIOR OF CSTS

Fig. 3 shows a floor framing plan and a framing elevation. For individual types, the shear force of the CSTS column and the axial force of the BRB at a story deformation angle of 1/100 rad. by static incremental analysis are plotted as shown in *Fig. 4*. In individual types, the shear force of the columns and the axial force of BRBs decrease as story level increases. The shear force of the CSTS columns and the axial force of BRBs of Type 1 are respectively comparable to Type 2. A similar tendency is observed in the shear force of the CSTS columns and the axial force of BRBs in the middle and lower stories. Thus, there are only small differences in the shear force of CSTS columns and the axial force of BRBs between Type 1 and Type 2.

For Type 1, the relationship between story shear force and relative story displacement obtained from the static incremental analysis results is plotted as shown in *Fig. 5*. The required horizontal load-carrying capacity of individual types is distributed within a story deformation angle of 1/100 rad. , and the horizontal load-carrying capacity of each layer exceeds the required horizontal load-carrying capacity.

CONCLUSIONS

This study examined the effects of whether or not tensile stress is transmitted between timber structural members on the behavior of the CSTS using a semi-rigid joint. The following findings were obtained.

1) There are only small differences in the shear force of CSTS columns and the axial force of BRBs regardless of whether or not tensile stress is transmitted between timber structural members.

2) A plastic hinge occurs at the beam ends in the CSTS column-to-beam connections after BRBs buckle.

3) The ratio of shear force carried by the timber of the CSTS columns gradually decreases after BRBs plasticize.

INVESTIGATION ON THE SEISMIC BEHAVIOR OF CONCRETE-FILLED STEEL PLATE COMPOSITE COUPLING BEAMS

Hong-Song Hu[a], Jian-Guo Nie[b]

[a] Disaster Prevention Research Institute(DPRI), Kyoto Univ., Gokasho, Uji, Kyoto 611-0011, Japan hhsong05@gmail.com

[b] Dept. of Civil Engineering, Tsinghua University, Beijing 100084, China
niejg@tsinghua.edu.cn

KEYWORDS: concrete-filled steel plate composite coupling beam; cyclic loading test; seismic behavior;

ABSTRACT

Coupled shear wall systems that consist of two or more shear walls connected by coupling beams are efficient lateral force resisting systems, which have been widely used in mid-rise and high-rise buildings(El-Tawil et al,2010). As the building height increases, the internal forces in the structural members increase significantly, which makes the conventional reinforced concrete(RC) coupled shear wall system uneconomic or even impossible to accomplish a design. To overcome this problem, a new coupled shear wall system referred to as concrete-filled steel plate(CFSP) composite coupled shear wall system was proposed by the authors, as shown in *Fig. 1*. In the proposed coupled shear wall system, CFSP composite wall piers (Nie et al, 2013) are coupled by CFSP composite coupling beams. Since the two structural members have similar configurations (both consist of surface steel plates and concrete infill), it is easy to proportion them with matched stiffness, load-carrying and deformation capacities. In this paper, six coupling beam specimens were tested under reversed cyclic loading to study the seismic behaviour of CFSP composite coupling beam.

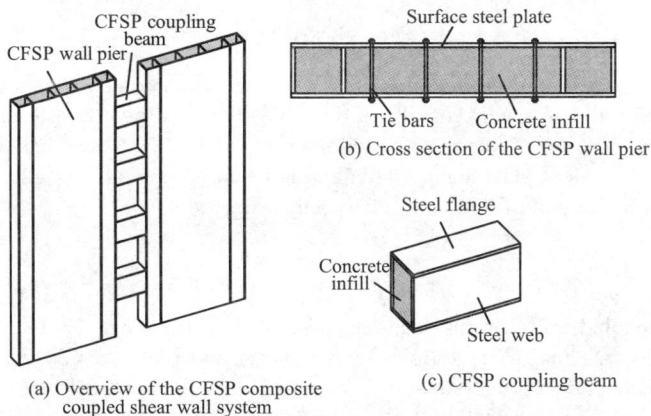

Fig 1. Details of the CFSP composite coupled shear wall system

Six specimens were designed with coupling beams between two shear wall piers. The wall piers were each rigidly attached to end plates that allowed the specimen to be bolted to the test setup. The parameters that were varied between specimens included the coupling beam span and steel plate thickness used for the coupling beam. All the specimens experienced a progression of limit states from crack initiation, shear buckling of the steel webs, and fracture propagation which ultimately caused a significant loss of load carrying capacity. The final failure modes of all the specimens are shown in *Fig. 2*. The fracture limit state was identified as critical for the composite coupling beam configurations tested in this study as the fracture propagation through the web and flange plates was associated with significant loss of shear capacity in all specimens. The fractures initiated at the corners of the coupling beam, and propagated through the steel webs and flanges at the interface of the coupling beam to the wall pier. Two local buckling phenomena were observed including compression local buckling at the beam ends and shear buckling of the steel webs. The occurrence and form of local buckling was affected by the steel plate thickness, span-to-height ratio, and fracture of steel plates. Concrete compression failure was not observed in the infill concrete for any of the specimens. The concrete crack patterns were consistent with the deformation of steel plates. For the longer span specimens, the infill concrete cracked at the beam ends, while for the shorter span specimens, the infill concrete cracked along the diagonal of coupling beam.

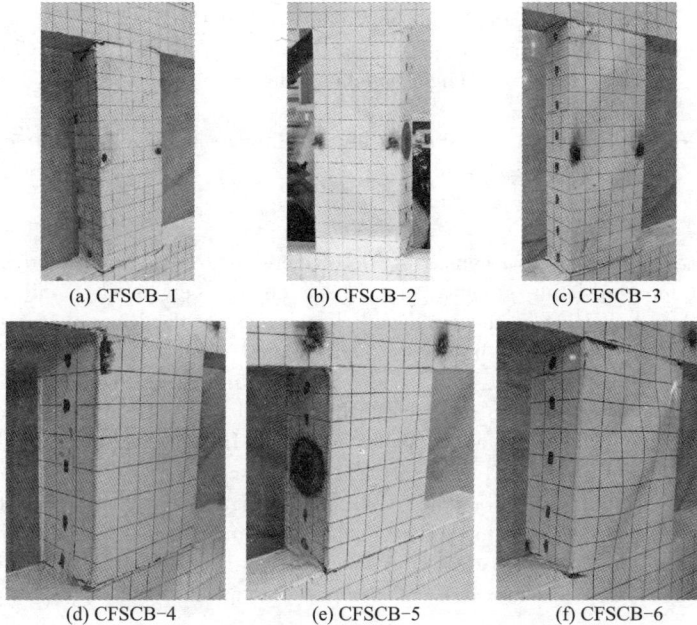

| (a) CFSCB-1 | (b) CFSCB-2 | (c) CFSCB-3 |
| (d) CFSCB-4 | (e) CFSCB-5 | (f) CFSCB-6 |

Fig 2. Failure modes

CONCLUSIONS

Since the specimen of improved detailing achieved large ultimate chord rotation, and the hysteretic loops were plump for all the specimens, it can be concluded that the concrete filled steel plate composite coupling beam has large deformation and energy dissipation capacities if fractures are delayed. The concrete filled steel plate composite coupling beam is therefore a viable option to use in the high rise buildings with concrete filled steel plate composite walls.

REFERENCES

[1] El-Tawil, S. , Harries, K. A. , Fortney, P. J. , Shahrooz, B. M. , and Kurama, Y. ,2010. "Seismic design of hybrid coupled wall systems: state of the art". J. Struct. Eng. ,136(7),755-769.
[2] Nie, J. G. , Hu, H. S. , Fan, J. S. , Tao, M. X. , Li, S. Y. , and Liu, F. J. ,2013. "Experimental study on seismic behavior of high-strength concrete filled double-steel-plate composite walls. " J. Constr. Steel Res. ,88,206-219.

Passive, Semi-active and Active Control

RESEARCH ON ADDITIONAL DAMPING EFFECT OF THE PENDULUM-TYPE TUNED MASS DAMPER

Deng Zhongliang[a], Fan Zhong[a], Liu Xianming[a]

[a]China Architectural Design & Research Group, China

dengzl@ cadg. cn, fanz@ cadg. cn, liuxm@ cadg. cn

KEYWORDS: Pendulum-type tuned mass dampers, mass ratios, frequency ratio, additional damping ratio.

ABSTRACT

The pendulum-type tuned mass damper(called P-TMD) had been widely applied in the wind vibration control of high-rise buildings and towering structures on account of being highly sensitive to environmental lateral loading, the structural dynamic response decreases consequently. *Table 1.* has listed parts of exsisting engineering projects with P-TMD.

Table 1. Engineering projects with P-TMD

Projects title	Location	Height(m)	Frequency of P-TMD(Hz)	Mass quality(t)
Sydney Tower	Sydney/Australia	305	0. 100 ~ 0. 500	220
Rokko island P & G	Kobe/Japan	117	0. 330-0. 620	270
Chifley Tower	Sydney/Australia	209	—	400
Kansai International Airport	Osaka /Japan	—	0. 800	10
MKD8 Hikarigaoka Office Building	Tokyo/Japan	100	0. 440	—
Shanghai World Financial Center(SWFC)	Shanghai/China	492	—	150
The Olympic Tower	Beijing/China	246. 8	0. 196	50
Canton Tower	GuangZhou/China	618	0. 105	1200
Taipei 101	Taiwan/China	509	—	660
Shanghai Tower	Shanghai/China	632	0. 106	1000

Additional damping effect had been studied in detail for SDOF with one pendulum under the following assumptions:

(1) The suspension point of the pendulum is ideal without no friction.

(2) The swing angle is rather small($\theta < 5°$), the swing line is always in the state of tesion.

(3) The potential energy of the pendulum ball is transmitted and obtained instantaneously from the kinetic energy of the main structure, which is:

$$\frac{1}{2}m\left[\dot{x}(t)\right]^2 = mgL[1 - \cos\theta(t)] \tag{1}$$

$$\dot{x}(t) = \sqrt{gL}\theta(t) \tag{2}$$

(4) The external motivation is transient.

Additional damping ratio is:

$$\xi_a = \frac{m\sqrt{\frac{g}{L}}}{2M\omega} = \frac{1}{2}\frac{m}{M}\frac{\omega_p}{\omega} = \frac{1}{2}\mu f \tag{3}$$

As shown in the *Fig. 2*, for mass ratio 2% usually applied in the engineering design of P-TMD, the additional damping ratio increases 0. 005 ~ 0. 010 approximately.

Analysis in this paper concluded that the swing of the pendulum ball could cause increasing of additional damping ratio, and decrease the dynamic structural response. Furthermore, the calculation

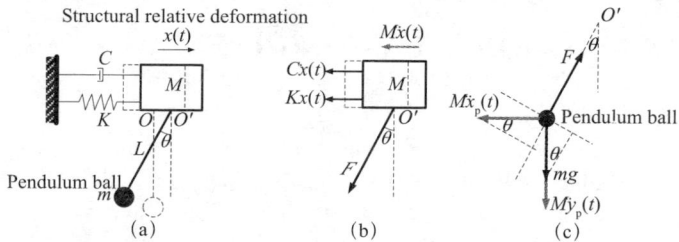

Fig 1. Single freedom degree structure system with the pendulum

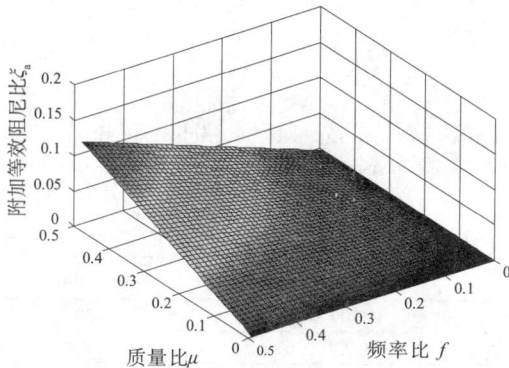

(a) 3D diagram of the additional damping ratio

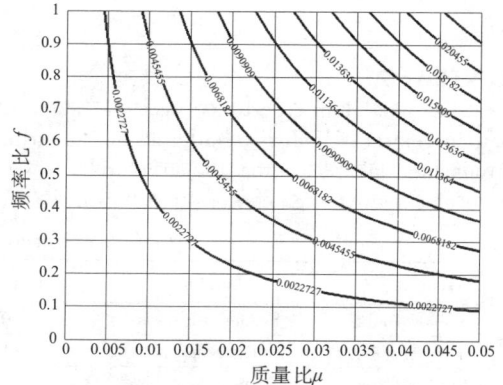

(b) Contour map

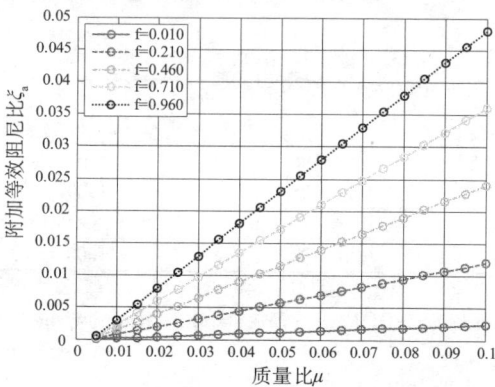

(c) $\xi_a - \mu$

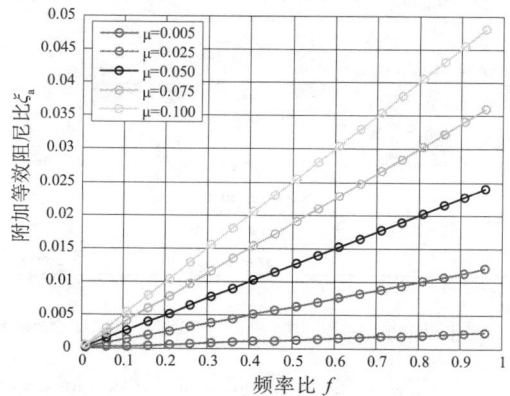

(d) $\xi_a - f$

Fig 2. Relationship of additional damping ratio ξ_a with mass ratio μ and frequency ratio f

method of additional damping ratio proposed could provide scientific guidance to the design and structural analysis. Finnaly, existing engineering projects were summarized to provide foundation for subsequent research and application of the P-TMD.

REFERENCES

[1] Ormondroyd J, Den Hartog J P. The theory of the dynamic vibration absorber[J]. Journal of Applied Mechanics, 1928, 50(7): 9-22.

[2] Den Hartog J P. Mechanical vibrations [M]. New York: Dover Publications, 1947: 126-131.

[3] Rana R, Soong T T. Parametric study and simplified design of tuned mass dampers [J]. Engineering Structures, 1998, 20(3): 193-204.

[4] G. B. Warburton, Optimum absorber parameters for variouscombinations of response and excitation parameters, Earthquake Engineering and Structural Dynamics, 1982, Vol. 10, pp. 381-401.

DAMAGE CONTROL OF COMPOSITE GYMNASIUM STRUCTURES WITH ENERGY-DISSIPATION ROOF BEARINGS

Yuki Terazawa[a], Toru Takeuchi[a], Kazuhiko Narita[b], Ryota Matsui[a] and Kou Maehara[a]

[a] Tokyo Institute of Technology, Japan

terazawa. y. aa@ m. titech. ac. jp, ttoru@ arch. titech. ac. jp, rmatui5@
mail. arch. titech. ac. jp, maehara. k. aa@ m. titech. ac. jp

[b] Ibaraki Prefectural Office, Japan

k3narita@ agate. plala. or. jp

KEYWORDS: seismic retrofit, gymnasium, cantilevered RC wall, out-of-plane response, roof bearing

ABSTRACT

A large numbers of steel roof bearings and roof braces in RC gymnasia were damaged during the 2011 Tohoku Earthquake, and the out-of-plane response of cantilevered RC walls supporting the roof frame on the gable side was identified as one of the main causes of the damage[1]. Typical damages and the collapse mechanism are shown in *Fig. 1*.

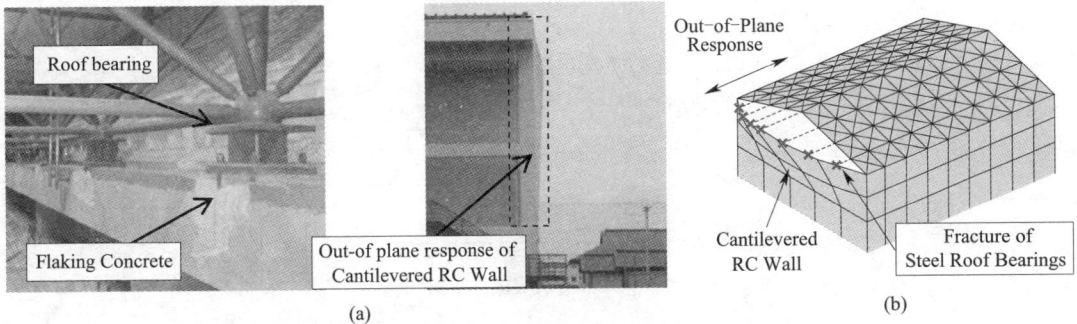

Fig 1. (a) Typical damages; (b) Damage mechanism of the steel roof bearing

In order to use those kinds of gymnasia as shelters after disasters, they are required seismic retrofit. However, a conventional retrofit method strengthening bearings with steel corner plates and replacing all the roof braces is uneconomical. In this paper, a seismic retrofit method inserting energy-dissipation elements between the roof bearings and the substructure is discussed and their response reduction effects on an actual gymnasium damaged in the 2011 Tohoku Earthquake is investigated. The energy-dissipation bearings including hysteresis dampers or viscos dampers are installed, and the performance including response reduction effect of cantilevered RC walls, displacement of the bearings, and their reaction force are compared with a non-controlled gymnasium through time history response analysis method. The analytical model with the energy-dissipation devices is shown in *Fig. 2*. Additionally, dynamic tests on a kind of friction dampers are carried out for examining the energy-dissipation performance as shown in *Fig. 3*.

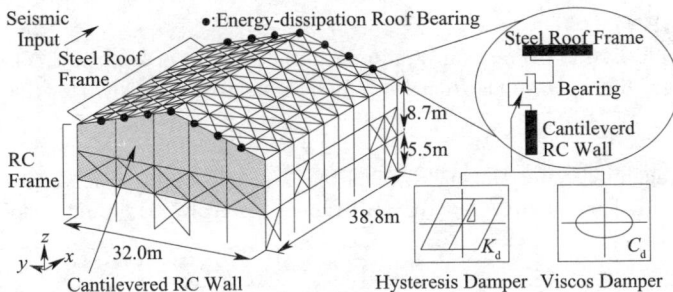

Fig 2. Analytical model

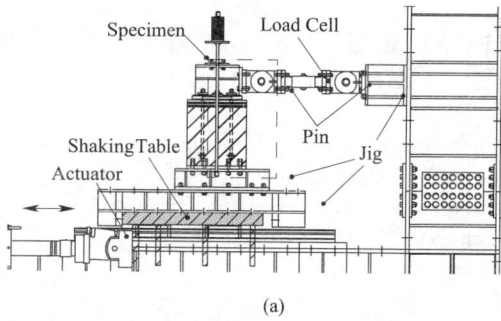

Fig 3. (a) Test Setup; (b) Specimen of the energy-dissipation bearing

On the basis ofthe analysis results and the dynamic test results, a seismic retrofit method using energy-dissipation devices at the steel roof bearings provides significant response reduction effects for the cantilevered RC walls and the bearings, and the friction damper bearing shows the stable hysteresis characteristic. Finally, a simplified response evaluation method using equivalent SDOF system shown in *Fig. 4* and an equivalent linearization technique are proposed to determine the optimum design of the energy-dissipation roof bearings, which by and large evaluates the response with safety margin.

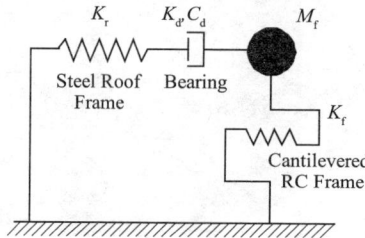

Fig 4. Equivalent SDOF system model

CONCLUSIONS

In conclusion, the following results are obtained.

1) A seismic retrofit method inserting energy-dissipation devicesbetween the steel roof and bearings provides a significant response reduction effect for the cantilevered RC walls, and the reaction force at the bearings can be maintained below the buckling strength of the steel roof beams.

2) A kind of friction dampers was developed to create a feasible design of the energy dissipation device at the roof bearings of RC gymnasia with steel roofs. The friction damper shows the stable hysteresis characteristics.

3) The simplified response evaluation method usingthe equivalent SDOF system and the equivalent linearization technique are proposed to determine the optimum design of the energy-dissipation devices. Although the proposed method overestimates the displacement response, it indicates an appropriate amount of dampers with safety margin.

ACKNOWLEDGMENT

The authors would like to express our gratitude to the support of Nippon Steel and Sumikin Engineering Co. , Ltd. , and the suggestions from Dr. Yuichi Matsuoka (NSENGI) for this study.

REFERENCES

[1] Narita, K. , Takeuchi, T. , and Matsui, R. (2014) , "Seismic Performance of school gymnasia with steel roofs supported by cantilevered RC wall frames", ACEE 2014 Conference Proceedings, Taipei Taiwan.

CONTROL OF STRUCTURAL RESPONSE WITH A NEW SEMI-ACTIVE VISCOUS DAMPING DEVICE

N. Khanmohammadi Hazaveh[a], S. Pampanin[a], J. G. Chase[b], G. W. Rodgers[b]

[a] University of Canterbury, Department of Civil Engineering, New Zealand
Nikoo. hazaveh@ pg. canterbury. ac. nz, stefano. pampanin@ canterbury. ac

[b] University of Canterbury, Department of Mechanical Engineering, New Zealand
geoff. chase@ canterbury. ac. nz, geoff. rodgers@ canterbury. ac. nz

KEYWORDS: Semi-active control devices, Semi-active viscous damper, reshape hysteretic structural response reduction factors.

ABSTRACT

Semi-active control devices can perform significantly better than passive devices, but also have the potential to achieve the performance approaching that of a fully active system. Semi-active devices offer significant promise for their ability to add supplemental damping and reduce seismic structural response in an easily controllable manner, and can be used in some modes to modify or reshape hysteretic structural response. However, many current semi-active devices are highly complex, limiting robustness, while those that can generate larger forces suffer from increased response lag time to do so. Thus, an ideal semi-active device would offer high forces, low complexity, and fast response. The semi-active viscous dampers could offer all these properties and could reduce not only the displacement response of a structure, but also the base shear. There are three semi-active viscous dampers, a 1-4, 1-3 and 2-4 device. In this study, a spectral analysis over periods of T = 0. 2-5. 0 sec under 20 design level earthquakes from the medium suite of the SAC project is used to compare three device control laws individually or in combination to sculpt structural hysteretic behaviour. Performance is assessed by evaluating reduction factors(RFs) compared to an uncontrolled structure for maximum displacement(Sd) and total base-shear(Fb), indicative of structural and foundation damage, respectively. Results show that combining the control laws to reshape the hysteresis loop can reduce the median value of both Sd and Fb by approximately 30% for periods less than 3. 0 sec and 20% for periods more than 3. 0. Thus, the results show that the proposed device and control laws have significant effect to reduction both structural response and base-shear. Overall, these results indicate the robustness of potentially very simple and robust semi-active viscous dampers to mitigate the risk of seismic damage to both the structure and foundation in a way that is economically suitable for either new designs or retrofit.

CONCLUSIONS

The semi-active viscous damper(*Fig. 1*) that allows a broader range of control laws than classical semi-active devices that has been investigated in this research. The most appropriate hysteresis loop is provided by the ability to semi-actively re-shape hysteretic behaviour via broader control laws. Based upon the investigation described herein, the following conclusions can be drawn:

• Minimizing base shear and structural response, depending on design requirements, can be achieved by sculpting the hysteretic behaviour semi-actively.

• Spectral displacement (sd) reduction factors showed considerable reductions achieved, with similar results between the 1-3 and 2-4 control law. The largest reductions were seen for the 1-4 device. However, these latter results come at the expensive of increased risk of increased base shear. Moreover, the results of RF of Sd for combination control law are closed to Sd RF of the 1-4 control laws.

The two control laws that reduce total base shear, as well as displacement response, are the 2-4 and combination device. However, the combination of the 1-4 and 2-4 control law could use the advantage of the both methods. Therefore, the combination method could reduce the structural response as well as the 1-4 control law and have a stable RF of Fb for all of the periods as same as the 2-4 control law.

Finally, the results presented provide initial insight that could lead to quantify risk and reward in term of foundational structural design parameters in a framework suitable for typical performance based design, to ease translation into practice.

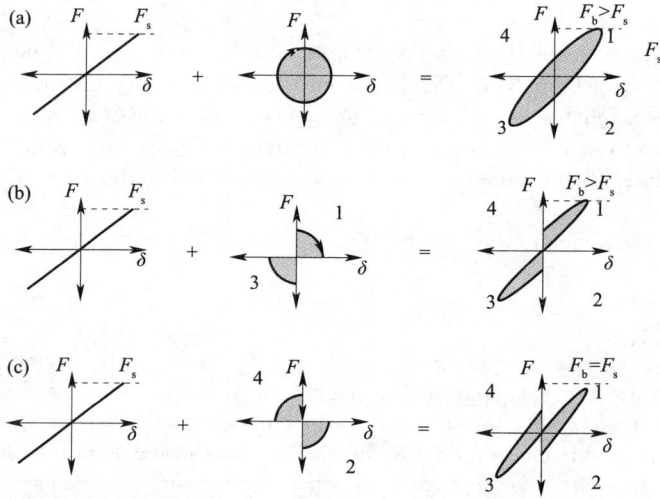

Fig. 1. Schematic hysteresis for (a)1-4 device, (b)1-3 device and (c)2-4 device[11]

ACKNOWLEDGMENT

This research is part of the SAFER concrete project funded by Natural Hazard Research Platform (NHRP).

REFERENCES

[1] Rodgers, G. W. , J. B. Mander, J. Geoffrey Chase, K. J. Mulligan, B. L. Deam, and A. Carr, Re-shaping hysteretic behaviour—spectral analysis and design equations for semi-active structures. Earthquake engineering & structural dynamics, 2007. 36(1): p. 77-100.

[2] Mulligan, K. , J. Chase, J. Mander, G. Rodgers, R. Elliott, R. Franco-Anaya, and A. Carr, Experimental validation of semi-active resetable actuators in a 1/5th scale test structure. Earthquake Engineering & Structural Dynamics, 2009. 38(4): pp. 517-536.

[3] N. Khanmohammadi Hazaveh, S. P. , J. G. Chase, G. W. Rodgers, Reshaping Structural Hysteresis Response with Semi-active Viscous Damping. bulletin of earthquake engineering, 2015. under review.

[4] N. Khanmohammadi Hazaveh, S. P. , G. W. Rodgers, J. G. Chase, Novel Semi-active Viscous Damping Device for Reshaping Structural Response. Conference: 6WCSCM(Sixth World Conference of the International Association for Structural Control and Monitoring), 2014.

EVALUATION OF DISSIPATIVE EFFECTIVENESS OF A HYBRID SYSTEM COMPOSED BY A BUCKLING RESTRAINED BRACE WITH A MAGNETO RHEOLOGICAL DAMPER

Norin Filip-Vacarescu[a], Aurel Stratan[a] and Dan Dubina[a], [b]

[a]Politehnica University of Timisoara, Department of Steel Structures and Structural Mechanics, Timisoara, Romania

norin. filip-vacarescu@ upt. ro, aurel. stratan@ upt. ro, dan. dubina@ upt. ro

[b] Romanian Academy, Fundamental and Advanced Technical Research Centre, Timisoara, Romania

KEYWORDS: Hybrid damping devices, Seismic, Dissipative, Performance.

ABSTRACT

A wealth of damping devices that function on various principles, have been studied and implemented in different structural configurations, for both new structures and for seismic retrofit of existing structures, to reduce seismic response and to limit damage. Research in the last years has started to inquire in the possibility to combine already established damping devices to create Hybrid Damping Systems (HDS). These hybrid systems are conceived to provide additional damping capacity in the structure and/or provide special advantages in the structural system such as recentering capabilities or even the replaceability of damaged components[1]. This can be achieved by the interaction between damping devices disposed in varied structural configurations. The purpose of this paper is to show the potential of a hybrid system made of buckling restrained brace(BRB) coupled with a magneto-rheological damper(MRD). The MR dampers can work either as passive devices or as semi-active ones, having a variable hysteretic behaviour that can be controlled by the variation of the magnetic field in the device with the change of current intensity as the MR liquids change their apparent viscosity when acted upon by a magnetic field. However, from technical point of view, these devices can still be considered limited in terms of capacity and can prove insufficient for high intensity seismic events. To tackle this issue the current research proposes the use of these dampers in a hybrid system with another secondary passive damping device more suitable for moderate to high levels of seismic action. The second passive device chosen is a BRB which is a well-known damping system that works on the concept of preventing the buckling of a ductile steel core that is introduced in a steel casing filled with concrete, obtaining a symmetric behaviour in tension and compression.

Numerical models were calibrated for the MRD working as a passive device(Fig. 1a) and for the BRB based on experimental test data found in literature.

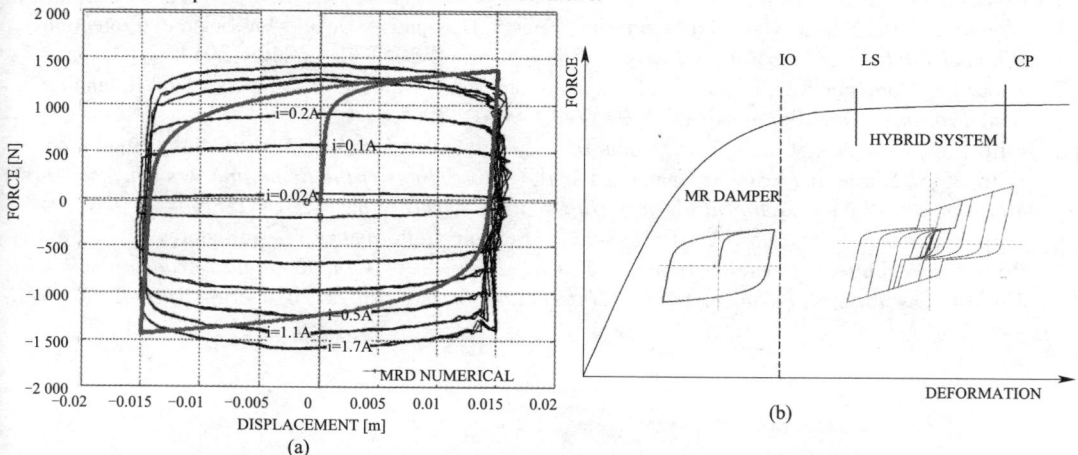

Fig 1. (a) MRD numerical model vs. experimental[2]; (b) Conceptual use of hybrid device

Using previous experience of a research program that studied the coupling of a friction damper with a steel brace in a design concept that led to a hybrid system[3,4], the damping system proposed was the combination of two different dissipative devices that can work together to answer in an optimal way the demands of both low level and moderate to high level intensity of seismic motions. The analytical models were simple numerical models used to study the multi-phase behaviour of the hybrid system. The aim of the numerical study was to prove that the hybrid system concept has the desired behaviour at element level and that it could be a viable solution to fulfilling the demands of the structure at different performance levels. The two elements were connected in series and modelled as 2 link elements. An interlock-out mechanism to make the transition between the two elements was also introduced in the numerical model as a gap-hook element connected in parallel with the MRD damper.

CONCLUSIONS

A hybrid dissipative system developed from two damping devices-i. e. BRD + MRD-is proposed and studied numerically. The results obtained for the hybrid system demonstrate the combined behaviour of the two passive devices. The MR damper is designed to provide damping and energy dissipation at a low level of seismic action and, after attainment of that level, it is able to engage a BRB to provide energy dissipation at higher levels of seismic action. In a performance based design approach this hybrid damper can be thought to provide energy dissipation through the MRD for levels of seismic action corresponding to immediate occupancy (IO) performance level and through the BRB for higher level seismic actions corresponding to life safety (LS), even up to collapse prevention (CP) performance levels (*Fig. 1*b). A simple numerical model was developed to demonstrate the concept of multi-stage behaviour of the hybrid damper concept by connecting in series two numerical models calibrated based on experimental data for the MR damper and for the BRB.

The paper summarise the first numerical trials for such a hybrid system in order to design an experimental testing program as part of on-going research.

ACKNOWLEDGMENT

This work was partially supported by the strategic grant POSDRU/159/1. 5/S/137070 (2014) of the Ministry of Labour, Family and Social Protection, Romania, co-financed by the European Social Fund-Investing in People, within the Sectorial Operational Programme Human Resources Development 2007-2013.

This work was partially supported by the research grant PN II nr. PCCA 77 / 2014-Protecţia seismică a structurilor cu sisteme de contravântuiri disipative echipate cu amortizoare cu fluid nano-micro magneto-reologic (SEMNAL-MRD). cknowledgments are optional. Numbered reference list shall be given as follows.

REFERENCES
[1] Wenke T, Eric ML. , "Hybrid Recentering Energy Dissipative Device for Seismic Protection", *Hindawi Publishing Corporation Journal of Structures*, vol 2014, ID 262409, 2014.
[2] Tudor S. , Gheorghe G. , Danut S. , "Magnetorheological fluids and dampers" (in Romanian), *Bren Printing House*, Bucuresti, ISBN 973-648-394-0. , 2005.
[3] Filip-Vacarescu N. , Stratan A. , Dubina D. , "Seismic Performance of Multistorey Steel Frames With Strain Hardening Friction Dampers-Part 1", *Proceedings of the Romanian Academy*, series A: *Mathematics, Physics, Technical Sciences, Information Science*, Vol. 1/2014, ISSN 1454-906, 2014.
[4] Filip-Vacarescu N. , Stratan A. , Dubina D. , "Seismic Performance of Multistorey Steel Frames With Strain Hardening Friction Dampers-Part 2", *Proceedings of the Romanian Academy*, series A: *Mathematics, Physics, Technical Sciences, Information Science*, Vol. 2/2014, ISSN 1454-906, 2014.

PERFORMANCE OF FIXED-PARAMETER CONTROL ALGORITHMS ON HIGH-RISE STRUCTURES EQUIPPED WITH SEMI-ACTIVE TUNED MASS DAMPERS

Demetris Demetriou[a], Nikolaos Nikitas[a], and Konstantinos Daniel Tsavdaridis[a]

[a]University of Leeds, School of Civil Engineering, United Kingdom
cn09dd@ leeds. ac. uk, N. Nikitas@ leeds. ac. uk, K. Tsavdaridis@ leeds. ac. uk

KEYWORDS: semi-active damper, control algorithm, wind excitation, high-rise buildings

ABSTRACT

The present study aims to investigate the relative performance of fixed parameter control (direct output and full state feedback) algorithms on the vibration mitigation of a wind excited high-rise structure equipped with a semi-active control device. The actual device opted is a tuned mass damper (TMD) where the active component is facilitated through variable structural damping provision. Obviously the ranges of maximum and minimum damping delivered along with the damper's stroke are essential parameters in the design of the whole semi-active system. Similar studies were performed before[1], yet the perspective taken within this study is new and attempts to determine the holistically most effective and robust solution in terms of a combined software-hardware analysis. An ensemble of some of the most commonly used algorithms in the structural and automotive control world are considered as a representative sample; namely five variants 1) the linear quadratic regulator (LQR), 2) the proportional integral derivative (PID), 3) the velocity (VBG) and 4) the displacement -based groundhook (DBG) and 5) the bang-bang control algorithm are realised for the case of the benchmark 76-floor wind-prone sway model building previously described by Yang et al[2] and here depicted in *Fig. 1*. Within the same publication the wind forcing, structural details and a number of performance limitations (e. g. max top floor acceleration) are clearly defined and similarly followed herein. Additionally a set of eleven performance indexes ($J_1, J_2, \cdots, J_{11}$) are introduced, relevant to different performance metrics of the vibration control efficacy. For them, the lower value indicates superior functioning. Out of the eleven indexes, J_{11} the one relating to the damper's maximum recorded stroke is critical for the analysis to follow.

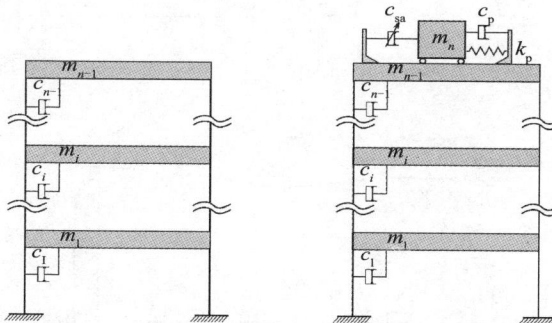

Fig 1. Sway n -storey sway frame with and without the semi-active TMD device

CONCLUSIONS

The analysis results indicate that there is an interesting interplay between the control algorithms opted and the semi-active damper design features of min-max critical damping ratio and stroke. *Table 1* depicts the superiority of the bang-bang based controller in terms of a minimal required travel length. Yet, this is partly the picture that pertains. It should be noted that these results derived when using identical critical damping ratio bands for all control cases (i. e. min 5%, max 20%)

Table 1. Performance index for different algorithms

Index	LQR	VBG	DBG	PID	BANG
J_{11}	2. 34	2. 05	2. 53	1. 89	1. 57

As *Fig. 1*a illustrates the vibration mitigation performance for different control scenarios is very close to each other, but the bang-bang option clearly is not the best. Twisting the problem and thresholding the allowable damper stroke to a conservative value while allowing different optimal critical damping ranges for different control algorithms results to *Figs. 1*b,c. There it becomes apparent that different algorithms identify in various ways the compromise between their design features and a truly optimised solution requires for an in-depth study of the compatibility between control hardware and software that is required.

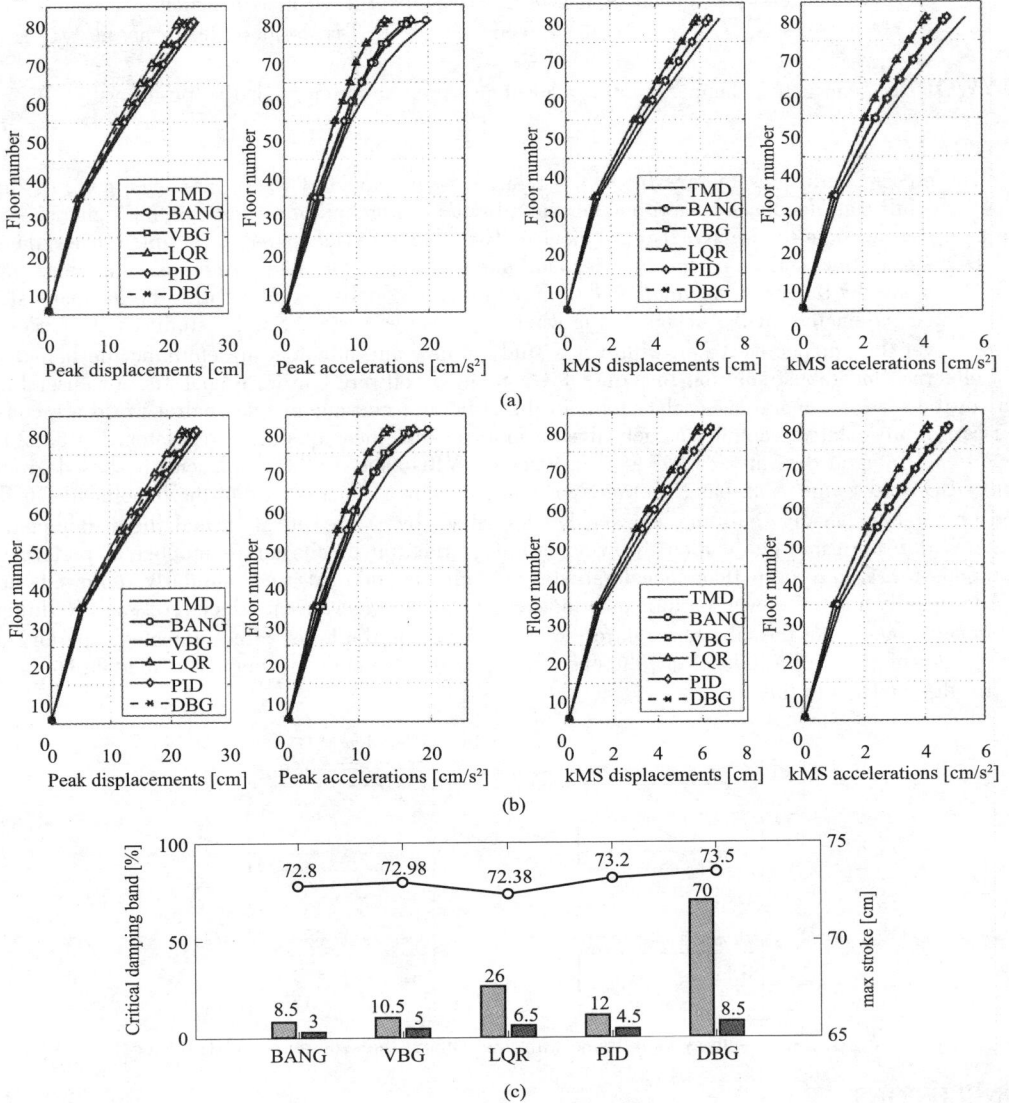

Fig 2. Indicative response results for the TMD and for the variously controlled STMD equipped structure when (a) same min-max critical damping ratios(5% ~20%)were used throughout, (b) maximum allowable strokes were kept similar and different min-max damping ratios was enabled in (c) the new damping ranges per algorithm)

REFERENCES

[1] Jansen LM, Dyke SJ. Semi-Active Control Strategies for MR Dampers: A Comparative Study. *Journal Engineering Mechanics.* 2000;126:795-803.

[2] Yang NY, Agrawal AK, Samali B, Wu JC. Benchmark Problem for Response Control of Wind-Excited Tall Buildings. *Journal of Engineering Mechanics.* 2004;130:437-46

EDDY CURRENT DAMPING AND ITS APPLICATION ON SEISMIC RESPONSES OF STEEL STRUCTURES: SOME NEW ADVANCES

Zheng-Qing Chen[a], Zhi-Wen Huang[a], Xu-Gang Hua[b], Yong-Kui Wen[b]

[a]Hunan University, College of Civil Engineering, China

zqchen@ hnu. edu. cn, zwhuang213@ hnu. edu. cn, cexghua@ hotmail. com

[b] Beijing Jiaotong University, School of   Civil Engineering, China

ykwen@ bjtu. edu. cn

KEYWORDS: steel structure, seismic response control, shaking table test, eddy current damping, ball screw

ABSTRACT

Two progresses in applyingeddy current damping(ECD)on structural vibration control have been made by the authors. One is the invention of plane ECD unit(PECDU)that provides damping force for tuned mass dampers(TMD). In China TMD with PECDU has been applied in several structural vibration mitigation programs including the Shanghai Center Building's TMD with an inertial mass block of 1000 t, being the largest TMD in the world. As shown in *Fig. 1*, A PECDU consists of a permanent magnets array, a conductive plane and two back irons. The damping force generated by the PECDU is given by,

$$F_{d} = \frac{n\ (\eta_{b}B)^{2}tS}{2\rho}v \tag{1}$$

Where n is the number of magnets, B is the magnetic flux density, η_{b} is the dimensionless modification coefficient for B decided by the air gap, the shape and topology of the permanent magnets and the characteristics of the magnetic circuit(to be determined experimentally), t is the thickness of the conductive plate, S is the pole projection area of a single magnet, ρ is the resistivity of the conductive plate, v is the relative velocity between the conductive plate and the magnets array.

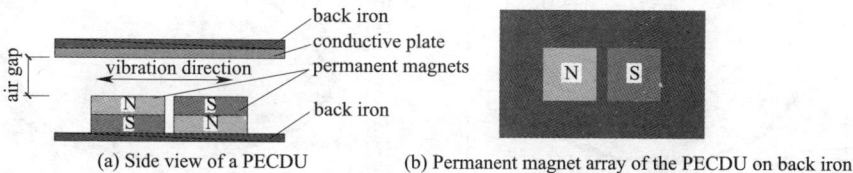

(a) Side view of a PECDU   (b) Permanent magnet array of the PECDU on back iron

Fig 1. Basic structure of a PECDU

Another progress is the invention of the axial EC Dampers based on Screw-Driven mechanism (ECDSD). ECDSD is capableof not only supplying very large damping force, but also generating very large inertial force like the EIMD presented by Yutaka Nakamura[1]. Compared with EIMD, ECDSD is simpler in configuration. *Figure 2* shows a conceptual sketch of an ECDSD. The ECDSD consists of a ball screw, a ball nut, two thrust bearings and a PECU, which is somewhat different from that in figure. 1. In this PECU the copper disk rotates between the two magnets and the magnetic circuit consists of the two magnets, the two back irons behind the magnets, the column made of soft iron, the copper disk and the air gap. When a relative motion is given between the top and the bottom rod end, the linear motion of the ball screw is converted into rotational motion of the ball nut as well as the copper disk connected to it by the ball screw mechanism. The copper disk rotates through the magnetic fluxes, so that eddy currents are generated in it, causing a damping torque in proportional to the angular velocity. The damping torque can be easily adjusted by varying the distance between the magnet and the ball screw. In additional, the rotating copper disk produces an inertial force.

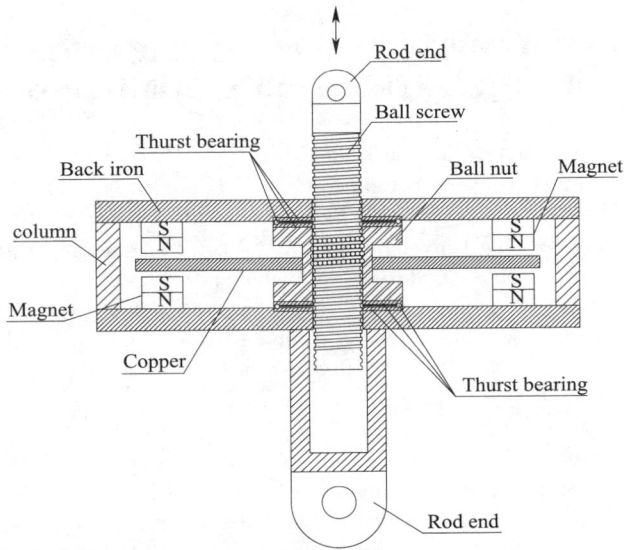

Fig. 2. Conceptual sketch of an ECDSD

The generated axial resisting force of an ECDSD, including the eddy current damping force, inertial force and friction force can be formulated like EIMD given by Nakamura[1],

$$F = m_e \ddot{u} + c_e \dot{u} + f_0 \cdot \text{sign}(\dot{u}) \tag{2}$$

Where m_e and c_e denotes the equivalent inertial mass and equivalent damping coefficient of the ECDSD, respectively. u is the stroke of the ECDSD, f_0 is the friction force.

Device performance tests of a reduced-scale were undertaken to verify the basic characteristics of the damper and the validity of the derived theoretical formulae.

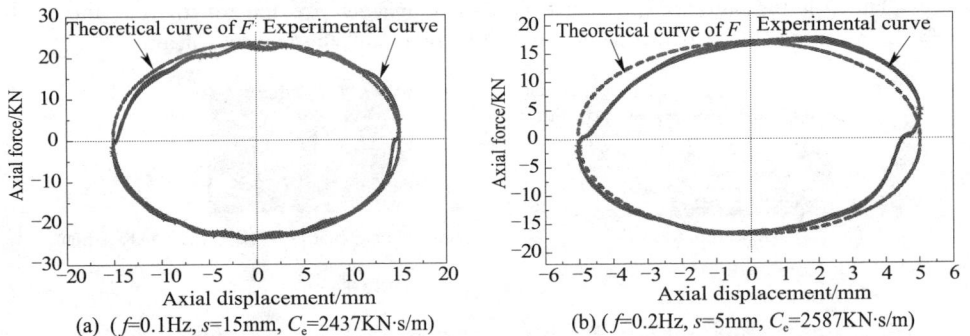

(a) (f=0.1Hz, s=15mm, C_e=2437KN·s/m) (b) (f=0.2Hz, s=5mm, C_e=2587KN·s/m)

Fig. 3. Force characteristics of the reduced-scale ECDSD

CONCLUSIONS

The Performance tests of the reduced-scale ECDSD validate the derived theoretical formulae for the damping force and inertial force of the ECDSD. In comparison to the EIMD with the same level of damping property and inertial force, the ECDSD has almost the same dimension with the EIMD, but its configuration is much simpler.

REFERENCES

[1] Yutaka Nakamura, etc. , Seismic response control using electromagnetic inertial mass dampers, Earthquake Engng Struct. Dyn. 2014; 43:507-527.
[2] Sodano H A, Bae J S. Eddy current damping in structures[J]. Shock and Vibration Digest, 2004, 36(6): 469-478.
[3] Weinberger M R. Drag force of an eddy current damper[J]. Aerospace and Electronic Systems, IEEE Transactions on,1977(2): 197-200.

SEISMIC RETROFIT OF A HIGH-RISE STEEL BUILDING USING FLUID VISCOUS DAMPERS

Shanshan Wang[a], Jiun-Wei Lai[b], Matthew Schoettler[b] and Stephen A. Mahin[a]

[a] Department of Civil and Environmental Engineering, University of California, Berkeley, U. S. A
shanonwang@ berkeley. edu, mahin@ berkeley. edu
[b] Pacific Earthquake Engineering Research Center, University of California, Berkeley, U. S. A
adrian. jwlai@ berkeley. edu, mschoettler@ berkeley. edu

KEYWORDS: Existing Tall Buildings, Retrofit, Fluid Viscous Damper, Optimization.

ABSTRACT

The Pacific Earthquake Engineering Research(PEER) Center has expanded its Tall Building Initiative project to include the seismic performance of existing tall buildings. A candidate 35-story steel building with representative details from that era was analyzed. Pushover analysis and nonlinear dynamic response history results for the as-built building are discussed in a companion paper[1]. In this paper, a potential retrofit approach using viscous dampers is presented. The performance criteria for the retrofit focused on reducing story drifts and accelerations to enhance safety in large events, and promote continued occupancy in more moderate ones.

Fluid Viscous Damper(FVD)is a velocity-dependent energy dissipation device, and both numerical and experimental investigations have shown the merits of such devices[2,3]. Considerable research in the 1990s resulted at least five code-oriented procedures for designing passive energy dissipation systems[4]. Those guidelines and codes such as ASCE 41[5] and ASCE 7[6] provide methods to account for the supplemental damping effects of these devices by modifying design spectrum. In this way, general damper characteristics can be selected. However, they do not prescribe specific methods for optimally placing dampers in a building, and damper placement could have large impacts on both structural behavior and the cost of using dampers[7]. A variety of optimal damper placement methods have been proposed. Among those, most are focused on designing FVD in low-to median-rise buildings, and intensive computation is needed. For a tall building with complex, three-dimensional complexity and nonlinear dynamic behavior, typical optimization algorithms are not efficient, especially during preliminary design.

In this paper, retrofit study starts by selecting damper locations within architectural constraints. Effective damping ratio is calculated in order to satisfy the target roof drift limit. Three conventional damper placement methods are proposed, and the structural responses including story drift ratio, floor acceleration, and damper responses are compared. Possible improved schemes are later presented taking into account performance-related and cost-related factors. A refined damper design scheme is proposed based on an overall cost-benefit analysis. Other design issues for FVD such as nonlinearity, bracing stiffness, and damper configurations are discussed briefly at the end.

CONCLUSIONS

FVDs are quite useful in upgrading high-rise buildings. They could reduce the peak drift ratio by up to 30% , and decrease the residual drift ratio by a similar amount. It also help suppress acceleration and lead to more rapid decay of vibrations. All those would greatly reduce structural and nonstructural damage as well as fear and discomfort of the occupants. However, the analysis results indicate the large damper forces are needed to achieve the desired responses, especially in the most deformed stories. Other strategies such as strengthening structural elements or using other energy dissipation device would be investigated in combination with FVD to achieve a more economic design without sacrificing structural performances. Seismic isolation above the base of the structure may also be a viable strategy.

REFERENCES

[1] Lai, J. -W. , Schoettler, M. , Wang, S. and Mahin, S. A. , "Seismic Performance Assessment of a Tall Building Having Pre-Northridge Moment-Resisting Connections", Proceedings, 8th International Conference on Behavior of Steel Structures in Seismic Areas, No. 162, Shanghai, China, July 1-3, 2015.

[2] Reinhorn, A. M. , Li, C. , & Constantinou, M. C. , "Experimental and analytical investigation of seismic retrofit of structures with supplemental damping, part I: fluid viscous damping devices", NCEER-95-0001, 1995.

[3] Symans, M. D. , & Constantinou, M. C. , "Passive fluid viscous damping systems for seismic energy dissipation", *ISET Journal of Earthquake Technology*, vol. 35, No. 4, p. 185-206, 1998.

[4] Ramirez, O. M. , Constantinou, M. C. , Kircher, C. A. , Whittaker, A. S. , Johnson, M. W. , Gomez, J. D. , & Chrysostomou, C. Z. , "Development and evaluation of simplified procedures for analysis and design of buildings with passive energy dissipation systems", MCEER- 00- 0010, 2001.

[5] ASCE, "Seismic Evaluation and Retrofit of Existing Buildings", American Society of Civil Engineers, ASCE/SEI 41-13, Reston, VA, 2014.

[6] ASCE, "Minimum Design Loadsfor Buildings and Other Structures", American Society of Civil Engineers, ASCE/SEI 7-10, Reston, VA, 2010.

[7] Soong, T. T. , &Dargush, G. F. , "Passive energy dissipation systems in structural engineering", Chichester, U. K. : John Wiley and Sons, 1997.

PERFORMANCE EVALUATION OF BUILDING FRAMES WITH ENERGY DISSIPATION SYSTEMS FUSEIS 1

GeorgiaDougka[a], Danai Dimakogianni[a] and Ioannis Vayas[a]

[a]Steel Structures Laboratory, National Technical University of Athens, Athens, Greece

e-mails: giouli@ dougka. gr, ddimakogianni@ hotmail. com, vastahl@ central. ntua. gr

KEYWORDS: seismic resistant system, energy dissipation fuses, self-centering

ABSTRACT

Experimental investigations on the innovative energy dissipation system "FUSEIS1" were presented during the STESSA 2012 Conference. This paper examines the seismic behavior of buildings where FUSEIS1 systems were applied. Non-linear static and dynamic analyses based on element properties as derived by the tests were performed and a relevant Design Guide was introduced to provide design recommendations and appropriate behavior factors. These analyses not only confirm the good seismic performance of the system but also indicate that the system can develop self-centering properties under certain conditions.

CONCLUSIONS

FUSEIS 1 is an innovative seismic resistant system with possibly self-centering capabilities that uses replaceable fuses to provide energy dissipation. This paper proceeds to the design of two case studies where linear, non-linear static and non-linear dynamic time history analyses were performed in order to investigate the system response. The results verified that inelastic deformations are restricted to the beams/pins leaving all other structural members of the main frame and the system (beams, columns) respond elastically. Also, it was proved that the behavior factors used, 5 for FUSEIS1-1 and 3 for FUSEIS1-2, meet the seismic performance objectives. Under real seismic excitations, the system exhibits a self-centering behavior with minimal residual drifts allowing for immediate occupancy after earthquake. For its confirmation more studies are needed.

ACKNOWLEDGMENT

Part of the presented research was done in the framework of the European Research Project RFSR-CT-2008-00032: "Dissipative devices for seismic resistant steel frames (Fuseis)". The financial contribution of the Research Fund for Coal and Steel of the European Community is gratefully acknowledged. The other partners of the project were RWTH Aachen University, Germany, Politecnico di Milano, Italy, Instituto Superior Tecnico Lisbon, Portugal and SIDENOR S. A, Greece.

REFERENCES

[1] Vayas I. , Karydakis Ph. , Dimakogianni D. , Dougka G. , Castiglioni C. A. , Kanyilmaz A. et al. "Dissipative devices for seismic-resistant steel frames (FUSEIS)", *Research Fund for Coal and Steel, European Commission*, EU 25901, EN 2013.

[2] Vayas I. , Karydakis Ph. , Dimakogianni D. , Dougka G. , Castiglioni C. A. , Kanyilmaz A. et al. "Dissipative devices for seismic resistant steel frames-The FUSEIS Project, Design Guide", *Research Programme of the Research Fund for Coal and Steel*, 2012.

[3] Dougka G. , Dimakogianni D. , Vayas I. , "Innovative energy dissipation systems (FUSEIS 1-1)- Experimental analysis", *Journal of Constructional Steel Research*, 96(5): 69-80, 2014.

[4] Dimakogianni D. , Dougka G. , Vayas I. , "Innovative seismic-resistant steel frames (FUSEIS 1-2) experimental analysis", *Steel Construction Design and Research*, 5(4): 212-221, 2012.

[5] SAP2000, CSI, Computers and Structures Inc. , www. csiberkeley. com

[6] EN1994-1-1, Design of composite steel and concrete structures. Part 1-1: General rules and rules for buildings, Comité Européen de Normalisation (CEN), 2005.

[7] EN 1993, Design of steel structures, Comité Européen de Normalisation (CEN), 2004.

［8］ EN1998-1-1,Design of structures for earthquake resistance. Part 1-1: General rules,seismic actions and rules for buildings. Comité Européen de Normalisation(CEN); 2003.

［9］ FEMA-356,Prestandard and Commentary for the seismic rehabilitation of Buildings,2000.

［10］ ATC- 40,Seismic Evaluation and Retrofit of Concrete Buildings,1996.

［11］ FEMA-P695,Quantification of building seismic performance factors,Washington,2009.

［12］ Seismomatch v. 2. 1. 0,Seismosoft,www. seismosoft. com

［13］ Vamvatsikos D. ,Cornell C. A. ,"The incremental dynamic analysis and its application to performance-based earthquake engineering",In: Proc. 12th European Conference on Earthquake Engineering,479,London,2002.

ENERGY BALANCE-BASED METHOD FOR RESPONSE CONTROL STRUCTURES WITH HYSTERETIC DAMPERS AND VISCOUS DAMPERS

Toshiaki Sato[a], Haruyuki Kitamura[a], Daiki Sato[b], Daisuke Sato[c],
Michio Yamaguchi[d], Naoya Wakita[d] and Yuta Watanuki[d]

[a] Dept. of Arch. , Faculty of Science and Technology, Tokyo Univ. of Science, Japan
sato_t@ rs. tus. ac. jp, kita-h@ rs. noda. tus. ac. jp

[b] Structural Engineering Research Center, Tokyo Institute of Technology, Japan
daiki-s@ serc. titech. ac. jp

[c] Kajima Corporation(Former Graduate Student, Tokyo Univ. of Science), Japan
satod@ kajima. com

[d] Nippon Steel & Sumikin Engineering Co. , LTD.
yamaguchi. michio@ eng. nssmc. com, wakita. naoya@ eng. nssmc. com,
watanuki. yuuta@ eng. nssmc. com

KEYWORDS: Response Control Structure, Response Prediction Method

ABSTRACT

After The Southern Hyogo prefecture earthquake in 1995, structural control devices have been widely used in order to decrease the seismic response of various buildings in Japan[1]. These devices are typically categorized as hysteretic and viscous dampers, according to their different characteristics as shown in *Fig. 1*.

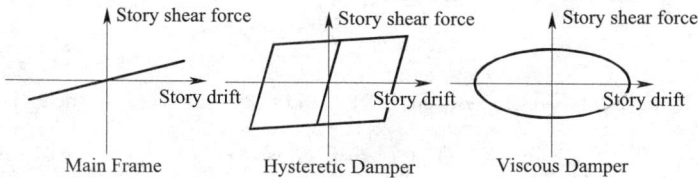

Fig 1. Hysteretic characteristics of main frame and the two types of dampers

Akiyama[2,3] proposed a prediction method based on energy balance, called the energy balance-based method, to quantitatively evaluate the seismic performance. The existing method adapts the response control structures with the respective types of dampers; however, the adaptability to predict the seismic response of the buildings, which are composed of both dampers, has not been clarified. This paper describes a method that can deal with these cases by rearranging the mathematical expression presented in past studies. The main concept of the proposed method is that the theory of vertical distribution governs energy distribution for each story. The proposed method is validated by comparing its results with time-history analyses. In addition, the effectiveness and applicability of using both dampers are verified using this method.

The energy balance equation, which is focused on the passive control structure composed of both hysteretic and viscous dampers, is expressed as

$$_fW_e(t) +_sW_p(t) +_hW_d(t) = E(t) \tag{1}$$

where $_fW_e$ denotes the elastic vibration energy of the main frame; $_sW_p$ denotes the dissipation energy by hysteretic dampers; $_hW_d$ denotes the damping energy by viscous dampers, and E represents the input energy that can be evaluated by the energy spectra V_E. An N-story structure is set up to discuss the evaluation of the respective energies in Eq. (1). The maximum story shear coefficient of the main frame is $_f\alpha_i$, that of hysteretic dampers is $_s\alpha_i$, that of viscous dampers is $_h\alpha_i$, and that of the sum of whole elements is α_i at the i-th story, as follows:

$$_f\alpha_i = \frac{_fQ_{\max i}}{\displaystyle\sum_{j=1}^{N} m_j g}, _s\alpha_{yi} = \frac{_sQ_{yi}}{\displaystyle\sum_{j=1}^{N} m_j g}, _h\alpha_i = \frac{_fQ_{\max i}}{\displaystyle\sum_{j=1}^{N} m_j g}, \alpha_i = \frac{Q_{\max i}}{\displaystyle\sum_{j=1}^{N} m_j g}$$

where m denotes mass, g denotes gravitational acceleration, and Q, shown in the numerator of the above formulas, is the maximum shear force of each element at the i-th story. The story shear coefficient α_0 and story drift δ_0 of the equivalent linear system are defined as following equations, which are the benchmarks for seismic response.

$$\alpha_0 = \frac{2\pi \cdot V_E}{T \cdot g}, \delta_0 = \frac{T \cdot V_E}{2\pi}$$

In order to investigate the seismic response determined by structural components such as the stiffness of the main frame and the quantity of dampers, this paper focuses on the response of the first story, which is described as the story drift and the story shear coefficient based on energy balance equation shown in *Eq.* (1).

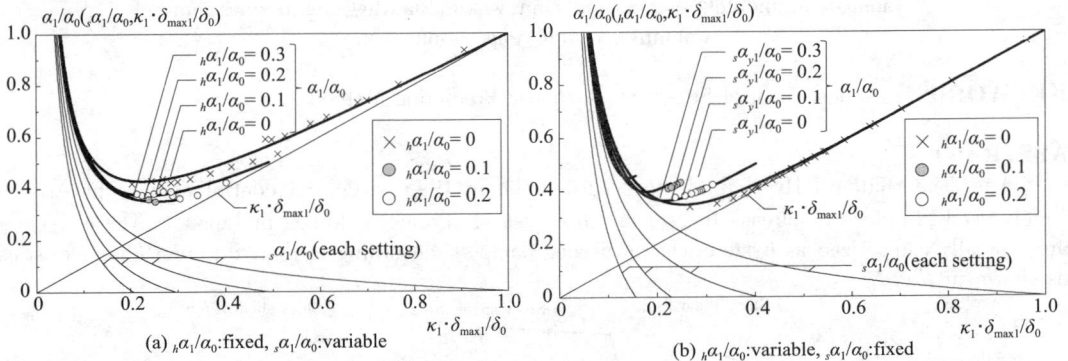

(a) $_h\alpha_1/\alpha_0$:fixed, $_s\alpha_1/\alpha_0$:variable

(b) $_h\alpha_1/\alpha_0$:variable, $_s\alpha_1/\alpha_0$:fixed

Fig 2. Relationship between story shear force and story drift based on energy balance

Fig. 2 (a) shows that the reduction of story drift corresponds to the increase in the quantities of viscous dampers when the story shear force is minimal. In contrast, the more the number of hysteretic dampers, the more is the decrease in the story drift. The evaluation results indicate that the synergistic effects of using both dampers are expected at the specific range, and the range is predictable based on energy balance.

ACKNOWLEDGMENT

This research was carried out under corroborative research of Tokyo Univ. of Science and Nippon Steel & Sumikin Engineering Co. , LTD. The collaboration is gratefully acknowledged.

REFERENCES

[1] Kitamura, H. , Kitamura, Y. , Ito, M. and Sakamoto, M. "Analysis of the present situation of response control systems in japan based on the survey of applied buildings", *Journal of Technology and Design*, AIJ, No. 18, pp. 55-60, 2003. 12

[2] Akiyama, H. "Earthquake resistant design to meet to diversified performance", *Journal of Structural and Construction Engineering*, AIJ, No. 472, pp. 85-90, 1995. 6

[3] Akiyama, H. "Earthquake-resistant design method for buildings based on energy balance", Gihodo Shuppan, Tokyo, Japan, 1999.

EQUIVALENT LINEARIZAED MODEL OF DAMPER RESPONSE FOR SEISMIC DESIGN OF STEEL STRUCTURES WITH NONLINER VISCOUS DAMPERS

Baiping Dong[a], Richard Sause[a] and James M. Ricles[a]

[a]ATLSS Engineering Research Center, Department of Civil and Environmental Engineering,
Lehigh University, USA

bad209@ lehigh. edu , rs0c@ lehigh. edu , jmr5@ lehigh. edu

KEYWORDS: Nonlinear viscous damper, equivalent linearization, damper-brace component, seismic design.

ABSTRACT

Damping devices are usually connected to other structural components to improve the seismic performance of steel structures. This paper presents a linearized model of the response of a nonlinear viscous damper in series with the associated bracing needed to connect the damper to the structure. The linearization of the damper-brace component is needed for preliminary seismic design of structures with nonlinear viscous dampers. The nonlinear viscous damper-brace component is represented as an equivalent linear elastic-viscous model by equating the energy dissipation per cycle of the damper to the energy dissipation per cycle of the equivalent linear elastic-viscous model. This linearization includes the effects of elastic flexibility of the structural components on the damper behaviour within a structure. To validate the equivalent linear elastic-viscous model, tests with predefined harmonic displacement time histories are conducted on a practical 0. 6-scale three-story structure with nonlinear viscous dampers. The test results are presented, and demonstrate that the equivalent linear elastic-viscous model for the damper-brace component is accurate for preliminary seismic design of a structure with nonlinear viscous dampers.

CONCLUSIONS

This paper focuses onan equivalent linearized model for nonlinear viscous damper-brace components for use in preliminary seismic design of structures with nonlinear viscous dampers. The equivalent linear elastic-viscous model, as shown in *Fig. 1*, includes a linear elastic spring and a linear viscous dashpot. The equivalent linearized model is obtained by equating the energy dissipation per cycle of the nonlinear viscous damper to the energy dissipation per cycle of the equivalent elastic-viscous model. With this linearization, the effects of the elastic flexibility of the brace (i. e. , the flexibility of the complete force path that connects the damper to the structure) on the damper behaviour and structural response can be directly taken into account in the preliminary design of structures with nonlinear viscous dampers. The equivalent linear elastic-viscous model for the damper-brace component was validated through tests with predefined harmonic floor displacement time histories on a 0. 6-scale three-story test structure with nonlinear viscous dampers which was designed and constructed under practical conditions. The results, as shown in *Fig. 2*, demonstrate that the equivalent linear elastic-viscous model is suitable for preliminary seismic design of structures with nonlinear viscous dampers. Analytical results using the equivalent elastic-viscous model show that a more flexible brace design is more likely to stiffen the structure and decrease the equivalent damping ratio by leading to a smaller story drift response.

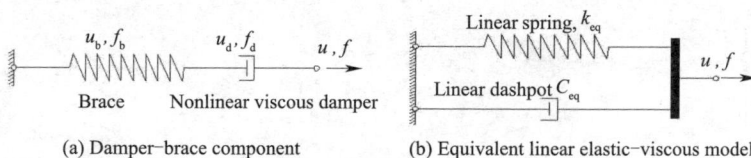

(a) Damper-brace component   (b) Equivalent linear elastic-viscous model

Fig 1. Equivalent linear elastic-viscous model for damper-brace component

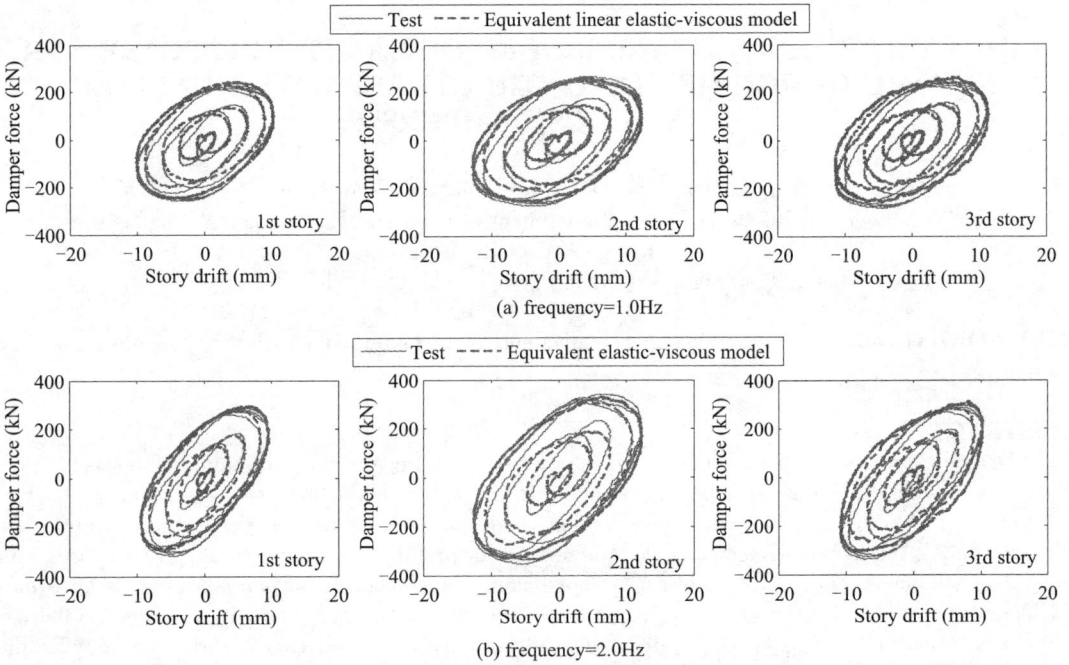

Fig 2. Comparison of damper force-story drift hysteresis behavior

REFERENCES

[1] Soong, T. T. , and Spencer, Jr. B. F. , "Supplemental energy dissipation: State-of-the-art and state-of-the-practice", *Engineering Structures*, 24(3), 243-259, 2002.

[2] Christopoulos, C. , and Filiatrault, A. , *Principles of passive supplemental damping and seismic isolation*, IUSS Press, 2006.

[3] ASCE 7-10, *Minimum design loads for buildings and other structures*. American Society of Civil Engineers, 2010.

[4] Fu, Y. , Kasai, K. , "Comparative study of frames using viscoelastic and viscous dampers", *Journal of Structural Engineering*; 124(5):513-522, 1998.

[5] Takewaki, I. , and Yoshitomi, S. , "Effects of support stiffness on optimal damper placement for a planar building frame", *The Structural Design of Tall Buildings*, 7:323-336, 1998.

[6] Singh, M. P. , Verma, N. P. , and Moreschi, L. M. , "Seismic analysis and design with Maxwell dampers", *Journal of Engineering Mechanics*; 129(3):273-282, 2003.

[7] Lin, W. , and Chopra, A. , "Earthquake Response of Elastic Single-Degree-of-Freedom Systems with Nonlinear Viscoelastic Dampers", *Journal of Engineering Mechanics*, 129(6), 597-606, 2003.

[8] Chen, Y. T. , and Chai, Y. H. , "Effects of brace stiffness on performance of structures with supplemental Maxwell model-based brace-damper systems", *Earthquake Engineering and Structural Dynamics*, 40(1), 75-92, 2011.

[9] Lehigh RTMD Users Guide. http://www. nees. lehigh. edu/resources/users-guide, 2013.

INTEGRATED OPTIMAL DESIGN FOR BELT TRUSS USING VISCOUS DAMPERS IN SUPER TALL BUILDINGS

Xin ZHAO[a], Tao SHI[b]

[a] Department of Structural Engineering, Tongji University, Shanghai, CHINA
Tongji Architectural Design (Group) Co. , Shanghai, CHINA
22zx@ tjadri. com

[b] Department of Structural Engineering, Tongji University, Shanghai, CHINA
ocean_shi@ 163. com

KEYWORDS: viscous damper, integrated optimal design, belt truss, super tall buildings.

ABSTRACT

Viscous dampers (VD) offer additional damping to the structure and dissipate the energy during earthquake. The structure will be safer with the reduction of response and internal forces due to increase of damping. The structural members can be further optimized for considering the reduction of earthquake actions due to additional damping introduced by the installation of viscous dampers. The integrated optimal design of belt trusses using viscous dampers is addressed in this paper. The virtual work principle is first to be employed to analyze the sensitivity of design parameters of belt truss members on the design constraints. An optimal design process applied to minimize the sizes of belt truss members is then introduced. A super tall building located in high seismicity area is applied in the final part of this paper to illustrate the effectiveness of the proposed method. Numerical analysis results show that the proposed methods are reasonable and it is effective to reduce the structure cost with the improvement of seismic performance.

CONCLUSIONS

A 250-meter-high super high-rise building with 63 floors and 3 mechanical floors is adopted in this study. The structural system is frame-core wall structure. Viscous dampers can dissipate energy under different earthquake intensity, including frequent earthquakes. Due to the energy dissipation capacity of the viscous dampers, the seismic response can be reduced and that increases additional damping ratio. To reduce the redundancies introduced by the viscous dampers, the following integrated optimal design process can be employed:

(1) Assess the contribution of different outrigger trusses on the structural lateral stiffness;

(2) Remove those outriggers which have small contribution on the lateral stiffness;

(3) Install damped outriggers with viscous dampers, and calculate the addition damping ratios;

(4) Determine the design constraints of components, such as story drift, period and stress ratio;

(5) Conduct sensitivity analysis and indicate those members which need to be optimized;

(6) Optimize those components of high redundancies, and check the design constraints to make sure they are below specified limits;

(7) After of the optimization of structural components, sensitivity analysis will be conducted using updated sizes of structural components. Those components of high redundancies will be optimized again like in step (6) until the margins of design constraints are as small as specified;

(8) Step (1) to (7) will be conducted for different structural schemes with different number of damped outrigger and stiff outrigger combinations;

(9) The overall costs of different structural schemes will be compared and the scheme with least cost will be the final scheme to be employed.

Table 6 is the optimization comparison with different number of viscous dampers, in which it can be clearly observed that the damped outriggers can be removed the inner outriggers that cost 365 ton steels. The outer outriggers of the scheme with 12 and 16 VDs cost more steels that other schemes. The price of every viscous damper is about 10,000 RMB. The price of steel is about 10,000 RMB per ton. In *Table 1*, the economy of scheme with 4VDs is the best.

Table 2 shows the period and story drift about original structure and scheme with 8 VDs. Period is a little larger than the previous scheme, and the story drift is reduced with the installation of viscous dampers.

Table 1. Optimization Scheme Comparison

	Original Structure	Scheme with 4 VDs	Scheme with 8 VDs	Scheme with 12 VDs	Scheme with 16 VDs
Outer Outrigger	149t	115t	115t	230t	230t
Inner Outrigger	365t	0	0	0	0
Belt Truss	1181t	1026t	976t	948t	934t
Viscous Damper	0	40,000RMB	80,000RMB	120,000RMB	160,000RMB
Total	16.9 million RMB	11.5 million RMB	10.9 million RMB	11.9 million RMB	11.8 million RMB

Table 2. Optimization Scheme Comparison of Global Design Criteria

	Original Structure	Scheme with 8 VDs
First period	5.83s	6.06s
Second Period	4.96s	5.00s
Third Period	3.96s	3.96s
Story Drift	1/674	1/692

This article discusses the optimization design method of viscous dampers for super high-rise building. This method is made use of sensitivity analysis, which is based on principle of virtual work. And the components can be optimized for the redundancies introduced by the viscous dampers. A 250-meter-high super high-rise building is addressed in this study, without reducing the safety of structure, the additional damping ratio is increased and the economy of the structure is improved.

We can draw the following conclusions:

1. The installation of the damped outriggers can increase additional damping ratio of structure, consume earthquake energy and reduce response under the earthquake. Meanwhile, there are some redundancies for component optimization.

2. In the process of integrated optimal design, sensitivity analysis, which is based on principle of virtual work, is used to judge the component's contribution to some criteria so that the optimal strategy can be obtained.

3. Optimization calculation results show that the integrated optimal design of viscous dampers is reasonable and can be guidance to other designs of super tall buildings.

ACKNOWLEDGMENT

The authors are grateful for the supports from the Shanghai Excellent Discipline Leader Program (No. 14XD1423900) and Key Technologies R & D Program of Shanghai (Grant No. 09dz1207704).

REFERENCES

[1] Rob J. Smith and Michael R. Willford. The damped outrigger concept for tall building [J]. The Structural Design of Tall and Special Buildings, 2007, 16: 501-517

[2] Rob Smith, Michael Willford. Damped outriggers for tall buildings [J]. The Arup Journal, 2008, 3: 15-21

[3] Zhengqing Chen and Zhihao Wang. A novel passive energy dissipation system for frame-core tube structure [A]. In: The Seventh Asia-Pacific Conference on Wind Engineering, Taipei: Taiwan, 2009.

[4] Baker, W. F. Stiffness Optimization Methods for Lateral Systems of Buildings: A Theoretical Basis. Electronic Computation. Indianapolis, Indiana: American Society of Civil Engineers, 269278, 1991.

THE LIFE CYCLE COST ASSESSMENT OF SUPER TALL BUILDINGS WITH VISCOUS DAMPING WALLS

Xi Zhan[a], Xin Zhao[b], Yimin Zheng[b]

a Department of Structural Engineering, Tongji University, Shanghai, CHINA

22zx@ tjadri. com

b Department of Structural Engineering, Tongji University, Shanghai, CHINA

Tongji Architectural Design(Group) Co. , Shanghai, CHINA

KEYWORDS: The life cycle cost assessment method, Viscous damping walls, Termination costs, Super tall buildings.

ABSTRACT

Life cycle cost assessment method not only considers the initial construction costs, but also considers the operation and termination costs during the life cycle of the super tall buildings. Viscous damping walls can effectively suppress the vibration amplitude of the super tall building structures under both earthquake and wind. Comparing with other viscous dampers, viscous damping walls can dissipate more energy due to shear type mechanism and larger contacting areas. By the introduction of viscous damping walls in super tall buildings, the material consumption of main structures can be reduced by the reduction of earthquake action, and the seismic rehabilitation efforts can be less due to the alleviation of structural damage during earthquakes. The above two aspects lead to the reduction of the life cycle cost of super tall buildings. A life cycle cost assessment method was proposed in this paper to consider the overall costs of the structure during the whole life cycle of the building with the consideration of the structural reliability during the building life cycle. A super tall building located in high seismicity area was applied in the last part of the paper to illustrate the proposed life cycle cost assessment method. Numerical analysis results show that the proposed method is reasonable and viscous damping walls can reduce the life cycle costs of super tall building.

CONCLUSIONS

Energy dissipation scheme may increase the investment cost in the initial period of construction, but it can reduce the loss cost of the structure for the device can decrease the destruction and damage under the earthquake. Not only initial cost of materials and construction cost but also the feature and the performance of the structure in the life cycle in the high-rise building structure initial design stage should be considered to assess and optimize diverse scheme in economy. A life cycle cost assessment method was introduced in this paper. A super tall building located in high seismicity area was applied in the last part of the paper to direct the design of high-rise building and to illustrate the proposed life cycle cost assessment method. Threestructure scheme-uncontrolled structure, controlled structure and integrated optimal structure were contrastive analyzed. Numerical analysis results show that the proposed method is reasonable and viscous damping walls can reduce the termination costs of super tall building.

Table 1. The life cycle cost for three plans

Design Method	Performance of Structure	Structural Costs Csb	Equipment Costs Cdb	Loss cost Cf	Construction Cost Cb	Life Cycle Cost CL
Primitive design	-	0	0	0	0	0
Performance improved Design	↑ ↑	0	320	-68. 74	320	251. 26
Integrated optimal design	↑	-1307. 41	320	-10. 89	-987. 41	-998. 3

(1) Although building structure with viscous damping walls and other energy dissipation scheme will increase the initial investment cost of building, energy dissipation device can reduce the seismic

damage to the main structure and the loss cost of main component under earthquake, in addition, with the additional damping provided by the viscous damping walls the structure components can be optimized. Therefore, with setting reasonable number and location of viscous damping walls, the whole cost of structure in the life cycle can be reduced by decreasing loss cost and the cost of structural members by optimizing.

(2) In this case, Integrated optimal design has a big advantage with minimum life cycle cost regardless of loss cost is not the minimum. The loss cost of the whole structure of the integrated optimal structure is less than loss cost of the uncontrolled structure, and viscous damping walls in the integrated optimal structure improve the performance of the structure and effectively protect the main structure component in the earthquake.

(3) Because the calculation of structure loss cost under all occurrence probability of earthquake intensity is complex, it can be simplified as taking several wave with typical peak acceleration for calculating and the relationship between the loss costs   under typical peak acceleration with its probability was plotted to get enclosed area of each curve with the coordinate axis. The simplification offers engineering practice with convenience and more point of the relationship graph is taking, the more accurate cost statistics are.

ACKNOWLEDGMENT

The authors are grateful for the supports from the Shanghai Excellent Discipline Leader Program (No. 14XD1423900) and Key Technologies R & D Program of Shanghai(Grant No. 09dz1207704).

REFERENCES

[1] Miyazaki, M. , Arima, F. , Kidata, Y. , Hristov, I. Earthquake Response Control Design of Buildings Using Viseous Damping Walls. Proceedings of First East Asia Pacific Conference, Bangkok, 1986. 1882 ~ 1891

[2] Arima, F. , Miyazaki, M. , Tanaka, H. , ect. A Study on Buildings with Large Damping Using Viscous Damping Walls. 9th World Conference on Earthquake Engineering. Tokyo-Kyoto: 1988. 821-826

[3] Fei, C. , Weiqing, L. , Shuguang, W. , ect. The research of Energy dissipation design of Jin Bainian square with viscous damping walls. Building Science, 2008, 24(9)56-59.

[4] Zhengcang, Z. , ect. Kagoshima airport terminal building aseismic reinforcement-three-dimensional elastoplastic analysis of structure adding viscous damping walls. Building Structures, 2000, (6) 20-23.

[5] Frangopol D M, Liu M. Maintenance and management of civil infrastructure based on condition, safety, optimization, and life-cycle cost[J]. Structure and Infrastructure Engineering. 2007, 3(1): 29-41

[6] Wen Y K, Kang Y J. Minimum building life-cycle cost design criteria. I: Methodology [J]. Journal of Structural Engineering-Asce. 2001, 127(3): 330-337.

[7] Hongwei, Z. Integrated optimal structural design for super tall buildings with bucking-restrained braces: Tongji University, 2015

[8] Standard of the People's Republic of China. GB/T 24335-2009 Classification of earthquake damage to buildings and special structures[S]. Beijing: China Architecture and Building Press, 2010 (in Chinese)

[9] FEMA356. Prestandard and Commentary for the Seismic Rehabilitation of Buildings [S]. Federal Emergency Management Agency, U. S, 2000.

OPTIMAL PLACEMENT OF VISCOELASTIC COUPLING DAMPERS IN SUPER TALL BUILDINGS

Xin ZHAO[a], Lang QIN[b]

[a] Department of Structural Engineering, Tongji University, Shanghai, CHINA
Tongji Architectural Design(Group)Co., Shanghai, CHINA
22zx@tjadri.com
[b] Department of Structural Engineering, Tongji University, Shanghai, CHINA
501332874@qq.com

KEYWORDS: Viscoelastic coupling damper, Coupling beam, Optimal placement, Super tall building.

ABSTRACT

Applied widely in super tall buildings, frame-core wall structures not only possess the advantages of frame and shear wall, but also excel in mechanical performance, seismic characteristic andeconomical efficiency. In high seismicity area, energy dissipation technology can be applied in frame-core wall structures to add damping and provide significant reductions to seismic responses. Viscoelastic coupling damper(VCD) is a kind of new damper, which replaces coupling beams in structural configurations, increasing the level of inherent damping of structures to control earthquake-induced dynamic vibrations. Optimal placement of VCDs can improve the energy dissipation capacity with limited number of dampers. Generally, viscoelastic dampers are set in locations where the largest displacement or the maximum shear force occurs without considering the effect of added controllers on the motion of buildings. The optimal placement of VCDs for the frame-core wall structures are studied in this paper and an effective optimal placement method is presented. A super tall building incorporating VCDs is presented as an example to illustrate the proposed optimal placement method. Numerical analysis results show that economical use of VCDs is possible by the method proposed.

CONCLUSIONS

The optimal placement of VCDs for the frame-core wall structures are studied in this paper andan effective optimal placement method is presented. A super tall building incorporating VCDs is presented as an example to illustrate the proposed optimal placement method. The number of VCDs to be installed in the structure is 30 and there are two available places where VCDs can be installed in a story. The optimal places to install VCDs through 5 design cycles are listed in *Table 1*. Then the total energy dissipated by 30 VCDs and the additional damping ratio can be calculated.

Table 1. The optimal stories to install VCDs.

Design Cycle	The optimal stories
1	32, 33, 34
2	18, 19, 20
3	17, 35, 36
4	16, 21, 31
5	22, 37, 64

To verify the optimal placement method proposed by the author, another several methods are used, such as the method based on shear forces in coupling beams, story drifts, and deformations of coupling beams which are non-sequential methods. The optimal locations of VCDs, energy dissipated introduced by VCDs and the additional damping ratio are listed in *Table 2*. It can be observed that the energy dissipated introduced by VCDs using the method proposed in the paper is higher than that of other methods. It is concluded that the method proposed in the paper is efficient to obtain the optimal locations for VCDs. It is also believed that the method can be applied for optimal design of other viscoelastic dampers.

Table 2. Comparison of different methods.

Method	The optimal stories	Energy dissipated (kN * m)	The additional damping ratio(%)
Shear forces in coupling beams	4-9,13-15,18-22,32-33	295. 84	1. 09
Story drifts	32-37,60-68	163. 15	0. 64
Deformations of coupling beams	19-21,29-37,61-63	274. 79	1. 10
Method proposed in the paper	16-22,31-37,64	352. 05	1. 34

REFERENCES

[1] Paulay, T. , Binney, J. R. , "Diagonally reinforced coupling beams of shear walls", ACI special publication,579-598,1974.

[2] Hindi, R. A. , Hassan, M. A. , "Nonlinear behavior of diagonally reinforced coupling beams", Proceedings of the 2004 Structures Congress-Building on the Past: Securing the Future, 137-146,2004.

[3] Bower, O. J. , Rassati, G. A. , "Axial Restraint in Diagonally Reinforced Concrete Coupling Beams: An Analytical Investigation", Proceedings of the 2008 Structures Congress-Structures Congress 2008: Crossing the Borders,1-11,2008.

[4] Shahrooz, B. M. , Remmetter, M. E. , Qin, F. , "Seismic response of composite coupled structural walls", Composite Construction in Steel and Concrete II,429-441,1992.

[5] Fortney, P. J. , Shahrooz, B. M. , Rassati, G. A. , "The next generation of coupling beams", Proceedings of the 5th International Conference on Composite Construction in Steel and Concrete V, 619-630,2006.

[6] Gong, B. , Shahrooz, B. M. , "Concrete-steel composite coupling beams. I: Component testing", Journal of Structural Engineering,127(6): 625-631,2001.

[7] Gong, B. , Shahrooz, B. M. , "Concrete-steel composite coupling beams. II: Subassembly testing and design verification", Journal of Structural Engineering,127(6): 632-638,2001.

[8] Harries, K. A. , Gong, B. , Shahrooz, B. M. , "Behavior and design of reinforced concrete, steel, and steel-concrete coupling beams", Earthquake Spectra,16(4): 775-799,2000.

[9] Chung, H. S. , Moon, B. W. , Lee, S. K. , et al. , "Seismic performance of friction dampers using flexure of RC shear wall system", Structural Design of Tall and Special Buildings,18(7): 807-822,2009.

[10] Teng,J. , Ma,B. T. ,Li,W. H. ,et al. ,"Pseudo-static test for coupling beam damper of coupled shear wall structures",Journal of Building Structures,31(12): 92-100,2010. (In Chinese)

[11] Teng,J. , Ma, B. T. , Zhou, Z. G. , et al. , "Key technique of energy dissipating damper on coupling beam to improve seismic resistance performance of coupling shear wall structures", Earthquake Resistant Engineering and Retrofitting,29(5): 1-6,2007. (In Chinese)

[12] Mao, C. X. , Wang,Z. Y. ,Zhang L. Q. ,et al. ,"Seismic performance of RC frame-shear wall structure with novel shape memory alloy dampers in coupling beams",15th World Conference on Earthquake Engineering, Lisbon, Portugal, Paper -ID: 4988, September 24-28,2012.

[13] Christopoulos, C. , Montgomery, M. , "Viscoelastic coupling dampers(VCDs)for enhanced wind and seismic performance of high-rise buildings", Earthquake Engineering & Structural Dynamics,42: 2217-2233,2013.

[14] Kim, H. J. , Choi, K. S. , Oh, S. H. , et al. , "Comparative study on seismic performance of conventional RC coupling beams and hybrid energy dissipative coupling beams used for RC shear walls",15th World Conference on Earthquake Engineering, Lisbon, Portugal, Paper -ID: 2254, September 24-28,2012.

[15] Zhang, R. H. , Song, T. T. , "Seismic design of viscoelastic dampers for structural application", Journal of Structural Engineering,118:1375-1392,1992.

[16] Lyons, R. M. , Christopoulos, C. , Montgomery, M. S. , "Enhancing the seismic performance of RC coupled wall high-rise buildings with viscoelastic coupling dampers",15th World Conference on Earthquake Engineering, Lisbon, Portugal, Paper -ID: 1573, September 24-28,2012.

[17] Chang, K. C. , Song, T. T. , Oh, S. , et al. , "Seismic response of a 2/5 scale steel structure with added viscoelastic damper", Technical Report NCEER-91-0012, National Center for Earthquake Engineering Research, Buffalo, NY,1991.

FUNDAMENTAL STUDY OF TUNED MASS DAMPER FLOOR SYSTEM TOWARDS SEISMIC RESPONSE CONTROL

Ping Xiang[a], Akira Nishitani[a]

[a]Waseda University, Department of Achitecture, Japan

xiangp. mail@ gmail. com, anix@ waseda. jp

KEYWORDS: Tuned mass damper, Floor isolation system, Seismic response control, Bending-shear type model, Optimum design, Stability maximization criterion, Multi-objective optimization.

ABSTRACT

A new design concept that main structural components such as beams and columns should remain elastic even during strong earthquakes receives more acknowledgements[1]. High strength steel has a larger elastic range compared with common mild steel. It is promising to employ new materials, such as high strength steels in new structural systems. An innovative scheme of seismic structural system, referred to as tuned mass damper floor system(TMDFS), was recently proposed by the authors, which is schematically illustrated in *Fig. 1*. The proposed scheme takes advantages of both the benefits of floor isolation system(FIS) and multiple tuned mass dampers(MTMDs). The floor components themselves serve as MTMDs in this scheme, and thus larger mass ratios of MTMDs can be achieved than those of conventional TMD systems, indicating higher control performance. Multi-vibration-modes can be controlled by installing such TMD floors in different storeys. Firstly, small scaled shaking table experiments are conducted with the experimental model as shown in *Fig. 2*. The control effect of TMDFS for a high-rise building which is simplified as a bending-shear model[2,3] is also investigated. A multi-objective optimization genetic algorithm[4] is employed to solve the multi-objective optimization problem with the combination of H_2 criterion and the stability maximization criterion(SMC)[3]. The favorable performance of TMDFS for high-rise buildings is demonstrated by comparing with a fixed floor building equipped with additional high-damping devices achieving 10% first modal damping ratio.

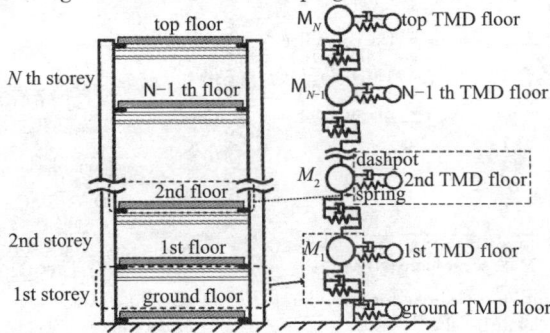

Fig. 1. Schematic model of TMDFS for an *N*-storey structure

CONCLUSIONS

Fig. 3 shows the experimental and simulated absolute accelerations of the main frame and TMD floor in TMDFS underEl Centro for instance, which compare well with each other, respectively. The accelerations of the TMD floor are smaller than those of the main frame in TMDFS. *Fig. 4* gives the experimental results comparison between the cases of fixed floor and TMD floor systems under JMA Kobe. *Figs. 5* and *6* are numerical responses of a 25-storey building under the 2011 Tohoku earthquake (TKY007, EW) with the normalized peak velocity of 50 cm/s(level 2), which indicate that TMDFS can remain in elastic range even under the long period ground motion of level 2 and free vibration responses decay the most rapidly if SMC TMDFS is employed.

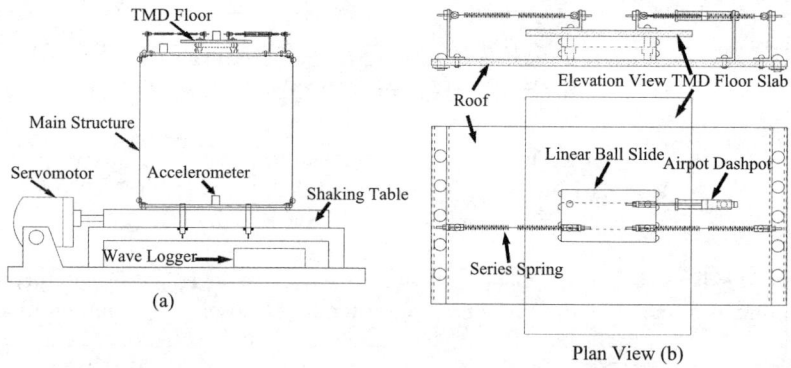

Fig. 2. Experiment model: (a) panoramic view and (b) close-up view of TMD floor

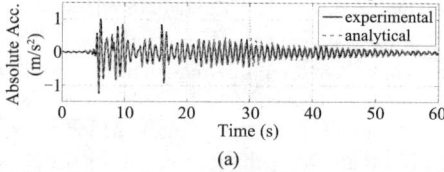

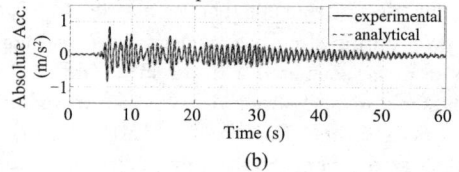

Fig. 3. Experimental and analytical absolute Acc. of TMDFS under El Centro: a) main frame and b) TMD floor

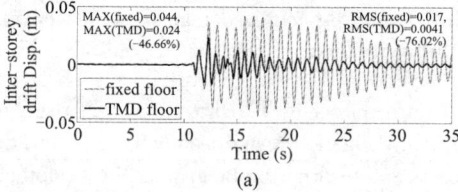

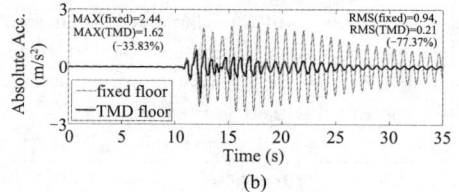

Fig. 4. Response comparison between fixed and TMD floor systems under JMA Kobe:
(a) inter-storey drift Disp. of main frame and (b) absolute Acc. of floor

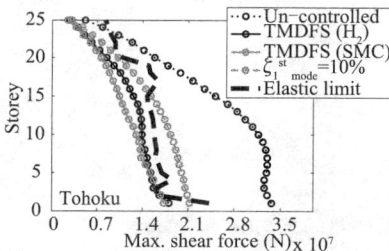

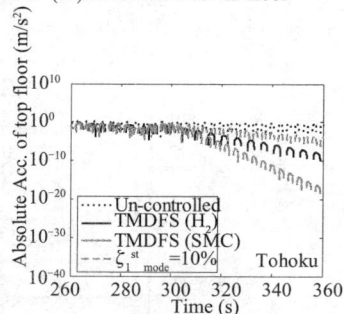

Fig. 5. Maximum storey shear force under Tohoku (Level 2)

Fig. 6. Absolute Acc. (log value) time history of top floor under Tohoku (Level 2)

REFERENCES

[1] Ohata M. , 2009. "New structural system constructed using new high-strength steel aiming for non-damage during a large earthquake", *Proceedings of* 10*th Korea-China-Japan Symposium on Structural Steel Construction*, 5-6 November, 2009, Seoul, South Korea, pp. 193-204.

[2] Niwa N. , Kobori T. , Takahashi M. , Hatada T. , Kurino H. , Tagami J. , 1995. "Passive seismic response controlled high-rise building with high damping device", *Earthquake Engineering and Structural Dynamics*, Vol. 24, No. 5, pp. 655-671.

[3] Xiang P. , Nishitani A. , 2014. "Optimum design of tuned mass damper floor system integrated into bending-shear type building based on H_∞, H_2, and stability maximization criteria", *Structural Control and Health Monitoring*, DOI: 10. 1002/stc. 1725.

[4] Deb K. , Pratap A. , Agarwal S. , Meyarivan T. , 2002. "A fast and elitist multiobjective genetic algorithm: NSGA-II", *IEEE transactions on Evolutionary Computation*, Vol. 2, No. 6, pp. 182-197.

Codification, Design, and Practice

AUSTRALIAN/NEW ZEALAND STANDARD FOR COMPOSITE STRUCTURES, AS/NZS 2327, SEISMIC PROVISIONS DEVELOPMENT

Kevin A. Cowie[a]

[a]Steel Construction New Zealand Inc.

kevin. cowie@ scnz. org

KEYWORDS: Standard, AS/NZS 2327, Composite construction, Composite structures.

ABSTRACT

Recent significant activity on the joint development of an Australian/New Zealand Standard for Bridge Structures in Steel and Composite Structures has highlighted the need and galvanised the industry to begin to develop a harmonised standard for steel-concrete composite structures for buildings. This project submitted to Standards Australia was initially titled "Suite of Standards for Composite Structures for use in Buildings and other non-bridge infrastructure incorporating existing AS2327. 1-2003, AS2327. 2-201X, AS2327. 3-201X and ASS2327. 4-201X" which received approval in November 2011 and commenced drafting in July 2012 and is due for completion by July 2015. The project is now referred to as the AS/NZS 2327, Composite Structures and will be a single document.

Composite structures using steel-concrete composite construction techniques generally tend to be the preferred method of construction for steel framed buildings in Australia, New Zealand and most developed countries. Composite construction tends to reduce the amount of structural steel being used and thus lends itself to greater sustainability benefits than other methods of construction. Furthermore, composite construction also provides steel-frames with the robustness of concrete frames by possessing the added advantages of light-weight which has many other potential benefits for the construction process.

Composite structures have been demonstrated to perform well in earthquakes. The 2010-2011 Christchurch earthquake sequence showed the performance of composite steel structures was generally better than anticipated especially given the levels of ground shaking which were more than twice that considered explicitly in design. Furthermore, unlike many reinforced concrete structures, a lot of damage was reparable. These earthquakes have resulted in composite steel being the construction material of choice in the Christchurch rebuild.

Currently in Australia, the design of composite structures for buildings is covered in a piecemeal fashion by three standards AS2327. 1-2003, AS3600-2009 and AS4100-2012 (Standards Australia 2003, 2009 and 2012) and in New Zealand by NZS 3404, (Standards New Zealand, 2007). The development of AS/NZS 2327 will improve harmonisation, will allow innovative aspects of composite construction and design to be incorporated in one standard and will also allow state of the art international research to be incorporated into a single document.

AS/NZS 2327 includes specific requirements for composite structures designed for seismic. The background behind the seismic provisions is outlined in this paper.

CONCLUSIONS

This paper has provided some background to the development of the seismic provision in a joint Australian/New Zealand standard on composite structures, namely AS/NZS 2327 Composite Structures. The structure and scope of the proposed seismic section in AS/NZS 2327 has been provided and innovations in the various sections have been highlighted.

ACKNOWLEDGMENT

This paper has been prepared by the author and is meant to be an informative exercise to outline the scope and structure of the proposed seismic requirements in AS/NZS 2327 Composite Structures Standard. The final standard is subject to committee decisions and the public comment phase and thus information herein is subject to change. The input of all the Australian and New Zealand nominating organisations has also been extremely valuable in ensuring that this project received the support in the proposal and drafting stages. We look forward to being able to communicate the progress of this project and hope for its completion in mid-late 2015.

REFERENCES

[1] SNZ (2004), Earthquake Actions Standard, NZS1170. 5-2004, Standards New Zealand, Wellington, New Zealand.

[2] SNZ (2007), NZS 3404-1997 + Amendment 2-2007, Steel Structures Standard, Standards New Zealand, Wellington, New Zealand.

[3] Standards Australia (1996) Australian Standard, AS 2327. 1-1996 Composite structures: simply-supported beams, Australia.

[4] Standards Australia (2003) Australian Standard, AS 2327. 1-2003 Composite structures: simply-supported beams, Australia.

[5] Standards Australia (2004) Australian Standard, AS5100. 6 Bridge design, Part 6 Steel and composite construction, Australia.

[6] Standards Australia (2009) Australian Standard, AS 3600-2009 Concrete Structures, Australia.

[7] Standards Australia (2012) Australian Standard, Steel Structures, AS4100-1998 (incorporating Amendment 1-2012), Australia.

[8] Standards Australia, (2007), Australian Standard, Structural Design Actions Part 4: Earthquake Actions in Australia, AS 1170. 4, Australia

DESIGN AND APPLICATION OF A MINIMAL-DISTURBANCE SEISMIC REHABILITATION TECHNIQUE COMPOSED OF LIGHT-WEIGHT STEEL ELEMENTS

Lei Zhang[a], Masahiro Kurata[a], Miho Sato[a], Oren Lavan[b], Masayoshi Nakashima[a]
[a]Kyoto University, Kyoto, Japan
zhanglei123gg@ gmail. com, kurata. masahiro. 5c@ kyoto-u. ac. jp,
sato. miho. 52w@ st. kyoto-u. ac. jp, nakashima@ archi. kyoto-u. ac. jp
[b]Technion-Israel Institute of Technology, Haifa, Israel
lavan@ cv. technion. ac. il

KEYWORDS: Seismic Rehabilitation Technique, Steel Moment-resisting Frame, Minimal-disturbance, Earthquake response analysis.

ABSTRACT

Seismic rehabilitation interrupts vision and physical space of building users in a large number of cases. Use of heavy construction equipment and hot work (welding / cutting) during rehabilitation is hard to be avoided. In response to these circumstances, this paper presents the development of a seismic rehabilitation technique for steel beam-column connections termed minimal-disturbance. The basic behavior of the rehabilitation technique is first verified through component-level testing and a simple model for a numerical analysis is constructed referring the test results. The effectiveness of the developed technique is examined numerically in application to a mid-rise steel-moment resisting frame.

INTRODUCTION

In steel moment-resisting buildings, deformation capacities of beam end often control the damage limits for serviceability, occupancy and collapse. With the presence of floor slab, which are designed to act together with steel beams, experience shows that strength and stiffness increase for the composite section compared to the bare steel beams but deterioration behavior under cyclic loading becomes asymmetric[1-3]. At states near ultimate, when the slab is in compression (i. e., positive bending), large tensile strains at bottom flanges result in ductile cracking and fracture.

This paper presents the design and application of a minimum-disturbance rehabilitation technique developed for improving seismic performance of steel beam-column connections. This technique is referred to asa MDAD (minimal-disturbance arm damper) hereinafter. First, the concept of a MDAD and the test result of a MDAD baseline model are presented. Then, a simple numerical model is developed based on the test result. Next, MDADs are applied to a four-story steel-moment resisting frame, and the changes in seismic demands, such as inter-story drift, roof drift and beam plastic rotation, were evaluated as measures of seismic capacity enhancement.

MINIMAL-DISTURBANCE ARM DAMPER

Fig. 1 (a) shows the configuration of a MDAD applied to a steel beam-column connection. The MDAD consists of several elements, such as tension-rods, energy dissipating plates, middle-connecting blocks, and other minor elements. The mid-span of beams and the upper part of a column (at one quarter of the story height from the joint) are connected with two tension rods at each side of the column and an energy dissipater. The latter consists of two steel bending plates that are attached to the two facing surfaces of the column using PC-bars [*Fig. 1* (a)]. For keeping equal deformation and yielding of two steel bending plates, rigid elements referred to as middle connecting blocks are used to connect the plates.

A simple model of a MDAD is constructed in a general-purpose structural analysis code, OpenSees [*Fig. 1* (b)]. Here, the yielding of bending plates is modelled by using zero-length spring elements that have a nonlinear material behavior. Under cyclic loading, the nonlinear spring provided stable hysteretic behavior.

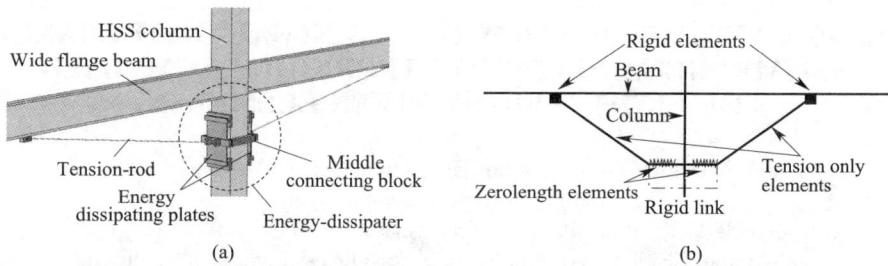

Fig. 1. (a)Schematic illustration of MDAD; (b)Modeling of MDAD

REHABILITATION PLAN OF FOUR-STORY FRAME

The MDAD is applied to a four story steel moment-resisting frame(*Fig.* 2a). In the application to the bare frame, the properties of the MDADs were selected as they yield at a force of 330 kN and a drift of 0. 36%. In nonlinear earthquake response analyses of the four-story frame, the MDADs contributed to reduce roof drifts and positive plastic rotations. The MDADs reduced the positive plastic rotations at beams-ends as high as 43%. This result illustrated the strong capability of MDAD in reducing the force demand at the bottom flange of beam. In addition, the negative plastic rotations slightly decreased as well since the increase in the negative bending moment with the addition of the MDADs were cancelled by the reduction in story drifts.

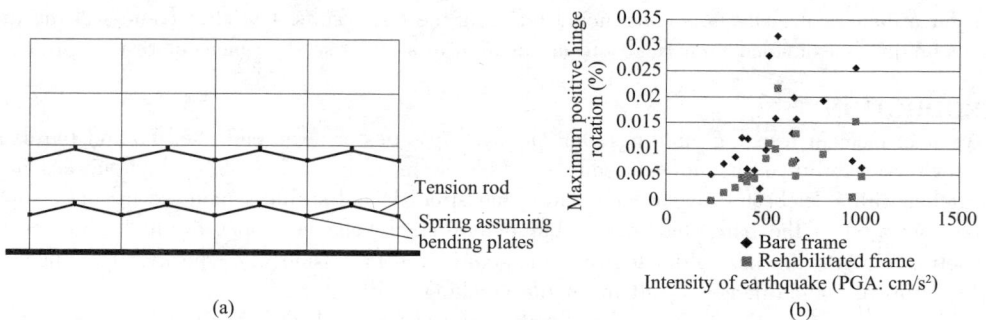

Fig. 2. Rehabilitation of four-story steel moment-resisting frame:
MDAD layout; reduction in positive plastic rotation.

CONCLUSIONS

The MDAD was designed to reduce plastic rotations in positive bending as well as inter-story drifts. The effectiveness of MDAD was verified through an application to a four-story steel moment-resisting frame. In nonlinear earthquake response analyses, the MDADs were very effective in reducing positive plastic rotations. The positive plastic rotations at beams-ends was reduced by as high as 43% in mean-plus standard deviation

REFERENCES

[1] Leon, R. T. , Hajjar, J. F. , and Gustafson, M. A. , 1998. "Seismic response of composite moment-resisting connections. I: Performance," J. Struct. Eng. , 1248, pp. 868-876.

[2] Chen, S-J. , Chao, Y. C. , 2001. "Effect of composite action on seismic performance of steel moment connections with reduced beam sections," J. Const. Steel Research, 57, pp. 417-434.

[3] Kim Y-J, Oh S-H, Moon T-S, 2004. "Seismic behavior and retrofit of steel moment connections considering slab effects," Eng. Struct. , 26(13): pp. 1993-2005.

SOME THOUGHTS FOR THE PREDICTION OF THE LOCAL INELASTIC CAPACITY OF MRF SUBJECTED TO SEISMIC ACTIONS

Anthimos Anastasiadis[a], Marius Mosoarca[b], Cristian Petrus[b] and Federico M. Mazzolani[c]

[a] Geostatic, Geotechnical & Structural Engineering, Greece
e-mail: anastasiadis@ geostatic. eu

[b] Politehnica University of Timisoara, Faculty of Architecture, Romania
e-mails: marius. mosoarca@ upt. ro, cristian. petrus@ student. upt. ro

[c] University of Naples Federico II, Dept. Structures for Engineering and Architecture, Italy
e-mail: fmm@ unina. it

KEYWORDS: Steel structures, Moment resisting frame, Seismic design, Local ductility.

ABSTRACT

Even 20 years after the Northridge and Kobe earthquakes, which represented a benchmark action particularly significant for the design and construction of steel structures, the prediction of the available local ductility of steel members still remains an open issue, [1]. Current design codes do not provide a clear procedure to evaluate the rotation capacity of a member. In order to verify directly the available ductility, based on the capacity-demand ratio, such a procedure should be set-up, [2]. The paper emphasizes the framework for the definition of the local ductility, considering that the component elements and their joints belong to a structural frame. Accordingly, it is important to distinguish different levels of influence, namely the available ductility under different loading conditions, monotonic, seismic (near-field, far-field), as well as under the effect of the conceptual detailing (strong column-weak beam, strengthening, weakening at the joint region) and finally under a given structural behavior.

CONCLUSIONS

Actually, the difficulties for the evaluation of the local inelastic capacity rise from the inherent variability of the loading, the material and the geometric parameters that can be evaluated in a post-elastic range. The experimental and theoretical background cumulated, particularly, in the last twenty five years creates the condition for a straightforward verification of the local ductility of steel moment resisting frames. The paper attempted to evidence the framework under which should be worked in order to obtain a ductile design based on the inelastic capacity of the element components. However, due to the fact that the monotonic ductility is relatively well studied, [3] and by using correction factors that take into account the reductive effect of the seismic action, we can obtain, for design purposes, the following conceptual relationship:

$$\frac{\text{Local Available}}{\text{Seismic Ductility}} = (\text{Correction Factors}) \times (\text{Available Local Monotonic Ductility})$$

According to the current philosophy of the ultimate limit state and the format of the Eurocodes we can provide: Seismic Required Ductility, $D_{req} \leq$ Seismic Available Ductility, D_{av}.

$$\gamma_{req}(\gamma_{fs};\gamma_{ns})D_{req}(D_{req. fs};D_{req. ns}) \leq \frac{D_{av}(D_{av. cyc.};D_{av. str})}{\gamma_{av}(\gamma_{cyc.};\gamma_{str.})} \tag{1}$$

where the required ductility could be evaluated directly from a time-history analysis, push-over or simplified relationships, while the available one by using a proper software (e. g DuctRot or other specialized computer program) or by simplified relationships suitably adjusted taking into account the seismic action, [2,3,4,5].

Moreover the available and the required ductility could be evaluated in terms of the ultimate rotation, θ_u, or in a non-dimensional format given by the rotation capacity, as a ratio of the ultimate to plastic rotation, $R = (\theta_u/\theta_p)-1$. In any case also the drift as well as the roof displacement could be used independently or in combination with the aforementioned in order to control the deformational capacity. Therefore, $D_{av} = (\theta_{u. av}; R_{av}; \mu_{\theta. av}; \mu_{\delta. av})$ and in the same direction $D_{req.} = (\theta_{u. req}; R_{req}; \mu_{\theta. req};$

$\mu_{\delta. req}$) could be defined. Focused on the available ductility we can distinguish the cyclic ductility, $D_{av. cyc}$, as well as the strain-rate ductility, $D_{av. str}$. In this direction, as a function of the main earthquake effect could be defined the followings:

$$D_{av. cyc} = (\text{factors introducing the cyclic effect}) \times D_{av. mon} \qquad (2a)$$
$$D_{av. str} = (\text{factors introducing the strain} - \text{rate effect}) \times D_{av. mon} \qquad (2b)$$

With respect to safety factors the γ_{av} should be determined taking into account the cyclic and strain-rate effect, while the γ_{req} could be evaluated taking into account the global frame behavior, the local soil conditions and also the characteristics of the action as defined from the far-source, γ_{fs}, and near-source, γ_{ns}, earthquake motion. Of paramount importance is to recognize the different effect on the available deformational capacity that has an earthquake of near-source with impulsive characteristics with very few inelastic cycles against a far-source earthquake with much more cyclic action, within the inelastic range, and longer duration.

Future editions of the codes (e. g. EC 8) must clearly specify, the joint and member available ductility, based on the past existing experimental and theoretical work. Furthermore, having in mind all the aforementioned, new experimental protocols should be defined, after extensive time-history analysis with the corresponding structural systems and accelerograms, in order to introduce both the main effect of the earthquake as well as to distinguish between low and high seismicity for any action of prequalification.

ACKNOWLEDGMENT

This paper is dedicated to Professor V. Gioncu who passed away on March 2013 and along with Prof. Mazzolani initiated the STESSA Conference early on 1994, in order to gather all the ideas regarding the design of steel structures in seismic areas. His contributions in the field of stability and ductility were recognized from the international engineering community by numerous publications and honors and further on we feel the need to acknowledge him for his warm friendship and not only scientific but also practical ideas shared with all of us.

REFERENCES

[1] Mazzolani FM. (ed). *Moment Resistant Connections of Steel Frames in Seismic Areas. Design and Reliability*, *RECOS*, E&FN SPON, Taylor & Francis Group, 2000.

[2] Gioncu V, Mazzolani FM. , *Ductility of Seismic Resistant Steel Structures*. London: Spon Press, 2002.

[3] Anastasiadis A, Mosorca M, Gioncu, V. , Prediction of available rotation capacity and ductility of wide flange beams. Part 2: Applications. *Journal of Constructional Steel Research*, 68: 176-191, 2012a.

[4] Anastasiadis A, Gioncu V, Mazzolani FM. , New trend in the evaluation of available ductility of steel members. In Proceedings of *Behaviour of Steel Structures in Seismic Areas*, STESSA 2000. Eds. Mazzolani & Tremblay, Rotterdam: Balkema, 3-10, 2000.

[5] Anastasiadis A, Gioncu V. , Ductility of IPE and HEA beam and beam-columns. In Proceeding of the 6th *International Colloquium on Stability and Ductility of Steel Structures*. Eds. Dubina D, Ivanyi M, London: Elsevier, 249-258, 1999.

DESIGN OF CONNECTIONS FOR COMPOSITE SPECIAL MOMENTFRAMES (C-SMF) WITH CONCRETE-FILLED STEEL TUBE (CFT) COLUMNS

Erica C. Fischer[a], Zhichao Lai[a] and Amit H. Varma[a]

[a]Purdue University, Lyles School of Civil Engineering, USA

fischere@ purdue. edu

laiz@ purdue. edu

ahvarma@ purdue. edu

KEYWORDS: Composite special moment frames, concrete-filled tube columns, moment connections.

ABSTRACT

The AISC *Seismic Provisions* (AISC 341-10)[1] provides design requirements and performance criteria for the beam-to-column connections in moment frames that serve as the primary seismic force resisting systems (SFRS) for steel building structures. The AISC *Prequalified Connections* (AISC 358-10)[2], however, do not include examples of beam-to-column connections applicable to composite moment frame construction. The engineer is required to present experimental results for the desired beam-to-column connection configuration, and demonstrate conformance to the performance requirements set forth in the AISC *Seismic Provisions*. This paper assists the engineer in this effort by providing an overview of previous research and testing performed on beam-to-column connections for composite moment frame construction, while focusing mainly on split-tee moment connections (for rectangular CFT columns) and through beam moment connections (for circular CFT columns), and their potential failure modes. This paper also includes comprehensive guidance and a complete example for the design and detailing these two types of connections in composite special moment frames. The connection details are based on the test results and the components are designed accordingly to provide the strength and ductility required by the performance criteria set forth in the AISC *Seismic Provisions*.

This paper along with the results of the test series performed by Peng[3] at Lehigh University, Schneider and Alostaz[4] at University of at University of Illinois at Urban-Champaign, and Elremaily[5] at University of Nebraska give structural engineers the tools to create a body of evidence to apply for prequalification of similar split-tee connections and through beam connections in C-SMF. The testing requirements outlined in Section K2 of the AISC *Seismic Provisions* are satisfied by the experimental testing program completed by Peng[3], Schneider and Alostaz[4], and Elremaily[5] discussed in this paper. This paper provides a comprehensive design procedure that satisfies the requirements of Section K1. 5 of the AISC *Seismic Provisions*. This paper also outlines many of the necessary sections, listed below, of the Prequalification Record required by Section K1. 6 of the AISC *Seismic Provisions*:

- General description of the prequalified connection
- Description of expected behavior of the connection in the elastic and inelastic ranges
- Definition of region of the connection that comprises the protected zone
- Detailed description of the design procedure for the connection
- List of references of test reports, research reports, and other publications that provide basis for prequalification

The experimental programs conducted by Schneider and Alostaz (1998), Peng (2001), andElremaily (2001) tested beam-to-column connections for C-SMFs. The experiments performed by Schneider and Alostaz (1998), and by Elremaily (2001) and the tests conducted by Peng (2001) showed through-beam connections and split-tee connections satisfy the requirements of the AISC *Seismic Provisions* for beam-to-column conections in C-SMFs. The AISC *Seismic Provisions* do not provide examples of moment connections for C-SMF construction. The engineer of record is required to create a "Prequalified Record" for the designed connection. This paper highlights many of the required sections for the "Prequalied Record" outlined in K1. 6 of the AISC *Seismic Provisions*.

The design of through-beam connections should account for the potential failure modes of the connection and organize them in a hierachal order from most ductile (desirable) to least: (i) plastic hinge formation in the beam, (ii) plastic hinge formation in the column, and (iii) panel zone failure of column. A similar hierachal order is outlined for the split-tee connection as : (i) plastic hinge formation in beam, (ii) stem yielding of split-tee, (iii) flange yielding of split-tee due to prying action, (iv) panel zone failure of column, and (v) bolt fracture in split-tee due to prying action of split-tee flange. Both connections are designed and detailed such that plastic hinge formation in the beams of the C-SMF frame is the most desirable failure mechanism.

CONCLUSIONS

The experimental programs conducted by Schneider and Alostaz(1998) , Peng(2001) , and Elremaily(2001) tested beam-to-column connections for C-SMFs. The experiments performed by Schneider and Alostaz(1998) , and by Elremaily(2001) and the tests conducted by Peng(2001) showed through-beam connections and split-tee connections satisfy the requirements of the AISC *Seismic Provisions* for beam-to-column conections in C-SMFs. The AISC *Seismic Provisions* do not provide examples of moment connections for C-SMF construction. The engineer of record is required to create a " Prequalified Record" for the designed connection. This paper highlights many of the required sections for the "Prequalied Record" outlined in K1. 6 of the AISC *Seismic Provisions*.

The design of through-beam connections should account for the potential failure modes of the connection and organize them in a hierachal order from most ductile(desirable) to least : (i) plastic hinge formation in the beam, (ii) plastic hinge formation in the column, and (iii) panel zone failure of column. A similar hierachal order is outlined for the split-tee connection as : (i) plastic hinge formation in beam, (ii) stem yielding of split-tee, (iii) flange yielding of split-tee due to prying action, (iv) panel zone failure of column, and (v) bolt fracture in split-tee due to prying action of split-tee flange. Both connections are designed and detailed such that plastic hinge formation in the beams of the C-SMF frame is the most desirable failure mechanism.

REFERENCES

[1] AISC ,2010. *Seismic provisions for structural steel buildings* , ANSI/ AISC 341-10. Chicago , IL.
[2] AISC ,2010. *Prequalified Connections for Special and Intermediate Steel Moment Frames for Seismic Applications* , ANSI/ AISC 358-10. Chicago , IL.
[3] Peng SW. ,2001. *Seismic resistant connection for concrete-filled tube column-to-WF beam moment resisting frames* [Dissertation]. Lehigh University.
[4] Schneider SP. ,1998. Alostaz YM. " Experimental behavior of connections to concrete-filled steel tubes". *J Constr Steel Res* ,45 :321-52.
[5] Elremaily A , Azizinamini A. ,2001. " Experimental behavior of steel beam to CFT column connections". *J Constr Steel Res* ,57 :1099-119.

A DESIGN APPROACH FOR COMPOSITE FRAMED STRUCTURES USING THE HYBRID FORCE/DISPLACEMENT (HFD) DESIGN METHOD

Konstantinos A. Skalomenos[a], George D. Hatzigeorgiou[b] and Dimitri E. Beskos[a]

[a]Department of Civil Engineering, University of Patras, Patras, GR-26500, Greece
skalomenos@ upatras. gr, beskos@ upatras. gr

[b]School of Science and Technology, Hellenic Open University, Patras, GR-26335, Greece
hatzigeorgiou@ eap. gr

KEYWORDS: Performance based design, Composite frames, Drift, Ductility, Behavior factor.

ABSTRACT

The seismic inelasticbehavior of regular planar composite steel/concrete MRFs consisting of I steel beams and concrete filled steel tube (CFT) columns is investigated. For this purpose, a family of 96 regular plane CFT-MRFs are subject to an ensemble of 100 ordinary ground motions scaled to different intensities in order to accommodate different performance levels and their response to these motions is recorded to form a response databank. Then, nonlinear regression analysis is performed and simple expressions for estimation of seismic displacements, drift and ductility demands are developed. More specifically, in this study the following objectives are examined: a) the relation between the behavior factor q and the maximum roof displacement ductility, μ; b) the relation that connects the maximum roof displacement $u_{r,max}$ with the maximum inter-storey drift ratio IDR_{max} and c) the relation that correlates μ with the maximum member rotation ductility μ_θ, along the height of the structure. The influence of specific parameters on the maximum structural response, such as the number of stories (n_s), the beam-to-column stiffness (ρ) and strength ratio (α), the level of inelastic deformation induced by the seismic excitation and the material strengths (e_s), is also taken into account. It is worth noticing that the proposed behavior factor q corresponds to the instant of the development of the first plastic hinge in the frame and for this reason, overstrength factors are implicitly considered, while additional static inelastic (pushover) analyses are not required. This essential aspect in conjunction with the ability of the proposed formulae to be adjusted to the framework of the hybrid force/displacement (HFD) seismic design method make possible the establishment of a design approach for CFT-MRFs using this method.

The HFD method combines the advantages of the well-known force-based[1] and displacement-based[2] seismic design methods in a hybrid (force/displacement) design scheme and works in a performance-based design (PBD) framework. The method has been proposed firstly in[3] and evolved by extensive parametric studies to reach its final stage dealing with plane steel frames of various kinds[4]. Herein, an effort is made to extend the HFD method to CFT-MRFs and apply it to a realistic design example of CFT-MRF using the proposed formulae. Furthermore, comparisons with a CFT-MRF designed according to the EC8 method are made on the basis of nonlinear time-history analyses of the designed frames under ten semi-artificial accelerograms, which are compatible with the EC8[1] elastic spectrum for three performance levels. The performance levels of (a) immediate occupancy (IO) under the frequently occurred earthquake (FOE), (b) life safety (LS) under the design basis earthquake (DBE) and (c) collapse prevention (CP) under the maximum considered earthquake (MCE) are assumed for seismic design of CFT-MRFs considered here. The corresponding limit values of local ductility and inter-storey drift ratio as defined by FEMA-273[5] for MRFs are also adopted in the framework of performance-based seismic design. The nonlinear time-history analysis results demonstrate the advantages of the proposed HFD method over the force-based seismic design procedure of EC8[1].

CONCLUSIONS

An efficient procedurebased on simple formulae for estimating the behavior factor, the seismic displacement, drift and ductility demands in regular plane CFT-MRF buildings subjected to ordinary (i. e. , without near-fault effects) earthquake ground motions has been presented. The proposed formulae were developed on the basis of the results of thousands of time history analyses, are simple, satisfy the physical constraints, express the central tendency with small dispersion and reflect the influence of several structural parameters(α, ρ, n_s, e_s). On the basis of the developments presented in this study, the following conclusions can be drawn:

(1) The q factor expression proposed here corresponds to the instant of first yielding in the frame and thus, pushover analysis and estimation of the overstrength factor are not needed. This important feature in conjunction with the ability of the proposed formulae to be adjusted to the framework of the HFD seismic design method make possible a design approach for CFT-MRFs using this method.

(2) Predefined values of the inter-storey drift ratio and local ductility for any limit state canbe incorporated to a target roof displacement in the initial design stage of the HFD method and then the appropriate behavior factor for limiting ductility demands can be calculated with the aid of the target roof displacement.

(3) All the performance metrics were satisfied for the designed frame by using the proposed HFD method, while the frame designed according to the EC8 method did notfulfill all the limit state values, as these defined in FEMA-273[5].

(4) Thenonlinear time-history analysis results(Table 1) revealed the consistency of the proposed relations to accurately estimate the seismic displacements and the adopted design criteria(drift and local member ductility) and the tendency of the EC8 method to overestimate the maximum roof displacement and underestimate the maximum inter-story drift ratio along the height of the frame, in all cases.

(5) Compared with the procedure adopted by current seismic design codes, the HFD methodwas found to be more rational and efficient for performance-based seismic design of regular plane CFT-MRFs buildings.

Table 1. Time history analyses results and comparison with design estimations of HFD and EC8 methods.

	HFD						EC8					
FOE		DBE		MCE			FOE		DBE		MCE	
	TH	EST	TH	EST	TH	EST	TH	EST	TH	EST	TH	EST
$IDR(\%)$	0.66	0.67	2.35	2.50	3.47	3.70	0.89	0.60	2.40	2.00	3.80	3.00
$u_{r,max(m)}$	0.086	0.086	0.265	0.28	0.395	0.43	0.086	0.094	0.23	0.313	0.34	0.47
μ_θ	0.87	–	2.56	2.20	3.80	3.30	0.85	–	2.65	–	4.07	–
q	–	–	2.82	2.80	4.23	4.20	–	6.5	2.95	6.5	4.10	6.5
μ_r	–	–	2.36	2.40	3.48	3.60	–	6.5	2.40	6.5	3.80	6.5

TH: time history analysis; EST: estimation of HFD; EC8: estimation using the equal-displacement rule

REFERENCES

[1] EC8, *Design of Structures for Earthquake Resistance-Part 1: General rules, seismic actions and rules for buildings*, European Committee for Standardization, Brussels, Belgium, 2004.

[2] Priestley MJN, Calvi GM, Kowalsky MJ, *Direct displacement-based seismic design*, Pavia, Italy: IUSS Press, 2007.

[3] Karavasilis TL, Bazeos N, Beskos DE, "A hybrid force/displacement seismic design method for plane steel frames", Behavior of steel structures in seismic area. In: Mazzolani F, Wada A, editors. *Proceedings of STESSA conference*, August. Yokohama, Japan: Taylor & Francis; 2006. p. 39-44.

[4] Tzimas AS, Karavasilis TL, Bazeos N, Beskos DE, "A hybrid force/displacement seismic design method for steel building frames", *Engineering Structures* 56:1452-1463, 2013.

[5] FEMA. *NEHRP Guidelines for the Seismic Rehabilitation of Buildings*, Report No. FEMA-273. Federal Emergency Management Agency, Washington, DC, 1997. A

SEISMIC DESIGN CRITERIA FOR STEEL MOMENT RESISTING FRAMES FOR COLLAPSE RISK MITIGATION

Ahmed Elkady[a], Dimitrios G. Lignos[a]

[a]McGill University, Department of Civil Engineering and Applied Mechanics, Canada
ahmed. elkady@ mail. mcgill. ca, dimitrios. lignos@ mcgill. ca

KEYWORDS: Dynamic overstrength, Collapse risk, steel SMF, Strong-column weak-beam ratio.

ABSTRACT

A number of analytical and experimental studies related to the seismic behaviour of steel frame buildings demonstrated the need to re-assess the lateral system overstrength factor that is currently used in the seismic design of frame buildings in highly seismic regions in North America[1]. Prior analytical studies on steel frame buildings with perimeter special moment frames (SMFs) showed that the overstrength factor calculated based on nonlinear static procedures can considerably vary from the one specified by ASCE/SEI 7-10[2] (i. e. , $\Omega_o = 3. 0$). Moreover, it is becoming increasingly important to quantify the collapse risk of steel frame buildings when subjected to extreme earthquakes beyond the design level consideration.

To properly quantify the system overstrength and collapse risk of steel SMFs, the lateral stiffness and strength contributions from structural elements other than the mainlateral force resisting system (LFRS) should be considered. In particular, the contributions of the composite concrete slab and the interior gravity framing system. These contributions have been typically ignored in past analytical studies as well as in the current seismic design practice in North America.

The flexural stiffness and strength contributions provided by the composite slab and the gravity framing system can either benefit or impair the overall seismic performance of a steel frame building. Prior analytical studies[3,4] showed that considering the gravity framing in the analytical model of a frame building can increase both its lateral stiffness and base shear strength as well as mitigate the lateral drifts of the LFRS. On the other hand, the amplified flexural strength of the steel beams due to the composite action may cause plastic hinging in the steel columns and consequently local story collapse mechanisms may develop[5,6].

To this end, a comprehensive analytical study is conducted on five archetype steel frame buildings with perimeter SMFs with heights ranging from 2 to 20 stories. The archetype frame buildings are designed as per the current seismic code in the US. State-of-the-art, 2-dimensional, analytical models are developed considering both the effects of the composite action and the interior gravity framing system on the lateral resistance and flexural stiffness of the archetypes. Rigorous nonlinear static and response history analysis through collapse are employed to quantify realistic overstrength values as well as the collapse risk of the considered archetype buildings. The effect of the strong-column-weak-beam (SCWB) ratio on the mitigation of collapse risk of archetype steel frame buildings with perimeter SMFs is also quantified.

CONCLUSIONS

The first goal of this paper was to propose more robust overstrength measures that may be potentially used as part of the seismic design process of steel SMFs. The second goal was to propose an effective value of the SCWB ratio that can be employed as part of the design process of steel SMFs in order to minimize their collapse risk over the archetype building life expectancy. The main conclusions of this paper are summarized as follows:

1. Bare steel SMFs(i. e. , B models) develop a dynamic overstrength factor, Ω_d, > 3. 0 When the composite action and the gravity framing are considered(i. e. , CG models), the dynamic overstrength is in average equal to 4. 0 without any period dependency. Unlike the static overstrength factor, Ω_s, measured from pushover analysis, the dynamic overstrength captures the inelastic force redistribution due to dynamic loading(see *Fig. 1*a); thus it can be used to reliably measure the system overstrength

and to evaluate the reliability of force-sensitive components in steel frame buildings subjected to earth-quake loading[7].

2. When both the composite and interior gravity framing action is considered as part of the analytical model, low-to mid-rise SMFs, designed as per[15] (i. e. , SCWB ratio $\geq$ 1. 0), achieve a probability of collapse in a return period of 50 years, P_c (50 years), larger than the 1% limit specified by the current seismic provisions[8] (see Fig. 1b). A P_c (50 years) < 1% is achieved only when the SMFs are designed with a SCWB ratio $\geq$ 1. 5. This ratio results in heavier column sections, however, the number of doubler plates required by the design provisions are significantly reduced and therefore the fabrication cost associated with the welding of these plates is also significantly reduced.

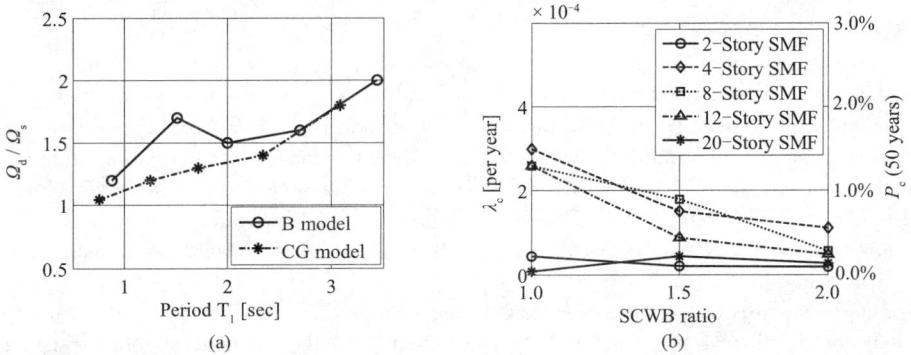

Fig. 1. (a) Ratio of dynamic to static overstrength factor versus first-mode period of different analytical models of the archetype frame buildings; (b) Mean annual frequency of collapse and the corresponding probability of collapse in 50 years versus SCWB ratio for all archetype frame building.

REFERENCES

[1] NIST. "Tentative framework for development of advanced seismic design criteria for new buildings". *NIST GCR* 12-917-20, NEHRP consultants Joint Venture, 2012.

[2] ASCE. *Minimum design loads for buildings and other structures*, ASCE/SEI 7-10. American Society of Civil Engineers: Reston, VA. ,2010.

[3] Ji, X. , Kato, M. , Wang, T. , Hitaka, T. , and Nakashima, M. "Effect of gravity columns on mitigation of drift concentration for braced frames". *Journal of Construc tional Steel Research*; 65(12): 2148-2156,2009. DOI: 10. 1016/j. jcsr. 2009. 07. 003.

[4] Gupta, A. and Krawinkler, H. "Seismic demands for the performance evaluation of steel moment resisting frame structures". *Report No.* 132, The John A. Blume Earthquake Engineering Center, Stanford University, CA ,1999.

[5] Elkady, A. and Lignos, D. G. "Modeling of the composite action in fully restrained beam-to-column connections: implications in the seismic design and collapse capacity of steel special moment frames". *Earthquake Engineering & Structural Dynamic*; 43(13): 1935-1954,2014. DOI: 10. 1002/eqe. 2430.

[6] Suita, K. , Yamada, S. , Tada, M. , Kasai, K. , Matsuoka, Y. , and Shimada, Y. "Collapse experiment on 4-story steel moment frame: Part 2 detail of collapse behavio r". *Proc. of The 14th World Conference on Earthquake Engineering*(*WCEE*). Beijing, China,2008.

[7] Elkady, A. and Lignos, D. G. "Effect of gravity framing o the overstrength and collapse capacity of steel frame buildings with perimeter special moment frames". *Earthquake Engineering & Structural Dynamic* ,2014. DOI: 10. 1002/eqe. 2519.

[8] AISC. *Seismic provisions for structural steel buildings*, ANSI/AISC 341-10. American Institute for Steel Construction: Chicago, IL ,2010.

LESSONS FROM STEEL STRUCTURES IN CHRISTCHURCH EARTHQUAKES

Gregory MacRae[a], G. Charles Clifton[b], Michel Bruneau[c], Amit Kanvinde[d] and Sean Gardiner[e]

[a] Dept of Civil and Natural Resources Engineering, Univ. of Canterbury, Christchurch, New Zealand gregory. macrae@ canterbury. ac. nz

[b] Dept of Civil and Environmental Engineering, University of Auckland, New Zealand c. clifton@ auckland. ac. nz

[c] Department of Civil Engineering, The University of Buffalo, New York, USA bruneau@ buffalo. edu

[d] Department of Civil Engineering, University of California, Davis, USA kanvinde@ ucdavis. edu

[e] Calibre Consulting, Christchurch, NZ sean. gardiner@ calibre. co. nz

Keywords: Christchurch earthquakes, Lessons learned, Steel, Seismic, Structures.

ABSTRACT

Lessons learned from the 2010/2011 Christchurch earthquake sequence about the behaviour of steel structures are described. Firstly, observed performance of steel structures is summarised. It is shown that many steel structures had very little damage. However, some structures suffered damage as a result of large foundation settlements. Yielding, buckling and fractures were observed in steel bridges and buildings. Reasons for observed damage are then described in the light of recent studies. It is shown that because of the lesser damage to steel structures and greater uncertainty over repair of reinforced concrete structures, that steel structures have become popular in the Christchurch rebuild. A number of these use low-damage systems.

CONCLUSIONS

During the 2010-2011 Canterbury earthquake sequence buildings within Christchurch were subjected tovery strong earthquake shaking. A number of lessons relating to the steel structures are:

1) Immediate reconnaissance after the Canterbury earthquake indicated very little damage, however various types of buckling and fracture occurred. The lack of damage may be a result of drift controlling the member sizes resulting in low expected ductility demands, and foundation/slab/non-structural element effects which meant the structure was twice as much as expected.

2) Subsequent studies indicate that:

(i) soft-soil and foundation flexibility effects may have minimized the distress at the foundation,

(ii) slab and non-structural element effects may have significantly decreased the response, and together with the likely member strengths mean that structures may be able to resist significantly greater levels of earthquake shaking than those for which they have been designed,

(iii) hardness has the possibility of being used to estimate remaining earthquake life of a damaged member, and

(iv) lack of redundancy may be able to be addressed using continuous columns, but a lack of redundancy does not always lead to collapse

3) Assessments of the behaviour of steel structures indicate that hardness has the possibility to be used to estimate the remaining earthquake life of some yielded elements.

4) Steel structures and steel-concrete composite structures have become systems of choice in the Christchurch rebuild. This is because of their reparability compared to reinforced concrete structures.

REFERENCES

[1] MacRae G. A. , Lessons from the February 2011 M6. 3 Christchurch Earthquake, Journal of Seismology

and Earthquake Engineering,14(3),2013,pp 81-93. http://www. iiees. ac. ir/jsee.

[2] Ainsworth,J. B. , Conlan, C. J. and Clifton G. C. "Influence of Column Base Rotational Stiffness and Non-Structural Elements on Moment Resisting Steel Frame Building Response to Severe Earthquakes",paper submitted for the 2015 Technical Conference of the NZSEE(2015).

[3] Bruneau M. , Anagnostopoulou M. ,MacRae G. , Clifton G. C. and Fussell A. ,Preliminary Report on Steel Building Damage from the Darfield Earthquake of September 4,2010,Bulletin of the New Zealand Society for Earthquake Engineering,December(2010)

[4] Clifton C, Bruneau M, MacRae GA, Leon R. and Fussell A. Steel Structure Damage from the Christchurch Earthquake Series of 2010 and 2011. Bulletin of the New Zealand Society for Earthquake Engineering 2011; 44(4): 279-318.

[5] Clifton G. C. ,Nashid H. , Ferguson G. ,Hodgson M. and Seal C. ,Bruneau M. ,MacRae G. A. and Gardiner S. ,"Performance of Eccentrically Braced Framed Buildings In The Christchurch Earthquake Series of 2010/2011",14 *World Conference on Earthquake Engineering*,Lisbon,Portugal, August 2012. Paper number 2502.

[6] Clifton G. C. ,"Lessons Learned for Steel Seismic Design from the 2010/2011 Canterbury Earthquake Series",Australian Eqke Eng Society 2013 Conference,Nov. 15-17,Tasmania.

[7] Gardiner S. ,Clifton G. C. , andMacRae G. A. ,2012. Australian Earthquake Engineering Society 2013 Conference,Nov. 15-17,Tasmania.

[8] Gardiner S. ,Clifton G. C. andMacRae G. A. "Performance, Damage Assessment and Repair of a Multistorey Eccentrically Braced Framed Building following the Christchurch Earthquake Series" ,Steel Innovations Conf. ,SCNZ,Wigram,Christchurch,21-22 Feb 2013,Paper 19.

[9] Kanvinde,A. ,2012. AISC Live Webinar Column Base Connections: What Several Years of Testing Has Taught Us. AISC.

[10] Clifton,G. C. and Ferguson. G. W. 2015. Peer Review DEE of HSBC Tower Version 9. 2a Confidential Document: University of Auckland,Faculty of Engineering.

[11] Pender,M. J. ,2012. Limit State Design of Foundations: Chapter 11 Elastic Models for Pile Stiffness. The University of Auckland,Auckland,New Zealand.

[12] Momtahan A. , Clifton G. C. Effects of Out of Plane Strength and Stiffness of Composite Floor Slabs on the Inelastic Response of Eccentrically Braced Frame Structures,8th International Conference on Behavior of Steel Structures in Seismic Areas, STESSA, Shanghai, China, July 1-3,2015.

[13] Imani R. ,and Bruneau M. ,"Effect of Link-Beam Stiffener and Brace Flange Alignment on Inelastic Cyclic Behavior of Eccentrically Braced Frames",AISC Engineering Journal

[14] Kanvinde A. M. ,Marshall K. S. ,Grilli,D. A. ; and Bomba,G. (2014),"Forensic Analysis of Link Fractures in Eccentrically Braced Frames during the February 2011 Christchurch Earthquake: Testing and Simulation",J. Struct. Eng. ,10. 1061/(ASCE)ST. 1943-541X. 0001043,04014146.

STRUCTURAL DESIGN ASPECTS OF NEXT GENERATION STEEL WIND ENERGY STRUCTURES

Evangelos Efthymiou[a],

[a]Lecturer Dr Civ. Eng. , Institute of Metal Structures, Department of Civil Engineering
Aristotle University of Thessaloniki, Thessaloniki, Greece
vefth@ civil. auth. gr

KEYWORDS: next generation wind energy structures, tall towers, steel construction, structural design considerations.

ABSTRACT

Towards the goal of making onshore wind energy the most competitive energy source, relative research efforts have been increasingly intensified in the last decade, while further advances in wind turbine technology are imperative. In response to this call, wind energy sector is focusing developing bigger structures that can support larger wind turbines, thus increasing energy yield, while addressing successfully cost and transportation challenges. The present paper introduces the so called next generation wind energy structures, namely towers with hub height over 100m, aiming to take advantage the greater and more stable wind speeds corresponding at this altitude, generating thus more power. The paper summarizes the latest developments concerning the structural forms of the towers, as well as modern innovative design concepts, so that both safety and material economy can result in cost-efficient power generators. Within the framework of the study, key features of these tall structures are examined and the special parameters that need to be considered at design stage are identified. Moreover, although wind loading is the primary action in such structures, wind profile at these heights combined with seismic actions compose a complex loading framework, particularly in earthquake prone areas. The herein presented work aims at highlighting the significant role that growth possesses regarding their structural behaviour.

In view of structural design, the increase of size has direct effects on the structural parts due to both the augmented wind pressure area and tower's height. Large values of bending moments arising at the tower's base must be safely undertaken by the structure's foundation. Besides the static stability of the tower, the dynamic behaviour of such large scale structures is also important. Constantly operating, the inherent dynamic loads create many issues that are not normally considered in static structures. Long-term loads are the normal operational wind loads that include dead, live, static wind pressure and standard operational overturning moments, shears and axial loads. Therefore, every aspect of the tower-foundation design must consider these provisions, assess dynamic characteristics and address fatigue-related issues.

CONCLUSIONS

As stated in literature, an often underestimated parameter is the consideration of the tower in the design process of a wind energy structure. However, it is of great relevance for the economic efficiency of a wind turbine, since it produces a significant portion of the wind turbines' initial costs, approximately 15% to 20% and also determines the costs for transport and erection. Moreover, examining the evolution of modern design of wind turbines, a trend of increasing the rotor's diameter is recorded. This fact leads to a proportionate increase in size of the structural parts of the wind turbine and especially of the height of the tower that supports its mechanical parts.

In pursuit of increased energy yield, wind energy sector introduced recently next generations wind turbine towers, namely tall towers that can support bigger rotor diameter and exploiting wind profiles at high altitudes, generating thus more energy. Hub-heights beyond 100m, lattice towers, hybrid towers, innovative design configurations, bolted friction joints, advanced dynamic loads, increased significance of soil-structure interaction compose this complex analysis framework. Additionally, various practical and constructional issues arise, preventing the necessary dimension enlargement of structural parts in order to undertake the increased stress resultants. Towards overcoming these barriers, there is a need for more R&D activities for innovative design and an adoption of an integrated approach, so that transportation, construction, maintenance and sustainability issues can be addressed successfully along with structural performance, providing a cost-benefit outcome.

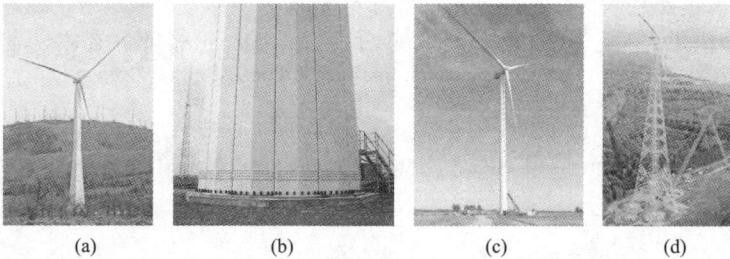

(a)　　　　　　　(b)　　　　　　　(c)　　　　　　　(d)

Fig. 1. Large diameter steel tower, Space frame design, Bolted steel shell, Lattice wind tower [17-19]

ACKNOWLEDGMENT

The author would like to express his gratitude to Aristotle University of Thessaloniki's Research Committee(ELKE AUTH)for the financial support provided regarding his research on the field of next generation wind energy structures.

REFERENCES

[1] Baniotopoulos CC, Borri C. , Stathopoulos Th. (Eds.). 2011. *Environmental Wind Engineering and Design of Wind Energy Structures*, 1st Edition, CISM(International Centre for Mechanical Sciences) Courses and Lectures, Vol. 531, SpringerWienNewYork, Italy.

[2] Lavassas I. , Nikolaidis G. , Zervas P. , Efthimiou E. , Doudoumis I. and Baniotopoulos C. C. , 2003. "Analysis and Design of the Prototype of an 1 MW Steel Wind Turbine Tower", *Engineering Structures*, 25: 1097-1106.

[3] Bazeos N, Hatzigeorgiou GD, Hondros ID, Karamaneas H, Karabalis DL, Beskos DE. ,2002. "Static, seismic and stability analyses of a prototype wind turbine steel tower", *Engineering Structures*. 24,8: 1015-1025.

[4] TPWind-European Wind Energy Technology Platform for Wind Energy, 2008. *Strategic Research Agenda-Market Deployment Strategy from 2008 to 2030*, Synopsis-Preliminary Discussion Document, TPWind Secretariat Brussels, Belgium.

[5] Cotrell J. , Stehly T. , Johnson J. , Roberts JO. , Parker Z. , Scott G. , 2014. Heimiller D. , *Analysis of Transportation and Logistics Challenges Affecting the Deployment of Larger Wind Turbines: Summary of Results*, NREL/TP-5000-61063, National Renewable Energy Laboratory, 2014.

[6] Stavridou N. , Efthymiou E. , Gerasimidis S. , Baniotopoulos CC. ,2013. "Modelling of the structural response of wind energy towers stiffened by internal rings", *Proc. of the 10th HSTAM International Congress on Mechanics*, 25-27 May 2013, Chania, Crete, Greece.

[7] Engström S. , Lyrner T. , Hassanzadeh M. , Stalin Th. & Johansson, J. ,2010. *Tall towers for large wind turbines*, Report from Vindforsk project V-342 Högatorn för vindkraftwerk, Elforsk rapport 10:48, Stockholm, 2010.

[8] Gasch R. , Twele J. ,2012. *Wind Energy Plants*, Springer, Heidelberg, New York.

A NEW STRATEGY TO PREVENT COLLAPSE OF COLUMNS IN BUILDINGS WITH STEEL CHEVRON BRACED STRUCTURE

Francesca Barbagallo[a], Melina Bosco[a], Edoardo M. Marino[a], Pier Paolo Rossi[a]

[a]Dept. of Civil Engineering and Architecture, University of Catania, v. le A, Doria 6, 95125 Catania

fbarbaga@ dica. unict. it, mbosco@ dica. unict. it, emarino@ dica. unict. it, prossi@ dica. unict. it

KEYWORDS: Chevron bracings, gravity columns, capacity design, seismic code.

ABSTRACT

Modern seismic codes define dissipative members, which should dissipate energy during the ground motion by means of their hysteretic behaviour, and non-dissipative members that should remain elastic. In concentrically braced frames, according to Eurocode 8 (EC8)[1], braces are the dissipative members, while all the other members (i. e. beams and columns) should remain elastic. Furthermore, the numerical models commonly used for the design of building structures with concentric bracings include only the braced frames and neglect the bending moment in structural members[2]. These numerical models are consistent with EC8, which defines concentrically braced frames as "those in which the horizontal forces are mainly resisted by members subjected to axial force s". However, the bending moments in columns (those of the braced frame as well as gravity columns) may achieve significant values because of the storey drift concentration that occurs when the braced frame overcomes the elastic limit. According to some recent studies[3-4], these bending moments combined with axial forces may cause yielding or buckling of columns before braces achieve their axial deformation capacity. Based on these considerations, in a previous paper[4], the authors proposed technological suggestions and design procedures to improve the seismic performance of columns of building structures with diagonal braces.

In this paper, these technological suggestions and design procedures are extended to chevron braced structures. In particular, the authors suggest the adoption of semi-rigid with low rotational stiffness or pinned connections at the ends of the columns of the braced frames[4]. This technological solution can reduce significantly or avoid the bending moment of these columns. With regards to gravity columns, it is considered that during the earthquake the value of gravity load is much smaller than that used for their design. Furthermore, the presence of continuous columns has a beneficial effect on the redistribution of the plastic deformation of braces[5,6]. Based on these considerations, it is suggested that gravity columns are continuous along the height of the building. However, the bending moments that arise in gravity columns during ground motions have to be determined and considered for the design of the gravity columns. These bending moments are predicted by a procedure that considers the behaviour of the whole structure (braced frame and gravity columns) under a seismic input increased up to the attainment of the brace failure. Three stages of behaviour, named "0", "1" and "2", are analysed. When the braces are in the elastic range of their behaviour (stage n. 0), the seismic force provides a displacement profile that is almost linear (*Fig. 1*a). Gravity columns basically rotate rigidly about their base and the bending moments are negligible. When the braces in compression of some storeys buckle, the stiffness of these storeys reduces and drops to zero when also the braces in tension yield. The horizontal displacement profile changes in these two stages of behaviour and storey drifts concentrate where the collapse mechanism has initiated. Fig. 1b shows the shape of the horizontal displacement profile when the compression braces of the three lower storeys have buckled (stage n. 1). Instead, the horizontal displacements demanded by the seismic force after the yielding of the tension braces of the three lower storeys are shown in Fig. 1c (stage n. 2). In the figures the horizontal displacements are scaled to obtain the same top displacement. In the stages n. 1 and 2, the gravity columns are deformed in flexure and sustain large bending moments.

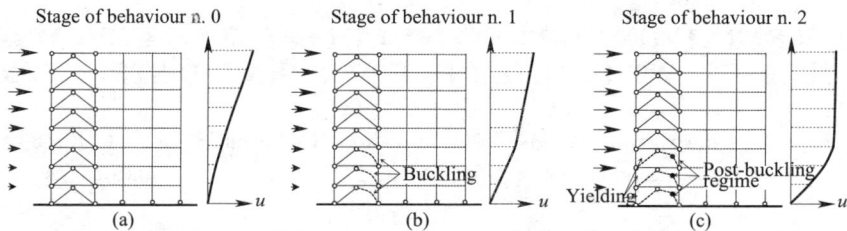

Fig. 1. Idealised stages of behaviour: (a) before buckling, (b) after buckling and (c) after yielding of braces.

CONCLUSIONS

Two steel building structures with chevron bracings have been designed according to EC8 and the proposed design procedure. The effect of the seismic force is determined by modal response spectrum analysis. The elastic spectrum proposed in EC8 for soil type C and characterised by a peak ground acceleration $a_{g,R}$ equal to 0.35 g is used. The q value equal to 2.5 stipulated in EC8 for high ductility chevron bracings is adopted. In order to separate the influence of each proposed modification, three design alternatives are analysed and compared.

2. In the first alternative, the frames are designed according to EC8. All the columns are continuous for the whole height of the building, but are considered pinned in the numerical model adopted in design. The gravity columns are designed to sustain gravity loads only.

2. In the second alternative, the cross-sections of all the members are those assigned in the previous design alternative. However, the columns of the braced frame are pinned at each storey.

3. In the third alternative, the braced frames are identical to those of the second alternative, but the gravity columns are designed to sustainalso bending moments according to the proposed procedure.

The seismic response of the case study structures is determined by nonlinear dynamic analysis for different seismic excitation levels. Then, the seismic excitation level that leads to the brace failure is evaluated. Finally, the seismic performance of columns for this seismic excitation level is determined. In case of the structures designed according to the first alternative (EC8), the columns of the braced frame exceed both the plastic resistance and the stability capacity. Instead, the seismic performance of the gravity columns is always suitable. The introduction of the pinned connections in the columns of the braced frames (second alternative) generally reduces their strength demand below their capacity. However, in this case, the response of the gravity columns is significantly aggravated and always exceeds the capacity, because bending moments are sustained only by the gravity columns. When the gravity columns are designed considering also the bending moments evaluated according to the proposed method (third alternative), the seismic performance significantly improves and both the resistance and stability requirements are generally fulfilled. Instead, the seismic performance of the columns of the braced frame remains virtually identical to that obtained in the case of the second design alternative.

REFERENCES

[1] Eurocode 8. Design of structures for earthquake resistance. European Committee for Standardisation 2003, prEN 1998-1-1/2/3.

[2] Brandonisio G, Toreno M, Grande E, Mele E, De Luca A, "Seismic design of concentric braced frames", *Journal of Constructional Steel Research*, 78: 22-37, 2012.

[3] Longo A, Montuori R, Piluso V, "Plastic design of seismic resistant V-braced frames", *Journal of Earthquake Engineering*, 12(8): 1246-66, 2008.

[4] Bosco M, Ghersi A, Marino EM, Rossi PP, "A capacity design procedure for columns of steel structures with diagonals braces", *The Open Construction and Building Technology Journal*, 8 (Supp 1: M2): 196-207, 2015.

[5] Ji X, Kato M, Wang T, Hitaka T, Nakashima M, "Effect of gravity columns on mitigation of drift concentration for braced frames", *Journal of Constructional Steel Research*, 65: 2148-56, 2009.

[6] Marino EM, "A unified approach for the design of high ductility steel frames with concentric braces in the framework of Eurocode 8", *Earthquake Engineering and Structural Dynamics*, 43 (1): 97-118, 2014.

OPTIMIZATION OF ENERGY – DISSIPATION DEVICES ARRANGEMENT
For Seismic Retrofit of Truss Tower Structures

Yusuke Kinouchi[a], Toru Takeuchi[a], Ryota Matsui[a], Toshiyuki Ogawa[a], Kazuhiro Fujishita[a]

[a] Tokyo Institute of Technology, Dept. of Arch. and Build. Eng. , Japan

kinouchi. yusuke@ gmail. com, ttoru@ arch. titech. ac. jp, matsui. r. aa@ m. titech. ac. jp,

togawa@ arch. titech. ac. jp, fujishita. k. aa@ m. titech. ac. jp

KEYWORDS: Truss Tower Structure, Seismic Control, Structural Optimization, Genetic Algorithm and Equivalent Linearization Method

ABSTRACT

It is important to assure theperformance of existing truss tower structures, due to the substantial role of telecommunication infrastructures after huge seismic disasters. In reference[1], response control retrofit of truss tower structures by replacing existing members with energy dissipation members was proposed and applied in a practical structure.

In this paper, the optimal design method for the arrangements of visco-elastic dampers(VED) with Genetic Algorithm(GA) is proposed. Objective functions are lateral displacement and member stress obtained by response spectrum analyses. The effectiveness of proposed optimization method based on the response evaluation is examined with time-history analysis and the optimal solutions are compared with the results of empirical damper design procedure.

The outline of the proposed optimal design method is shown in *Fig. 1*. The equivalent damping ratio of m-th mode $_mh_{eq}$ is calculated based on the assumption of a series relationship between provided VEDs and the main structure. As analysis model, 3D truss tower model placed on SRC building(R model) and self-supported tower(G model) are considered. The equivalent damping effect is evaluated from the elastic structural stiffness and each hysteresis loop of VED assuming vibration mode as constant in spite of added damping. Responses on each mode are synthesized with Complete Quadratic Combination(CQC) method and maximum responses of each analysis model are evaluated. These evaluated values are optimized with GA.

Objective function of Structural Optimization Problem(SOP) 1 and SOP2 is minimizing the lateral displacement at the top of the structure, δ_{top} and that of SOP3 is maximizing the buckling risk index. In SOP1, all members including chord members can be replaced, while only diagonal members are replaced in SOP2 and SOP3. Obtained optimal shapes and design indexes are shown in *Fig. 2* and *Fig. 3*. Damper arrangement of SOP1 is concentrated on lower part of truss tower and that of SOP2 is distributed into two parts, lower part and middle part of truss tower. On the same optimization case, the optimal solutions are similar for varying seismic input waves. Maximum response of optimal solution for SOP1 is shown in *Fig. 4*. Proposed response evaluation results show good correspondence with analysis results.

Fig. 5 shows the two empirical approaches by elastic analysis of non-retrofit buildings. One is replacing large relative displacement members to dampers on time-history analysis (*Method-A*), the other is replacing high buckling risk members (*Method-B*). *Method-B* produced only an upper arrangement without base zones, and *Method-A* produced relatively similar

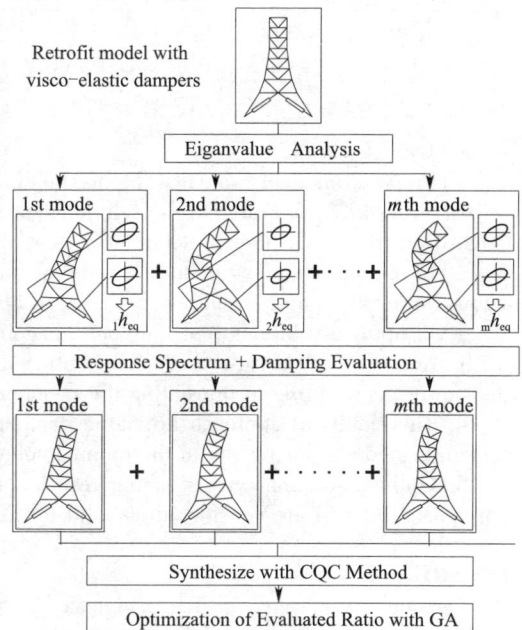

Fig. 1. Outline of optimal design method

results to the optimization analyses.

Upper column: Proposed Evaluation Lower Column: Time History Analysis

HACHINOHE-EW		JMAKOBE-NS		RANDOM	
1R	1G	2R	2G	3R	3G
21. 5cm	19. 0 cm	21. 2 cm	19. 3 cm	22. 4 cm	18. 6 cm
18. 3 cm	14. 7 cm	16. 2 cm	15. 2 cm	20. 1 cm	16. 4 cm

Fig. 2. Optimal design of SOP1

		HACHINOHE-EW		JMAKOBE-NS		RANDOM	
		1R	1G	2R	2G	3R	3G
SOP2							
	δ_{top}	30. 0 cm	26. 5 cm	31. 9 cm	28. 0 cm	30. 5 cm	25. 7 cm
		26. 5 cm	23. 5 cm	28. 7 cm	23. 5 cm	28. 2 cm	22. 8 cm
	σ_i/f_{ci}	0. 823	0. 429	0. 732	0. 450	0. 829	0. 474
		0. 615	0. 424	0. 614	0. 416	0. 630	0. 459
SOP3							
	δ_{top}	30. 0 cm	27. 0 cm	31. 9 cm	28. 4 cm	30. 6 cm	25. 7 cm
		26. 4 cm	24. 4 cm	28. 6 cm	24. 0 cm	28. 3 cm	22. 8 cm
	σ_i/f_{ci}	0. 664	0. 412	0. 650	0. 435	0. 674	0. 417
		0. 624	0. 402	0. 608	0. 400	0. 649	0. 402

Fig. 3. Optimal design of SOP2 and SOP3

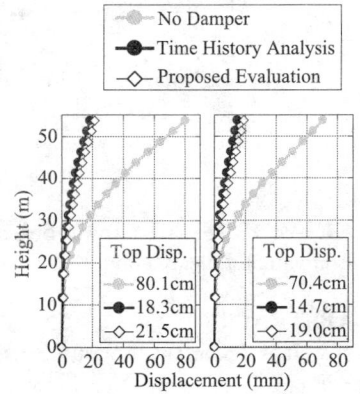

Fig. 4. Maximum response

(a)1R Case (b)1G Case

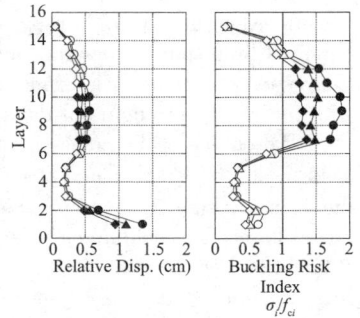

Fig. 5. Maximum analysis value
(1R, non-retrofit model)

CONCLUSIONS

1) Proposed method evaluates the maximum lateral displacement of retrofit model in good accuracy. It is proved by full search of the solution space of SOP2 and SOP3 that proposed optimization method can find the global optimal solution in most analysis cases.

2) In optimal solutions including code members, columns and diagonal members of the lower part of truss tower are mainly replaced to VEDs on optimization of SOP1.

3) In optimal solutions replacing only diagonal members, VEDs are arranged in mainly two parts, which are lower part of truss tower contributing to displacement response reduction and middle part of truss tower contributing to preventing the member buckling.

4) The empirical approach arranging dampers with connecting points of larger relative displacement indicated similar results to the optimal solution of SOP2 and 3, which proves its effectiveness.

5) Optimal solutions yields similar results for various input waves, therefore, the optimal solutions of proposed method are considered as valid for large varieties of earthquake waves.

REFERENCES

[1] Ookouchi Y. , Takeuchi T. , Uchiyama T. , Suzuki K. , Sugiyama T. , Ogawa T. , Kato S. , 2006. "Experimental Studies of Tower Structures with Hysteretic Dampers", *Journal of the International Association for Shell and Spatial Structures*, Vol. 47, No. 3, pp. 229-236.

DEVELOPMENT OF R_Y, R_T FACTORS AND PROBABLE BRACE RESISTANCE AXIAL LOADS FOR THE SEISMIC DESIGN OF BRACING CONNECTIONS AND OTHER MEMBERS

Steven Cerri[a], Harrison Moir[b], Dimitrios G. Lignos[a]

[a]McGill University, Montreal, Quebec, Canada

steven. cerri@ mail. mcgill. ca, dimitrios. lignos@ mcgill. ca

[b] Atkins & Van Groll Inc. Consulting Engineering, Toronto, Ontario, Canada

harrison. moir@ mail. mcgill. ca

KEYWORDS: probable brace resistance, bracing connections, braced frames, post-buckling probable brace resistance, R_y, R_t factors

ABSTRACT

In the North America limit state design approach, steel brace connections must be designed to resist brace axial loads that correspond to the probable(expected) buckling strength and tensile yielding of the steel braces. In order to ensure that within these elements, the desired ductile mode of yielding governs and other undesirable brittle failure modes are precluded from the "capacity design" methodology, the true strength of the yielding members is required. As a general requirement for brace connections subjected to seismic loading, current seismic provisions[1,2] require that brace connections must resist the maximum tension load that the brace member can attain when its gross cross sectional area (A_g) yields. This requires a reliable estimate of the R_y factor, which represents the ratio of expected yield stress to the specified minimum yield stress(F_y) of the steel material. In order to account for the shear lag phenomenon in typical brace connections and avoid connection failure by circumferential fracture we need to enforce the steel brace net area(A_n) at the connection. To this end, the R_y and R_t factors of various steel materials that are commonly used in steel construction to design brace connections for seismic applications need to be better defined. These parameters also affect the probable compressive and post-buckling compressive resistance(C_u and C'_u, respectively) of braces subjected to inelastic cyclic loading. Therefore, more robust seismic design criteria need to be developed for beams of chevron bracing under maximum bending moment due to C'_u exterior and interior columns in brace configurations.

To this end, a steel brace database was developed in a consistent format and the measured material properties($F_{y,m}$, $F_{u,m}$) including the available compressive resistance(C_u), post-buckling resistance (C'_u) from various brace shapes were documented. This paper discusses the summary of key statistics for the development of comprehensive R_y and R_t factors for bracing connections designed for seismic applications including the mean, standard deviation, and the coefficient of variation, of actual-to-specified minimum yield and ultimate stresses per material type and shape. Furthermore, relationships are proposed that express the probable compressive resistance C_u as well as the probable post-buckling compressive resistance C'_u of compressive elements as a function of their global slenderness given a pre-defined ductility level(see *Fig. 1*). The proposed relationships are categorized based on the steel brace shape that is typically used in the North America steel design practice. The proposed relationships can be directly employed for the seismic design of beams, columns and connections other than brace connections as part of steel braced frames designed for moderate or high seismicity.

CONCLUSIONS

This paper summarizes a comprehensive assessment of an extensive steel brace database[3,4] of 295 bracing members from 24 different experimental programs in order to propose improved values for the expected yield stress to the specified minimum yield stress(R_y) and the expected tensile stress to the specified minimum ultimate stress(R_t) for various steel material types that are used in the current seismic design practice in North America with emphasis in Canada. Recommended equations for the probable post-buckling resistance of bracing connections and other members for various ductility levels are also proposed based on the brace shape that is employed in the design process of concentrically braced frames. The main findings are summarized as follows:

- For commonly used steel material types (i. e. , A500 Gr. B) for the fabrication of HSS steel braces in the North America the recommended R_y value by[1] is satisfactory. These values should also be considered in a future revision of[2].

- The compressive strength of various steel braces at first buckling C_u, generally agrees with the predicted values based on the compressive equation per CSA-S16 seismic provisions[2] when the proposed R_y, R_t values are employed.

- Equations have been proposed to predict the minimum probable brace compressive strength C'_u, at ductility levels of 2 and 3 depending on the brace shape that is used as part of the seismic design process of concentrically braced frames. The proposed equations match reasonably well the experimental data and illustrate that brace members with moderate global slenderness (i. e. , $0.5 \leqslant \lambda \leqslant 1.0$) have a post-buckling strength of at least 50% of their first buckling strength C_u, regardless of the brace shape.

- The current CSA-S16 seismic provisions[2] underestimate by more than 50% the post-buckling compressive resistance of steel braces for moderate global slenderness values.

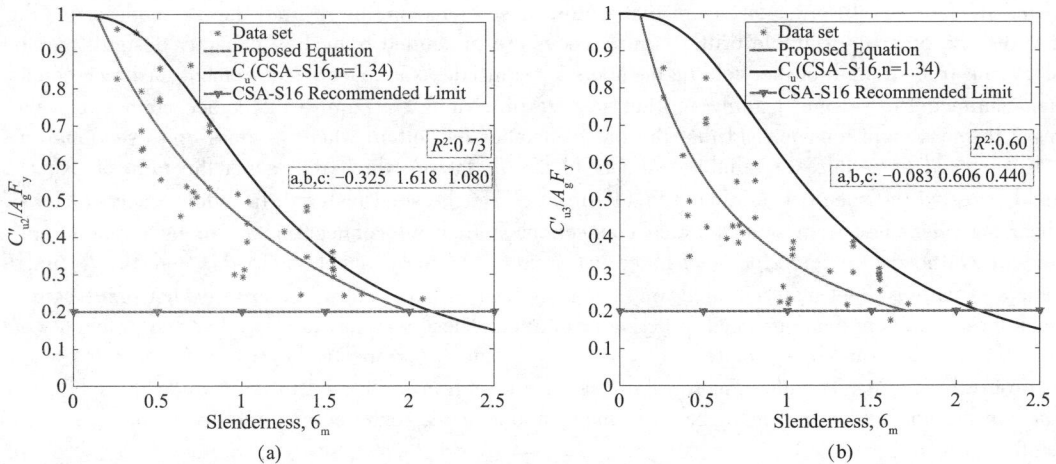

Fig. 1. Post-buckling compressive resistance for ductility levels 2 and 3
for round HSS and wide flange steel braces

ACKNOWLEDGMENT

This study is based on work supported by the Steel Structures Education Foundation (SSEF). This financial support is gratefully acknowledged. Any opinions, findings, and conclusions or recommendations expressed in this paper are those of the authors and do not necessarily reflect the views of sponsors.

REFERENCES

[1] AISC, "Seismic provisions for structural steel buildings", ANSI/ASIC 341-10, ANSI/AISC 341-10, Chicago, Illinois, 2010.

[2] S16-09, "Design of steel structures," Canadian Standards Association (CSA), Mississauga, Ontario, Canada, 2009.

[3] Lignos, DG, Karamanci E, Martin G. , "A steel database for modeling post-buckling behaviour and fracture of concentrically braced frames under earthquakes", *Proceedings 15th World Conference of Earthquake Engineering* (15WCEE), Lisbon, Portugal, 2012.

442

COST COMPARISON OF MRF, CBF AND EBF MID-HEIGHT STEEL BUILDINGS IN BOGOTÁ

Miguel Ángel Montaña[a], Francisco López-Almansa[b]

[a] PhD, Technical University of Catalonia, Barcelona, e-mail: miguelmontaa@ yahoo. es

[b] Professor, Technical University of Catalonia, Barcelona, e-mail:
francesc. lopez-almansa@ upc. edu

KEYWORDS: Colombia, Steel Structures, Seismic Performance, Pushover analysis, Construction Cost.

ABSTRACT

A number of mid-height steel buildings have been recently erected in Bogotá (Colombia). Their seismic risk might be high, given the new current microzonation of Bogotá and the lack of comprehensive previous studies. This study is carried out on eighteen representative prototype buildings. These eighteen buildings are generated by combining the values of three parameters: span-length (6 and 8 m), number of floors (5, 10 and 15) and earthquake-resistant systems (moment-resistant frames MRF, concentrically-braced frames CBF, and eccentrically-braced frames with chevron braces EBF). Each building is designed for ten seismic zones in Bogotá; therefore, 10 steel building are studied in this paper. In the framework of the Performance-Based Design, the seismic vulnerability of the buildings is assessed through 2-D pushover analyses. The structural properties of the buildings are compared, in terms of construction cost, to investigate the practical repercussions of the new regulation and of the considered parameters.

Fig. 1 displays plan views of the prototype buildings and Figure represents elevations of the 5-story buildings.

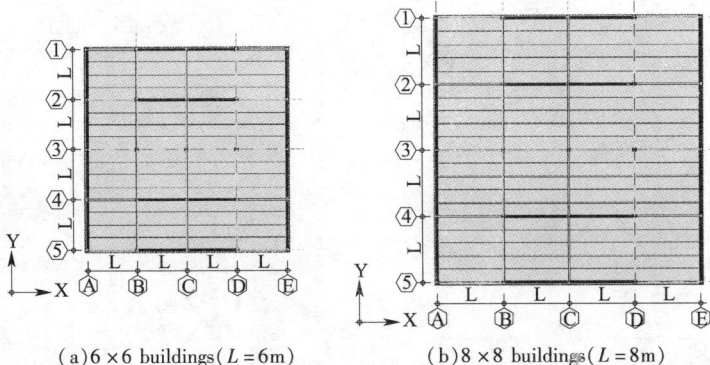

(a) 6 ×6 buildings ($L = 6$m)　　　(b) 8 ×8 buildings ($L = 8$m)

Fig. 1. Floor slab layout

CONCLUSIONS

Seismic zone. For the 5-and 10-story buildings, the construction cost in soft soil ("Lacustre" zones) is lower than in stiff soil ("Piedemonte" zones). The softer the soil, the lower the price. This trend can be explained by the higher spectral amplitudes in the range of periods of interest.

Building height. With few exceptions, for any span-length and structural type (MRF, CBF and EBF), the unit cost grows significantly with the height of the buildings. This effect can be attributed to the higher price of the elevation machinery and operations and, mainly, to the heavier structural members.

Span-length. Except for few cases (under the former microzonation) the unit cost for the 8 mspan-length buildings is higher than for the 6 m ones. This fact is mainly generated by the heavier beams and joists.

Structural type. In all the cases, the cost for EBF is significantly lower than for the other solutions; the price for EBF is about half the one for CBF. In general, the MRF option is cheaper than the CBF one.

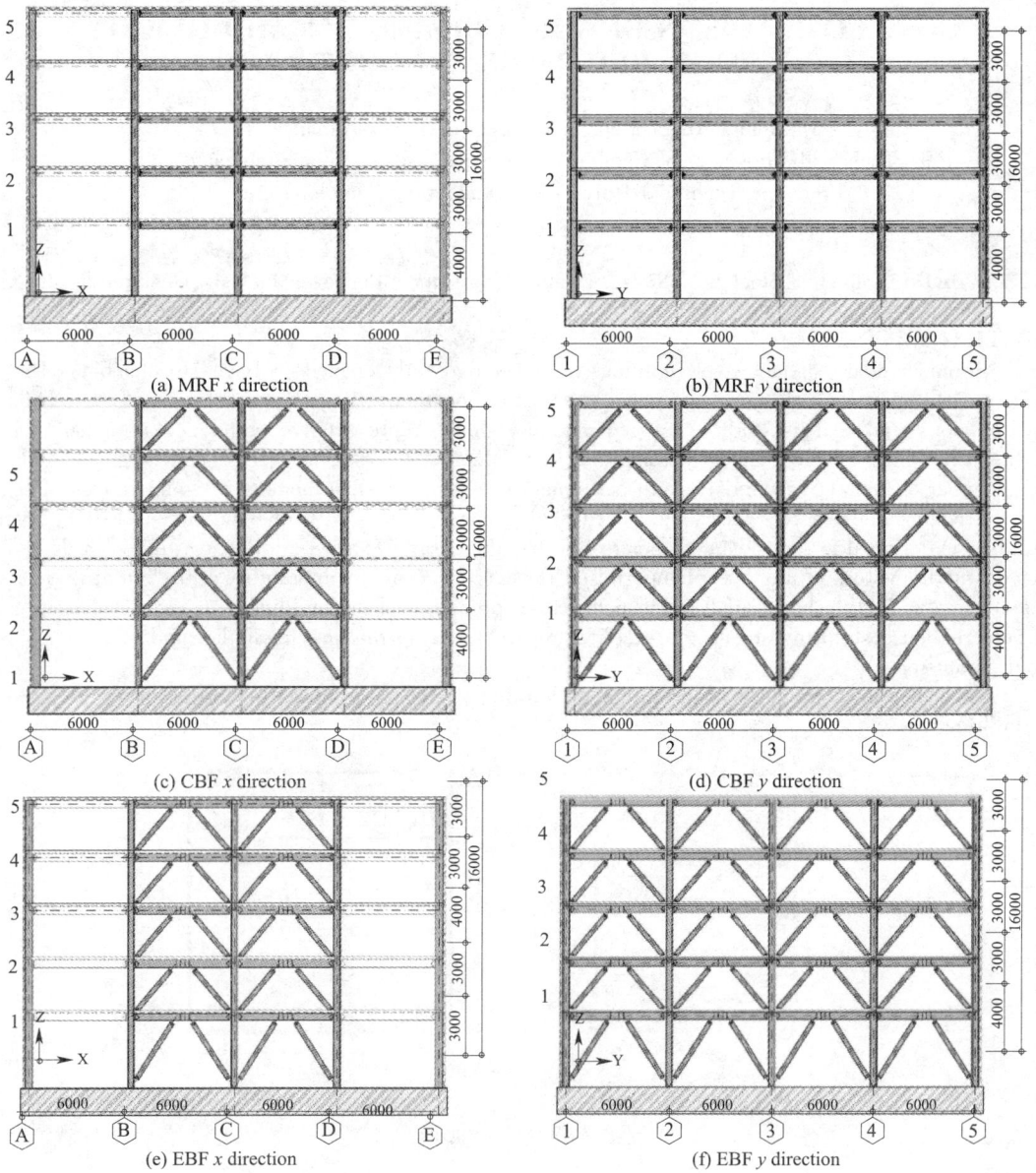
Fig. 2. Elevation views of the 5-story buildings

ACKNOWLEDGMENT

This work has received financial support from the Spanish Government under projects CGL2011-23621 and BIA2011-26816 and from the European Union(Feder).

REFERENCES

[1] López Almansa F, Montaña MA, 2014. "Numerical seismic vulnerability analysis of steel buildings in Bogotá, Colombia", *Journal of Constructional Steel Research*, Vol. 9(1), pp. 1-14.

[2] Montaña MA, 2014. "Seismic vulnerability analysis of mid-height steel buildings in Bogotá", Doctoral dissertation. Technical University of Catalonia.

AN APROACH FOR SEISMIC DESIGN OF BUILDINGS STRUCTUREDWITH ECCENTRICALLY BRACED FRAMES IN MEXICO

Alonso Gómez-Bernal[a], Antonio Gascón-Ramírez[b], Luis Aguilar Ugarte[b], Hugón Juárez-García[a]

[a]Universidad Autónoma Metropolitana Azcapotzalco, Departament of Materials, México
agb@ correo. azc. uam. mx, hjg@ correo. azc. uam. mx

[b]Universidad Autónoma MetropolitanaAzcapotzalco, División CBI, México
antonio_gascon@ hotmail. com, luisaguilarugarte@ gmail. com

KEYWORDS: Eccentrically Braced Frames; Link Overstrength Factor, Mexico earthquakes.

ABSTRACT

This paper provides a study that evaluates the behavior under seismic loading of steel buildings structured with Eccentrically Braced Frames(EBF). The evaluation was performed both by static non-linear analysis and nonlinear dynamic analysis. The research consists of a parametric study of EBF models with Shear Link of 3-story, 6-story and 12-story, designed according with the Mexican Construction Code[1]

Brace and column sizes in EBFs depend on the capacity of the Links. Braces are designed for the maximum deliverable forces from the adjacent link. Column demands, however, stem from links in all the stories above. The extent of simultaneous yielding and hardening of links determines the maximum axial force that can develop in the columns. Richards andUang[2] investigated the effect of axial loads in EBF columns using results from time-history analyses. They found that simultaneous link hardening and corresponding column demands exceeded values prescribed by the AISC Provisions[3] for columns at the base of the low rise frames and those located in the upper stories of high rise frames. Based on the results, suggestions to determine the loads more appropriate for columns of frames were made.

Inelastic analyses were performed using DRAIN2DX. Structure Models were subjected to different types of strong ground motions. Fourteen records of horizontal acceleration from earthquakes recorded in Mexico(10) and other countries(4) were used; the acceleration records were classified according to themotion type. Transform fault movement records: Mexicali Valley(CPE045 and DELS7910), and Northdridge(JEN292); with subduction source and containing high local effects: Chilpancingo (CHI1591), Mexico City(SCT1 B and TLHB8509); besides records close to the source: Zacatula (ZACA), SICARTSA (SICC), Manzanillo(MZ01), Caleta de Campos(CALE), Chile(Llolleo10), Taiwan. These ground motions were elected by contain the most damage potential recorded in Mexico.

The main objective of this article is to study the effect of the type of strong ground motion in Mexico on steel buildings with EBF. The impact of thebehavior of each frame element in the global response is analized. In the other hand, the variation of Link shear forces is evaluated, and axial loads in the columns are verified by performing analysis with some selected accelerograms from Mexican Earthquakes. In basis on the results it is possible give recommendations for design of EBF in Mexico.

CONCLUSIONS

The design of all buildings used in this study with EBFs and SMFs(ductility factor, $Q = 4$), was made with the requirements marked on the AISC seismic Provisions 2010, relative to EBF, as well as for the another frames without braces. Besides, the seismic design forces, were stablished by the Mexican Construction Code, RCDF; none of the buildings exceeded the limits of drift story ratios, required by the Mexican Code. The design capacity philosophy was applied in the design; and, ductile requirements for EBF and SMF($Q = 4$) was verify.

In the Nonlinear static analysis, when a distortion of 0. 037 rad was used on the pushover analysis, all buildings exceeded the ductility which were designed. 12-story models presented plastic joints in other elements in addition to the Links, as well as the rest of the not braced frames; this due to greater displacement specified for analysis. Maximum parameters values were presented when constant load pattern was applied. Displacements carried the same pattern in each step of the analysis. Maximum Drift Story usually arose at the first levels of each building.

In the Dynamic Analysis, no inelastic deformations were observed outside the potential plastic zones(links and the base ends of the first floor columns), except in 12-story models and for some small deformations in the central column and in some beams outside the link from the upper stories. During the analysis plastic deformations were registered in links of the first seven stories.

In all models, beam segments outside the links remained elastic during the analysis, except in the seventh and eighth stories of 12-story models, because they have smaller cross-sections than similar members of upper stories. Generally the greatest values of forces were reached at the link end(not at the column end) of the beam segment. Except for the second story central column, no inelastic deformations were registered in the columns outside the potential plastic zones near the base.

We note in models of 12-story, that when were applied acceleration records with epicenter close to the source(which is the case of external records), these earthquakes cause greater demands in the first two stories of the frame as the rest of the frame(are the biggest story drifts); besides, the shape of the curves from this earthquakes is very different respect the Mexican accelerograms, which are more or less uniform with regard to height. In addition, there is much difference between the values in the drifts story between the two groups, external earthquakes generate the greater values of displacement and distortion.

Simultaneous link yielding and hardening was common in the 3-story models. A hardening factor of at least 1.25 should be used for computing column demands in low-rise EBFs with short links, as was proposed by Richards andUang[2]. This hardening factor should also be considered for computing column demands in the top stories of a high rise EBF(Total number of stories/3). Lower stories of tall EBFs can be designed based on less link hardening in the stories above, since simultaneous yielding becomes less likely when more stories are considered, values of the hardening factor of 1.1 is appropriate for computing demands for the lower stories defined by 2/3 (Total number of stories), as is proposed by AISC2010[3].

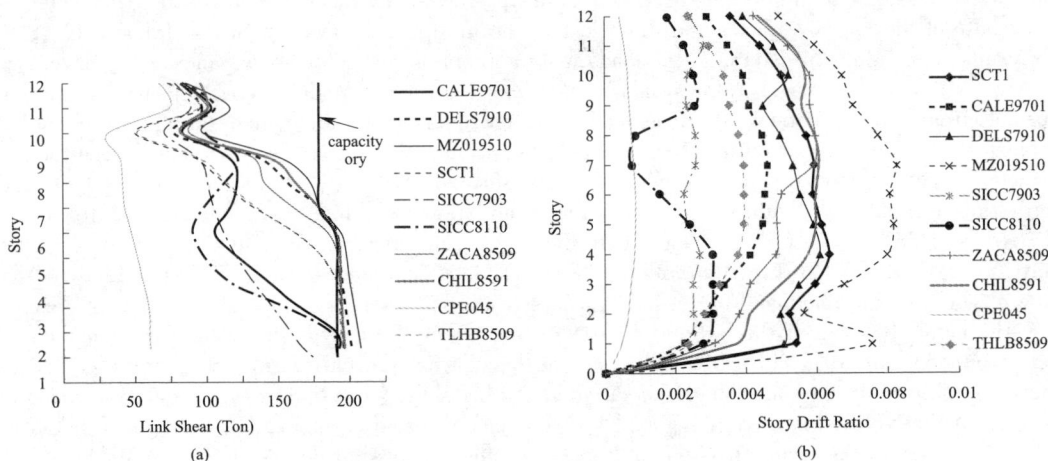

Fig. 1. (a) Maximum Shear Forces for Links in 12-story models;
(b) Maximum Drift Story ratio in 12-Story models,

REFERENCES

[1] Gaceta Oficial del Distrito Federal(2004), "Reglamento de Construcciones para el Distrito Federal", México.

[2] Richards, P., and Uang, C-M (2006). Capacity Based Design of Eccentrically Braced Frames. *Proc. 8th U.S. National Conf. on Earthquake Engineering*, Abril, 18-22, p 1100, San Fco., CA., USA.

[3] American Institute of Steel Construction, AISC(2010), "Seismic Provisions for Structural Steel Buildings," Chicago Il, USA.

SEISMIC LOSS ESTIMATION FOR EFFICIENT DECISION MAKING TO DESIGN MOMENT RESISTING FRAMES: EUROCODE 8 VERSUS TPMC

A. Longo[a] and V. Piluso[a]

[a]Department of Civil Engineering-University of Salerno

alongo@ unisa. it, v. piluso@ unisa. it

KEYWORDS: Mechanism Control, Moment Resisting Frames, Seismic Hazard, Fragility Curves.

ABSTRACT

In this paper a loss estimation methodology which provides various measures of seismic risk has been applied aiming to compare the seismic performance of Moment Resisting Steel Frames. A procedure for evaluating the seismic loss is herein briefly summarised and applied to MRFs designed according to different criteria. In particular, two design approaches are examined. The first one corresponds to the provisions suggested by Italian Seismic Code, while the second approach is based on a rigorous application of Theory of Plastic Mechanism Control which is devoted to the design of structures able to exhibit, in the ultimate conditions, a collapse mechanism of global type. The aim of the presented work is to focus on the use of seismic loss estimation for decision making and on the seismic assessment of the considered design procedures. A brief state of the art on seismic loss's estimation is given with reference to a specific study case of 5-storey's commercial office building which is designed by means of the two above mentioned methodologies.

CONCLUSIONS

Most of the current seismic design procedures are based on defining a "design earthquake" with prescribed probabilities of exceedance in a given time period. One of the main drawbacks of this procedure is that the probabilistic nature of the seismic excitations is only "accounted for" in the definition of the design response spectra, and not in the statement of performance objectives. In the recent years, specifically after the Northridge Earthquake, considerable research effort has been focused on defining probabilistic performance objectives that balance desirable structural performance and life cycle costs. As a consequence, the new trend in Performance Based Seismic Design (PBSD) is aimed at the evaluation of the seismic risk, defining a performance objective, in terms of (mean) annual frequency of exceeding a specified limit state, LS. In general, probabilistic PBSD should account for all the sources of uncertainty. In particular, in engineering problems, the sources of uncertainty are classified into two major groups known as aleatory uncertainty and epistemic uncertainty. The first identifies the more familiar "natural variability" such as the time and the magnitudes of future earthquakes in a region, record-to-record variability in acceleration time history amplitudes and phases, etc; the second group identifies the limitation of knowledge and data that professionals currently have about, for example, the modelling of structural systems in the highly non-linear range and the exact numerical values of parameters of physical and random (stochastic) models. In order to rigorously assess the seismic risk of a structure, all of the above uncertainties should be accounted for. Thus it becomes necessary that the problem of seismic risk is cast into a probabilistic framework which can propagate such uncertainties in each of the input variables and give a probabilistic output useful for decision making processes.

As an example, regarding the seismic input, a source of aleatory randomness is the variation in the level of ground motion observed at a site due to different ground motions resulting from the same rupture magnitude and source-to-site distance, while epistemic uncertainty results from the ground motion prediction equation used to estimate the level of a given seismic intensity measure at the site. Aleatory randomness is deemed to be an inherent property of complex phenomena, therefore it cannot be reduced; conversely, epistemic uncertainty is knowledge-based and, as a consequence, can be reduced provided that better knowledge of phenomena is acquired. As these two different uncertainties are related to different aspects of the considered problem they deserve a separate treatment within a decision making process.

This paper presents an application dealing with the use of a estimation method of seismic losses. It can be applied either for decision making at various stages of design or as a seismic assessment method. Reference is made to moment resisting frames designed according to two different design solutions (Theory of Plastic Mechanism Control proposed by[1] and Seismic Provision suggested by Italian Seismic Code[2]). A brief state of the art on seismic loss's estimation is given with reference to a specific study case of a 5-storey commercial office building. Using a case study structure, a full loss assessment is performed and discussion is given to each of the possible outputs for decision making. Some simplifying assumptions are made in the loss assessment in order to accomplish the goal of this manuscript which is to present the interpretation of loss assessment results for use in decision making. For sake of simplicity, this paper is primarily concerned with aleatory randomness only.

The seismic loss estimation methodology allows to compare the design approaches by means of a quantification of the seismic risk of engineered structures thus allowing consistent communication and rational decision making regarding the acceptance or mitigation of the seismic risk by means of different design criteria. The results of the application show that the structure designed according to Italian Seismic Provisions provide a major structural loss associated to the damage in the structural components when compared with the structure designed according to TPMC.

REFERENCES

[1] Mazzolani, F. M. , Piluso, V. , "Plastic Design of Seismic Resistant Steel Frames", Earthquake Engineering and Structural Dynamics, Vol. 26, pp. 167-191, 1997.

[2] NTC 2008-D. M. 14 Gennaio 2008: 《Nuove Norme Tecniche per le Costruzioni》(Italian Seismic Code).

[3] Aslani, H. "Probabilistic Earthquake Loss Estimation and Loss Disaggregation in Buildings". Ph. D. Thesis, John A. Blume Earthquake Engineering Centre, Dept. of Civil and Environmental Engineering Stanford University, 2005.

[4] Mitrani-Reiser, J. "An Ounce of Prevention: Probabilistic Loss Estimation for Performance-based Earthquake Engineering". Ph. D. Thesis, California Institute of technology, 2007.

[5] CSI 2007. SAP 2000: Integrated Finite Element Analysis and Design of Structures. Analysis Reference. Computer and Structure Inc. University of California, Berkeley.

[6] Longo A, Montuori R, Piluso V. (2008c). Influence of design criteria on the seismic reliability of X-braced frames. Journal of Earthquake Engineering; 12, 406-431, DOI: 10.1080/13632460701457231;

[7] Longo A, Montuori R, Piluso V. (2009). Seismic reliability of V-braced frames: influence of design methodologies. Earthquake Engineering and Structural Dynamics, vol. 38: 1587-1608, DOI: 10.1002/eqe.919.

[8] Pacific Earthquake Engineering Research Center, PEER Strong Motion Database, http://peer.berkeley.edu/smcat.

[9] Buchan, J. "Internal report of non-structural and contents inventories for the Christchurch Council Chambers building". Christchurch, New Zealand, 2007.

[10] Bradley, B. A. "SLAT: Seismic Loss Assessment Tool, version 1.12 user manual". Department of Civil Engineering, University of Canterbury, Christchurch, New Zealan, 2008.

[11] Porter, K. , Kennedy, R. and Bachman, R. "Creating Fragility Functions for Performance-based Earthquake Engineering". Earthquake Spectra. 23(2): 471-489, 2007.

SEISMIC DESIGN OF CFT-MRF AND BRBF STRUCTURAL SYSTEMS FOR STEEL BUILDINGS IN ECUADOR

Pedro P. Rojas[a], Mario E. Aguaguiña[a] and Ricardo A. Herrera[b]

[a]Escuela Superior Politécnica del Litoral, Facultad de Ingenierí
a en Ciencias de la Tierra, Ecuador
pprojas@ espol. edu. ec, maguagui@ espol. edu. ec

[b]Universidad de Chile, Facultad de Ciencias Físicas y Matemáticas, Chile
riherrer@ ing. uchile. cl

KEYWORDS: Seismic design, Structural system, Composite moment resisting frames, Concrete – filled steel tubes, Buckling-restrained braced frames.

ABSTRACT

Reinforced concrete is the preferred material used in Ecuador for most buildings. Moment Resisting Frames (MRFs) is the most common structural system chosen for reinforced concrete construction. However, steel construction has gained some popularity during the last 15 years. Most of the steel buildings constructed in this period are low-rise structures composed mainly of MRFs. This paper presents the initial steps of an ambitious research program to provide the local steel industry with other alternatives for structural systems to taller buildings. The four phases of this program are outlined. Composite Moment Resisting Frames consisting of concrete-filled steel tube (CFT) columns and steel beams (CFT-MRFs) as well as Buckling-Restrained Braced Frames (BRBFs) are the structural systems proposed in this paper to be considered in the seismic design of mid-rise and high-rise buildings. A CFT-MRF system has greater lateral stiffness compare to conventional steel systems and the interaction between steel and concrete results in members with larger axial and flexural capacity. On the other hand, BRBF System has shown stable hysteretic response and large energy dissipation capacity when subjected to cyclic loading since buckling-restrained braces (BRBs) are able to achieve yielding under tension or compression without buckling. Extensive analytical and experimental research has been carried out on these systems in several countries, mostly in the United States and Japan. In contrast, Ecuadorian seismic code does not include design provisions for CFT-MRFs or BRBFs due to a lack of knowledge relative to the analysis, design and construction of buildings using this type of structural systems. This paper reports the partial results from the first phase of the research program. A review of the most relevant aspects related to the analysis and design of CFT-MRF and BRBF Systems is provided. A design procedure was developed and detailing requirements were proposed for a 24-story prototype building, where CFT-MRFs and BRBFs constituted the main lateral load resisting system, based on the seismic design provisions and recommendations from American codes and related research. The paper also describes the recommendations for testing and qualification of CFT moment and BRB connections.

CONCLUSIONS

A research program to study analytically and experimentally the CFT-MRF and BRBF structural systems has been outlined. The research program seeks to provide the Ecuadorian steel industry with other alternatives for structural systems to mid-rise and high-rise buildings different from those of conventional steel and reinforced concrete construction. The research program has been divided in four phases. The main results of the first phase are presented in this paper. A 24-story prototype building, where the lateral load resisting system was comprised of CFT-MRFs and CFT-BRBFs, was designed. The results of the three-dimensional finite element model of the prototype building show that the structure satisfies the strength and drift requirements included in the codes. Future work shall address the evaluation of the seismic performance of the prototype building.

ACKNOWLEDGMENT

This research is based upon work to be presented to the Department of Civil Engineering atEscuela Superior Politécnica del Litoral(ESPOL) , Guayaquil, Ecuador in candidacy for the degree of Civil Engineer of Mister Mario E. Aguaguiña. The research reported herein was done with the participation of the ESPOL and the University of Chile. The findings, opinions and conclusions expressed in this paper are those of the authors and do not necessary reflects the views of the sponsors.

REFERENCES

[1] Cassagne A. ,2009. "Estado de la Práctica del Diseño y Construcción de Edificios Existentes de Acero Resistentes a Momento y Recomendaciones para la Construcción de Edificios Nuevos de Acero en la Ciudad de Guayaquil". *Tesis de Grado, Facultad de Ingeniería en Ciencias de la de la Tierra, Escuela Superior Politécnica del Litoral, Guayaquil, Ecuador.*

[2] Fahnestock L. A. , Ricles J. M. , R. Sause, 2006. "Analytical and Large-Scale Experimental Studies of Earthquake-Resistant Buckling-Restrained Braced Frame Systems". *ATLSS Reports, Paper 71, Bethlehem, PA, United States.*

[3] Herrera R. , 2005. "Seismic Behavior of Concrete Filled Tube Column-Wide Flange Beam Frames". *Ph. D. Dissertation, Department of Civil and Environmental Engineering, Lehigh University, Bethlehem, PA, United States.*

[4] Morino S. , Tsuda K. , 2003. "Design and Construction of Concrete-Filled Steel Tube Column System in Japan". *Earthquake Engineering and Engineering Seismology, Vol. 4, No. 1, pp. 51-73.*

[5] Muhummud T. , 2004. "Seismic Behavior and Design of Composite SMRFs with Concrete Filled Steel Tubular Columns and Steel Wide Flange Beams". *Ph. D. Dissertation, Department of Civil and Environmental Engineering, Lehigh University, Bethlehem, PA, United States.*

[6] Sabelli R. , López W. , 2004. "Design of Buckling-Restrained Braced Frames". *Proceedings of the North American Steel Construction Conference(NASCC) , Long Beach, CA, United States.*

[7] Hussain S. , Benschoten P. V. , Al Satari M. , Lin S. , 2006. "Buckling Restrained Braced Frame (BRBF) Structures: Analysis, Design and Approval Issues". *Proceedings of the 75th SEAOC Annual Convention, Long Beach, CA, United States.*

[8] American Institute of Steel Construction, 2010. "Seismic Provisions for Structural Steel Buildings" . *ANSI/AISC 341-10, Chicago, IL, United States.*

[9] American Society of Civil Engineers, 2010. "Minimum Design Loads for Buildings and Other Structures". *ASCE Standard ASCE/SEI 7-10, Reston, VA, United States, ISBN 978-0-7844-1085-1.*

[10] Habibullah A. , Wilson E. , 1997. "SAP2000: Integrated Finite Element Analysis and Design of Structures". *Computers and Structures, Inc. , Berkeley, CA, United States.*

[11] American Institute of Steel Construction, 2010. "Specification for Structural Steel Buildings" . *ANSI/AISC 360-10, Chicago, IL, United States.*

[12] American Institute of Steel Construction, 2010. "Prequalified Connections for Special and Intermediate Steel Moment Frames for Seismic Applications". *ANSI/AISC 358-10, Chicago, IL, United States.*

[13] Watanabe A. , Hitomi A. , Saeki E. , Wada A. , Fujimoto M. , 1988. "Properties of Brace Encased in Buckling-Restraining Concrete and Steel Tube". *Proceedings of the Ninth World Conference on Earthquake Engineering, Vol. IV, pp. 719-724, August 2-9, Tokyo-Kyoto, Japan.*

[14] Christopulos A. S. , 2005. "Improved Seismic Performance of Buckling Restrained Braced Frames". *M. S. Thesis, Department of Civil and Environmental Engineering, University of Washington, Seattle, WA, United States.*

[15] Palazzo G. , López-Almansa F. , Cahís X. , Crisafulli F. , 2009. "A low-tech dissipative buckling restrained brace. Design, analysis, production and testing" . *Engineering Structures, Vol. 31, pp. 2152-2161.*

SEISMIC BEHAVIOR OF TWO STEEL SOLUTIONS FOR APARTMNT EXTENSIONS IN THE CASE OF LARGE PREFABRICATED REINFORCED CONCRETE COLLECTIVE DWELLINGS

Miodrag Popov[a], Daniel Grecea[a,b], Adrian Dogariu[a], Viorel Ungureanu[a,b]

[a] Politehnica University of Timişoara, Department of Steel Structures and Structural Mechanics, Timişoara, ROMANIA

[b] Romanian Academy, Timisoara Branch, Timişoara, ROMANIA

popov. mio@ gmail. com, daniel. grecea@ upt. ro, adrian. dogariu@ upt. ro, viorel. ungureanu@ upt. ro

KEYWORDS: collective dwellings, prefabricated reinforced concrete panels, steel cantilevered extensions, seismic behaviour.

ABSTRACT

Almost 60 percent of the urban population of Romania lives in collective dwellings made out of large reinforced concrete panels and erected during the Communist regime. Nowadays, along with theincrease of life standards, the spatial conformation of the apartments no longer serves the contemporary living conditions, mainly because the small living surfaces and the enclosed and inflexible constructive systems. Solutions for horizontally extending the individual apartments have been applied, although these cannot provide, in many cases, any forms of long term structural guarantee.

This paper describesa solution of prefabricated steel-frame extension modules and two structural solutions that allow the flexible addition of these boxes to an existing collective dwelling for an increase of the apartments' interior surface(*Fig. 1*).

At the same time, a seismicstructural analysis of the existing building with several extensions is conducted, establishing the structural impact of the interventions and the appropriate conditions the two solutions can be applied.

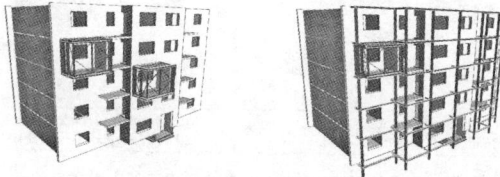

Fig. 1. Two structural solutions for attaching the prefabricated extensions: independent and global

CONCLUSIONS

The prefabricated extension modules described in this paperproves a reliable solution in response to the need of increasing apartment surfaces.

As demonstrated by the structural analysis taking into consideration an eccentric placement of the modules, none of theattaching solutions induces significant mass in comparison to the existing buildings. The additional foundation reactions are below the tolerable + 2% accepted for these buildings.

Due to the very small extension mass, the dynamic characteristics of the buildings remain almost unchanged. The first and second vibration modes are longitudinal and transversal with a period of 0. 103 sec for the initial building, and slightly increase at 0. 104 sec for the building with independent extensions(*Fig. 2*). When an entire frame, discreetly connected with the main building is used for supporting the extensions 0. 11 sec first period is obtained. Even so, the structure Eigen modes do not change. These facts underline the insignificant changes, introduced by the extension, in the dynamic characteristics of the buildings, extremely important features in case of an earthquake.

Further studies need to be undertaken referring axial forces that appear locally, in the connection region. For additional axial forces, reinforcement solutions are recommended.

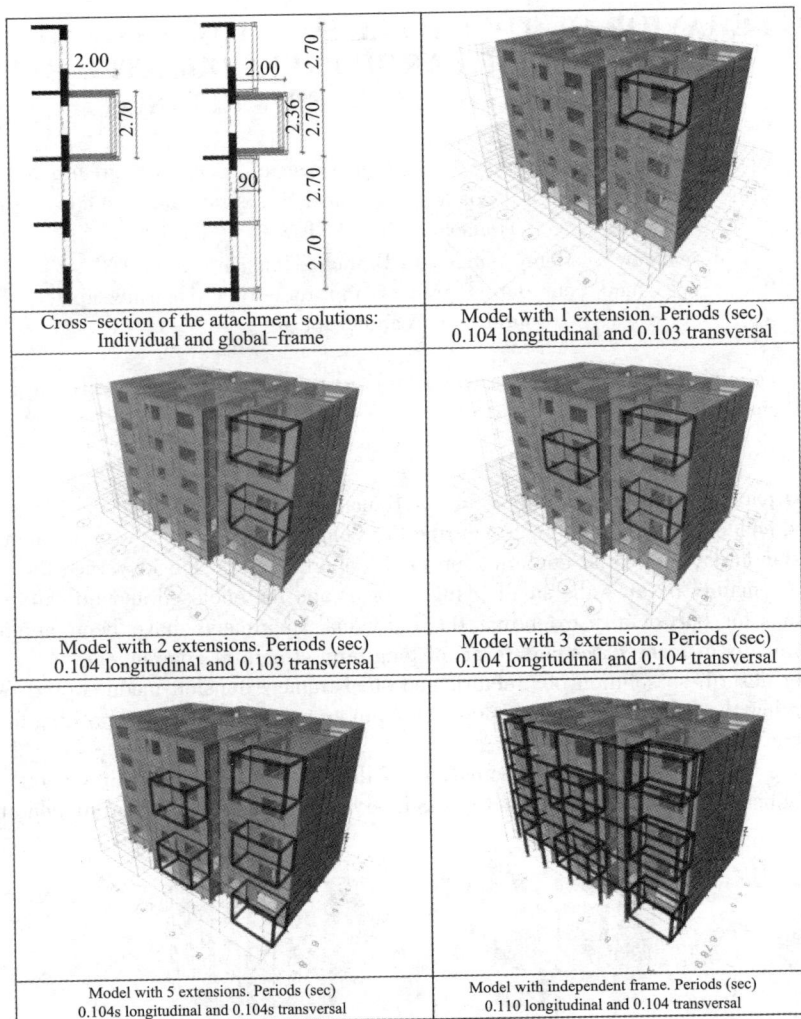

Cross-section of the attachment solutions: Individual and global-frame	Model with 1 extension. Periods (sec) 0.104 longitudinal and 0.103 transversal
Model with 2 extensions. Periods (sec) 0.104 longitudinal and 0.103 transversal	Model with 3 extensions. Periods (sec) 0.104 longitudinal and 0.104 transversal
Model with 5 extensions. Periods (sec) 0.104s longitudinal and 0.104s transversal	Model with independent frame. Periods (sec) 0.110 longitudinal and 0.104 transversal

Fig. 2. Two structural concepts for cantilevering the prefabricated extensions: independent and global

REFERENCES

[1] Popov M. , Szitar M. , Sămănță M. , "An Integrated Approach-Retrofitting the blocks of flats made of prefabricated panels" , *Romanian-Finnish Seminar on Opportunities in Sustainably Retrofitting the Large Panel Reinforced Concrete Building Stock-January 28 , 2013 , Timişoara , Romania* , edited by Ungureanu V. and Fülöp L. , Timişoara-Orizonturi Universitare , 159-175 , 2014.

[2] M. Economidou et al. , "Europe's buildings under the microscope-A country-by-country review of the energy performance of buildings" , Building Performance Institute of Europe , 39-50 , 2011.

[3] Botici A. A. , Ungureanu V. , Ciutina A. , Botici A. , Dubina D. , "Structural interventions for rehabilitation of precast large concrete panels for residential buildings" , *Proceedings of the 2nd International Conference on Protection of Historical Constructions. 7-9 May 2014 , Antalya , Turkey* , 133-139 , 2014.

BUCKLING RESTRAINED BRACE RETROFIT TECHNIQUE FOR EXISTING ELECTRIC POWER TRANSMISSION TOWERS

Marco Trovato[a], Li Sun[a], Bozidar Stojadinovic[a]

[a]Department of Civil, Environmental and Geomatic Engineering,
Swiss Federal Institute of Technology (ETH) Zurich, Zurich, Switzerland
marco. trovato@ gmx. ch, sun@ ibk. baug. ethz. ch, stojadinovic@ ibk. baug. ethz. ch

KEYWORDS: Transmission tower, Steel angle, Resilience, Retrofit, Buckling Restrained Bracing.

ABSTRACT

Structures that support high-voltage electrical power transmission cables are, mostly, light truss towers. These three-dimensional latticed structures are made using steel or aluminum bars, angles and/ or tubes. Buckling is the limit state that often controls the sizes and the strengths of the members, as well as the member bracing requirements. During wind, ice and earthquake disasters in the recent past in China and many other countries, transmission tower structures were seriously damaged resulting in expensive repairs and prolonged periods of electric power outages. This study was undertaken to improve the resilience of existing transmission tower to extreme natural hazard loads.

A retrofit strategy for the existing latticed transmission tower structures proposed in this study is somewhat different with respect to the conventional retrofit approach. First, the mechanical objective is not to increase the compressive strength of the restrained member but to provide it with more deformation capacity under compression. This is done to protect the existing tower connections and gusset plates. Second, the restraining casing needs to be placed on the existing transmission towers that may be in inaccessible locations. Thus, the restraining tube consists of two relatively light halves that are connected by in-situ bolting, and is not filled in with a filler material. This enables a single construction worker to climb the tower and perform this retrofit using only light battery-powered tools.

The test program included an unrestrained base-line specimen and three restrained specimens. A typical restrained specimen is showing in *Fig. 1*. Specimen data is shown in *Table 1* (the target gap means is the intended distance between the core angle and its restraining tube, in the direction of weak axis). The restraining casing consisted of cold-formed U-shapes or angles. The casing elements were connected using thread-forming screws that were installed using a light battery-operated screwdriver. The casing was secured against sliding along the braced angle using a stopper installed at midlength. The specimens were tested by monotonically compressing them in a uniaxial compression machine until failure.

Table 1. Dimensions of the specimen element.

No.	Section (mm)	Slenderness	b/t ratio	Casing	Target Gap (mm)
T-0	LNP 70 ×7	73.5	7.7		
T-1	LNP100 ×10	51.3	7.8	U 165/80 ×6	3
T-3	LNP 50 ×5	102.8	7.6	U 85/45 ×3	2.5
T-6	LNP 70 ×7	73.5	7.7	FLA105 ×8, LNP 75/50 ×7 FLB 205 ×8	2.5

The force-deformation response of the tested specimens is shown in *Fig.* 2. Compared to the baseline Specimens T-0, the retrofitted specimens exhibited significantly improved behavior. After reaching the buckling load, the core angle of the retrofitted specimens buckled and contacted the casing. Three contact points formed after some additional deformation of the core. This enabled compression load transfer from the core to the casing and effectively engaged the casing in carrying the compression load. The combined action of the core and the casing made it possible for the retrofitted specimens to maintain the compression load at the core buckling load level while undergoing a substantial shortening.

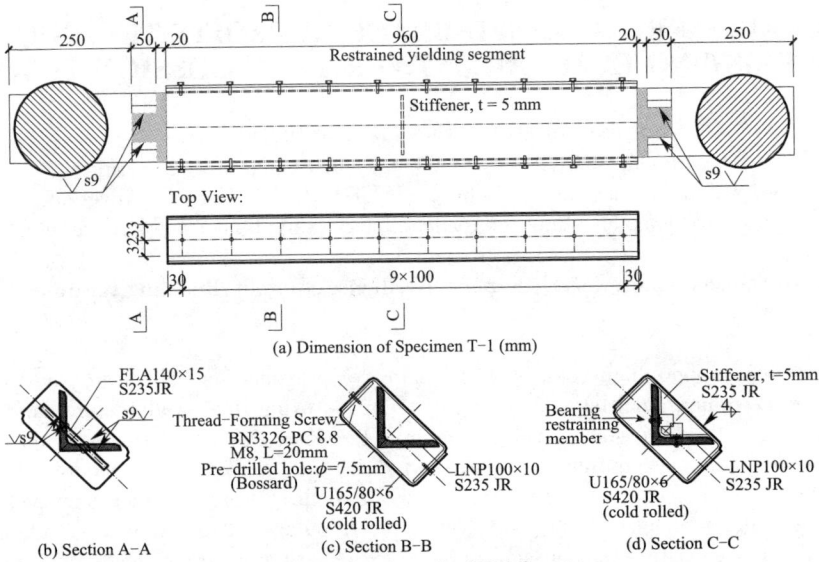

Top View:

(a) Dimension of Specimen T-1 (mm)

(b) Section A-A

(c) Section B-B

(d) Section C-C

Fig. 1. Design drawing of a BRB Specimen.

CONCLUSIONS

The test results on representative specimens show that:

1. The proposed casing can restrain buckling of the existing single angle truss members.

2. The thickness of the casing and the gap between the casing and the existing single angle truss member can be designed to maintain the resistance of the braced member at the buckling load of the core while enabling a substantial axial deformation.

3. A single worker can construct the proposed solution using battery-operated tools.

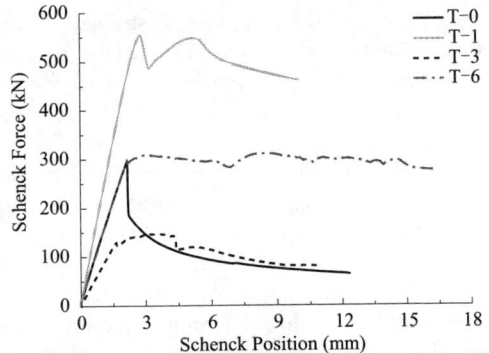

Fig. 2. Force-deformation response of the tested specimens.

ACKNOWLEDGMENT

The support and advice of the technicians in the Structural Testing Laboratory of the Institute of Structural Engineering(IBK), D-BAUG, ETH Zurich is greatly appreciated.

REFERENCES

[1] Inoue K., Sawaizumi S., Higashibata Y., 2001. "Stiffening requirements for unbounded braces encased in concrete panels". *Journal of Structural Engineering*, Vol. 127, No. 6, pp. 712-719.

[2] Sabelli R., Mahin S., Chang C., 2003. "Seismic demands on steel braced frame buildings with buckling restrained braces". *Engineering Structures*, Vol. 25, pp. 655-666.

[3] Usami T., Lu Z., Ge H., 2005. "A seismic upgrading method for steel arch bridges using buckling-restrained braces". *Earthquake Engineering and Structural Dynamics*, Vol. 34, pp. 471-496.

OUT-OF-PLANE SEISMIC DESIGN BY TESTING OF KNAUF DRYWALL PARTITIONS

Luigi Fiorino[a], Dominik Herfurth[b], Hans U. Hummel[b], Ornella Iuorio[a], Raffaele Landolfo[a], Vincenzo Macillo[a], Tatiana Pali[a], Maria Teresa Terracciano[a]

[a]Department of Structure for Engineering and Architecture, University of Naples "Federico II", Naples, Italy lfiorino@ unina. it, ornella. iuorio@ unina. it, landolfo@ unina. it, vincenzo. macillo@ unina. it, tatianapali@ unina. it, mariateresa. terracciano@ unina. it
[b]KnaufGips KG, Iphofen, Germany
herfurth. dominik@ knauf. de, Hummel. Hans-Ulrich@ knauf. de

KEYWORDS: Design by testing, drywall partitions, Knauf systems, non-structural components, out-of-plane monotonic tests, out-of-plane step-relaxation tests.

ABSTRACT

Past earthquakes have shown that the damage to non-structural elements can severely limit the functionality of most affected buildings and cause substantial economic losses compared to structural damage. Among the non-structural components, the ceiling-partition systems represent a significant investment in construction sector. Nevertheless, their seismic performance is poorly understood, because information and specific guidance are very limited. Since the behaviour of these systems cannot be easily simulated with traditional structural analysis, seismic design by testing is an effective procedure. In this framework, an important collaboration between KnaufGips KG Company and University of Naples "Federico II" is started in the last years, in order to overcome the lack of information about the behaviour and the design of lightweight drywall systems under seismic actions. In particular, this collaboration has led to the planning of a detailed research consisting in an extended experimental campaign, which is currently ongoing. The main research objective is to investigate the seismic performance of drywall components produced by KnaufGips KG, i. e. lightweight steel (interior and exterior) partition walls and suspended ceilings. The current paper presents the experimental program, with particular reference to the out-of-plane monotonic tests and step-relaxation tests on lightweight steel partitions, illustrating the wall specimens, the test set-up and instrumentation and the main experimental outcomes.

The experimental campaign is principally focused on the seismic behaviour of lightweight steel gypsum board partitions and on their interaction with other non-structural components, i. e. exterior walls and suspended continuous plasterboard ceilings, and structural elements. More information about the experimental activity are provided in[1]. In order to provide an answer to the Eurocode 8 Part 1[2] requirements, in terms of verification of non-structural components and systems and their connections to the building structure, out-of-plane tests on full-scale single drywall partitions are planned. The partition walls can be schematized as a simply supported beam with a concentrated force acting at mid-span. Therefore, in order to perform the codified verification, two parameter of the walls are necessary: its resistance and its fundamental vibration period. In order to evaluate the resistance of the investigated systems, three-point bending tests under quasi-static monotonic loads are carried out on walls. In addition, the dynamic tests, namely step-relaxation tests, are performed for the evaluation of the fundamental vibration period and the damping ratio of these systems.

Wall specimens are tested in horizontal position by using a set-up structure for three point bending tests (*Fig. 1*). The load is applied at wall mid-span through a hydraulic actuator. In particular, the actuator transfers the load to the wall by means of two beams: an upper beam, directly connected to the actuator, and a lower beam, that is connected to the upper beam and transfers the load to the specimen bottom face. The upper and lower beams are connected between them at both their ends by means of two different restraint systems, depending on test type to be carried out. In particular, in the case of step-relaxation tests, the restraint system is made of electromagnetic devices, which can be deactivated to a given load/displacement value for releasing the lower beam, in such a way to allow the wall free vibration.

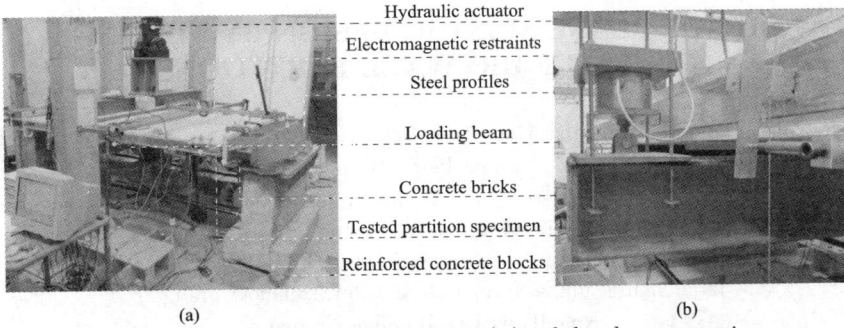

Fig. 1. Test set-up for out-of-plane tests(a) and the electromagnetic device used for step-relaxation tests(b).

The results of quasi-static monotonic tests show that the specimen initial response is not influenced by the joint types(fixed or sliding joints) and dowel types(plastic or steel dowels). In fact, the elastic strength and stiffness depend only on the stud spacing, which strongly increase (doubling the values) passing from the spacing of 600 mm to 300 mm(*Fig. 2*). The type of joints and dowels at the wall ends influences significantly the post peak response. In the case of fixed joints, after the occurrence of stud local buckling, the peak load corresponds at the flexural breaking of the gypsum panels and the ultimate behaviour depends on the dowel types. Walls with sliding joints exhibit lower deformation capacity than those with fixed joints(*Fig. 3*). The preliminary elaborations of the step-relaxation test results showed that the fundamental vibration period is about 0.07 s, whereas the damping ratio range between 3% and 8%.

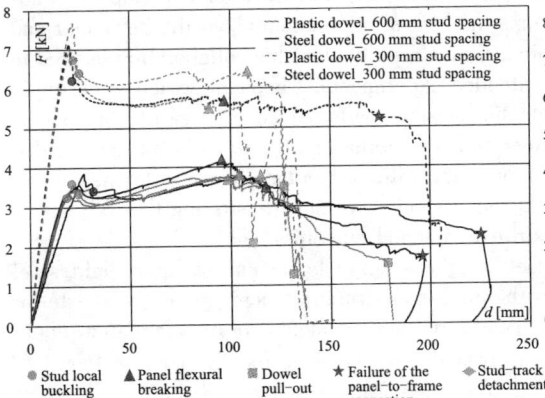

Fig. 2. F-d curve for walls with fixed joints.

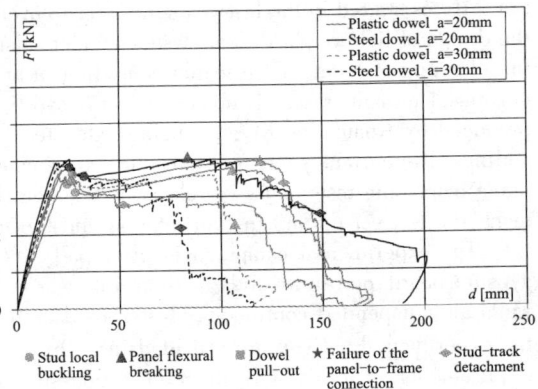

Fig. 3. F-d curve for walls with sliding joints.

CONCLUSIONS

An experimental campaign involving out-of-plane quasi-static monotonic tests and step-relaxation tests on Knauf drywall partition walls has been presented and discussed in the current paper. The obtained results can be considered the starting point for developing the out-of-plane seismic design assisted by testing of drywall partitions walls according to Eurocode 8 prescriptions.

REFERENCES

[1] Fiorino,L. ,Iuorio,O. ,Macillo,V. ,Terracciano,M. T. ,Pali,T. ,Landolfo,R. ,"Seismic response evaluation of non-structural drywall building components: planning of an experimental campaign", in proc. of 7th European conference on steel and composite structures Eurosteel 2014,2014.
[2] CEN,EN 1998-1-1,Eurocode 8,Design of structures for earthquake resistance-Part 1-1: General rules, seismic actions and rules for buildings,European Committee for Standardization,2005.

PERFORMANCE ASSESSMENT OF X-CBF DESIGNED ACCORDING TO AN IMPROVED (EC8 HT-BASED) APPROACH

Melina Bosco[a], Giuseppe Brandonisio[b], Edoardo M. Marino[a] and Elena Mele[b]

[a] Department of Civil and Environmental Engineering, University of Catania, Catania, Italy
mbosco@ dica. unict. it, emarino@ dica. unict. it

[b] Department of Structures for Engineering and Architecture, University of Naples, Naples, Italy
giuseppe. brandonisio@ unina. it, elenmele@ unina. it

KEYWORDS: concentric braced frames, capacity design, slenderness ratio, overstrength, dynamic response.

ABSTRACT

The paper analyses the seismic performance of steel structures with X Concentric Braced Frames (X-CBF) designed both according to Eurocode 8 and according to a new design method proposed by the authors in previous works. In detail, for validating the proposed design methodology, non-linear dynamic analyses have been carried out by using the computer code CLAP, with reference to several buildings characterized by different numbers of stories, numbers of braces in plan and by different structural grid dimensions. The results of the numerical analyses, expressed in terms of several parameters of structural performances, show that the proposed approach leads to: (i) reduce the unnecessary global overstrength due to the application of EC8 design rules; (ii) obtain an optimized structural design; (iii) ensure a good local and global ductility of braces under cyclic actions; (iv) have a greater "flexibility" in the design process.

CONCLUSIONS

The paper presents the results of non-linear dynamic analyses carried out by using the computer code CLAP, with the aim of validating the "Ω^* method" proposed by the authors in[2] for seismic design of steel buildings with X concentric braced frames, which is based on the EC8 design procedure, but "rationalized" thanks to the introduction some minor changes and the possibility of selecting the number of diagonals that contribute to the dissipative mechanism.

To this aim, boththe EC8 and the Ω^* design procedures are applied to 30 case studies, consisting of three and six storey buildings with the same floor plan (*Figs. 1*a and *1*b), and characterised by different structural solutions coming out from the different number of X-CBFs located along each principal direction of the building plan.

The non-linear dynamic analyses have been performed using a 2D numerical model, by considering the three seismic excitation levels corresponding to the probability of exceedance of 50%, 10% and 2% in 50 years. The results have been analysed in terms of structural performances indexes, appointed as *buckling index* (I-Sta) (*Fig. 1*c) and *yielding index* (I-Pla) (*Fig. 1*c), and in terms of ductility demands (m) of the braced frames and in terms of maximum storey drift (D_{max}).

The critical analysis of these performance parameters has confirmed what was underlined in[2] about the ability of the Ω^* method for obtaining X-CBFs characterised by a satisfactory seismic behaviour without introducing the "unnecessary" overstrength requirements emerged by applying the EC8 design procedure.

In fact, the flexibility of Ω^* method in the design process allows for optimized structural solutions (the structural weights are generally lower than the ones coming from the EC8 procedure, which does not give feasible solutions in several cases), and allows for obtaining a more uniform distribution of overstrength than the EC8, with a good local and global ductility of braced frames under cyclic actions comparable to the EC8 structural solutions.

Therefore, the possibility of Ω^* method of selecting the number of diagonals to be involved in the dissipative mechanism of the braced frame cannot lead to high level of damage and to potential damage concentrations (soft-storey mechanism), being the ductility demand generally comparable with the EC8 counterpart and always lesser that the limit values provided by the EC8[8].

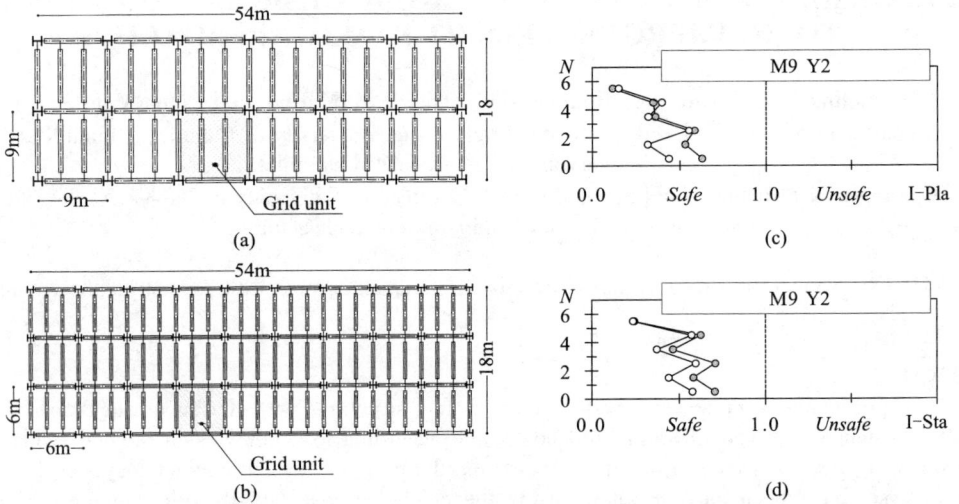

Fig. 1. Floor framing plan：(a)9 ×9m grid unit and(b)6 ×6m grid unit(M6)；
Yielding(I-Pla) (c) andbuckling(I-Sta) (d) indexes of columns for M9Y2
systems subjected to the 10%／50 years ground motions

REFERENCES

［1］ CEN. Eurocode 8(EC8)：Design of Structures for Earthquake Resistance-Part 1：General Rules，
Seismic Actions and Rules for Buildings，EN 1998-1. European Committee for Standardization：
Bruxelles，Belgium，2005.

［2］ Brandonisio G，Toreno M，Grande E，Mele E，De Luca A. Seismic design of concentric braced
frames. *Journal of Constructional Steel Research*，78：22-37，2012.

［3］ Ogawa K，Tada M，Computer program for static and dynamic analysis of steel frames considering
the deformation of joint panel. *Proceedings of the Seventeenth Symposium on Computer Technology
of Information*，*Systems and Applications*，Tokyo，1994. (in Japanese)

［4］ Somerville P et al. Development of ground motion time histories for phase 2 of the FEMA／SAC
steel project. SAC background document. Report No. SAC／BD-99-03，SAC Joint Venture，555 U-
niversity Ave.，Sacramento，1997.

［5］ Marino EM. A unified approach for the design of high ductility steel frames with concentric braces
in the framework of Eurocode 8. *Earthquake Engineering and Structural Dynamics*，2013.

［6］ Marino EM，Nakashima M. Seismic performance and new design procedure for chevron-braces
frames. *Earthquake Engineering and Structural Dynamics*，35：433-452，2006.

［7］ Tremblay R，Archambault M-H，Filiatrault A. Seismic response of concentrically braced steel
frames made with rectangular hollow bracing members. *Journal of Structural Engineering*，ASCE；
129(12)：1626-1636，2003.

［8］CEN. Eurocode 8：Design of structures for earthquake resistance-Part 3：Assessment and retrofitting of
buildings，EN 1998-3. European Committee for Standardization：Bruxelles，Belgium，2005.

［9］CEN. Eurocode 3(EC3)：Design of steel structures-Part. 1. 1：General rules and rules for build-
ings，UNI ENV 1993-1-1. August 2005.

［10］Brandonisio G.，De Luca A.，Grande E.，Mele E. Non-linear response of concentric braced
frames. Proceedings of the 5th International Conference on Behaviour of Steel Structures in Seismic Areas-*Stessa* 2006，399-405，2006.

Seismic, Wind and Exceptional Loads

STRENGTH AMPLIFICATION OF STRUCTURAL STEEL UNDER DYNAMIC CYCLIC LOADING DUE TO HIGH STRAIN – RATE

Yuko Shimada[a] ,Yu Jiao[b] ,and Satoshi Yamada[c]

[a] Chiba University,Japan
yshimada@ faculty. chiba-u. jp
[b] Tokyo University of Science,Japan
yujiao@ rs. tus. ac. jp
[c] Tokyo Institute of Technology,Japan
naniwa@ serc. titech. ac. jp

KEYWORDS : Steel , Strain-rate , Stress Increment , Cyclic Behavior , Dynamic Test

ABSTRACT

In the field of seismic engineering,it is known that high strain-rate might increase stress of steel. Considering real response of buildings under earthquakes,it is necessary to evaluate the characteristics of structural steel under high strain-rate. Previous studies related to strain-rate include high-speed tension-tests and cyclic loading tests with limited amplitudes. However,the detail of stress increasing affected by high strain-rate in theelasto-plastic behavior of structural steel under cyclic loading is not clear.

This study investigates the rise of stress due to high strain-rate by a series of high-speed cyclic loading test of structural steel used in general steel structures. In the test,the range of strain-rate is set in between semi-static to dynamic range,which is equivalent to the strain-rate of beams when earthquake occurs. Obtained results clarified that stress increasing effect of structural steel in dynamic hystereticbehaviors can be modeled by two functions; One is a function of stress rising according to strain-rate. The stress of the first yield point has linear increasing ratio with high strain-rate. And the other is a function of stress change ratio according to the cumulative plastic strain from the starting point of loading. It is based on that the stress rising during cyclic behavior after yielding caused by heat after making plastic.

Proposed stress rise model of structural steel based on high strain-rate is easy to adapt in the evaluation of seismic performance of steel buildings. The valid scope ofthise model is within the practical range of both cumulative plastic strain and strain-rate.

CONCLUSIONS

Based on the high-speed cyclic loading test,this study was obtained the tendency andaffector of stress increment of steel by strain-rate under not only monotonic loading but also cyclic loading. Stress of steel increased as strain-rate rises in spite of loading protocol,strain amplitude,and steel material.

Analysing detail of test results,stress increment ratio in the first yield point keep 1. 0 till 0. 1%/ s,and increase to the strain-rate after 0. 1%/s. And stress inelasto-plastic area decreased by heat of plastic behaviour and recovered its value by heat radiation during interval. Also,stress increment ratio just after beginning loading set through interval was stable in spite of cumulative strain. Therefore,it is clarified that strain-rate dependent component and plastic strain(heat) dependent component are independent in stress increasing in elasto-plastic area. Stress increment ratio considered strain-rate is expressed in a product between stress increment ratio in the first yield point by strain-rate and changing ratio of stress increment ratio by plastic strain specimen experienced as shown in *equation*(1).

$$\frac{\sigma_{dyn}}{\sigma_{sta}} = \alpha_y \beta \{0. 06 (1. 0 + \log_{10} + \dot{\varepsilon}) + 1. 0 \} \frac{-0. 4 (\sum_t \varepsilon_{prel} - 0. 2) + 1}{1. 08} \tag{1}$$

Here,σ_{sta} : first yield point by strain-rate under semi-static area,σ_{dyn} : first yield point by strain-rate in dynamic area, ε :strain-rate, $\sum_t \varepsilon_{prel}$: cumulative plastic relative strain after a loading set started

Coverages of this experimental equation are practical as 10%/s in strain-rate and 20% in cumulative plastic relative strain. This equation is able to obtain stress increment ratio even if strain-rate changes from moment to moment as spreading plastic area under random load such as earthquake motion. Moreover, behavior of steel member affected by strain-rate is represented easily and accurately on analysis and seismic performance is evaluated steel member not only static behavior but also effect of dynamic behaviour. It is a great contribution to seismic engineering.

ACKNOWLEDGMENT

This work was supported by Grant-in-Aid for Research Activity Start-up No. 22860014. The use of dynamic test faculty was supported by the collaborative research at Structural Engineering Research Center(S. E. R. C) in Tokyo Institute of Technology.

REFERENCES

[1] Manjoine M. J. , "Influence of strain rate and temperature on yieldstresses of mild steel" , Transactions of the ASME. Journal of applied mechanics, 11 : A211-A218 , 1944

[2] NagarajaRao N. , Lohrmann M. , Tall L. , "Effect of strain rate on the yield stress of structural steel" , ASTM Journal of Materials, 1 : 1, Publication No. 293, Fritz Lab. Reports Paper 1684 , 1966.

[3] Trembley R. , Timler P. , Bruneau M. , and Filiatrault A. , "Performance of steel structures during the 1994 Northridge earthquake" , Can. J. Civ. Eng. 22 : 338-360 , 1995.

[4] Saeki, E. , Sugisawa, M. , Yamaguchi, T. , Mochizuki, H. , and Wada, A. , "A study on low cycle fatigue characteristics of low yield strength steel " , J. Struct. Constr. Eng. , AIJ, 472 : 139-147 , 1995.

[5] Suzuki, T. , Sawaizumi, S. , Yamaguchi, T. , and Ikebe, T. , "Strain-Rate Effect on the Mechanical Properties of Low-Yield Steel" , Summaries of Technical Papers of Annual Meeting Architectural Institute of Japan, C-1 : 501-502 , 1997.

[6] Yamada, S. , Yamaguchi, M. , Takeuchi, Y. , Takeuchi, T. , and Wada, A. , "Hysteresis Characteristics of Steel Material for Dampers based on Dynamic and Cyclic Loading Tests : Experimental study on cyclic behavior of steel material for dampers considering a strain rate dependence Part 1 " , J. Struct. Constr. Eng. , AIJ , 553 : 121-128 2002.

[7] Suita, K. , Kaneta, K. , Kohzu, I. , and Yasutomi, M. , "Effect of Strain Rates on Hysteresis Characteristics of Steel Structural Joints due to High-Speed Cyclic Loadings" , J. Struct. Constr. Eng. , AIJ , 463 : 95-104 , 1994.

[8] Shimada Y. , "Database on steel hysteretic characteristic of steel affected by strain-rate" , Proceedings of Constructional Steel , JSSC , 20 : 1-8 , 2012.

[9] Osakada K. , "Applied Plastic Dynamics-Dynamics of plastic deformation and finite element analysis" , Baifukan, Co. LTD , ISBN 978-4-563-06737-3 , 200430 September-2 October, 2014 , Torino, Italia , ISBN 978-88-905870-0-9 , Vol. 2 , pp. 1015-1024.

INFLUENCE OF EARTHQUAKE DAMAGE ON PASSIVE FIRE PROTECTION AND STRUCTURAL FIRE BEHAVIOUR

Markus Knobloch[a] and Mario Fontana[b]

[a] Ruhr-Universität Bochum, Institute of Steel, Lightweight and Composite Structures, Germany
markus. knobloch@ rub. de

[b] ETH Zürich, Institute of Structural Engineering, Switzerland
mario. fontana@ ethz. ch

KEYWORDS: Fire after earthquake, Fire protection, Robustness

ABSTRACT

Structural fire protection should ensure an adequate fire safety of tall buildings. Passive fire protection is applied to structural members and forms an encasement to reduce the heating of the load-bearing structure by insulating the member. In particular, steel structures usually demand fire protection to avoid structural failure in case of fire. Earthquakes may cause partial damage or even loss of fire protection as well as subsequent fires. Particularly, lightweight fire protections, like spray applied protection or intumescent coatings may be vulnerable if exposed to earthquakes. Local damage of the fire protection leads to increased heating of load-bearing members in case of a subsequent fire and decreases the fire resistance of protected structural members. Current fire design methods usually do not consider damage of fire protection caused by earthquakes. Additionally, design rules often neglect causal relations of extreme events, like fire after earthquake. However, fire after earthquake is neither independent from the preceding action nor uncommon. Many fires have been reported after previous earthquakes e. g. in Japan, New Zealand or California.

The vulnerability and robustness of fire protection are important for the structural safety of buildings. The use of fire protection possessing low vulnerability helps avoiding damages that may cause increased heating and failure of structural members as well as potential progressive structural collapse and follow-up failures as business interruptions etc.

The vulnerabilities depend on the different types of fire protection. Steel structures can be protected by fire protective sprays and intumescent coatings directly applied to the members, by fire protective claddings forming an encasement, or by steel-concrete composite sections. Steel-concrete composite sections obtain a high resistance against damage due tothe mechanical action caused by an earthquake. Hence, the vulnerability of steel-concrete composite sections is lower compared to sprays, coatings and claddings.

Experimental and numerical studies[1,2] on the fire resistance of steel columns with partial loss of fire protection showed that even small areas of damaged fire protection might lead to a significant reduction of structural fire resistance. The failure of the steel columns with damaged fire protection either by flexural buckling or by yielding is strongly depending on the size of the local damage.

The fire safety strategy for tall buildings has to fulfil the general fire safety objectives:

- Safety of occupants and fire brigade,
- Safety of neighbours and their property,
- Limitation of financial loss(building, contents and follow-up costs), and
- Protection of the environment in case of fire.

To fulfil these objectives, it is of particular importance to prevent progressiveor global collapse. Global collapse of tall buildings has large indirect consequences, like a large number of fatalities, property loss, business interruption etc. Fire following an earthquake has much greater potential of causing progressive collapse than fire alone. Earthquakes may

1. Decrease the remaining structural resistance due to large inelastic deflections and partial damage of the load-bearing structure,

2. Cause partial loss of fire protection decreasing the fire resistance of the remaining structural members and

3. Cause subsequent fire.

Furthermore, alternative load paths may be destroyed due to the damaged structure caused by the earthquake. All these effects should be considered for the realistic analysis of steel structures in fire following an earthquake.

If the robustness of the fire protected structural steel member(member level) is insufficient to prevent the failure of additional structural members, the failure of one or several stories of a multi-storey building may occur(global structure level). If the lower part of the building is not able to resist the additional large amount of energy due to the failure of one or more stories, a progressive collapse of the entire structurewill follow causing enormous consequences not only for the owner and occupants of the building but for society in general(business interruption etc.).

The following options exist to prevent progressive collapse of exceptional building structures subjected to fire caused by an earthquake:

■ The remaining structure(lower part) can be designed to carry the additional dynamic impact load caused by the fall of the upper part resulting from the failure of one storey, or

■ The structure can be designed to prevent the failure of several structural members after earthquake and subsequent fire to prevent the failure of a storey.

A case study[3] analysed these concepts for reducing the risk of progressive collapse of building structures. The study considered fires caused by accidental actions, like earthquakes, initial damage of the load-bearing structure and a partial damage of the fire protection. The study showed benefits of the concepts with a higher resistance against initial damage. Additionally, it may be more cost-effective to design the structure to prevent the failure of structural members than to design the structure to withstand the failure of an entire storey. The additional costs for a structural design considering the causal relation of earthquakes and fires is often justified by the enormous consequences caused by the failure of tall building structures.

CONCLUSIONS

Earthquakes may cause a partial damage of both the load-bearing structure and the fire protection. Furthermore, earthquakes have a high potential to cause subsequent fires. Local damage of fire protection leads to increased heating of load-bearing members in case of fires after earthquake and may distinctly decrease the fire resistance of protected structural steel members, thus increasing the risk of structural failure and progressive collapse of building structures. Hence, the use of fire protection resistant to both the mechanical impact of an earthquake and the thermal impact of a fire is important for the structural fire safety. For tall building structures, it is particularly important to prevent structural failure and progressive collapse to fulfil the fire safety objectives and avoid serious consequences, like fatalities, property loss and business interruption.

REFERENCES

[1] Wong WW, Li GQ. , "Behavior of steel columns in a fire with partial damage to the fire protection", Journal of Constructional Steel Research, 65:1392-1400, 2009.

[2] Fontana M, Knobloch M. , "Fire resistance of steel column with partial loss of fire protection", Proceedings of the IABSE Symposium Metropolitan Habitats and Infrastructure, Shanghai, pp. 352-354, IABSE, Zurich 2004.

[3] Faber MH, Kübler O, Fontana M, Knobloch M. , "Failure Consequences and Reliability Acceptance Criteria for Exceptional Building Structures ", IBK Report No. 285, ETH Zürich, Zurich, 2004.

THE EFFECT OF EARTHQUAKE CHARACTERISTICS ON THE LOCALIZED BEHAVIOR OF MOMENT CONNECTIONS UNDER FIRE

Hussam Mahmoud[a], Mehrdad Memari[a], Collin Turbert[b]

[a]Colorado State University, Department of Civil and Environmental Engineering, USA
hussam. mahmoud@ colostate. edu, memarim@ engr. colostate. edu
[b]Nelson Engineering, USA
cturbert@ nelsonengineering. net

KEYWORDS: Seismic, Fire, Multi-Hazard, Local Performance

ABSTRACT

The occurrence of fires igniting during and immediately following a seismic event is a significant issue in highly active seismic regions. Structures which have been exposed to earthquake-induced damage are particularly vulnerable to the effects of fire loading and the combination of events represents a unique scenario that is not often considered during the design process. Records from historical earthquakes show that given the right conditions, the damage caused by the subsequent fire can be much worse than thedamage caused by the seismic action itself, this being true for both single buildings and whole regions. Therefore, there is a need to understand and quantify the response of steel frames to the combined events of earthquake and fire in high seismic regions. The steel structure presented in this paper include line-element model of a 16-story frames with RBS connections, which allowed for assessing the global behavior of the frames as well as the constraint provided by the frame to specific beam-column subassembly for analysing the local behavior of the connections. Exterior connection at 2nd floor of the frame was chosen to be analysed in detail to assess the local behavior of the joints. Commercial FE analysis software, ABAQUS, is utilized to conduct the analyses.

CONCLUSIONS

The occurrence of fire following earthquakes can lead to substantial structural damage [1]. This study utilized nonlinear time-history analysis to evaluate the response of steel reduced beam section connections under the cascading hazards of earthquake and fire. This was realized through the development of detailed 3D finite element models of the connections with realistic boundary conditions representing the restraints provided by the remainder of the frame connected to the connection. The temperature-dependent material properties were adopted from ASCE 7-10[2] and density was assumed to be constant. A multi-step sequential analysis was developed to simulate the seismic response of the beam-column subassembly and subsequent thermal loading during post-earthquake fire exposure. The procedure was designed to transfer the imposed stresses and strains from the previous analysis step (earthquake) to the next (fire) in order to capture the effect of residual damage during the fire simulation. The beam-column subassembly model was exposed to thermal loading in its undamaged state in order to determine the expected temperature distributions within the nodes. Fire proofing was taken into consideration during this step by only directly exposing the steel members to the ambient fire temperature where damage to fire proofing was expected to have occurred during the earthquake. This was based on observed damage patterns from experimental studies conducted by Braxtan[3], which found that the SFRM insulation in moment frame beam hinge regions debonded, cracked, and spalled during inelastic seismic response, leaving the bare steel exposed to the ambient temperature.

In order to accurately capture the demands imposed by the surrounding frame on the subassembly, a flexible restraint model was developed to provide a more realistic estimate of the changing mechanical restraint provided by the frame during the fire analysis. For this approach, additional finite element models were developed for the adjoining steel framework and examined throughout the entirety of the fire simulation for each earthquake scenario. The simulation was terminated at twenty representative locations throughout the fire simulation and a series of unit deformation-controlled analyses were performed to evaluate the force-displacement relationship of the subassembly surrounding framework. The response data was then used to calibrate nonlinear restraint elements for the subassembly boundary conditions imposed during the post-earthquake fire simulation. These elements were continually updated throughout the fire simulations to capture the change in restraint provided by the frame throughout the fire. A depiction of this procedure can be seen in *Fig. 1*.

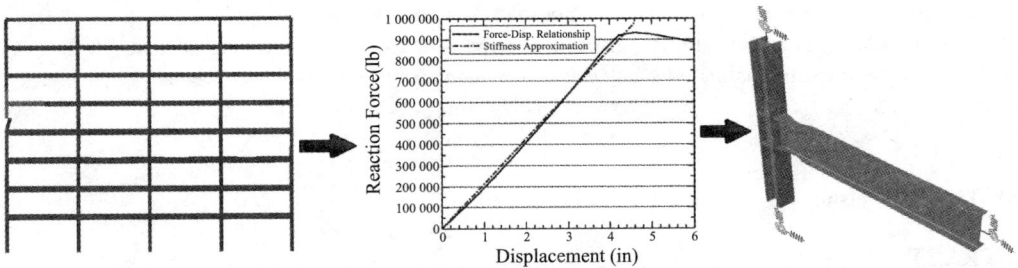

Fig. 1. ABAQUS stiffness model configuration

The demand imposed on the subassembly during the earthquake simulation was very variable. Yielding occurred in several key locations for all of the models, including the weld access hole and the reduced beam section. In addition, permanent deformation was observed and comprised of web and flange local buckling in the reduced beam section. An example of such is depicted in *Fig. 2*(a). For the beam-column subassembly, the demand imposed during the subsequent fire analysis did produce significant deformation in all of the models, particularly in the reduced beam section. Both lateral torsional buckling as well as localized web buckling was generated during the fire simulations. This behavior is shown in *Fig. 2*(b).

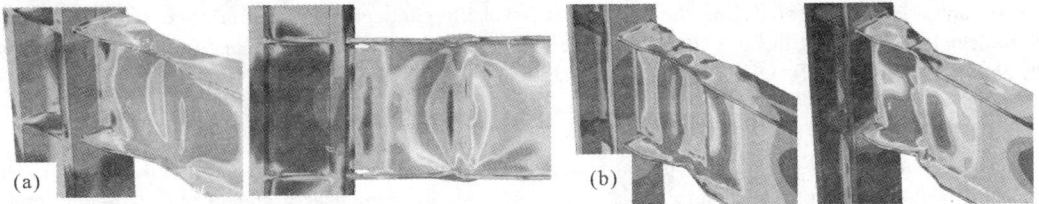

(a)   (b)

Fig. 2. (a)Observed damage during the earthquake simulation;
(b)Example of observed damage during the fire portion of the FFE simulations for the beam-column subassembly

The results showed that although reductions in strength do not occur until temperatures of 400 degree Celsius are reached, thermal expansion starts immediately with the increase of temperature. The combination of thermal expansion and high axial stiffness in the early stages of a fire results in large compressive forces early in the heating. The analyses showed that high axial compressive forces often occur at temperatures as low as 200 to 400°C. The magnitude of the compression force appears to be limited by the local buckling capacity of the beam at elevated temperatures. During the cooling phase of the fire and after complete cool down, very large tensile forces can develop in the beam, column and the connection, which could potentially cause connection failure. Significant residual forces and deformations may remain in the beam and column after a fire.

REFERENCES
[1] Scawthorn C. R. ,2008. "The shakeout scenario supplemental study: fire following earthquake", *SPA Risk LLC*.
[2] ASCE/SEI 7-10,2010. "Minimum design loads for buildings and other structures", *American Society of Civil Engineers*, Reston, VA.
[3] Braxtan N. L. ,2009. "Post-earthquake fire performance of steel moment-frame building columns". *Ph. D. Dissertation*, *Department of Civil and Environmental Engineering*, Lehigh University, Bethlehem, PA.

DESIGN CONSTRAINTS FOR THE OPTIMAL STRUCTURAL DESIGN OF SUPER TALL BUILDINGS UNDER EARTHQUAKE AND WIND

Xin Zhao[a], Xiang Jiang[b], Yaomin Dong[b]

[a] Department of Structural Engineering, Tongji University, Shanghai, CHINA

Tongji Architectural Design(Group) Co. , Shanghai, CHINA

22zx@ tjadri. com

[b] Department of Structural Engineering, Tongji University, Shanghai, CHINA

786379623@ qq. com, dongluoyaomin@ 163. com

KEYWORDS : Optimal structural design constraints, seismic design, wind design, super tall buildings.

ABSTRACT

Earthquake actions and wind loads for super tall buildings are significant due to the great building height and huge building mass, and the structural design of a super tall building is commonly controlled by horizontal loads. The optimal seismic design constraints are different from the optimal wind design constraints in many aspects, such as design criteria, minimum design measures and structural responses. The optimal structural design constraints for super tall buildings has to consider the structural performance requirements under both earthquake actions and wind loads. This paper addresses the differences between the optimal seismic and optimal wind design constraints for super tall buildings under different structural performance requirements, say stiffness, strength, stability and human comfort. The controlling factors for different structural performance requirements are thoroughly discussed for both seismic and wind design. Based on the above discussion, this paper obtained a comprehensive comparison of design constraints for optimal structural design by considering both seismic and wind design requirements for the reference of super tall buildings. A real super tall building of more than 500m high is applied as an example in the final part of this paper to show the specific constraints requirements for the optimal structural design of super tall buildings under both earthquake actions and wind loads.

CONCLUSIONS

Through constraints analysis of both the seismic and wind performance design constraints for super tall buildings under different structural performance requirements, say stiffness, strength, stability and human comfort, we can obtain a comprehensive comparison of design constraints for optimal structural design by considering both seismic and wind design requirements. The constraints considered for seismic and wind performance design are listed respectively in *Table 1* and 2.

Table 1. Seismic performance constraints.

Seismic performance	Design constraints	Component constraints	Assembly constraints	Global constraints
Strength	Capacity	√		
Stability	Stability of structural members	√		
Ductility	Compressive axial ratio of columns and walls	√		
Multiple guard lines	Proportion of frame shear to story shear		√	
Overall stiffness	Story drift			√
Overall strength	Shear-weight ratio			√
	Torsion displacement ratio			√
Overall regularity	Lateral stiffness ratio			√
	Shear capacity ratio			√

Table 2. Wind performance constraints.

Wind performance	Design constraints	Component constraints	Assembly constraints	Global constraints
Strength	Capacity	√		
Stiffness	Shear deformation	√		
Overall stiffness	Story drift			√
Overall Comfort	Wind induced acceleration			√

We can find thatcapacity of structural members and story drifts are checked in both the seismic and wind performance design. Extra performance constraints considered in seismic design are stability of structural members, compressive axial ratio of columns and walls, proportion of frame shear to story shear, shear-weight ratio, torsion displacement ratio, lateral stiffness ratio and shear capacity ratio. Constraint of wind induced acceleration is only considered in wind performance design.

Case study in this paper shows the specific constraints requirements for the optimal structural design performance design constraints of super tall buildings under both earthquake actions and wind loads and gives a good reference to the similar projects. We can come to the conclusion that:

1) Capacity of structural members and story drifts need to be checked in both the seismic performance design and wind performance design. The differences lies in load combination factors. The capacity of structural calculation should confirm to relevant codes or computing method;

2) Extra performance constraints considered in seismic design than wind design are stability of structural members, compressive axial ratio of columns and walls, proportion of frame shear to story shear, shear-weight ratio, torsion displacement ratio, lateral stiffness ratio and shear capacity ratio;

3) Constraint of wind induced acceleration is only considered in wind performance design;

4) The constraints mentioned can all be taken as the optimal design constraints in the computational optimization. Once the relationships between constraints and design variables are established, the optimal design scheme of tall buildings can be easily obtained;

5) Last but not least, the determination of optimal constraints vary a lot due to the differences of performance design objectives in the optimal design. Specific analysis of each specific design problem is necessary.

ACKNOWLEDGMENT

The authors are grateful for the support from the Shanghai excellent discipline leader program (No. 14XD1423900)and Key Technologies R & D Program of Shanghai(Grant No. 09dz1207704).

REFERENCES

[1] Xu, P. F. , Dai, G. Y. , Performance-based seismic design of tall building structures beyond the code-specification [J]. *Chinese journal of civil engineering*, 38(1): 1-10, 2005.

[2] J G J 3-2010. Technical specification for concrete structures of tall buildings Beijing, Chinese building industry press, 2010 .

[3] Bertero, R. D. , Bertero, V. V. , Performance-based Seismic Engineering: the Need for a Reliable Conceptual Comprehensive Approach, *Earthquake Engineering and Structural Dynamics*, 31:627-652, 2002.

[4] Ding, J. M. , Zhao, X. , Chao, S. , Performance based seismic design for the Shanghai Tower, *7th International Conference on Urban Earthquake Engineering*(*7CUEE*)& *5th International Conference on Earthquake Engineering*(*5ICEE*), 2010

[5] Wang, M. P. , Zhou, X. Y. , Performance-based seismic design of building structures [J]. *Building Structure*, 33(3): 59-61. , 2003.

[6] Chan, C. M. , Optimal lateral stiffness design of tall buildings of mixed steel and concrete construction. *The Structural Design of Tall Buildings*, 10(3), 155-177, 2001.

[7] Chan, C. M. , Huang, M. F. , Kwok, K. C. S. , Stiffness optimization for wind-induced dynamic serviceability design of tall buildings [J] . *Journal of Structural Engineering*, 135 (8): 985-997, 2009.

[8] Huang, M. F. , Wind-induced vibration performance-based optimal structural design of tall buildings [J]. *Engineering Mechanics*, 30(2): 240-246, 253, 2012.

COMBINED TUNED DAMPER BASED WIND AND EARTHQUAKE VIBRATION CONTROL FOR SUPER TALL BUILDINGS

Lilin Wang[a], Yimin Zheng[b], Xin Zhao[b]

[a]Department of Structural Engineering, Tongji University, Shanghai, CHINA

22zx@ tjadri. com

[b]Department of Structural Engineering, Tongji University, Shanghai, CHINA

Tongji Architectural Design(Group) Co. , Shanghai, CHINA

KEYWORDS: Seismic performance, wind-induced responses, combined tuned damper, super tall buildings.

ABSTRACT

The wind-induced vibrations of super tall buildings become excessive due to strong wind loads, super building height and high flexibility. Tuned mass damper(TMD) and tuned liquid column damper (TLCD) have been widely used to control wind-induced vibration for super tall buildings for decades. This paper conducts an implementation assessment and then proposes a combined supplemental damping system which contains both TMD and TLCD. First of all, therationality and correctness of the CTD system are illustrated in theory. Then, the formulation of the CTD system is obtained based on the equations of TLCD and TMD. Finally, when it comes to the design of CTD, there are two aspects to be considered, one of which called human comfort performance design under fluctuating wind and the other called safety performance design under strong wind and earthquake. A case study is conducted for the recommended damping system installed in a real super tall building project, which focuses on the predicted performance and responses under varying conditions, specifically various seismic and wind events. Performing this analysis on the damping system ensures that it will perform as intended throughout a range of operating conditions. Meanwhile, the analysis results suggested that by installing the recommended damping system, the building accelerations can be reduced to the desired levels. Considering both the economical advantage of the TLCD system and the high efficiency of the TMD system, the CTD system is a competitive option for wind-induced vibration control of super tall buildings.

CONCLUSIONS

The parameters of CTD for the case are shown in *Table 1*.

Table 1. Damping system basic parameters.

Classification	Value	Note
Target Frequency	0. 11	Hz
Location of TMD	Lever 138	Elevation 623
Location of TLCD	Lever M04	Elevation 623
TMD Mass	500	Tone
TLCD effective Mass	500	Tone
Generalized Mass Ratio of CTD	1. 3	%
Frequency Ratio	99. 1	%
Tuned Range	95% to 115% of Normal Frequency	
Viscous Damping Devices(VDDs)	$F_{VDD} = C_{VDD} * V^2$	
Snubbing Devices	$F_{SVDD} = C_{snubberVDD} * V^2$	

The TMD was designed to provide a desired amount of acceleration reduction under wind excitation, and the performance is optimized for moderate winds(i. e. wind events with return intervals 10 years). However, knowledge of the loads developed within the TMD system and those between the TMDs and local structure are necessary for design against all possible types of strong excitation. In this project, different hazard level winds and seismic excitations were considered. For this TMD system, 2500-year seismic loads were found to dominate the ultimate strength design.

1. *Wind responses*

Acceleration Reduction: The acceleration reduction has been optimized for the 10-year return period winds. Due to the limited space, TMD is designed to have best possible performance within the allocated space. The analysis results suggest that the implementation of this damping system(TMD + TLCD) reduces peak 10-year acceleration at the top occupied floor of the tower by approximately 25%, i. e. from 28 milli-g to 21 milli-g for top occupied floor; from 18 milli-g to 14 milli-g for top residential floor in the critical wind direction, as shown in below.

Table 2. Acceleration Reduction.

Maximum 10-year Acceleration(m/sec2)	Without Damper	With Damper	Criteria
Top Occupied Floor(Hotel)	27. 6	20. 7	25
Top Residential Floor	18. 4	13. 8	15

Loads: To determine the forces that need to be resisted by the damping system and local structure, analyses are conducted at 500-year return period wind levels. The corresponding responses will be applied for the ultimate strength design of the TMD and structural design of the building.

Table 3. Load Check of wind response.

500-year wind events	TMD Amplitude(m)	Cable Force/each group(2 cables)	Column Force(kN)/ each	VDD Force(kN)/ each
TMD Response/Loads	1. 3	1200	1000	140

2. Seismic responses

Loads: The damping system has been designed to avoid damage to either the structure or itself during seismic events. Analyses are performed at earthquake levels of probabilistic 2%-in-50-year PGA(200gal) values. The corresponding responses will be applied for the ultimate strength design of the TMD and structural design of the building.

Table 4. Load Check of seismic response.

2500-year seismic events	TMD Amplitude(m)	Cable Force/each group(2 cables)	Column Force(kN)/ each	VDD Force(kN)/ each
TMD Response/Loads (with snubbing)	2. 0	2400	2200	1100

3. Sunbbing system

Larger TMD displacements during higher return period wind or seismic events must be controlled to ensure that local damage is not caused. The snubbing system is specified to arrest excessive TMD motions with the use of eight shock absorbers. The current design is not expected to snub due to wind action at return periods up to and including 500 years. Seismic response strongly governs most of the displacement and force response for most of the TMD components. The snubbing system is designed to engage at the estimated TMD displacement corresponding to a 2500-year return period earthquake event and beyond. To incorporate the snubbing forces into the TMD system, dynamic models of the TMD with highly detailed geometryneed be established and the new model accounts for the mass moment of inertia of the TMD, the force applied by the snubbing system, the stiffness of the cables, etc.

ACKNOWLEDGMENT

The authors are grateful for the support from the Shanghai Excellent Discipline Leader Program (No. 14XD1423900) and Key Technologies R & D Program of Shanghai(Grant No. 09dz1207704).

REFERENCES

[1] Simiu E, Scanlan R. H. Wind effects on structures(2nd Edition). John Willey & Sons, 1986.
[2] Y. L. Xu, B. Samali and K. C. S. Kwok. Control of Along-wind Response of Structures by Mass and Liquid Dampers. ASCE Journal of Engineering Mechanics. 1992, 118(1): 20 ~ 39.
[3] MOTIONEERING INC. (Canada). Simple Pendulum with Variable Restoring Force: CA, 2391683[P] , 2003-12-26.
[4] Wang Z M. Vibration Control of High rising Structures [M]. Shanghai: Tongji University Press, 1997. (in Chinese).

THE BEHAVIOR OF SPHERICAL DOMESUNDER WIND AND EARTHQUAKE ACTION

Shuai Xu[a], Zhihua Chen[b] and Federico M. Mazzolani[c]

[a]Department of Civil Engineering, Tianjin University, Tianjin, China
sxu. tju@ hotmail. com, zhchen@ tju. edu. cn
[b] Key Laboratory of Coast Civil Structure Safety, Ministry of Education, Tianjin, China
[c]Department of Structures for Engineering and Architecture,
University of Naples "Federico II", Naples, Italy
fmm@ unina. it

KEYWORDS: Dome structure; Wind load; Earthquake; Numerical analysis.

INTRODUCTORY REMARKS

Due to their outstanding advantages, large-span latticed domes are widely used both in civil and industrial fields. Due to the high sensitive of these kind of structures to buckling phenomena, the effects of non symmetrical loading conditions due to wind and earthquake represent a key issue in the design of this kind of structures. Serious damage of domes was caused by extreme weather conditions, mainly due to strong wind. Examples of collapse of domes due to earthquake were not found in the literature, but in principle it is not correct to state 'a priori' that the wind load is always more severe than earthquake. In fact it is possible, for a given shape and material of the dome, to identify the regions, characterized by different intensities of the wind and earthquake actions, where the seismic effects are the severest ones and, therefore, must be considered in the design process.

In the technical literature, the behaviour of domes under wind and earthquake action has been investigated separately, without comparing both effects. The scope of the present paper is to fill this gap and to identify some behavioural zones for spherical domes subjected to wind load and horizontal earthquake actions, considering various dome shapes with different height-to-diameter ratios and made of two constructional materials(steel and aluminium).

RESULTS OF THE ANALYSIS

Different symbols are used to presentall the analyzed results in order to compare the degree of danger of different loading conditions as shown in *Fig. 1*. The effect of wind and earthquake on the six domes as summarized in *Fig. 2.* , where the horizontal axis is the region of several accelerations and the vertical axis is the region with different wind pressures.

	Dome 1 H/D =1/6, AL	Dome 2 H/D =1/6, S	Dome 3 H/D =1/4, AL	Dome 4 H/D=1/4, S	Dome 5 H/D =1/2, AL	Dome 6 H/D =1/2, S
Earthquake is more severe than wind	●	■	▲	◆	⬟	▼
Wind is more severe than earthquake	○	□	△	◇	⬠	▽

Fig. 1. Case studies

According to the symbols of *Fig. 1*, the risk map coming from the analysis is shown in the following *Fig. 2*. The main result is that in the low seismicity zones wind represents the most severe loading condition. As far as the degree of seismicity increases, the wind becomes less important. For a given combination of wind and earthquake, the increasing of self-weigh going from aluminium to steel produces more severe conditions in case of earthquake.

CONCLUSIONS

Based on the analyzed results, it is obvious that the dome is more sensitive to wind load than earthquake action as far as the shape becomes shallower. For many loading cases discussed in this paper, the wind load is more severe than the earthquake action. For high height-to-diameter steel domes,

2.0
1.6
wind load
(kN/m²) 1.2
0.8
0.4

0.05　0.1　0.2　0.3　0.4　0.5　0.6　0.7　0.8　0.9　1.0
earthquake acceleration (a/g)

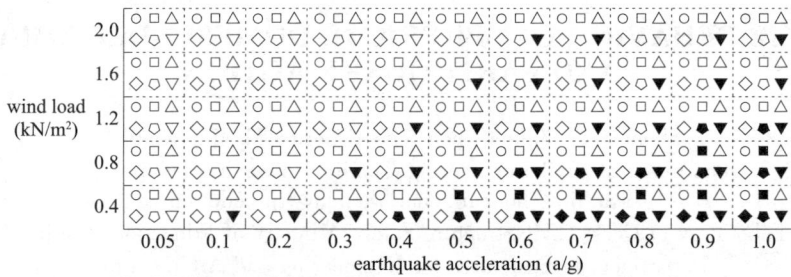

Fig. 2. Different degree of danger due to wind and earthquake

earthquake becomes more severe than wind in the regions of high seismic activity. As expected, the effect of earthquake on steel domes is always more significant than the one on aluminium domes, which means that aluminum domes are more suitable to be used in zones of high seismic activity.

REFERENCES

[1] YasushiUematsu, Osamu Kuribara, Motohiko Yamada, Akihiro Sasaki, Takeshi Hongo. "Wind-induced dynamic behavior and its load estimation of a single-layer latticed dome with a long span", *Journal of Wind Engineering and Industrial Aerodynamics*, 89:1671-1687,2001.

[2] Beles, A. A. , and Soare, M. V. "Some Observations on the Failure of a Dome of Great Span". In R. M. Davies(editor), *Space Structures*, Proc. 1st Int. Conf. Space Struct. , University of Surrey, 419-423,1966.

[3] Mazzolani, F. M. "3D aluminium structures", *Thin-Walled Structures*,61(SI):258-266,2012.

[4] Li Y. Q. , Tamura Y. "Equivalent static wind load estimation in wind-resistant design of single-layer reticulated shells", *Wind Structure*,8(6):443-454,2005.

[5] Kim J. Y. , Yu E. , Kim D. Y. , Tamura Y. "Long-term monitoring of wind-induced responses of a large-span roof structure", *Journal of Wind Engineering and Industrial Aerodynamics*,99, Issue 9: 955-963,2011.

[6] Li Y. Q. , Tamura Y. , Yoshida A. , Katsumura A. , Cho K. "Wind loading and its effects on single-layer reticulated cylindrical shells", *Journal of Wind Engineering and Industrial Aerodynamics*, Vol. 94, Issue 12: 949-973,2006.

[7] Shiro Kato, Takashi Ueki, Yoichi Mukaiyama. "Study of dynamic collapse of single layer reticular domes subjected to earthquake motion and the estimation of statically equivalent seismic forces", *International Journal of Space Structures*: 191-204,1997.

[8] Koichiro Ishikawa, Shiro Kato. "Elastic-plastic dynamic buckling analysis of reticular domes subjected to earthquake motion", *International Journal of Space Structures* (3/4): 205-215,1997.

[9] Shen S. Z. "The dynamic stability problem of reticular shells", Invited Lecture, *IASS-APCS Symposium*, Taibei, China: 44-46,2003.

[10] Mazzolani, F. M. "Two Twin Aluminium Domes of the Enel Plant in Civitavecchia (Italy)", *Proc. of the 11th International Aluminium Conference INALCO 2010 ' New Frontiers in Light Metals'*, Eindhoven,23-25,2010.

[11] "SAP2000 Tutorial Manual", *Computers and Structures*,2009.

[12] "EN 1991-1-3: Eurocode 1: Actions on structures-Part 1-3: General actions-Snow loads", *European Committee for Standardization*,2003.

[13] "EN 1991-1-4: Eurocode 1: Actions on structures-Part 1-4: General actions-Wind actions", *European Committee for Standardization*,2005.

[14] "EN 1998: Eurocode 8: Design of structures for earthquake resistance", *European Committee for Standardization*,2004.

EXPERIMENTAL STUDY OF HIGH-PERFORMANCE STRUCTURAL STEEL Q345GJ UNDER CYCLIC LOADING

Gang Xiong[a], Bo Yang[a], Le Shen[a], Ying Hu[a], Shidong Nie[a] and Guoxing Dai[a]

[a]Mountain town construction and the ministry of education key laboratory of new technology;
School of civil engineering, Chongqing university, China
e-mails: xionggang76@ 126. com, yang0206@ cqu. edu. cn, ryun2010@ 163. com,
y. hu@ cqu. edu. cn, 380999766@ qq. com, dgx1018@ 126. com

KEYWORDS: High-performance structural steel Q345GJ; Cyclic loading; Stress-strain skeleton curves; Material constitutive model

ABSTRACT

High-performance structural steel has been utilized in some steel-framed structures. The investigation about high performance steel has already been conducted in the literature, but its constitutive relationship under cyclic loading has not been studied sufficiently. Fifteen coupon specimens of high performance steel(Q345GJ) were tested under monotonic and cyclic loading with eleven different loading conditions, and the corresponding stress-strain relationships under these different loadingprotocols have been obtained. The high-performance steel constitutive model, mechanical properties, deformation and ductility have been studied. The stress-strain skeleton curves of high-performance structural steel Q345GJ under cyclic loading are achieved by curve fitting using the modified Ramberg-Osgood model. The results are important for analyzing the behaviour of high-performance structural steel under seismic loading.

CONCLUSIONS

Table 1. Summary of the material characteristics from test data

Type	E_s(GPa)	f_y(MPa)	f_u(MPa)	N_p	E_n(N·mm)
Monotonic tensile(average)	201	451	590	-	545327
Monotonic compression(average)	206	451	587	-	29382
H3	193	449	580	9	335188
H4	183	433	573	20	688969
H5	198	438	605	8	298621
H6	197	442	579	9	319173
H7	213	456	554	16	314925
H8	199	457	616	18	530642
H9	202	454	579	8	44616
H11	208	510	580	102	2311375

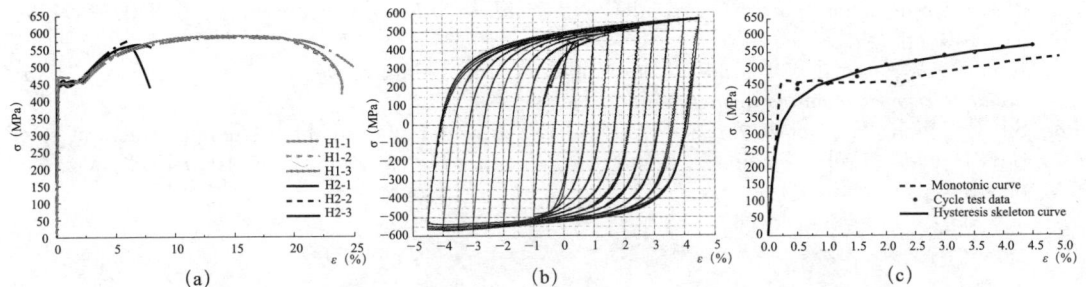

Fig. 1. (a) Monotonic tests; (b) Hysteresis curve of H4; (c) Hysteresis skeleton curve of H4

Table 1 and Fig. 1 show some material parameters and curves from test data. It can be seen from Fig. 1(c) that the skeleton curve is above the Monotonic curve, due to cyclic hardening. And Fig. 2

shows that there are two main failure modes, one is fracture and other one is buckling. Under cyclic loading, the specimens gradually accumulated fatigue damage and finally fractured, so the specimens had no significant voice and necking when fractured, which is quite different from the failure mode of monotonic tensile specimens.

Some conclusions can be drawn from this study:

(1) Q345GJ steel has no significant voice and necking phenomenon when fatigue fracture occurs.

(2) The skeleton curve of Q345GJ steel under cyclic loading is obviously different from the tensile curve under monotonic load. It also suggests that the material parameters obtained from monotonic tensile curves cannot be used for the Q345GJ steel performance analysis under the cyclic loading.

(3) The degradation characteristics of Q345GJ steel under cyclic loading contain macro buckling and micro damage.

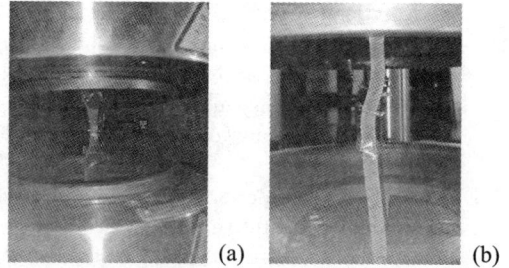

(a)   (b)

Fig. 2. (a) Failed by fracture; (b) Failed by buckling

ACKNOWLEDGMENT

The authors gratefully acknowledge the financial support provided by the National Natural Science Foundation of China (No. 51078368), the Fundamental Research Funds for the Central Universities (No. 0218005208011), and the Start Up Grant of Chongqing University (No. 0218001104414 & No. 0218001104410).

REFERENCES

[1] Nie S. D. , Dai G. X. , Yang B. , et. Al, 2009. "Series of high-performance GJ structural steel development and application in engineering construction". *Industrial Construction ISSN*, 1671-5799, Vol. S1, pp. 135-141. (in chinese)

[2] Shi Y. J. , Wang M. , Wang Y. Q. , 2011. "Experimental and constitutive model study of structural steel under cyclic loading". *Journal of Constructional Steel Research*, ISSN 0143-974X, Vol. 67, pp. 1185-1197.

[3] Shi G. , Wang M. , Bai Y. , et al, 2012. "Experimental and modeling study of high-strength structural steel under cyclic loading". *Engineering Structures*, ISSN 0141-0296, Vol. 37, pp. 1-13.

[4] Zhou F. , Chen Y. , Wu Q. , 2015. "Dependence of the cyclic response of structural steel on loading history under large inelastic strains". *Journal of Constructional Steel Research*, ISSN 0143-974X, Vol. 104, pp. 64-73.

[5] Wang Y. B. , Li G. Q. , Cui W. , et al, 2015. "Experimental investigation and modeling of cyclic behavior of high strength steel". *Journal of Constructional Steel Research*, ISSN 0143-974X, Vol. 104, pp. 37-48.

[6] Jia L. J. , Kuwamura Hitoshi, M. ASCE, 2013. "Prediction of cyclic behaviors of mild steel at large plastic strain using coupon test results". *Journal of Structural Engineering*, ISSN 0733-9445, Vol. 140, pp. 1-9.

[7] Ramberg W, Osgood W R, 1943. "Description of stress-strain curves by three paramet ers". *National advisory committee for aeronautics*.

[8] Shi G. , Wang F. , Dai G. X. , et al, 2012. "Experimental study of high strength structural steel Q460D under cyclic loading". *China Civil Engineering Journal*, ISSN 1000-131X, Vol. 45, pp. 48-55. (in chinese)

Author Index

A

Abbas Syed Murtuza 313
Abu Anthony 181
Abubakar S. 177
Aguaguiña Mario E. 449
Aguilar-Ugarte Luis 445
Ahmed Aziz 213
Akkad Nader 223
Amadio Claudio 223
An Yuwei 279
An Dong-Ya 361
Anastasiadis Anthimos 425
Araki Keita 195, 197
Araújo Miguel 343, 355
Aribert Jean-Marie 303, 347
Asada Hayato 135, 227, 241
Augusto Hugo 211, 341
Avossa A. M. 293
Ayrumyan Eduard 229

B

Bai Yong-Tao 327
Barbagallo Francesca 339, 437
Barnwell Nicholas 185
Beato Alexis Rafael Ovalle 87
Beetham Tessa 181
Benavent-Climent Amadeo 65
Berman Jeffrey W. 77
Beskos Dimitri E. 429
Bezabeh Matiyas Ayalew 73
Bian G. 143
Borzouie J. 177, 179
Bosco Melina 271, 337, 339, 437, 457
Bradley Cameron R. 89
Brandonisio Giuseppe 457
Bruneau Michel 77, 433
Bucciero Bianca 367
Bull Desmond 255, 323
Buonopane S. G. 143
Buru Stefan Marius 287

C

Cai Yun-Zhu 363
Calado Luis 153
Cassiano David 295
Castaldo Carmine 373

Castro José Miguel 211, 341, 343, 355, 155
Cerri Steven 441
Charney Finley 117, 251
Chase J. Geoffery 93, 179, 209, 253, 255, 371, 395
Chaudhari Tushar D. 255
Chen Ying-Zhi 187
Chen Xingchen 261
Chen Zheng-Qing 401
Chen Su-Wen 131
Chen Yi-Yi 61, 187, 325
Chen Yiyi 63, 215, 377
Chen Zhihua 471
Chou Chung-Che 87
Chuan Guang-Hong 61
Chuang Ming-Chieh 161
Chung Ping-Ting 87
Ciutina Adrian 263, 265
Clayton Patricia M. 77
Clifton G. Charles 44, 93, 179, 181, 189, 191, 209, 255, 321, 369, 383, 433
Connor Robert J. 127
Corte Gaetano Della 149
Costanzo Silvia 237, 239
Cowie Kevin A. 421
Crisan Andrei 267
Cui Yao 217
Cui Jia-Chun 361

D

D'Aniello Mario 237, 239, 273, 295
Dai Guoxing 473
Dan Daniel 125
Demetriou Demetris 399
Deng Zhong-Liang 391
Deng Hong-Zhou 129, 165
Dhakal Rajesh R. 319
Di Hao 243
Dimakogianni Danai 405
Dinu Florea 207, 269
Djojo Gary S. 369
Dogariu Adrian 451
Domínguez David 65
Dong Baiping 91, 409
Dong Yaomin 467